AF695485

Functional Analysis

GRADUATE STUDIES IN MATHEMATICS 191

Functional Analysis

Theo Bühler
Dietmar A. Salamon

American Mathematical Society
Providence, Rhode Island

EDITORIAL COMMITTEE

Dan Abramovich
Daniel S. Freed (Chair)
Gigliola Staffilani
Jeff A. Viaclovsky

2010 *Mathematics Subject Classification.* Primary 46-01, 47-01.

For additional information and updates on this book, visit
www.ams.org/bookpages/gsm-191

Library of Congress Cataloging-in-Publication Data

Names: Bühler, Theo, 1978– author. | Salamon, D. (Dietmar), author.
Title: Functional analysis / Theo Bühler, Dietmar A. Salamon.
Description: Providence, Rhode Island : American Mathematical Society, [2018] | Series: Graduate studies in mathematics ; volume 191 | Includes bibliographical references and index.
Identifiers: LCCN 2017057159 | ISBN 9781470441906 (alk. paper)
Subjects: LCSH: Functional analysis. | Spectral theory (Mathematics) | Semigroups of operators. | AMS: Functional analysis – Instructional exposition (textbooks, tutorial papers, etc.). msc | Operator theory – Instructional exposition (textbooks, tutorial papers, etc.). msc
Classification: LCC QA320 .B84 2018 | DDC 515/.7—dc23
LC record available at https://lccn.loc.gov/2017057159

Copying and reprinting. Individual readers of this publication, and nonprofit libraries acting for them, are permitted to make fair use of the material, such as to copy select pages for use in teaching or research. Permission is granted to quote brief passages from this publication in reviews, provided the customary acknowledgment of the source is given.

Republication, systematic copying, or multiple reproduction of any material in this publication is permitted only under license from the American Mathematical Society. Requests for permission to reuse portions of AMS publication content are handled by the Copyright Clearance Center. For more information, please visit www.ams.org/publications/pubpermissions.

Send requests for translation rights and licensed reprints to reprint-permission@ams.org.

© 2018 by the author. All rights reserved.
Printed in the United States of America.

∞ The paper used in this book is acid-free and falls within the guidelines
established to ensure permanence and durability.
Visit the AMS home page at http://www.ams.org/

10 9 8 7 6 5 4 3 2 1 23 22 21 20 19 18

Contents

Preface

These are notes for the lecture course *"Functional Analysis I"* held by the second author at ETH Zürich in the fall semester 2015. Prerequisites are the first year courses on *Analysis* and *Linear Algebra*, and the second year courses on *Complex Analysis*, *Topology*, and *Measure and Integration*.

The material of Section 1.4 on elementary Hilbert space theory, Subsection 5.4.2 on the Stone–Weierstraß Theorem, and the appendix on the Lemma of Zorn and Tychonoff's Theorem was not covered in the lectures. These topics were assumed to have been covered in previous lecture courses. They are included here for completeness of the exposition.

The material of Subsection 2.4.4 on the James space, Section 5.5 on the functional calculus for bounded normal operators, and Chapter 6 on unbounded operators was not part of the lecture course (with the exception of some of the basic definitions in Chapter 6 that are relevant for infinitesimal generators of strongly continuous semigroups). From Chapter 7 only the basic material on strongly continuous semigroups in Section 7.1, on their infinitesimal generators in Section 7.2, and on the dual semigroup in Section 7.3 were included in the lecture course.

23 October 2017

Theo Bühler
Dietmar A. Salamon

Introduction

Classically, functional analysis is the study of function spaces and linear operators between them. The relevant function spaces are often equipped with the structure of a Banach space and many of the central results remain valid in the more general setting of bounded linear operators between Banach spaces or normed vector spaces, where the specific properties of the concrete function space in question only play a minor role. Thus, in the modern guise, functional analysis is the study of Banach spaces and bounded linear operators between them, and this is the viewpoint taken in the present book. This area of mathematics has both an intrinsic beauty, which we hope to convey to the reader, and a vast number of applications in many fields of mathematics. These include the analysis of PDEs, differential topology and geometry, symplectic topology, quantum mechanics, probability theory, geometric group theory, dynamical systems, ergodic theory, and approximation theory, among many others. While we say little about specific applications, they do motivate the choice of topics covered in this book, and our goal is to give a self-contained exposition of the necessary background in abstract functional analysis for many of the relevant applications.

The book is addressed primarily to third year students of mathematics or physics, and the reader is assumed to be familiar with first year analysis and linear algebra, as well as complex analysis and the basics of point set topology and measure and integration. For example, this book does not include a proof of completeness and duality for L^p spaces.

There are naturally many topics that go beyond the scope of the present book, such as Sobolev spaces and PDEs, which would require a book on its own and, in fact, very many books have been written about this subject; here we just refer the interested reader to [**19**, **28**, **30**]. We also restrict the discussion to linear operators and say nothing about nonlinear functional analysis. Other topics not covered include the Fourier transform (see [**2**, **48**, **79**]), maximal regularity for semigroups (see [**76**]), the space of Fredholm operators on an infinite-dimensional Hilbert space as a classifying space for K-theory (see [**5**–**7**, **42**]), Quillen's determinant line bundle over the space of Fredholm operators (see [**71**, **77**]), and the work of Gowers [**31**] and Argyros–Haydon [**4**] on Banach spaces on which every bounded linear operator is the sum of a scalar multiple of the identity and a compact operator. Here is a description of the contents of the book, chapter by chapter.

Chapter 1 discusses some basic concepts that play a central role in the subject. It begins with a section on metric spaces and compact sets which includes a proof of the Arzelà–Ascoli Theorem. It then moves on to establish some basic properties of finite-dimensional normed vector spaces and shows, in particular, that a normed vector space is finite-dimensional if and only if the unit ball is compact. The first chapter also introduces the dual space of a normed vector space, explains several important examples, and contains an introduction to elementary Hilbert space theory. It then introduces Banach algebras and shows that the group of invertible elements is an open set. It closes with a proof of the Baire Category Theorem.

Chapter 2 is devoted to the three fundamental *principles of functional analysis*. They are the *Uniform Boundedness Principle* (a pointwise bounded family of bounded linear operators on a Banach space is bounded), the *Open Mapping Theorem* (a surjective bounded linear operator between Banach spaces is open), and the *Hahn–Banach Theorem* (a bounded linear functional on a linear subspace of a normed vector space extends to a bounded linear functional on the entire normed vector space). An equivalent formulation of the Open Mapping Theorem is the *Closed Graph Theorem* (a linear operator between Banach spaces is bounded if and only if it has a closed graph) and a corollary is the *Inverse Operator Theorem* (a bijective bounded linear operator between Banach spaces has a bounded inverse). A slightly stronger version of the Hahn–Banach Theorem, with the norm replaced by a quasi-seminorm, can be reformulated as the geometric assertion that two convex subsets of a normed vector space can be separated by a hyperplane whenever one of them has nonempty interior. The chapter also discusses reflexive Banach spaces and includes an exposition of the James space.

The subjects of Chapter 3 are the weak topology on a Banach space X and the weak* topology on its dual space X^*. With these topologies X and X^* are locally convex Hausdorff topological vector spaces and the chapter begins with a discussion of the elementary properties of such spaces. The central result of the third chapter is the Banach–Alaoglu Theorem which asserts that the unit ball in the dual space is compact with respect to the weak* topology. This theorem has important consequences in many fields of mathematics. The chapter also contains a proof of the Banach–Dieudonné Theorem which asserts that a linear subspace of the dual space of a Banach space is weak* closed if and only if its intersection with the closed unit ball is weak* closed. A consequence of the Banach–Alaoglu Theorem is that the unit ball in a reflexive Banach space is weakly compact, and the Eberlein–Šmulyan Theorem asserts that this property characterizes reflexive Banach spaces. The Kreĭn–Milman Theorem asserts that every nonempty compact convex subset of a locally convex Hausdorff topological vector space is the closed convex hull of its extremal points. Combining this with the Banach–Alaoglu Theorem, one can prove that every homeomorphism of a compact metric space admits an invariant ergodic Borel probability measure. Some properties of such ergodic measures can be derived from an abstract functional analytic ergodic theorem which is also established in this chapter.

The purpose of Chapter 4 is to give a basic introduction to Fredholm operators and their indices including the stability theorem. A Fredholm operator is a bounded linear operator between Banach spaces that has a finite-dimensional kernel, a closed image, and a finite-dimensional cokernel. Its Fredholm index is the difference of the dimensions of kernel and cokernel. The stability theorem asserts that the Fredholm operators of any given index form an open subset of the space of all bounded linear operators between two Banach spaces, with respect to the topology induced by the operator norm. It also asserts that the sum of a Fredholm operator and a compact operator is again Fredholm and has the same index as the original operator. The chapter includes an introduction to the dual of a bounded linear operator, a proof of the Closed Image Theorem which asserts that an operator has a closed image if and only if its dual does, an introduction to compact operators which map the unit ball to pre-compact subsets of the target space, a characterization of Fredholm operators in terms of invertibility modulo compact operators, and a proof of the stability theorem for Fredholm operators.

The purpose of Chapter 5 is to study the spectrum of a bounded linear operator on a real or complex Banach space. A first preparatory section discusses complex Banach spaces and the complexifications of real Banach spaces, the integrals of continuous Banach space valued functions on compact intervals, and holomorphic operator valued functions. The chapter then introduces the spectrum of a bounded linear operator, examines its elementary properties, discusses the spectra of compact operators, and establishes the holomorphic functional calculus. The remainder of this chapter deals exclusively with operators on Hilbert spaces, starting with a discussion of complex Hilbert spaces and the spectra of normal and self-adjoint operators. It then moves on to C* algebras and the continuous functional calculus for self-adjoint operators, which takes the form of an isomorphism from the

C* algebra of complex valued continuous functions on the spectrum to the smallest C* algebra containing the given operator. The next topic is the Gelfand representation and the extension of the continuous functional calculus to normal operators. The chapter also contains a proof that every normal operator can be represented by a projection valued measure on the spectrum, and that every self-adjoint operator is isomorphic to a direct sum of multiplication operators on L^2 spaces.

Chapter 6 is devoted to unbounded operators and their spectral theory. The domain of an unbounded operator on a Banach space is a linear subspace. In most of the relevant examples the domain is dense and the operator has a closed graph. The chapter includes a discussion of the dual of an unbounded operator and an extension of the Closed Image Theorem to this setting. It then examines the basic properties of the spectra of unbounded operators. The remainder of the chapter deals with unbounded operators on Hilbert spaces and their adjoints. In particular, it extends the functional calculus and the spectral measure to unbounded self-adjoint operators.

Strongly continuous semigroups of operators are the subject of Chapter 7. They play an important role in the study of many linear partial differential equations such as the heat equation, the wave equation, and the Schrödinger equation, and they can be viewed as infinite-dimensional analogues of the exponential matrix $S(t) := e^{tA}$. In all the relevant examples the operator A is unbounded. It is called the *infinitesimal generator* of the strongly continuous semigroup in question. A central result in the subject is the Hille–Yosida–Phillips Theorem which characterizes the infinitesimal generators of strongly continuous semigroups. The dual semigroup is not always strongly continuous. It is, however, strongly continuous whenever the Banach space in question is reflexive. The chapter also includes a basic treatment of analytic semigroups and their infinitesimal generators. It closes with a study of Banach space valued measurable functions and of solutions to the inhomogeneous equation associated to a semigroup.

Each of the seven chapters ends with a problem section, which we hope will give the interested reader the opportunity to deepen their understanding of the subject.

Chapter 1

Foundations

This introductory chapter discusses some of the basic concepts that play a central role in the subject of *Functional Analysis*. In a nutshell, functional analysis is the study of normed vector spaces and bounded linear operators. Thus it merges the subjects of linear algebra (vector spaces and linear maps) with that of point set topology (topological spaces and continuous maps). The topologies that appear in functional analysis will in many cases arise from metric spaces. We begin in Section 1.1 by recalling the basic definitions and list several examples of Banach spaces that will be used to illustrate the theory throughout the book. The central topic is the study of compact sets and the main results are the characterization of sequentially compact subsets of a metric space in terms of open covers and the Arzelà–Ascoli Theorem which gives a compactness criterion for subsets of the space of continuous functions on a compact metric space. Section 1.2 moves on to the study of finite-dimensional normed vector spaces. It shows that any two norms on a finite-dimensional vector space are equivalent, and that a normed vector space is finite-dimensional if and only if the unit ball is compact. The section also contains a brief introduction to bounded linear operators and to product and quotient spaces. Section 1.3 introduces the dual space of a normed vector space and explains several important examples. Section 1.4 contains a brief introduction to elementary Hilbert space theory, including a proof of the Cauchy–Schwarz inequality and the Riesz Representation Theorem. Section 1.5 examines some basic properties of power series in Banach algebras. It shows, via the geometric series, that the space of invertible operators on a Banach space is open and that the map that assigns to an invertible operator its inverse is continuous. The Baire Category Theorem is the subject of Section 1.6.

1.1. Metric Spaces and Compact Sets

This section begins by recalling the basic definitions of a metric space and a Banach space and gives several important examples of Banach spaces. It then moves on to the study of compact subsets of a metric space and shows that sequential compactness is equivalent to the condition that every open cover has a finite subcover (Theorem 1.1.4). The second main result of this section is the Arzelà–Ascoli Theorem, which characterizes the precompact subsets of the space of continuous functions from a compact metric space to another metric space, equipped with the supremum metric, in terms of equi-continuity and pointwise precompactness (Theorem 1.1.11).

1.1.1. Banach Spaces.

Definition 1.1.1 (Metric Space). A **metric space** is a pair (X, d) consisting of a set X and a function $d : X \times X \to \mathbb{R}$ that satisfies the following axioms.

(M1) $d(x, y) \geq 0$ for all $x, y \in X$, with equality if and only if $x = y$.

(M2) $d(x, y) = d(y, x)$ for all $x, y \in X$.

(M3) $d(x, z) \leq d(x, y) + d(y, z)$ for all $x, y, z \in X$.

A function $d : X \times X \to \mathbb{R}$ that satisfies these axioms is called a **distance function** and the inequality in (M3) is called the **triangle inequality**. A subset $U \subset X$ of a metric space (X, d) is called **open** (or d-**open**) if, for every $x \in U$, there exists a constant $\varepsilon > 0$ such that the **open ball**

$$B_\varepsilon(x) := B_\varepsilon(x, d) := \{y \in X \mid d(x, y) < \varepsilon\}$$

(centered at x with radius ε) is contained in U. The set of d-open subsets of X will be denoted by

$$\mathscr{U}(X, d) := \{U \subset X \mid U \text{ is } d\text{-open}\}.$$

It follows directly from the definitions that the collection $\mathscr{U}(X, d) \subset 2^X$ of d-open sets in a metric space (X, d) satisfies the axioms of a **topology** (i.e. the empty set and the set X are open, arbitrary unions of open sets are open, and finite intersections of open sets are open). A subset F of a metric space (X, d) is closed (i.e. its complement is open) if and only if the limit point of every convergent sequence in F is itself contained in F.

Recall that a sequence $(x_n)_{n \in \mathbb{N}}$ in a metric space (X, d) is called a **Cauchy sequence** if, for every $\varepsilon > 0$, there exists an $n_0 \in \mathbb{N}$ such that any two integers $n, m \geq n_0$ satisfy the inequality $d(x_n, x_m) < \varepsilon$. Recall also that a metric space (X, d) is called **complete** if every Cauchy sequence in X converges.

The most important metric spaces in the field of functional analysis are the normed vector spaces.

Definition 1.1.2 (Banach Space). A **normed vector space** is a pair $(X, \|\cdot\|)$ consisting of a real vector space X and a function

$$X \to \mathbb{R} : x \mapsto \|x\|$$

satisfying the following axioms.

(N1) $\|x\| \geq 0$ for all $x \in X$, with equality if and only if $x = 0$.

(N2) $\|\lambda x\| = |\lambda| \, \|x\|$ for all $x \in X$ and $\lambda \in \mathbb{R}$.

(N3) $\|x + y\| \leq \|x\| + \|y\|$ for all $x, y \in X$.

Let $(X, \|\cdot\|)$ be a normed vector space. Then the formula

$$d(x,y) := \|x - y\| \tag{1.1.1}$$

for $x, y \in X$ defines a distance function on X. The resulting topology is denoted by $\mathscr{U}(X, \|\cdot\|) := \mathscr{U}(X, d)$. X is called a **Banach space** if the metric space (X, d) is **complete**, i.e. if every Cauchy sequence in X converges.

Here are six examples of Banach spaces.

Example 1.1.3. **(i)** Fix a real number $1 \leq p < \infty$. Then the vector space $X = \mathbb{R}^n$ of all n-tuples $x = (x_1, \dots, x_n)$ of real numbers is a Banach space with the norm-function

$$\|x\|_p := \left(\sum_{i=1}^{n} |x_i|^p \right)^{1/p}$$

for $x = (x_1, \dots, x_n) \in \mathbb{R}^n$. For $p = 2$ this is the Euclidean norm. Another norm on $\mathbb{R}^n$ is given by $\|x\|_\infty := \max_{i=1,\dots,n} |x_i|$ for $x = (x_1, \dots, x_n) \in \mathbb{R}^n$.

(ii) For $1 \leq p < \infty$ the set of p-summable sequences of real numbers is denoted by

$$\ell^p := \left\{ x = (x_i)_{i \in \mathbb{N}} \in \mathbb{R}^{\mathbb{N}} \,\middle|\, \sum_{i=1}^{\infty} |x_i|^p < \infty \right\}.$$

This is a Banach space with the norm $\|x\|_p := (\sum_{i=1}^{\infty} |x_i|^p)^{1/p}$ for $x \in \ell^p$. Likewise, the space $\ell^\infty \subset \mathbb{R}^{\mathbb{N}}$ of bounded sequences is a Banach space with the supremum norm $\|x\|_\infty := \sup_{i \in \mathbb{N}} |x_i|$ for $x = (x_i)_{i \in \mathbb{N}} \in \ell^\infty$.

(iii) Let $(M, \mathcal{A}, \mu)$ be a measure space, i.e. M is a set, $\mathcal{A} \subset 2^M$ is a σ-algebra, and $\mu : \mathcal{A} \to [0, \infty]$ is a measure. Fix a constant $1 \leq p < \infty$. A measurable function $f : M \to \mathbb{R}$ is called **p-integrable** if $\int_M |f|^p \, d\mu < \infty$ and the space of p-integrable functions on M will be denoted by

$$\mathcal{L}^p(\mu) := \left\{ f : M \to \mathbb{R} \,\middle|\, f \text{ is measurable and } \int_M |f|^p \, d\mu < \infty \right\}.$$

The function $\mathcal{L}^p(\mu) \to \mathbb{R} : f \mapsto \|f\|_p$ defined by

$$\|f\|_p := \left(\int_M |f|^p \, d\mu\right)^{1/p} \tag{1.1.2}$$

is nonnegative and satisfies the triangle inequality (Minkowski's inequality). However, in general it is not a norm, because $\|f\|_p = 0$ if and only if f vanishes **almost everywhere** (i.e. on the complement of a set of measure zero). To obtain a normed vector space, one considers the quotient

$$L^p(\mu) := \mathcal{L}^p(\mu)/\sim,$$

where

$$f \sim g \quad \overset{\text{def}}{\Longleftrightarrow} \quad f = g \text{ almost everywhere.}$$

The function $f \mapsto \|f\|_p$ descends to this quotient space and, with this norm, $L^p(\mu)$ is a Banach space (see [**75**, Theorem 4.9]). In this example it is often convenient to abuse notation and use the same letter f to denote a function in $\mathcal{L}^p(\mu)$ and its equivalence class in the quotient space $L^p(\mu)$.

(iv) Let $(M, \mathcal{A}, \mu)$ be a measure space, denote by $\mathcal{L}^\infty(\mu)$ the space of bounded measurable functions, and denote by

$$L^\infty(\mu) := \mathcal{L}^\infty(\mu)/\sim$$

the quotient space, where the equivalence relation is again defined by equality almost everywhere. Then the formula

$$\|f\|_\infty := \operatorname{ess\,sup}|f| = \inf\left\{c \geq 0 \,\middle|\, f \leq c \text{ almost everywhere}\right\} \tag{1.1.3}$$

defines a norm on $L^\infty(\mu)$, and $L^\infty(\mu)$ is a Banach space with this norm.

(v) Let M be a topological space. Then the space $C_b(M)$ of bounded continuous functions $f : M \to \mathbb{R}$ is a Banach space with the supremum norm

$$\|f\|_\infty := \sup_{p \in M} |f(p)|$$

for $f \in C_b(M)$.

(vi) Let $(M, \mathcal{A})$ be a measurable space, i.e. M is a set and $\mathcal{A} \subset 2^M$ is a σ-algebra. A **signed measure** on $(M, \mathcal{A})$ is a function $\mu : \mathcal{A} \to \mathbb{R}$ that satisfies $\mu(\emptyset) = 0$ and is σ-additive, i.e. $\mu(\bigcup_{i=1}^\infty A_i) = \sum_{i=1}^\infty \mu(A_i)$ for every sequence of pairwise disjoint measurable sets $A_i \in \mathcal{A}$. The space $\mathcal{M}(M, \mathcal{A})$ of signed measures on $(M, \mathcal{A})$ is a Banach space with the norm given by

$$\|\mu\| := |\mu|(M) := \sup_{A \in \mathcal{A}} \left(\mu(A) - \mu(M \setminus A)\right) \tag{1.1.4}$$

for $\mu \in \mathcal{M}(M, \mathcal{A})$ (see [**75**, Exercise 5.34]).

1.1.2. Compact Sets. Let (X, d) be a metric space and let $K \subset X$. Then the restriction of the distance function d to $K \times K$ is a distance function, denoted by $d_K := d|_{K \times K} : K \times K \to \mathbb{R}$, so (K, d_K) is a metric space in its own right. The metric space (X, d) is called **(sequentially) compact** if every sequence in X has a convergent subsequence. The subset K is called **(sequentially) compact** if (K, d_K) is compact, i.e. if every sequence in K has a subsequence that converges to an element of K. It is called **precompact** if its closure is sequentially compact. Thus K is compact if and only if it is precompact and closed. The subset K is called **complete** if (K, d_K) is a complete metric space, i.e. if every Cauchy sequence in K converges to an element of K. It is called **totally bounded** if it is either empty or, for every $\varepsilon > 0$, there exist finitely many elements $\xi_1, \ldots, \xi_m \in K$ such that

$$K \subset \bigcup_{i=1}^{m} B_\varepsilon(\xi_i).$$

The next theorem characterizes the compact subsets of a metric space (X, d) in terms of the open subsets of X. It thus shows that compactness depends only on the topology $\mathscr{U}(X, d)$ induced by the distance function d.

Theorem 1.1.4 (Characterization of Compact Sets). *Let (X, d) be a metric space and let $K \subset X$. Then the following are equivalent.*

(i) *K is sequentially compact.*

(ii) *K is complete and totally bounded.*

(iii) *Every open cover of K has a finite subcover.*

Proof. See page 7. □

Let $(X, \mathscr{U})$ be a topological space. Then condition (iii) in Theorem 1.1.4 is used to define compact subsets of X. Thus a subset $K \subset X$ is called **compact** if every open cover of K has a finite subcover. Here an open cover of K is a collection $(U_i)_{i \in I}$ of open subsets $U_i \subset X$, indexed by the elements of a nonempty set I, such that $K \subset \bigcup_{i \in I} U_i$, and a finite subcover is a finite collection of indices $i_1, \ldots, i_m \in I$ such that $K \subset U_{i_1} \cup \cdots \cup U_{i_m}$. Thus Theorem 1.1.4 asserts that a subset of a metric space (X, d) is sequentially compact if and only if it is compact as a subset of the topological space $(X, \mathscr{U})$ with $\mathscr{U} = \mathscr{U}(X, d)$. A subset of a topological space is called **precompact** if its closure is compact. Elementary properties of compact sets include the fact that every compact subset of a Hausdorff space is closed, that every closed subset of a compact set is compact, and that the image of a compact set under a continuous map is compact (see [**45**, **61**]).

We give two proofs of Theorem 1.1.4. The first proof is more straightforward and uses the axiom of dependent choice. The second proof is taken from Herrlich [**34**, Prop. 3.26] and only uses the axiom of countable choice.

The **axiom of dependent choice** asserts that, if $\mathbf{X}$ is a nonempty set and $\mathbf{A} : \mathbf{X} \to 2^{\mathbf{X}}$ is a map that assigns to each element $\mathbf{x} \in \mathbf{X}$ a nonempty subset $\mathbf{A}(\mathbf{x}) \subset \mathbf{X}$, then there exists a sequence $(\mathbf{x}_k)_{k\in\mathbb{N}}$ in $\mathbf{X}$ such that $\mathbf{x}_{k+1} \in \mathbf{A}(\mathbf{x}_k)$ for all $k \in \mathbb{N}$. In the axiom of dependent choice the first element of the sequence $(\mathbf{x}_k)_{k\in\mathbb{N}}$ can be prescribed. To see this, let $\mathbf{x}_1 \in \mathbf{X}$, define $\widetilde{\mathbf{X}}$ as the set of all tuples of the form $\widetilde{\mathbf{x}} = (n, \mathbf{x}_1, \dots, \mathbf{x}_n)$ with $n \in \mathbb{N}$ and $\mathbf{x}_k \in \mathbf{A}(\mathbf{x}_{k-1})$ for $k = 2, \dots, n$, and for $\widetilde{\mathbf{x}} = (n, \mathbf{x}_1, \dots, \mathbf{x}_n) \in \widetilde{\mathbf{X}}$ define the set $\widetilde{\mathbf{A}}(\widetilde{\mathbf{x}}) := \{(n+1, \mathbf{x}_1, \dots, \mathbf{x}_n, \mathbf{x}) \,|\, \mathbf{x} \in \mathbf{A}(\mathbf{x}_n)\}$. Then $\widetilde{\mathbf{X}}$ is nonempty and $\widetilde{\mathbf{A}}(\widetilde{\mathbf{x}})$ is nonempty for every $\widetilde{\mathbf{x}} \in \widetilde{\mathbf{X}}$. Now apply the axiom of dependent choice to $\widetilde{\mathbf{A}}$.

The **axiom of countable choice** asserts that, if $(\mathbf{A}_k)_{k\in\mathbb{N}}$ is a sequence of nonempty subsets of a set $\mathbf{A}$, then there exists a sequence $(\mathbf{a}_k)_{k\in\mathbb{N}}$ in $\mathbf{A}$ such that $\mathbf{a}_k \in \mathbf{A}_k$ for all $k \in \mathbb{N}$. This follows from the axiom of dependent choice by taking $\mathbf{X} := \mathbb{N}\times\mathbf{A}$ and $\mathbf{A}(k, \mathbf{a}) := \{k+1\}\times\mathbf{A}_{k+1}$ for $(k, \mathbf{a}) \in \mathbb{N}\times\mathbf{A}$.

Lemma 1.1.5. *Let (X, d) be a metric space and let $K \subset X$. Then the following are equivalent.*

(i) *Every sequence in K has a Cauchy subsequence.*

(ii) *K is totally bounded.*

Proof of (ii) $\Longrightarrow$ (i) in Lemma 1.1.5. The argument only uses the axiom of countable choice. Assume that K is totally bounded and let $(x_n)_{n\in\mathbb{N}}$ be a sequence in K. We prove that there exists a sequence of infinite subsets $\mathbb{N} \supset T_1 \supset T_2 \supset \cdots$ such that, for all $k, m, n \in \mathbb{N}$,

$$m, n \in T_k \qquad \Longrightarrow \qquad d(x_m, x_n) < 2^{-k}. \tag{1.1.5}$$

Since K is totally bounded, it follows from the axiom of countable choice that there exists a sequence of ordered finite subsets

$$S_k = \{\xi_{k,1}, \dots, \xi_{k,m_k}\} \subset K$$

such that

$$K \subset \bigcup_{i=1}^{m_k} B_{2^{-k-1}}(\xi_{k,i}) \qquad \text{for all } k \in \mathbb{N}.$$

Since $x_n \in K$ for all $n \in \mathbb{N}$, there must exist an index $i \in \{1, \dots, m_1\}$ such that the open ball $B_{1/4}(\xi_{1,i})$ contains infinitely many of the elements x_n. Let i_1 be the smallest such index and define the set

$$T_1 := \{n \in \mathbb{N} \,|\, x_n \in B_{1/4}(\xi_{1,i_1})\}.$$

This set is infinite and satisfies $d(x_n, x_m) \le d(x_n, \xi_{1,i_1}) + d(\xi_{1,i_1}, x_m) < 1/2$ for all $m, n \in T_1$. Now fix an integer $k \ge 2$ and suppose, by induction, that T_{k-1} has been defined. Since T_{k-1} is an infinite set, there must exist an index $i \in \{1, \dots, m_k\}$ such that the ball $B_{2^{-k-1}}(\xi_{k,i})$ contains infinitely many of the elements x_n with $n \in T_{k-1}$. Let i_k be the smallest such index and define

$$T_k := \{n \in T_{k-1} \,|\, x_n \in B_{2^{-k-1}}(\xi_{k,i_k})\}.$$

This set is infinite and satisfies $d(x_n, x_m) \le d(x_n, \xi_{k,i_k}) + d(\xi_{k,i_k}, x_m) < 2^{-k}$ for all $m, n \in T_k$. This completes the induction argument and the construction of a decreasing sequence of infinite sets $T_k \subset \mathbb{N}$ that satisfy (1.1.5).

We prove that $(x_n)_{n \in \mathbb{N}}$ has a Cauchy subsequence. By (1.1.5) there exists a sequence of positive integers $n_1 < n_2 < n_3 < \cdots$ such that $n_k \in T_k$ for all $k \in \mathbb{N}$. Such a sequence can be defined by the recursion formula

$$n_1 := \min T_1, \qquad n_{k+1} := \min \left\{ n \in T_{k+1} \,\middle|\, n > n_k \right\}$$

for $k \in \mathbb{N}$. It follows that $n_k, n_\ell \in T_k$ and hence

$$d(x_{n_k}, x_{n_\ell}) < 2^{-k} \qquad \text{for } \ell \ge k \ge 1.$$

Thus the subsequence $(x_{n_k})_{k \in \mathbb{N}}$ is a Cauchy sequence. This shows that (ii) implies (i) in Lemma 1.1.5. The converse will be proved on page 9. □

First proof of Theorem 1.1.4. We prove that (i) implies (iii) using the axiom of dependent choice. Assume that the set K is nonempty and sequentially compact, and let $\{U_i\}_{i \in I}$ be an open cover of K. Here I is a nonempty index set and the map $I \to 2^X : i \mapsto U_i$ assigns to each index i an open set $U_i \subset X$ such that $K \subset \bigcup_{i \in I} U_i$. We prove in two steps that there exist indices $i_1, \dots, i_m \in I$ such that $K \subset \bigcup_{j=1}^m U_{i_j}$.

Step 1. *There exists a constant $\varepsilon > 0$ such that, for every $x \in K$, there exists an index $i \in I$ such that $B_\varepsilon(x) \subset U_i$.*

Assume, by contradiction, that there is no such constant $\varepsilon > 0$. Then

$$\forall \varepsilon > 0 \;\exists x \in K \;\forall i \in I \;\; B_\varepsilon(x) \not\subset U_i.$$

Take $\varepsilon = 1/n$ for $n \in \mathbb{N}$. Then the set $\{x \in K \,|\, B_{1/n}(x) \not\subset U_i \text{ for all } i \in I\}$ is nonempty for every $n \in \mathbb{N}$. Hence the axiom of countable choice asserts that there exists a sequence $x_n \in K$ such that

$$B_{1/n}(x_n) \not\subset U_i \qquad \text{for all } n \in \mathbb{N} \text{ and all } i \in I. \tag{1.1.6}$$

Since K is sequentially compact, there exists a subsequence $(x_{n_k})_{k \in \mathbb{N}}$ that converges to an element $x \in K$. Since $K \subset \bigcup_{i \in I} U_i$, there exists an $i \in I$ such that $x \in U_i$. Since U_i is open, there is an $\varepsilon > 0$ such that $B_\varepsilon(x) \subset U_i$. Since $x = \lim_{k \to \infty} x_{n_k}$, there is a $k \in \mathbb{N}$ such that $d(x, x_{n_k}) < \frac{\varepsilon}{2}$ and $\frac{1}{n_k} < \frac{\varepsilon}{2}$. Thus $B_{1/n_k}(x_{n_k}) \subset B_{\varepsilon/2}(x_{n_k}) \subset B_\varepsilon(x) \subset U_i$ in contradiction to (1.1.6).

Step 2. *There exist indices $i_1, \dots, i_m \in I$ such that $K \subset \bigcup_{j=1}^m U_{i_j}$.*

Assume, by contradiction, that this is wrong. Let $\varepsilon > 0$ be the constant in Step 1. We prove that there are sequences $x_n \in K$ and $i_n \in I$ such that

$$B_\varepsilon(x_n) \subset U_{i_n}, \qquad x_n \notin U_{i_1} \cup \cdots \cup U_{i_{n-1}}, \tag{1.1.7}$$

for all $n \in \mathbb{N}$ (with $n \geq 2$ for the second condition). Choose $x_1 \in K$. Then, by Step 1, there exists an index $i_1 \in I$ such that $B_\varepsilon(x_1) \subset U_{i_1}$. Now suppose, by induction, that $x_1, \dots, x_k$ and $i_1, \dots, i_k$ have been found such that (1.1.7) holds for $n \leq k$. Then

$$K \not\subset U_{i_1} \cup \cdots \cup U_{i_k}.$$

Choose an element $x_{k+1} \in K \setminus (U_{i_1} \cup \cdots \cup U_{i_k})$. By Step 1, there exists an index $i_{k+1} \in I$ such that $B_\varepsilon(x_{k+1}) \subset U_{i_{k+1}}$. Thus the existence of sequences x_n and i_n that satisfy (1.1.7) follows from the axiom of dependent choice. More precisely, let $\mathbf{X}$ be the set of all pairs $\mathbf{x} = (x, J)$ such that J is a finite subset of I and $x \in K \setminus \bigcup_{j \in J} U_j$. For $\mathbf{x} = (x, J) \in \mathbf{X}$ let $\mathbf{A}(\mathbf{x})$ be the set of all pairs $\mathbf{x}' = (x', J') \in \mathbf{X}$, where $J' = J \cup \{i'\}$, $i' \in I$, and $B_\varepsilon(x') \subset U_{i'}$. Then $\mathbf{A}(\mathbf{x}) \neq \emptyset$ for all $\mathbf{x} \in \mathbf{X}$ by assumption and the choice of ε in Step 1. Thus there is a sequence $\mathbf{x}_n = (x_n, J_n) \in \mathbf{X}$ such that $\mathbf{x}_{n+1} \in \mathbf{A}(\mathbf{x}_n)$ for all n. So $J_n \setminus J_{n-1} = \{i_n\}$ is a singleton such that $B_\varepsilon(x_n) \subset U_{i_n}$ for each $n \in \mathbb{N}$. Moreover $i_1, \dots, i_{n-1} \in J_n$ and so $x_n \in K \setminus \bigcup_{k=1}^{n-1} U_{i_k}$ for each integer $n \geq 2$. Thus the sequences $(x_n)_{n \in \mathbb{N}}$ and $(i_n)_{n \in \mathbb{N}}$ satisfy (1.1.7) as claimed.

By (1.1.7) we have $d(x_n, x_k) \geq \varepsilon$ for $k \neq n$, so $(x_n)_{n \in \mathbb{N}}$ does not have a convergent subsequence, contradicting (i). This shows that (i) implies (iii).

We prove that (iii) implies (ii) without using any version of the axiom of choice. Thus assume that every open cover of K has a finite subcover. Assume that K is nonempty and fix a constant $\varepsilon > 0$. Then the sets $B_\varepsilon(\xi)$ for $\xi \in K$ form a nonempty open cover of K. Hence there exist finitely many elements $\xi_1, \dots, \xi_m \in K$ such that $K \subset \bigcup_{i=1}^m B_\varepsilon(\xi_i)$. This shows that K is totally bounded.

We prove that K is complete. Let $(x_n)_{n \in \mathbb{N}}$ be a Cauchy sequence in K and suppose, by contradiction, that $(x_n)_{n \in \mathbb{N}}$ does not converge to any element of K. Then no subsequence of $(x_n)_{n \in \mathbb{N}}$ can converge to any element of K. Thus, for every $\xi \in K$, there is an $\varepsilon > 0$ such that $B_\varepsilon(\xi)$ contains only finitely many of the x_n. For $\xi \in K$ let $\varepsilon(\xi) > 0$ be half the supremum of the set of all $\varepsilon \in (0, 1]$ such that $\#\{n \in \mathbb{N} \,|\, x_n \in B_\varepsilon(\xi)\} < \infty$. Then the set $\{n \in \mathbb{N} \,|\, x_n \in B_{\varepsilon(\xi)}(\xi)\}$ is finite for every $\xi \in K$. Thus $\{B_{\varepsilon(\xi)}(\xi)\}_{\xi \in K}$ is an open cover of K that does not have a finite subcover, in contradiction to (iii). This shows that (iii) implies (ii).

That (ii) implies (i) was shown in Lemma 1.1.5, using the axiom of countable choice, and this completes the first proof of Theorem 1.1.4. □

The above proof of Theorem 1.1.4 requires the axiom of dependent choice and only uses the implication (ii) $\Longrightarrow$ (i) in Lemma 1.1.5. The second proof follows [**34**, Prop. 3.26] and only requires the axiom of countable choice.

Proof of (i) $\Longrightarrow$ (ii) in Lemma 1.1.5. The proof follows [**34**, Prop. 3.26] and only uses the axiom of countable choice. We argue indirectly and assume that K is not totally bounded and hence also nonempty. Then there exists a constant $\varepsilon > 0$ such that K does not admit a finite cover by balls of radius ε, centered at elements of K. We prove in three steps that there exists a sequence in K that does not have a Cauchy subsequence.

Step 1. *For $n \in \mathbb{N}$ define the set*

$$K_n := \left\{ (x_1, \dots, x_n) \in K^n \,\middle|\, \begin{array}{l} \text{if } i,j \in \{1,\dots,n\} \text{ and } i \neq j, \\ \text{then } d(x_i,x_j) \geq \varepsilon \end{array} \right\}.$$

There is a sequence $(x_k)_{k\in\mathbb{N}}$ in K such that $(x_{n(n-1)/2+1}, \dots, x_{n(n+1)/2}) \in K_n$ for every integer $n \geq 1$.

We prove that K_n is nonempty for every $n \in \mathbb{N}$. For $n = 1$ this holds because K is nonempty. If it is empty for some $n \in \mathbb{N}$ then there exists an integer $n \geq 1$ such that $K_n \neq \emptyset$ and $K_{n+1} = \emptyset$. In this case, choose an element $(x_1, \dots, x_n) \in K_n$. Since $K_{n+1} = \emptyset$, this implies $K \subset \bigcup_{i=1}^n B_\varepsilon(x_i)$, contradicting the choice of ε. Since $K_n \neq \emptyset$ for all $n \in \mathbb{N}$, the existence of a sequence $(x_k)_{k\in\mathbb{N}}$ as in Step 1 follows from the axiom of countable choice.

Step 2. *For every collection of $n-1$ elements $y_1, \dots, y_{n-1} \in K$, there is an integer i such that $\frac{(n-1)n}{2} < i \leq \frac{n(n+1)}{2}$ and $d(y_j, x_i) \geq \frac{\varepsilon}{2}$ for $j = 1, \dots, n-1$.*

Otherwise, there exists a map $\nu : \{\frac{(n-1)n}{2} + 1, \dots, \frac{n(n+1)}{2}\} \to \{1, \dots, n-1\}$ such that $d(x_i, y_{\nu(i)}) < \frac{\varepsilon}{2}$ for all i. Since the target space of ν has smaller cardinality than the domain, there is a pair $i \neq j$ in the domain with $\nu(i) = \nu(j)$ and so $d(x_i, x_j) \leq d(x_i, y_{\nu(i)}) + d(y_{\nu(j)}, x_j) < \varepsilon$, in contradiction to Step 1.

Step 3. *There exists a subsequence $(x_{k_n})_{n\in\mathbb{N}}$ such that $k_1 = 1$ and*

$$\frac{(n-1)n}{2} < k_n \leq \frac{n(n+1)}{2}, \qquad d(x_{k_m}, x_{k_n}) \geq \frac{\varepsilon}{2} \qquad \textit{for } m < n. \tag{1.1.8}$$

Define $k_1 := 1$, fix an integer $n \geq 2$, and assume, by induction, that the integers $k_1, k_2, \dots, k_{n-1}$ have been found such that (1.1.8) holds with n replaced by any number $n' \in \{2, \dots, n-1\}$. Then, by Step 2, there exists a unique smallest integer k_n such that $\frac{(n-1)n}{2} < k_n \leq \frac{n(n+1)}{2}$ and $d(x_{k_m}, x_{k_n}) \geq \frac{\varepsilon}{2}$ for $m = 1, \dots, n-1$. This proves the existence of a subsequence $(x_{k_n})_{n\in\mathbb{N}}$ that satisfies (1.1.8). The sequence $(x_{k_n})_{n\in\mathbb{N}}$ in Step 3 does not have a Cauchy subsequence. This shows that (i) implies (ii) in Lemma 1.1.5. $\square$

Second proof of Theorem 1.1.4. A sequentially compact metric space is complete, because a Cauchy sequence converges if and only if it has a convergent subsequence. Hence the equivalence of (i) and (ii) in Theorem 1.1.4 follows directly from Lemma 1.1.5.

We prove that (ii) implies (iii), following the argument in [**34**, Prop. 3.26]. Assume that K is complete and totally bounded. Suppose, by contradiction, that there is an open cover $\{U_i\}_{i\in I}$ of K that does not have a finite subcover. Then $K \neq \emptyset$. For $n, m \in \mathbb{N}$ define

$$A_{n,m} := \left\{ (x_1, \dots, x_m) \in K^m \,\middle|\, K \subset \bigcup_{j=1}^{m} B_{1/n}(x_j) \right\}.$$

Then, for every $n \in \mathbb{N}$, there exists an $m \in \mathbb{N}$ such that $A_{n,m} \neq \emptyset$, because K is totally bounded and nonempty. For $n \in \mathbb{N}$ let $m_n \in \mathbb{N}$ be the smallest positive integer such that $A_{n,m_n} \neq \emptyset$. Then, by the axiom of countable choice, there is a sequence $a_n = (x_{n,1}, \dots, x_{n,m_n}) \in A_{n,m_n}$ for $n \in \mathbb{N}$.

Next we construct a sequence $(y_n)_{n\in\mathbb{N}}$ in K such that $\bigcap_{\nu=1}^{n} B_{1/\nu}(y_\nu) \cap K$ cannot be covered by finitely many of the sets U_i for any $n \in \mathbb{N}$. For $n = 1$ define $y_1 := x_{1,k}$, where

$$k := \min \left\{ j \in \{1, \dots, m_1\} \,\middle|\, \begin{array}{l} \text{the set } B_1(x_{1,j}) \cap K \text{ cannot} \\ \text{be covered by finitely many } U_i \end{array} \right\}.$$

Assume, by induction, that $y_1, \dots, y_{n-1}$ have been chosen such that the set $\bigcap_{\nu=1}^{n-1} B_{1/\nu}(y_\nu) \cap K$ cannot be covered by finitely many of the U_i and define $y_n := x_{n,k}$, where

$$k := \min \left\{ j \in \{1, \dots, m_n\} \,\middle|\, \begin{array}{l} \text{the set } B_{1/n}(x_{n,j}) \cap \bigcap_{\nu=1}^{n-1} B_{1/\nu}(y_\nu) \cap K \\ \text{cannot be covered by finitely many } U_i \end{array} \right\}.$$

This completes the construction of the sequence $(y_n)_{n\in\mathbb{N}}$. It satisfies

$$d(y_n, y_m) < \frac{1}{m} + \frac{1}{n} \leq \frac{2}{m} \qquad \text{for } n > m \geq 1,$$

because $B_{1/n}(y_n) \cap B_{1/m}(y_m) \neq \emptyset$. Hence $(y_n)_{n\in\mathbb{N}}$ is a Cauchy sequence in K. Since K is complete, the limit $y^* := \lim_{n\to\infty} y_n$ exists and is an element of K. Choose an index $i^* \in I$ such that $y^* \in U_{i^*}$ and choose a constant $\varepsilon^* > 0$ such that $B_{\varepsilon^*}(y^*) \subset U_{i^*}$. Then

$$B_{1/n}(y_n) \subset B_{\varepsilon^*}(y^*) \subset U_{i^*}$$

for n sufficiently large in contradiction to the choice of y_n. This proves that (ii) implies (iii).

That (iii) implies (ii) was shown in the first proof without using the axiom of choice. This completes the second proof of Theorem 1.1.4. □

It follows immediately from Theorem 1.1.4 that every compact metric space is separable. Here are the relevant definitions.

Definition 1.1.6. Let X be a topological space. A subset $S \subset X$ is called **dense in** X if its closure is equal to X or, equivalently, every nonempty open subset of X contains an element of S. The space X is called **separable** if it admits a countable dense subset. (A set is called **countable** if it is either finite or countably infinite.)

Corollary 1.1.7. *Every compact metric space is separable.*

Proof. Let $n \in \mathbb{N}$. Since X is totally bounded by Theorem 1.1.4, there exists a finite set $S_n \subset X$ such that $X = \bigcup_{\xi \in S_n} B_{1/n}(\xi)$. Hence $S := \bigcup_{n\in\mathbb{N}} S_n$ is a countable dense subset of X by the axiom of countable choice. □

Corollary 1.1.8. *Let (X, d) be a metric space and let $A \subset X$. Then the following are equivalent.*

(i) *A is precompact.*

(ii) *Every sequence in A has a subsequence that converges in X.*

(iii) *A is totally bounded and every Cauchy sequence in A converges in X.*

Proof. That (i) implies (ii) follows directly from the definitions.

We prove that (ii) implies (iii). By (ii) every sequence in A has a Cauchy subsequence and so A is totally bounded by Lemma 1.1.5. If $(x_n)_{n\in\mathbb{N}}$ is a Cauchy sequence in A, then by (ii) there exists a subsequence $(x_{n_i})_{i\in\mathbb{N}}$ that converges in X, and so the original sequence converges in X because a Cauchy sequence converges if and only if it has a convergent subsequence.

We prove that (iii) implies (i). Let $(x_n)_{n\in\mathbb{N}}$ be a sequence in the closure $\overline{A}$ of A. Then, by the axiom of countable choice, there exists a sequence $(a_n)_{n\in\mathbb{N}}$ in A such that $d(x_n, a_n) < 1/n$ for all $n \in \mathbb{N}$. Since A is totally bounded, it follows from Lemma 1.1.5 that the sequence $(a_n)_{n\in\mathbb{N}}$ has a Cauchy subsequence $(a_{n_i})_{i\in\mathbb{N}}$. This subsequence converges in X by (iii). Denote its limit by a. Then $a \in \overline{A}$ and $a = \lim_{i\to\infty} a_{n_i} = \lim_{i\to\infty} x_{n_i}$. Thus $\overline{A}$ is sequentially compact. This proves Corollary 1.1.8. □

Corollary 1.1.9. *Let (X, d) be a complete metric space and let $A \subset X$. Then the following are equivalent.*

(i) *A is precompact.*

(ii) *Every sequence in A has a Cauchy subsequence.*

(iii) *A is totally bounded.*

Proof. This follows directly from the definitions and Corollary 1.1.8. □

1.1.3. The Arzelà–Ascoli Theorem. It is a recurring theme in functional analysis to understand which subsets of a Banach space or topological vector space are compact. For the standard Euclidean space $(\mathbb{R}^n, \|\cdot\|_2)$ the Heine-Borel Theorem asserts that a subset of $\mathbb{R}^n$ is compact if and only if it is closed and bounded. This continues to hold for every finite-dimensional normed vector space and, conversely, every normed vector space in which the closed unit ball is compact is necessarily finite-dimensional (see Theorem 1.2.11 below). For infinite-dimensional Banach spaces this leads to the problem of characterizing the compact subsets. Necessary conditions are that the subset is closed and bounded, however, these conditions can no longer be sufficient. For the Banach space of continuous functions on a compact metric space a characterization of the compact subsets is given by a theorem of Arzelà and Ascoli which we explain next.

Let (X, d_X) and (Y, d_Y) be metric spaces and assume that X is compact. Then the space

$$C(X,Y) := \{f : X \to Y \,|\, f \text{ is continuous}\}$$

of continuous maps from X to Y is a metric space with the distance function

$$d(f,g) := \sup_{x\in X} d_Y(f(x), g(x)) \qquad \text{for } f,g \in C(X,Y). \tag{1.1.9}$$

This is well defined because the function $X \to \mathbb{R} : x \mapsto d_Y(f(x), g(x))$ is continuous and hence is bounded because X is compact. That (1.1.9) satisfies the axioms of a distance function follows directly from the definitions. When X is nonempty, the metric space $C(X,Y)$ with the distance function (1.1.9) is complete if and only if Y is complete, because the limit of a uniformly convergent sequence of continuous functions is again continuous.

Definition 1.1.10. A subset

$$\mathscr{F} \subset C(X,Y)$$

is called **equi-continuous** if, for every $\varepsilon > 0$, there exists a constant $\delta > 0$ such that, for all $x, x' \in X$ and all $f \in \mathscr{F}$,

$$d_X(x,x') < \delta \qquad \Longrightarrow \qquad d_Y(f(x), f(x')) < \varepsilon.$$

It is called **pointwise compact** if, for every element $x \in X$, the set

$$\mathscr{F}(x) := \{f(x) \,|\, f \in \mathscr{F}\}$$

is a compact subset of Y. It is called **pointwise precompact** if, for every element $x \in X$, the set $\mathscr{F}(x)$ has a compact closure in Y.

Since every continuous map defined on a compact metric space is uniformly continuous, every finite subset of $C(X,Y)$ is equi-continuous.

Theorem 1.1.11 (**Arzelà–Ascoli**). *Let (X, d_X) and (Y, d_Y) be metric spaces such that X is compact and let $\mathscr{F} \subset C(X, Y)$. Then the following are equivalent.*

(i) *$\mathscr{F}$ is precompact.*

(ii) *$\mathscr{F}$ is pointwise precompact and equi-continuous.*

Proof. We prove that (i) implies (ii). Thus assume $\mathscr{F}$ is precompact. That $\mathscr{F}$ is pointwise precompact follows from the fact that the evaluation map $C(X, Y) \to Y : f \mapsto \mathrm{ev}_x(f) := f(x)$ is continuous for every $x \in X$. Since the image of a precompact set under a continuous map is again precompact (Exercise 1.7.1), it follows that the set $\mathscr{F}(x) = \mathrm{ev}_x(\mathscr{F})$ is a precompact subset of Y for every $x \in X$.

It remains to prove that $\mathscr{F}$ is equi-continuous. Assume $\mathscr{F}$ is nonempty and fix a constant $\varepsilon > 0$. Since the set $\mathscr{F}$ is totally bounded by Lemma 1.1.5, there exist finitely many maps $f_1, \dots, f_m \in \mathscr{F}$ such that

$$\mathscr{F} \subset \bigcup_{i=1}^{m} B_{\varepsilon/3}(f_i). \tag{1.1.10}$$

Since X is compact, each function f_i is uniformly continuous. Hence there exists a constant $\delta > 0$ such that, for all $i \in \{1, \dots, m\}$ and all $x, x' \in X$,

$$d_X(x, x') < \delta \qquad \Longrightarrow \qquad d_Y(f_i(x), f_i(x')) < \frac{\varepsilon}{3}. \tag{1.1.11}$$

Now let $f \in \mathscr{F}$ and let $x, x' \in X$ such that

$$d_X(x, x') < \delta. \tag{1.1.12}$$

By (1.1.10) there exists an index $i \in \{1, \dots, m\}$ such that $d(f, f_i) < \frac{\varepsilon}{3}$. Thus

$$d_Y(f(x), f_i(x)) < \frac{\varepsilon}{3}, \qquad d_Y(f(x'), f_i(x')) < \frac{\varepsilon}{3}.$$

Moreover, it follows from (1.1.11) and (1.1.12) that

$$d_Y(f_i(x), f_i(x')) < \frac{\varepsilon}{3}.$$

Putting these last three inequalities together and using the triangle inequality, we find

$$\begin{aligned} d_Y(f(x), f(x')) &\leq d_Y(f(x), f_i(x)) + d_Y(f_i(x), f_i(x')) + d_Y(f_i(x'), f(x')) \\ &\leq \frac{\varepsilon}{3} + \frac{\varepsilon}{3} + \frac{\varepsilon}{3} \\ &= \varepsilon. \end{aligned}$$

This shows that $\mathscr{F}$ is equi-continuous, and thus we have proved that (i) implies (ii).

We prove that (ii) implies (i). Let $(f_n)_{n\in\mathbb{N}}$ be a sequence in $\mathscr{F}$ and let $(x_k)_{k\in\mathbb{N}}$ be a dense sequence in X (Corollary 1.1.7). We prove in three steps that $(f_n)_{n\in\mathbb{N}}$ has a convergent subsequence.

Step 1. *There exists a subsequence $(g_i)_{i\in\mathbb{N}}$ of $(f_n)_{n\in\mathbb{N}}$ such that the sequence $(g_i(x_k))_{i\in\mathbb{N}}$ converges in Y for every $k\in\mathbb{N}$.*

Since $\mathscr{F}(x_k)$ is precompact for each k, it follows from the axiom of dependent choice (page 6) that there is a sequence of subsequences $(f_{n_{k,i}})_{i\in\mathbb{N}}$ such that, for each $k\in\mathbb{N}$, the sequence $(f_{n_{k+1,i}})_{i\in\mathbb{N}}$ is a subsequence of $(f_{n_{k,i}})_{i\in\mathbb{N}}$ and the sequence $(f_{n_{k,i}}(x_k))_{i\in\mathbb{N}}$ converges in Y. Thus the diagonal subsequence $g_i := f_{n_{i,i}}$ satisfies the requirements of Step 1.

Step 2. *The subsequence $(g_i)_{i\in\mathbb{N}}$ in Step 1 is a Cauchy sequence in $C(X,Y)$.*

Fix a constant $\varepsilon>0$. Then, by equi-continuity, there exists a constant $\delta>0$ such that, for all $f\in\mathscr{F}$ and all $x,x'\in X$,

$$d_X(x,x')<\delta \qquad\Longrightarrow\qquad d_Y(f(x),f(x'))<\frac{\varepsilon}{3}. \tag{1.1.13}$$

Since the balls $B_\delta(x_k)$ form an open cover of X, there exists an $m\in\mathbb{N}$ such that $X=\bigcup_{k=1}^m B_\delta(x_k)$. Since $(g_i(x_k))_{i\in\mathbb{N}}$ is a Cauchy sequence for each k, there exists an $N\in\mathbb{N}$ such that, for all $i,j,k\in\mathbb{N}$, we have

$$1\le k\le m,\quad i,j\ge N \qquad\Longrightarrow\qquad d_Y(g_i(x_k),g_j(x_k))<\varepsilon/3. \tag{1.1.14}$$

We prove that $d(g_i,g_j)<\varepsilon$ for all $i,j\ge N$. To see this, fix an element $x\in X$. Then there exists an index $k\in\{1,\dots,m\}$ such that $d_X(x,x_k)<\delta$. This implies $d_Y(g_i(x),g_i(x_k))<\varepsilon/3$ for all $i\in\mathbb{N}$, by (1.1.13), and so

$$\begin{aligned} d_Y(g_i(x),g_j(x)) &\le d_Y(g_i(x),g_i(x_k))+d_Y(g_i(x_k),g_j(x_k))+d_Y(g_j(x_k),g_j(x))\\ &<\frac{\varepsilon}{3}+\frac{\varepsilon}{3}+\frac{\varepsilon}{3}\\ &=\varepsilon \end{aligned}$$

for all $i,j\ge N$ by (1.1.14). Hence $d(g_i,g_j)=\max_{x\in X}d_Y(g_i(x),g_j(x))<\varepsilon$ for all $i,j\ge N$ and this proves Step 2.

Step 3. *The subsequence $(g_i)_{i\in\mathbb{N}}$ in Step 1 converges in $C(X,Y)$.*

Let $x\in X$. By Step 2, $(g_i(x))_{i\in\mathbb{N}}$ is a Cauchy sequence in $\mathscr{F}(x)$. Since $\mathscr{F}(x)$ is a precompact subset of Y, the sequence $(g_i(x))_{i\in\mathbb{N}}$ has a convergent subsequence and hence converges in Y. Denote the limit by $g(x):=\lim_{i\to\infty}g_i(x)$. Then the sequence g_i converges uniformly to g by Step 2 and so $g\in C(X,Y)$.

Step 3 shows that every sequence in $\mathscr{F}$ has a subsequence that converges to an element of $C(X,Y)$. Hence $\mathscr{F}$ is precompact by Corollary 1.1.8. This proves Theorem 1.1.11. □

Corollary 1.1.12 (Arzelà–Ascoli). *Let (X, d_X) be a compact metric space, let (Y, d_Y) be a metric space, and let $\mathscr{F} \subset C(X, Y)$. Then the following are equivalent.*

(i) *$\mathscr{F}$ is compact.*

(ii) *$\mathscr{F}$ is closed, pointwise compact, and equi-continuous.*

(iii) *$\mathscr{F}$ is closed, pointwise precompact, and equi-continuous.*

Proof. That (i) implies (ii) follows from Theorem 1.1.11, because every compact subset of a metric space is closed, and the image of a compact set under a continuous map is compact. Here the continuous map in question is the evaluation map $C(X, Y) \to Y : f \mapsto f(x)$ associated to $x \in X$. That (ii) implies (iii) is obvious. That (iii) implies (i) follows from Theorem 1.1.11, because a subset of a metric space is compact if and only if it is precompact and closed. This proves Corollary 1.1.12. □

When the target space Y is the Euclidean space $(\mathbb{R}^n, \|\cdot\|_2)$ in part (i) of Example 1.1.3, the Arzelà–Ascoli Theorem takes the following form.

Corollary 1.1.13 (Arzelà–Ascoli). *Let (X, d) be a compact metric space and let $\mathscr{F} \subset C(X, \mathbb{R}^n)$. Then the following hold.*

(i) *$\mathscr{F}$ is precompact if and only if it is bounded and equi-continuous.*

(ii) *$\mathscr{F}$ is compact if and only if it is closed, bounded, and equi-continuous.*

Proof. Assume $\mathscr{F}$ is precompact. Then $\mathscr{F}$ is equi-continuous by Theorem 1.1.11, and is bounded, because a sequence whose norm tends to infinity cannot have a convergent subsequence. Conversely, assume $\mathscr{F}$ is bounded and equi-continuous. Then, for each $x \in X$, the set $\mathscr{F}(x) \subset \mathbb{R}^n$ is bounded and therefore is precompact by the Heine–Borel Theorem. Hence $\mathscr{F}$ is precompact by Theorem 1.1.11. This proves (i). Part (ii) follows from (i) and the fact that a subset of a metric space is compact if and only if it is precompact and closed. This proves Corollary 1.1.13. □

Exercise 1.1.14. This exercise shows that the hypothesis that X is compact cannot be removed in Corollary 1.1.13. Consider the Banach space $C_b(\mathbb{R})$ of bounded continuous real valued functions on $\mathbb{R}$ with the supremum norm. Find a closed bounded equi-continuous subset of $C_b(\mathbb{R})$ that is not compact.

There are many versions of the Arzelà–Ascoli Theorem. For example, Theorem 1.1.11, Corollary 1.1.12, and Corollary 1.1.13 continue to hold, with the appropriate notion of equi-continuity, when X is any compact topological space. This is the content of the following exercise.

Exercise 1.1.15. Let X be a compact topological space and let Y be a metric space. Then the space $C(X,Y)$ of continuous functions $f : X \to Y$ is a metric space with the distance function (1.1.9). A subset $\mathscr{F} \subset C(X,Y)$ is called **equi-continuous** if, for every $x \in X$ and every $\varepsilon > 0$, there exists an open neighborhood $U \subset X$ of x such that $d_Y(f(x), f(x')) < \varepsilon$ for all $x' \in U$ and all $f \in \mathscr{F}$.

(a) Prove that the above definition of equi-continuity agrees with the one in Definition 1.1.10 whenever X is a compact metric space.

(b) Prove the following variant of the Arzelà–Ascoli Theorem for compact topological spaces X.

Arzelà–Ascoli Theorem. *Let X be a compact topological space and let Y be a metric space. A set $\mathscr{F} \subset C(X,Y)$ is precompact if and only if it is pointwise precompact and equi-continuous.*

Hint 1: If $\mathscr{F}$ is precompact, use the argument in the proof of Theorem 1.1.11 to show that $\mathscr{F}$ is pointwise precompact and equi-continuous.

Hint 2: Assume $\mathscr{F}$ is equi-continuous and pointwise precompact.

Step 1. The set $F := \{f(x) \,|\, x \in X,\ f \in \mathscr{F}\} \subset Y$ is totally bounded.
Show that F is precompact (Exercise 1.7.1) and use Corollary 1.1.8.

Step 2. The set $\mathscr{F}$ is totally bounded.
Let $\varepsilon > 0$. Cover F by finitely many open balls $V_1, \dots, V_n$ of radius $\varepsilon/3$ and cover X by finitely many open sets $U_1, \dots, U_m$ such that

$$\sup_{x,x' \in U_i} \sup_{f \in \mathscr{F}} d_Y(f(x), f(x')) < \varepsilon/3 \qquad \text{for } i = 1, \dots, m.$$

For any function $\alpha : \{1, \dots, m\} \to \{1, \dots, n\}$ define

$$\mathscr{F}_\alpha := \left\{ f \in \mathscr{F} \,\middle|\, f(U_i) \cap V_{\alpha(i)} \neq \emptyset \text{ for } i = 1, \dots, m \right\}.$$

Prove that $d(f,g) = \sup_{x \in X} d_Y(f(x), g(x)) < \varepsilon$ for all $f, g \in \mathscr{F}_\alpha$. Let A be the set of all α such that $\mathscr{F}_\alpha \neq \emptyset$. Prove that $\mathscr{F} = \bigcup_{\alpha \in A} \mathscr{F}_\alpha$ and choose a collection of functions $f_\alpha \in \mathscr{F}_\alpha$, one for each $\alpha \in A$.

Step 3. The set $\mathscr{F}$ is precompact.
Use Lemma 1.1.5 and Step 3 in the proof of Theorem 1.1.11 to show that every sequence in $\mathscr{F}$ has a subsequence that converges in $C(X,Y)$.

In contrast to what one might expect from Exercise 1.1.14, there is also a version of the Arzelà–Ascoli Theorem for the space of continuous functions from an arbitrary topological space X to a metric space Y. This version uses the compact-open topology on $C(X,Y)$ and is explained in Exercise 3.7.5.

1.2. Finite-Dimensional Banach Spaces

The purpose of the present section is to examine finite-dimensional normed vector spaces with an emphasis on those properties that distinguish them from infinite-dimensional normed vector spaces, which are the main subject of functional analysis. Finite-dimensional normed vector spaces are complete, their linear subspaces are closed, linear functionals on them are continuous, and their closed unit balls are compact. Theorem 1.2.11 below shows that this last property characterizes finite-dimensionality. Before entering into the main topic of this section, it is convenient to first introduce the concept of a bounded linear operator.

1.2.1. Bounded Linear Operators. The second fundamental concept in functional analysis, after that of a Banach space, is the notion of a bounded linear operator. In functional analysis it is common practice to use the term *linear operator* instead of *linear map*, although both terms have the exact same meaning, namely that of a map between vector spaces that preserves addition and scalar multiplication. The reason lies in the fact that the relevant normed vector spaces in applications are often function spaces and then the elements of the space on which the operator acts are themselves functions. If domain and target of a linear operator are *normed* vector spaces, it is natural to impose continuity with respect to the norm topologies. This underlies the following definition.

Definition 1.2.1 (Bounded Linear Operator). Let $(X, \|\cdot\|_X)$ and $(Y, \|\cdot\|_Y)$ be real normed vector spaces. A linear operator

$$A : X \to Y$$

is called **bounded** if there exists a constant $c \geq 0$ such that

$$\|Ax\|_Y \leq c \|x\|_X \qquad \text{for all } x \in X. \tag{1.2.1}$$

The smallest constant $c \geq 0$ that satisfies (1.2.1) is called the **operator norm of** A and is denoted by

$$\|A\| := \|A\|_{\mathcal{L}(X,Y)} := \sup_{x \in X \setminus \{0\}} \frac{\|Ax\|_Y}{\|x\|_X}. \tag{1.2.2}$$

A bounded linear operator with values in $Y = \mathbb{R}$ is called a **bounded linear functional** on X. The space of bounded linear operators from X to Y is denoted by[1]

$$\mathcal{L}(X, Y) := \left\{ A : X \to Y \,\middle|\, A \text{ is linear and bounded} \right\}.$$

Then $(\mathcal{L}(X, Y), \|\cdot\|_{\mathcal{L}(X,Y)})$ is a normed vector space. The resulting topology on $\mathcal{L}(X, Y)$ is called the **uniform operator topology**.

[1]Many authors use the notation $\mathcal{B}(X, Y)$ for the space of bounded linear operators.

Theorem 1.2.2. *Let $(X, \|\cdot\|_X)$ and $(Y, \|\cdot\|_Y)$ be real normed vector spaces and let $A : X \to Y$ be a linear operator. The following are equivalent.*

(i) *A is bounded.*

(ii) *A is continuous.*

(iii) *A is continuous at $x = 0$.*

Proof. We prove that (i) implies (ii). If A is bounded, then

$$\begin{aligned} \|Ax - Ax'\|_Y &= \|A(x - x')\|_Y \\ &\le \|A\| \, \|x - x'\|_X \end{aligned}$$

for all $x, x' \in X$ and so A is Lipschitz-continuous. Since every Lipschitz-continuous function is continuous, this shows that (i) implies (ii). That (ii) implies (iii) follows directly from the definition of continuity.

We prove that (iii) implies (i). Thus assume A is continuous at $x = 0$. Then it follows from the ε-δ definition of continuity with $\varepsilon = 1$ that there exists a constant $\delta > 0$ such that, for all $x \in X$,

$$\|x\|_X < \delta \qquad \Longrightarrow \qquad \|Ax\|_Y < 1.$$

This implies $\|Ax\|_Y \le 1$ for every $x \in X$ with $\|x\|_X = \delta$. Now let $x \in X \backslash \{0\}$. Then $\|\delta \|x\|_X^{-1} x\|_X = \delta$ and so $\|A(\delta \|x\|_X^{-1} x)\|_X \le 1$. Multiply both sides of this last inequality by $\delta^{-1}\|x\|_X$ to obtain the inequality

$$\|Ax\|_Y \le \delta^{-1} \, \|x\|_X$$

for all $x \in X$. This proves Theorem 1.2.2. □

Recall that the **kernel** and **image** of a linear operator $A : X \to Y$ between real vector spaces are the linear subspaces defined by

$$\begin{aligned} \ker(A) &:= \{x \in X \mid Ax = 0\} \subset X, \\ \operatorname{im}(A) &:= \{Ax \mid x \in X\} \subset Y. \end{aligned}$$

If X and Y are normed vector spaces and $A : X \to Y$ is a bounded linear operator, then the kernel of A is a closed subspace of X by Theorem 1.2.2. However, its image need not be a closed subspace of Y.

Definition 1.2.3 (Equivalent Norms). Let X be a real vector space. Two norms $\|\cdot\|$ and $\|\cdot\|'$ on X are called **equivalent** if there is a constant

$$c \ge 1$$

such that

$$\frac{1}{c} \, \|x\| \le \|x\|' \le c \, \|x\|$$

for all $x \in X$.

Exercise 1.2.4. **(i)** Definition 1.2.3 yields an equivalence relation on the set of all norm functions on X.

(ii) Two norms $\|\cdot\|$ and $\|\cdot\|'$ on X are equivalent if and only if the identity maps $\mathrm{id} : (X, \|\cdot\|) \to (X, \|\cdot\|')$ and $\mathrm{id} : (X, \|\cdot\|') \to (X, \|\cdot\|)$ are bounded linear operators.

(iii) Two norms $\|\cdot\|$ and $\|\cdot\|'$ on X are equivalent if and only if they induce the same topologies on X, i.e. $\mathscr{U}(X, \|\cdot\|) = \mathscr{U}(X, \|\cdot\|')$.

(iv) Let $\|\cdot\|$ and $\|\cdot\|'$ be equivalent norms on X. Show that $(X, \|\cdot\|)$ is complete if and only if $(X, \|\cdot\|')$ is complete.

1.2.2. Finite-Dimensional Normed Vector Spaces.

Theorem 1.2.5. *Let X be a finite-dimensional real vector space. Then any two norms on X are equivalent.*

Proof. Choose an ordered basis $e_1, \dots, e_n$ on X and define

$$\|x\|_2 := \sqrt{\sum_{i=1}^{n} |x_i|^2} \qquad \text{for } x = \sum_{i=1}^{n} x_i e_i, \qquad x_i \in \mathbb{R}.$$

This is a norm on X. We prove in two steps that every norm on X is equivalent to $\|\cdot\|_2$. Fix any norm function $X \to \mathbb{R} : x \mapsto \|x\|$.

Step 1. *There is a constant $c > 0$ such that $\|x\| \le c \|x\|_2$ for all $x \in X$.*

Define $c := \sqrt{\sum_{i=1}^{n} \|e_i\|^2}$ and let $x = \sum_{i=1}^{n} x_i e_i$ with $x_i \in \mathbb{R}$. Then, by the triangle inequality for $\|\cdot\|$ and the Cauchy–Schwarz inequality on $\mathbb{R}^n$, we have

$$\|x\| \le \sum_{i=1}^{n} |x_i| \|e_i\| \le \sqrt{\sum_{i=1}^{n} |x_i|^2} \sqrt{\sum_{i=1}^{n} \|e_i\|^2} = c \|x\|_2 .$$

This proves Step 1.

Step 2. *There is a constant $\delta > 0$ such that $\delta \|x\|_2 \le \|x\|$ for all $x \in X$.*

The set $S := \{x \in X \mid \|x\|_2 = 1\}$ is compact with respect to $\|\cdot\|_2$ by the Heine–Borel Theorem, and the function $S \to \mathbb{R} : x \mapsto \|x\|$ is continuous by Step 1. Hence there exists an element $x_0 \in S$ such that $\|x_0\| \le \|x\|$ for all $x \in S$. Define

$$\delta := \|x_0\| > 0.$$

Then every nonzero vector $x \in X$ satisfies $\|x\|_2^{-1} x \in S$, hence $\| \|x\|_2^{-1} x\| \ge \delta$, and hence $\|x\| \ge \delta \|x\|_2$. This proves Step 2 and Theorem 1.2.5. □

Theorem 1.2.5 has several important consequences that are special to finite-dimensional normed vector spaces and do not carry over to infinite dimensions.

Corollary 1.2.6. *Every finite-dimensional normed vector space is complete.*

Proof. This holds for the Euclidean norm on $\mathbb{R}^n$ by a theorem in first year analysis, which follows rather directly from the completeness of the real numbers. Hence, by Theorem 1.2.5 and part (iv) of Exercise 1.2.4, it holds for every norm on $\mathbb{R}^n$. Thus it holds for every finite-dimensional normed vector space. □

Corollary 1.2.7. *Let $(X, \|\cdot\|)$ be a normed vector space. Then every finite-dimensional linear subspace of X is a closed subset of X.*

Proof. Let $Y \subset X$ be a finite-dimensional linear subspace and denote by $\|\cdot\|_Y$ the restriction of the norm on X to the subspace Y. Then $(Y, \|\cdot\|_Y)$ is complete by Corollary 1.2.6 and hence Y is a closed subset of X. □

Corollary 1.2.8. *Let $(X, \|\cdot\|)$ be a finite-dimensional normed vector space and let $K \subset X$. Then K is compact if and only if K is closed and bounded.*

Proof. This holds for the Euclidean norm on $\mathbb{R}^n$ by the Heine–Borel Theorem. Hence it holds for every norm on $\mathbb{R}^n$ by Theorem 1.2.5. Hence it holds for every finite-dimensional normed vector space. □

Corollary 1.2.9. *Let $(X, \|\cdot\|_X)$ and $(Y, \|\cdot\|_Y)$ be normed vector spaces and suppose* $\dim X < \infty$. *Then every linear operator $A : X \to Y$ is bounded.*

Proof. Define the function $X \to \mathbb{R} : x \to \|x\|_A$ by

$$\|x\|_A := \|x\|_X + \|Ax\|_Y \qquad \text{for } x \in X.$$

This is a norm on X. Hence, by Theorem 1.2.5, there exists a constant $c \geq 1$ such that $\|x\|_A \leq c\,\|x\|_X$ for all $x \in X$. Hence A is bounded. □

The above four corollaries spell out some of the standard facts in finite-dimensional linear algebra. The following four examples show that in none of these four corollaries can the hypothesis of finite-dimensionality be dropped. Thus in functional analysis one must dispense with some of the familiar features of linear algebra. In particular, linear subspaces need no longer be closed subsets and linear maps need no longer be continuous.

Example 1.2.10. (i) Consider the space $X := C([0, 1])$ of continuous real valued functions on the closed unit interval $[0, 1]$. Then the formulas

$$\|f\|_\infty := \sup_{0\leq t\leq 1} |f(t)|, \qquad \|f\|_2 := \left(\int_0^1 |f(t)|^2\right)^{1/2}$$

for $f \in C([0,1])$ define norms on X. The space $C([0,1])$ is complete with $\|\cdot\|_\infty$ but not with $\|\cdot\|_2$. Thus the two norms are not equivalent. (**Exercise:** Find a sequence of continuous functions $f_n : [0,1] \to \mathbb{R}$ that is Cauchy with respect to the L^2-norm and has no convergent subsequence.)

(ii) The space $Y := C^1([0,1])$ of continuously differentiable real valued functions on the closed unit interval is a dense linear subspace of $C([0,1])$ with the supremum norm and so is not a closed subset of $(C([0,1]), \|\cdot\|_\infty)$.

(iii) Consider the closed unit ball

$$B := \{f \in C([0,1]) \mid \|f\|_\infty \le 1\}$$

in the Banach space $C([0,1])$ with the supremum norm. This set is closed and bounded, but not equi-continuous. Hence it is not compact by the Arzelà–Ascoli Theorem (Corollary 1.1.13). More explicitly, for $n \in \mathbb{N}$ define the function $f_n \in B$ by $f_n(t) := \sin(2^n \pi t)$ for $0 \le t \le 1$. Then $\|f_n - f_m\| \ge 1$ for $n \neq m$ and hence the sequence $(f_n)_{n\in\mathbb{N}}$ does not have any convergent subsequence. Theorem 1.2.11 below shows that the compactness of the unit ball characterizes the finite-dimensional normed vector spaces.

(iv) Let $(X, \|\cdot\|)$ be an infinite-dimensional normed vector space and choose an unordered basis $E \subset X$ such that $\|e\| = 1$ for all $e \in E$. Thus every nonzero vector $x \in X$ can be uniquely expressed as a finite linear combination $x = \sum_{i=1}^{\ell} x_i e_i$ with $e_1, \dots, e_\ell \in E$ pairwise distinct and $x_i \in \mathbb{R} \setminus \{0\}$. By assumption E is an infinite set. (The existence of an unordered basis requires the Lemma of Zorn or, equivalently, the axiom of choice by Theorem A.1.3.) Choose any unbounded function $\lambda : E \to \mathbb{R}$ and define the linear map $\Phi_\lambda : X \to \mathbb{R}$ by $\Phi_\lambda(\sum_{i=1}^{\ell} x_i e_i) := \sum_{i=1}^{\ell} \lambda(e_i) x_i$ for all $\ell \in \mathbb{N}$, all pairwise distinct ℓ-tuples of basis vectors $e_1, \dots, e_\ell \in E$, and all $x_1, \dots, x_\ell \in \mathbb{R}$. Then $\Phi_\lambda : X \to \mathbb{R}$ is an unbounded linear functional.

Theorem 1.2.11. *Let $(X, \|\cdot\|)$ be a normed vector space and denote the closed unit ball and the closed unit sphere in X by*

$$B := \{x \in X \mid \|x\| \le 1\}, \qquad S := \{x \in X \mid \|x\| = 1\}.$$

Then the following are equivalent.

(i) $\dim X < \infty$.

(ii) *B is compact.*

(iii) *S is compact.*

Proof. That (i) implies (ii) follows from Corollary 1.2.8 and that (ii) implies (iii) follows from the fact that a closed subset of a compact set in a topological space is compact.

We prove that (iii) implies (i). We argue indirectly and show that if X is infinite-dimensional, then S is not compact. Thus assume X is infinite-dimensional. We claim that there exists a sequence $x_i \in X$ such that

$$(1.2.3) \qquad \|x_i\| = 1, \qquad \|x_i - x_j\| \geq \tfrac{1}{2} \qquad \text{for all } i, j \in \mathbb{N} \text{ with } i \neq j.$$

This is then a sequence in S that does not have any convergent subsequence and so it follows that S is not compact.

To prove the existence of a sequence in X satisfying (1.2.3) we argue by induction and use the axiom of dependent choice. For $i = 1$ choose any element $x_1 \in S$. If $x_1, \dots, x_k \in S$ satisfy $\|x_i - x_j\| \geq \frac{1}{2}$ for $i \neq j$, consider the subspace $Y \subset X$ spanned by the vectors $x_1, \dots, x_k$. This is a closed subspace of X by Corollary 1.2.7 and is not equal to X because $\dim X = \infty$. Hence Lemma 1.2.12 below asserts that there exists a vector $x = x_{k+1} \in S$ such that $\|x - y\| \geq \frac{1}{2}$ for all $y \in Y$ and hence, in particular, $\|x_{k+1} - x_i\| \geq \frac{1}{2}$ for $i = 1, \dots, k$. This completes the induction step and shows, by the axiom of dependent choice (see page 6), that there exists a sequence $x_i \in X$ that satisfies (1.2.3) for $i \neq j$.

More precisely, take $\mathbf{X} := \bigsqcup_{k \in \mathbb{N}} S^k$ and, for every $\mathbf{x} = (x_1, \dots, x_k) \in S^k$, define $\mathbf{A}(\mathbf{x})$ as the set of all $k+1$-tuples $\mathbf{y} = (x_1, \dots, x_k, x) \in S^{k+1}$ such that $\|x - x_i\| \geq \frac{1}{2}$ for $i = 1, \dots, k$. The above argument shows that this set is nonempty for all $\mathbf{x} \in \mathbf{X}$ and so the existence of the required sequence $(x_i)_{i \in \mathbb{N}}$ follows from the axiom of dependent choice. This proves Theorem 1.2.11. □

Lemma 1.2.12 (Riesz Lemma). *Let $(X, \|\cdot\|)$ be a normed vector space and let $Y \subset X$ be a closed linear subspace that is not equal to X. Fix a constant $0 < \delta < 1$. Then there exists a vector $x \in X$ such that*

$$\|x\| = 1, \qquad \inf_{y \in Y} \|x - y\| \geq 1 - \delta.$$

Proof. Let $x_0 \in X \setminus Y$. Then $d := \inf_{y \in Y} \|x_0 - y\| > 0$ because Y is closed. Choose $y_0 \in Y$ such that

$$\|x_0 - y_0\| \leq \frac{d}{1 - \delta}$$

and define $x := \|x_0 - y_0\|^{-1}(x_0 - y_0)$. Then $\|x\| = 1$ and

$$\|x - y\| = \frac{\|x_0 - y_0 - \|x_0 - y_0\| \, y\|}{\|x_0 - y_0\|} \geq \frac{d}{\|x_0 - y_0\|} \geq 1 - \delta$$

for all $y \in Y$. This proves Lemma 1.2.12. □

Theorem 1.2.11 leads to the question of how one can characterize the compact subsets of an infinite-dimensional Banach space. For the Banach space of continuous functions on a compact metric space with the supremum norm this question is answered by the Arzelà–Ascoli Theorem (Corollary 1.1.13). The Arzelà–Ascoli Theorem is the source of many other compactness results in functional analysis.

1.2.3. Quotient and Product Spaces.

Quotient Spaces. Let $(X, \|\cdot\|)$ be a real normed vector space and let $Y \subset X$ be a closed subspace. Define an equivalence relation $\sim$ on X by

$$x \sim x' \qquad \Longleftrightarrow \qquad x' - x \in Y.$$

Denote the equivalence class of an element $x \in X$ under this equivalence relation by $[x] := x + Y := \{x + y \,|\, y \in Y\}$ and denote the quotient space by

$$X/Y := \{x + Y \,|\, x \in X\}.$$

For $x \in X$ define

$$\|[x]\|_{X/Y} := \inf_{y \in Y} \|x + y\|_X \,. \tag{1.2.4}$$

Then X/Y is a real vector space and the formula (1.2.4) defines a norm function on X/Y. (**Exercise:** Prove this.) The next lemma is the key step in the proof that if X is a Banach space so is the quotient space X/Y for every closed linear subspace $Y \subset X$.

Lemma 1.2.13. *Let X be a normed vector space and let $Y \subset X$ be a closed linear subspace. Let $(x_i)_{i\in\mathbb{N}}$ be a sequence in X such that $([x_i])_{i\in\mathbb{N}}$ is a Cauchy sequence in X/Y with respect to the norm (1.2.4). Then there exist a subsequence $(x_{i_k})_{k\in\mathbb{N}}$ and a sequence $(y_k)_{k\in\mathbb{N}}$ in Y such that $(x_{i_k} + y_k)_{k\in\mathbb{N}}$ is a Cauchy sequence in X.*

Proof. Choose $i_1 := 1$ and let $i_2 > i_1$ be the smallest integer bigger than i_1 such that $\inf_{y\in Y} \|x_{i_1} - x_{i_2} + y\|_X < 2^{-1}$. Once $i_1, \dots, i_k$ have been constructed, choose $i_{k+1} > i_k$ to be the smallest integer bigger than i_k such that $\inf_{y\in Y} \left\|x_{i_k} - x_{i_{k+1}} + y\right\|_X < 2^{-k}$. This completes the inductive construction of the subsequence $(x_{i_k})_{k\in\mathbb{N}}$. Now use the axiom of countable choice to find a sequence $(\eta_k)_{k\in\mathbb{N}}$ in Y such that $\left\|x_{i_k} - x_{i_{k+1}} + \eta_k\right\|_X < 2^{-k}$ for all $k \in \mathbb{N}$. Define

$$y_1 := 0, \qquad y_k := -\eta_1 - \cdots - \eta_{k-1} \qquad \text{for } k \geq 2.$$

Then

$$\left\|x_{i_k} + y_k - x_{i_{k+1}} - y_{k+1}\right\|_X = \left\|x_{i_k} - x_{i_{k+1}} + \eta_k\right\|_X < 2^{-k}$$

for all $k \in \mathbb{N}$ and hence $(x_{i_k} + y_k)_{k\in\mathbb{N}}$ is a Cauchy sequence. This proves Lemma 1.2.13. □

Theorem 1.2.14 (Quotient Space). *Let X be a normed vector space and let $Y \subset X$ be a closed linear subspace. Then the following hold.*

(i) *The map $\pi : X \to X/Y$ defined by $\pi(x) := [x] = x + Y$ for $x \in X$ is a surjective bounded linear operator.*

(ii) *Let $A : X \to Z$ be a bounded linear operator with values in a normed vector space Z such that $Y \subset \ker(A)$. Then there exists a unique bounded linear operator $A_0 : X/Y \to Z$ such that $A_0 \circ \pi = A$.*

(iii) *If X is a Banach space, then X/Y is a Banach space.*

Proof. Part (i) follows directly from the definitions.

To prove part (ii) observe that the operator $A_0 : X/Y \to Z$, given by

$$A_0[x] := Ax \qquad \text{for } x \in X,$$

is well defined whenever $Y \subset \ker(A)$. It is obviously linear and it satisfies

$$\|A_0[x]\|_Z = \|A(x+y)\|_Z \le \|A\| \, \|x+y\|_X$$

for all $x \in X$ and all $y \in Y$. Take the infimum over all $y \in Y$ to obtain the inequality $\|A_0[x]\|_Z \le \inf_{y \in Y} \|A\| \, \|x+y\|_X = \|A\| \, \|[x]\|_{X/Y}$ for all $x \in X$. This proves part (ii).

To prove part (iii), assume X is complete and let $(x_i)_{i \in \mathbb{N}}$ be a sequence in X such that $([x_i])_{i \in \mathbb{N}}$ is a Cauchy sequence in X/Y with respect to the norm (1.2.4). By Lemma 1.2.13 there exist a subsequence $(x_{i_k})_{k \in \mathbb{N}}$ and a sequence $(y_k)_{k \in \mathbb{N}}$ in Y such that $(x_{i_k} + y_k)_{k \in \mathbb{N}}$ is a Cauchy sequence in X. Since X is a Banach space, there exists an element $x \in X$ such that

$$\lim_{k \to \infty} \|x - x_{i_k} - y_k\|_X = 0.$$

Hence $\lim_{k \to \infty} \|[x - x_{i_k}]\|_{X/Y} = \lim_{k \to \infty} \inf_{y \in Y} \|x - x_{i_k} + y\|_X = 0$. Thus the subsequence $([x_{i_k}])_{k \in \mathbb{N}}$ converges to $[x]$ in X/Y. Since a Cauchy sequence converges whenever it has a convergent subsequence, this proves Theorem 1.2.14. □

Product Spaces. Let X and Y be normed vector spaces. Then the product space $X \times Y$ admits the structure of a normed vector space. However, there is no canonical norm on this product space although it has a canonical product topology (page 110). Examples of norms that induce the product topology are

$$\|(x,y)\|_p := \left(\|x\|_X^p + \|y\|_Y^p\right)^{1/p}, \qquad 1 \le p < \infty, \tag{1.2.5}$$

and

$$\|(x,y)\|_\infty := \max\left\{\|x\|_X, \|y\|_Y\right\} \tag{1.2.6}$$

for $x \in X$ and $y \in Y$.

Exercise 1.2.15. **(i)** Show that the norms in (1.2.5) and (1.2.6) are all equivalent and induce the product topology on $X \times Y$.

(ii) Show that the product space $X \times Y$, with any of the norms in (1.2.5) or (1.2.6), is a Banach space if and only if X and Y are Banach spaces.

1.3. The Dual Space

1.3.1. The Banach Space of Bounded Linear Operators. This section returns to the normed vector space $\mathcal{L}(X,Y)$ of bounded linear operators from X to Y introduced in Definition 1.2.1. The next theorem shows that the normed vector space $\mathcal{L}(X,Y)$ is complete whenever the target space Y is complete, even if X is not complete.

Theorem 1.3.1. *Let X be a normed vector space and let Y be a Banach space. Then $\mathcal{L}(X,Y)$ is a Banach space with respect to the operator norm.*

Proof. Let $(A_n)_{n\in\mathbb{N}}$ be a Cauchy sequence in $\mathcal{L}(X,Y)$. Then

$$\|A_nx - A_mx\|_Y = \|(A_n - A_m)x\|_Y \le \|A_n - A_m\|\,\|x\|_X$$

for all $x \in X$ and all $m, n \in \mathbb{N}$. Hence $(A_nx)_{n\in\mathbb{N}}$ is a Cauchy sequence in Y for every $x \in X$. Since Y is complete, this implies that the limit

$$Ax := \lim_{n\to\infty} A_nx \tag{1.3.1}$$

exists for all $x \in X$. This defines a map $A : X \to Y$. That it is linear follows from the definition, the fact that the limit of a sum of two sequences is the sum of the limits, and the fact that the limit of a product of a sequence with a scalar is the product of the limit with the scalar.

It remains to prove that A is bounded and that $\lim_{n\to\infty}\|A - A_n\| = 0$. To see this, fix a constant $\varepsilon > 0$. Since $(A_n)_{n\in\mathbb{N}}$ is a Cauchy sequence with respect to the operator norm, there exists an integer $n_0 \in \mathbb{N}$ such that

$$m, n \in \mathbb{N}, \qquad m, n \ge n_0 \qquad \Longrightarrow \qquad \|A_m - A_n\| < \varepsilon.$$

This implies

$$\begin{aligned} \|Ax - A_nx\|_Y &= \lim_{m\to\infty}\|A_mx - A_nx\|_Y \\ &\le \limsup_{m\to\infty}\|A_m - A_n\|\,\|x\|_X \\ &\le \varepsilon\,\|x\|_X \end{aligned} \tag{1.3.2}$$

for every $x \in X$ and every integer $n \ge n_0$. Hence

$$\|Ax\|_Y \le \|Ax - A_{n_0}x\|_Y + \|A_{n_0}x\|_Y \le (\varepsilon + \|A_{n_0}\|)\,\|x\|_X$$

for all $x \in X$ and so A is bounded. It follows also from (1.3.2) that, for each $\varepsilon > 0$, there is an $n_0 \in \mathbb{N}$ such that $\|A - A_n\| \le \varepsilon$ for every integer $n \ge n_0$. Thus $\lim_{n\to\infty}\|A - A_n\| = 0$ and this proves Theorem 1.3.1. □

1.3.2. Examples of Dual Spaces. An important special case is where the target space Y is the real axis. Then Theorem 1.3.1 asserts that the space

$$X^* := \mathcal{L}(X, \mathbb{R}) \tag{1.3.3}$$

of bounded linear functionals $\Lambda : X \to \mathbb{R}$ is a Banach space for every normed vector space X (whether or not X is itself complete). The space of bounded linear functionals on X is called the **dual space of** X. The dual space of a Banach space plays a central role in functional analysis. Here are several examples of dual spaces.

Example 1.3.2 (Dual Space of a Hilbert Space). Let H be a Hilbert space, i.e. H is a Banach space and the norm on H arises from an inner product $H \times H \to \mathbb{R} : (x, y) \mapsto \langle x, y\rangle$ via $\|x\| = \sqrt{\langle x, x\rangle}$. Then every element $y \in H$ determines a linear functional $\Lambda_y : H \to \mathbb{R}$ defined by

$$\Lambda_y(x) := \langle x, y\rangle \qquad \text{for } x \in H. \tag{1.3.4}$$

It is bounded by the Cauchy–Schwarz inequality (Lemma 1.4.2) and the Riesz Representation Theorem asserts that the map

$$H \to H^* : y \mapsto \Lambda_y$$

is an isometric isomorphism (Theorem 1.4.4).

Example 1.3.3 (Dual Space of $L^p(\mu)$). Let $(M, \mathcal{A}, \mu)$ be a measure space and fix a constant $1 < p < \infty$. Define the number $1 < q < \infty$ by

$$\frac{1}{p} + \frac{1}{q} = 1. \tag{1.3.5}$$

The **Hölder inequality** asserts that the product of two functions $f \in \mathcal{L}^p(\mu)$ and $g \in \mathcal{L}^q(\mu)$ is μ-integrable and satisfies

$$\left|\int_M fg\, d\mu\right| \le \|f\|_p \|g\|_q. \tag{1.3.6}$$

(See [**75**, Thm. 4.1].) This implies that every $g \in L^q(\mu)$ determines a bounded linear functional $\Lambda_g : L^p(\mu) \to \mathbb{R}$ defined by

$$\Lambda_g(f) := \int_M fg\, d\mu \qquad \text{for } f \in L^p(\mu). \tag{1.3.7}$$

It turns out that

$$\|\Lambda_g\|_{\mathcal{L}(L^p(\mu),\mathbb{R})} = \|g\|_q$$

for all $g \in L^q(\mu)$ (see [**75**, Thm. 4.33]) and that the map

$$L^q(\mu) \to L^p(\mu)^* : g \mapsto \Lambda_g$$

is an isometric isomorphism (see [**75**, Thm. 4.35]). The proof relies on the Radon–Nikodým Theorem (see [**75**, Thm. 5.4]).

Example 1.3.4 (Dual Space of $L^1(\mu)$). The assertion of Example 1.3.3 extends to the case $p = 1$ and shows that the natural map

$$L^\infty(\mu) \to L^1(\mu)^* : g \mapsto \Lambda_g$$

is an isometric isomorphism if and only if the measure space $(M, \mathcal{A}, \mu)$ is localizable. In particular, the dual space of $L^1(\mu)$ is isomorphic to $L^\infty(\mu)$ whenever $(M, \mathcal{A}, \mu)$ is a σ-finite measure space. (See [**75**, Def. 4.29] for the relevant definitions.) However, the dual space of $L^\infty(\mu)$ is in general much larger than $L^1(\mu)$, i.e. the map

$$L^1(\mu) \to L^\infty(\mu)^* : g \mapsto \Lambda_g$$

in (1.3.7) is an isometric embedding but is typically far from surjective.

Example 1.3.5 (Dual Space of ℓ^p). Fix a number $1 < p < \infty$ and consider the Banach space ℓ^p of p-summable sequences of real numbers, equipped with the norm

$$\|x\|_p := \left(\sum_{i=1}^{p} |x_i|^p\right)^{1/p} \qquad \text{for } x = (x_i)_{i\in\mathbb{N}} \in \ell^p.$$

(See part (ii) of Example 1.1.3.) This is the special case of the counting measure on $M = \mathbb{N}$ in Example 1.3.3 and so the dual space of ℓ^p is isomorphic to ℓ^q, where $1/p + 1/q = 1$. Here is a proof in this special case.

Associated to every sequence $y = (y_i)_{i\in\mathbb{N}} \in \ell^q$ is a bounded linear functional $\Lambda_y : \ell^p \to \mathbb{R}$, defined by

$$\Lambda_y(x) := \sum_{i=1}^{\infty} x_i y_i \tag{1.3.8}$$

for $x = (x_i)_{i\in\mathbb{N}} \in \ell^p$. It is well defined by the Hölder inequality (1.3.6). Namely, in this case the Hölder inequality takes the form

$$\sum_{i=1}^{\infty} |x_i y_i| \le \|x\|_p \|y\|_q$$

for $x = (x_i)_{i\in\mathbb{N}} \in \ell^p$ and $y = (y_i)_{i\in\mathbb{N}} \in \ell^q$ and hence the limit

$$\sum_{i=1}^{\infty} x_i y_i = \lim_{n\to\infty} \sum_{i=1}^{n} x_i y_i$$

in (1.3.8) exists. Thus, for each $y \in \ell^q$, the map $\Lambda_y : \ell^p \to \mathbb{R}$ in (1.3.8) is well defined and linear and satisfies the inequality

$$|\Lambda_y(x)| \le \|x\|_p \|y\|_q \qquad \text{for all } x \in \ell^p.$$

Thus Λ_y is a bounded linear functional on ℓ^p for every $y \in \ell^q$ with norm

$$\|\Lambda_y\| = \sup_{x \in \ell^p \setminus \{0\}} \frac{|\Lambda_y(x)|}{\|x\|_p} \le \|y\|_q .$$

Hence the formula (1.3.8) defines a bounded linear operator

$$\ell^q \to (\ell^p)^* : y \mapsto \Lambda_y. \tag{1.3.9}$$

In fact, it turns out that $\|\Lambda_y\| = \|y\|_q$ for all $y \in \ell^q$. To see this, fix a nonzero element $y = (y_i)_{i\in\mathbb{N}} \in \ell^q$ and consider the sequence $x = (x_i)_{i\in\mathbb{N}}$, defined by $x_i := |y_i|^{q-1}\mathrm{sign}(y_i)$ for $i \in \mathbb{N}$, where $\mathrm{sign}(y_i) := 1$ when $y_i \ge 0$ and $\mathrm{sign}(y_i) := -1$ when $y_i < 0$. Then $|x_i|^p = |y_i|^{(q-1)p} = |y_i|^q$ and thus

$$\|x\|_p = \left(\sum_{i=1}^\infty |y_i|^q\right)^{1-1/q} = \|y\|_q^{q-1}, \qquad \Lambda_y(x) = \sum_{i=1}^\infty x_i y_i = \sum_{i=1}^\infty |y_i|^q = \|y\|_q^q .$$

This shows that

$$\|\Lambda_y\| \ge \frac{|\Lambda_y(x)|}{\|x\|_p} = \frac{\|y\|_q^q}{\|y\|_q^{q-1}} = \|y\|_q$$

and so $\|\Lambda_y\| = \|y\|_q$. Thus the map (1.3.9) is an isometric embedding.

We prove that it is surjective. For $i \in \mathbb{N}$ define

$$e_i := (\delta_{ij})_{j\in\mathbb{N}}, \tag{1.3.10}$$

where δ_{ij} denotes the **Kronecker symbol**, i.e. $\delta_{ij} := 1$ for $i = j$ and $\delta_{ij} := 0$ for $i \ne j$. Then $e_i \in \ell^p$ for every $i \in \mathbb{N}$ and the subspace $\mathrm{span}\{e_i \,|\, i \in \mathbb{N}\}$ of all (finite) linear combinations of the e_i is dense in ℓ^p. Let $\Lambda : \ell^p \to \mathbb{R}$ be a nonzero bounded linear functional and define $y_i := \Lambda(e_i)$ for $i \in \mathbb{N}$. Since $\Lambda \ne 0$ there is an $i \in \mathbb{N}$ such that $y_i \ne 0$. Consider the sequences

$$\xi_n := \sum_{i=1}^n |y_i|^{q-1}\mathrm{sign}(y_i) e_i \in \ell^p, \qquad \eta_n := \sum_{i=1}^n y_i e_i \in \ell^q \qquad \text{for } n \in \mathbb{N}.$$

Since $(q-1)p = q$, they satisfy

$$\|\xi_n\|_p = \left(\sum_{i=1}^n |y_i|^q\right)^{1-1/q} = \|\eta_n\|_q^{q-1}$$

and $\Lambda(\xi_n) = \sum_{i=1}^n |y_i|^q = \|\eta_n\|_q^q$, and so

$$\left(\sum_{i=1}^n |y_i|^q\right)^{1/q} = \|\eta_n\|_q = \frac{\Lambda(\xi_n)}{\|\xi_n\|_p} \le \|\Lambda\|$$

for $n \in \mathbb{N}$ sufficiently large. Thus $y = (y_i)_{i\in\mathbb{N}} \in \ell^q$. Since $\Lambda_y(e_i) = \Lambda(e_i)$ for all $i \in \mathbb{N}$ and the linear subspace $\mathrm{span}\{e_i \,|\, i \in \mathbb{N}\}$ is dense in ℓ^p, it follows that $\Lambda_y = \Lambda$. This proves that the map (1.3.8) is an isometric isomorphism.

Example 1.3.6 (Dual Space of ℓ^1). The discussion of Example 1.3.5 extends to the case $p = 1$ and shows that the natural map

$$\ell^\infty \to (\ell^1)^* : y \mapsto \Lambda_y$$

defined by (1.3.8) is a Banach space isometry. Here $\ell^\infty \subset \mathbb{R}^{\mathbb{N}}$ is the space of bounded sequences of real numbers equipped with the supremum norm. (**Exercise:** Prove this by adapting Example 1.3.5 to the case $p = 1$.)

There is an analogous map $\ell^1 \to (\ell^\infty)^* : y \mapsto \Lambda_y$. This map is again an isometric embedding of Banach spaces, however, it is far from surjective. The existence of a linear functional on ℓ^∞ that cannot be represented by a summable sequence can be established via the Hahn–Banach Theorem.

Example 1.3.7 (Dual Space of c_0). Consider the closed linear subspace of ℓ^∞ which consists of all sequences of real numbers that converge to zero. Denote it by

$$c_0 := \left\{ x = (x_i)_{i\in\mathbb{N}} \in \mathbb{R}^{\mathbb{N}} \,\middle|\, \lim_{i\to\infty} x_i = 0 \right\} \subset \ell^\infty. \tag{1.3.11}$$

This is a Banach space with the supremum norm

$$\|x\|_\infty := \sup_{i\in\mathbb{N}} |x_i|.$$

Every sequence $y = (y_i)_{i\in\mathbb{N}} \in \ell^1$ determines a linear functional $\Lambda_y : c_0 \to \mathbb{R}$ via (1.3.8). It is bounded and $\|\Lambda_y\| \le \|y\|_1$ because

$$|\Lambda_y(x)| \le \sum_{i=1}^\infty |x_i y_i| \le \|x\|_\infty \sum_{i=1}^\infty |y_i| = \|x\|_\infty \|y\|_1$$

for all $x \in c_0$. Thus the map

$$\ell^1 \to c_0^* : y \mapsto \Lambda_y \tag{1.3.12}$$

is a bounded linear operator. In fact, it is an isometric isomorphism of Banach spaces. To see this, let $y = (y_i)_{i\in\mathbb{N}} \in \ell^1$ and define $\varepsilon_i := \mathrm{sign}(y_i)$ for $i \in \mathbb{N}$. Thus $\varepsilon_i = 1$ when $y_i \ge 0$ and $\varepsilon_i = -1$ when $y_i < 0$. For $n \in \mathbb{N}$ define $\xi_n := \sum_{i=1}^n \varepsilon_i e_i \in c_0$, where $e_i \in c_0$ is defined by (1.3.10). Then

$$\Lambda_y(\xi_n) = \sum_{i=1}^n |y_i|, \qquad \|\xi_n\|_\infty = 1.$$

Thus $\|\Lambda_y\| \ge \sum_{i=1}^n |y_i|$ for all $n \in \mathbb{N}$, hence

$$\|\Lambda_y\| \ge \sum_{i=1}^\infty |y_i| = \|y\|_1 \ge \|\Lambda_y\|,$$

and so $\|\Lambda_y\| = \|y\|_1$. This shows that the linear map (1.3.12) is an isometric embedding and, in particular, is injective.

We prove that the map (1.3.12) is surjective. Let $\Lambda : c_0 \to \mathbb{R}$ be a nonzero bounded linear functional and define the sequence $y = (y_i)_{i\in\mathbb{N}} \in \mathbb{R}^{\mathbb{N}}$ by $y_i := \Lambda(e_i)$ for $i \in \mathbb{N}$ where $e_i \in c_0$ is the sequence in (1.3.10). As before, define $\xi_n := \sum_{i=1}^n \operatorname{sign}(y_i)e_i \in c_0$ for $n \in \mathbb{N}$. Then $\|\xi_n\| = 1$ for n sufficiently large and therefore

$$\sum_{i=1}^n |y_i| = \Lambda(\xi_n) \leq \|\Lambda\| \qquad \text{for all } n \in \mathbb{N}.$$

This implies $\|y\|_1 = \sum_{i=1}^\infty |y_i| \leq \|\Lambda\|$ and so $y \in \ell^1$. Since $\Lambda_y(e_i) = y_i = \Lambda(e_i)$ for all $i \in \mathbb{N}$ and the linear subspace $\operatorname{span}\{e_i \,|\, i \in \mathbb{N}\}$ is dense in c_0 (prove this!), it follows that $\Lambda_y = \Lambda$. Hence the map (1.3.12) is a Banach space isometry and so $c_0^* \cong \ell^1$.

Example 1.3.8 (Dual Space of $C(M)$). Let M be a second countable compact Hausdorff space, so M is metrizable [**61**]. Denote by $\mathcal{B} \subset 2^M$ its **Borel σ-algebra**, i.e. the smallest σ-algebra containing the open sets. Consider the Banach space $C(M)$ of continuous real valued functions on M with the supremum norm and denote by $\mathcal{M}(M)$ the Banach space of signed Borel measures $\mu : \mathcal{B} \to \mathbb{R}$ with the norm in equation (1.1.4) (see Example 1.1.3). Every signed Borel measure $\mu : \mathcal{B} \to \mathbb{R}$ determines a bounded linear functional $\Lambda_\mu : C(M) \to \mathbb{R}$ defined by

$$\Lambda_\mu(f) := \int_M f\, d\mu \qquad \text{for } f \in C(M). \tag{1.3.13}$$

The Hahn Decomposition Theorem asserts that for every signed Borel measure $\mu : \mathcal{B} \to \mathbb{R}$ there exists a Borel set $P \subset M$ such that $\mu(B \cap P) \geq 0$ and $\mu(B \setminus P) \leq 0$ for every Borel set $B \subset M$ (see [**75**, Thm. 5.19]). Since every Borel measure on M is regular (see [**75**, Def. 3.1 and Thm. 3.18]) this can be used to show that $\|\Lambda_\mu\|_{\mathcal{L}(C(M),\mathbb{R})} = \|\mu\|$. Now every bounded linear functional $\Lambda : C(M) \to \mathbb{R}$ can be expressed as the difference of two positive linear functionals $\Lambda^\pm : C(M) \to \mathbb{R}$ (see [**75**, Ex. 5.35]). Hence it follows from the Riesz Representation Theorem (see [**75**, Cor. 3.19]) that the linear map

$$\mathcal{M}(M) \to C(M)^* : \mu \mapsto \Lambda_\mu$$

is an isometric isomorphism.

Exercise 1.3.9. Let X be an infinite-dimensional normed vector space and let $\Lambda : X \to \mathbb{R}$ be a nonzero linear functional. The following are equivalent.

(i) Λ is bounded.

(ii) The kernel of Λ is a closed linear subspace of X.

(iii) The kernel of Λ is not dense in X.

1.4. Hilbert Spaces

This section introduces some elementary Hilbert space theory. It shows that every Hilbert space is isomorphic to its own dual space.

Definition 1.4.1 (Inner Product). Let H be a real vector space. A bilinear map

$$H \times H \to \mathbb{R} : (x, y) \mapsto \langle x, y \rangle \tag{1.4.1}$$

is called an **inner product** if it is **symmetric**, i.e. $\langle x, y \rangle = \langle y, x \rangle$ for all $x, y \in H$, and **positive definite**, i.e. $\langle x, x \rangle > 0$ for all $x \in H \setminus \{0\}$. The **norm** associated to an inner product (1.4.1) is the function

$$H \to \mathbb{R} : x \mapsto \|x\| := \sqrt{\langle x, x \rangle}. \tag{1.4.2}$$

Lemma 1.4.2 (Cauchy–Schwarz Inequality). *Let H be a real vector space equipped with an inner product (1.4.1) and the norm (1.4.2). The inner product and norm satisfy the* **Cauchy–Schwarz inequality**

$$|\langle x, y \rangle| \le \|x\| \, \|y\| \tag{1.4.3}$$

and the **triangle inequality**

$$\|x + y\| \le \|x\| + \|y\| \tag{1.4.4}$$

for all $x, y \in H$. Thus (1.4.2) is a norm on H.

Proof. The Cauchy–Schwarz inequality is obvious when $x = 0$ or $y = 0$. Hence assume $x \neq 0$ and $y \neq 0$ and define

$$\xi := \frac{x}{\|x\|}, \qquad \eta := \frac{y}{\|y\|}.$$

Then $\|\xi\| = \|\eta\| = 1$. Hence

$$0 \le \|\eta - \langle \xi, \eta \rangle \xi\|^2 = \langle \eta, \eta - \langle \xi, \eta \rangle \xi \rangle = 1 - \langle \xi, \eta \rangle^2.$$

This implies $|\langle \xi, \eta \rangle| \le 1$ and hence $|\langle x, y \rangle| \le \|x\| \, \|y\|$. In turn it follows from the Cauchy–Schwarz inequality that

$$\begin{aligned} \|x + y\|^2 &= \|x\|^2 + 2\langle x, y \rangle + \|y\|^2 \\ &\le \|x\|^2 + 2\|x\| \, \|y\| + \|y\|^2 \\ &= (\|x\| + \|y\|)^2 . \end{aligned}$$

This proves the triangle inequality (1.4.4) and Lemma 1.4.2. □

Definition 1.4.3 (Hilbert Space). An inner product space $(H, \langle \cdot, \cdot \rangle)$ is called a **Hilbert space** if the norm (1.4.2) is complete.

Theorem 1.4.4 (Riesz). *Let H be a Hilbert space and let $\Lambda : H \to \mathbb{R}$ be a bounded linear functional. Then there exists a unique element $y \in H$ such that*

$$\Lambda(x) = \langle y, x \rangle \qquad \text{for all } x \in H. \tag{1.4.5}$$

This element $y \in H$ satisfies

$$\|y\| = \sup_{0 \neq x \in H} \frac{|\langle y, x \rangle|}{\|x\|} = \|\Lambda\|. \tag{1.4.6}$$

Thus the map $H \to H^* : y \mapsto \langle y, \cdot \rangle$ is an isometry of normed vector spaces.

Proof. See page 33. □

Theorem 1.4.5. *Let H be a Hilbert space and let $K \subset H$ be a nonempty closed convex subset of H. Then there exists a unique element $x_0 \in K$ such that $\|x_0\| \le \|x\|$ for all $x \in K$.*

Proof. Define

$$\delta := \inf \left\{ \|x\| \,\middle|\, x \in K \right\} \ge 0.$$

We prove existence. Choose a sequence $x_i \in K$ with $\lim_{i\to\infty} \|x_i\| = \delta$. We prove that x_i is a Cauchy sequence. Fix a constant $\varepsilon > 0$. Then there exists an integer $i_0 \in \mathbb{N}$ such that

$$i \in \mathbb{N}, \quad i \ge i_0 \qquad \Longrightarrow \qquad \|x_i\|^2 < \delta^2 + \frac{\varepsilon^2}{4}.$$

Let $i, j \in \mathbb{N}$ such that $i \ge i_0$ and $j \ge i_0$. Then $\frac{1}{2}(x_i + x_j) \in K$ because K is convex and hence $\|x_i + x_j\| \ge 2\delta$. This implies

$$\begin{aligned} \|x_i - x_j\|^2 &= 2\|x_i\|^2 + 2\|x_j\|^2 - \|x_i + x_j\|^2 \\ &< 4\left(\delta^2 + \frac{\varepsilon^2}{4}\right) - 4\delta^2 = \varepsilon^2. \end{aligned}$$

Thus x_i is a Cauchy sequence. Since H is complete the limit $x_0 := \lim_{i\to\infty} x_i$ exists. Moreover $x_0 \in K$ because K is closed and $\|x_0\| = \delta$ because the norm function (1.4.2) is continuous.

We prove uniqueness. Fix two elements $x_0, x_1 \in K$ with

$$\|x_0\| = \|x_1\| = \delta.$$

Then $\frac{1}{2}(x_0 + x_1) \in K$ because K is convex and so $\|x_0 + x_1\| \ge 2\delta$. Thus

$$\|x_0 - x_1\|^2 = 2\|x_0\|^2 + 2\|x_1\|^2 - \|x_0 + x_1\|^2 = 4\delta^2 - \|x_0 + x_1\|^2 \le 0$$

and therefore $x_0 = x_1$. This proves Theorem 1.4.5. □

Proof of Theorem 1.4.4. We prove existence. If $\Lambda = 0$ then the vector $y = 0$ satisfies (1.4.5). Hence assume $\Lambda \neq 0$ and define

$$K := \{x \in H \mid \Lambda(x) = 1\}.$$

Then $K \neq \emptyset$ because there exists an element $\xi \in H$ such that $\Lambda(\xi) \neq 0$ and hence $x := \Lambda(\xi)^{-1}\xi \in K$. The set K is closed because $\Lambda : H \to \mathbb{R}$ is continuous, and it is convex because Λ is linear. Hence Theorem 1.4.5 asserts that there exists an element $x_0 \in K$ such that

$$\|x_0\| \le \|x\| \qquad \text{for all } x \in K.$$

We prove that

$$x \in H, \quad \Lambda(x) = 0 \qquad \Longrightarrow \qquad \langle x_0, x\rangle = 0. \tag{1.4.7}$$

To see this, fix an element $x \in H$ such that $\Lambda(x) = 0$. Then $x_0 + tx \in K$ for all $t \in \mathbb{R}$. This implies

$$\|x_0\|^2 \le \|x_0 + tx\|^2 = \|x_0\|^2 + 2t\langle x_0, x\rangle + t^2 \|x\|^2 \qquad \text{for all } t \in \mathbb{R}.$$

Thus the differentiable function $t \mapsto \|x_0 + tx\|^2$ attains its minimum at $t = 0$ and so its derivative vanishes at $t = 0$. Hence

$$0 = \left.\frac{d}{dt}\right|_{t=0} \|x_0 + tx\|^2 = 2\langle x_0, x\rangle$$

and this proves (1.4.7).

Now define $y := \|x_0\|^{-2}x_0$. Fix an element $x \in H$ and define $\lambda := \Lambda(x)$. Then $\Lambda(x - \lambda x_0) = \Lambda(x) - \lambda = 0$. Hence it follows from (1.4.7) that

$$0 = \langle x_0, x - \lambda x_0\rangle = \langle x_0, x\rangle - \lambda\|x_0\|^2.$$

This implies $\langle y, x\rangle = \|x_0\|^{-2}\langle x_0, x\rangle = \lambda = \Lambda(x)$. Thus y satisfies (1.4.5).

We prove (1.4.6). Assume $y \in H$ satisfies (1.4.5). If $y = 0$, then $\Lambda = 0$ and so $\|y\| = 0 = \|\Lambda\|$. Hence assume $y \neq 0$. Then

$$\|y\| = \frac{\|y\|^2}{\|y\|} = \frac{\Lambda(y)}{\|y\|} \le \sup_{0 \neq x \in H} \frac{|\Lambda(x)|}{\|x\|} = \|\Lambda\|.$$

Conversely, it follows from the Cauchy–Schwarz inequality that

$$|\Lambda(x)| = |\langle y, x\rangle| \le \|y\|\|x\|$$

for all $x \in H$ and hence $\|\Lambda\| \le \|y\|$. This proves (1.4.6).

We prove uniqueness. Assume $y, z \in H$ satisfy $\langle y, x\rangle = \langle z, x\rangle = \Lambda(x)$ for all $x \in H$. Then $\langle y - z, x\rangle = 0$ for all $x \in H$. Take $x := y - z$ to obtain

$$\|y - z\|^2 = \langle y - z, y - z\rangle = 0$$

and so $y - z = 0$. This proves Theorem 1.4.4. □

We will see, as we proceed, that Hilbert spaces have several features that are not shared by general Banach spaces. One of these is that every closed subspace of a Hilbert space has a *complement*, i.e. another closed subspace whose direct sum with the original subspace is the entire Hilbert space. To explain this, we define the **orthogonal complement** of a subset $S \subset H$ by

$$S^\perp := \{ y \in H \,|\, \langle x, y\rangle = 0 \text{ for all } x \in S \}. \tag{1.4.8}$$

It follows directly from the definitions that $S^\perp$ is a closed subspace of H.

Corollary 1.4.6. *Let H be a Hilbert space and let $E \subset H$ be a closed subspace. Then $H = E \oplus E^\perp$.*

Proof. If $x \in E \cap E^\perp$, then $\|x\|^2 = \langle x, x\rangle = 0$ and hence $x = 0$. If $x \in H$, then the set $K := x + E = \{x + \xi \,|\, \xi \in E\}$ is a closed convex subset of H. Hence Theorem 1.4.5 asserts that there exists an element $\xi \in E$ such that $\|x - \xi\| \le \|x - \eta\|$ for all $\eta \in E$. Hence, for all $\eta \in E$, we have

$$0 = \left.\frac{d}{dt}\right|_{t=0} \frac{\|x - \xi + t\eta\|^2}{2} = \langle x - \xi, \eta\rangle.$$

Thus $x - \xi \in E^\perp$ and so $x \in E \oplus E^\perp$. This proves Corollary 1.4.6. □

In Chapter 2 we will encounter closed subspaces of Banach spaces that are not *complemented* (see Subsection 2.3.5).

Example 1.4.7. Let $(M, \mathcal{A}, \mu)$ be a measure space. Then $H := L^2(\mu)$ is a Hilbert space. The inner product is induced by the bilinear map

$$\mathcal{L}^2(\mu) \times \mathcal{L}^2(\mu) \to \mathbb{R} : (f, g) \mapsto \langle f, g\rangle := \int_M fg \, d\mu. \tag{1.4.9}$$

It is well defined because the product of two L^2-functions $f, g : M \to \mathbb{R}$ is integrable by the Cauchy–Schwarz inequality. That it is bilinear and symmetric follows directly from the properties of the Lebesgue integral. In general, it is not positive definite. However, it descends to a positive definite symmetric bilinear form on the quotient space

$$L^2(\mu) = \mathcal{L}^2(\mu)/\sim,$$

where the equivalence relation is defined by equality almost everywhere as in part (iii) of Example 1.1.3. The inner product on $L^2(\mu)$ induced by (1.4.9) is called the L^2 **inner product**. The norm associated to this inner product is the L^2 norm in (1.1.2) with $p = 2$. By [**75**, Thm. 4.9] the space $L^2(\mu)$ is complete with this norm and hence is a Hilbert space.

Special cases are the Euclidean space $(\mathbb{R}^n, \|\cdot\|_2)$ in part (i) of Example 1.1.3, associated to the counting measure on the set $M = \{1, \dots, n\}$, and the space ℓ^2 in part (ii) of Example 1.1.3, associated to the counting measure on the set $M = \mathbb{N}$.

1.5. Banach Algebras

We begin the discussion with a result about convergent series in a Banach space. It extends the basic assertion in first year analysis that every absolutely convergent series of real numbers converges. We will use Lemma 1.5.1 to study power series in a Banach algebra.

Lemma 1.5.1 (Convergent Series). *Let $(X, \|\cdot\|)$ be a Banach space and let $(x_i)_{i\in\mathbb{N}}$ be a sequence in X such that*

$$\sum_{i=1}^{\infty} \|x_i\| < \infty.$$

Then the sequence $\xi_n := \sum_{i=1}^{n} x_i$ in X converges. Its limit is denoted by

$$\sum_{i=1}^{\infty} x_i := \lim_{n\to\infty} \sum_{i=1}^{n} x_i. \tag{1.5.1}$$

Proof. Define $s_n := \sum_{i=1}^{n} \|x_i\|$ for $n \in \mathbb{N}$. This sequence is nondecreasing and converges by assumption. Moreover, for every pair of integers $n > m \geq 1$, we have

$$\|\xi_n - \xi_m\| = \left\| \sum_{i=m+1}^{n} x_i \right\| \leq \sum_{i=m+1}^{n} \|x_i\| = s_n - s_m.$$

Hence $(\xi_n)_{n\in\mathbb{N}}$ is a Cauchy sequence in X. Since X is complete, this sequence converges, and this proves Lemma 1.5.1. □

Definition 1.5.2 (Banach Algebra). A real (respectively complex) **Banach algebra** is a pair consisting of a real (respectively complex) Banach space $(\mathcal{A}, \|\cdot\|)$ and a bilinear map $\mathcal{A} \times \mathcal{A} \to \mathcal{A} : (a, b) \mapsto ab$ (called the **product**) that is associative, i.e.

$$(ab)c = a(bc) \qquad \text{for all } a, b, c \in \mathcal{A}, \tag{1.5.2}$$

and satisfies the inequality

$$\|ab\| \leq \|a\| \, \|b\| \qquad \text{for all } a, b \in \mathcal{A}. \tag{1.5.3}$$

A Banach algebra $\mathcal{A}$ is called **commutative** if $ab = ba$ for all $a, b \in \mathcal{A}$. It is called **unital** if there exists an element $\mathbb{1} \in \mathcal{A} \setminus \{0\}$ such that

$$\mathbb{1}a = a\mathbb{1} = a \qquad \text{for all } a \in \mathcal{A}. \tag{1.5.4}$$

The **unit** $\mathbb{1}$, if it exists, is uniquely determined by the product. An element $a \in \mathcal{A}$ of a unital Banach algebra $\mathcal{A}$ is called **invertible** if there exists an element $b \in \mathcal{A}$ such that $ab = ba = \mathbb{1}$. The element b, if it exists, is uniquely determined by a, is called the **inverse of** a, and is denoted by $a^{-1} := b$. The invertible elements form a group $\mathcal{G} \subset \mathcal{A}$.

Example 1.5.3. **(i)** The archetypal example of a Banach algebra is the space $\mathcal{L}(X) := \mathcal{L}(X, X)$ of bounded linear operators from a Banach space X to itself with the operator norm (Definition 1.2.1 and Theorem 1.3.1). This Banach algebra is unital whenever $X \neq \{0\}$ and the unit is the identity. It turns out that the invertible elements of $\mathcal{L}(X)$ are the bijective bounded linear operators from X to itself. That the inverse of a bijective bounded linear operator is again a bounded linear operator is a nontrivial result, which follows from the Open Mapping Theorem (see Theorem 2.2.5 below).

(ii) An example of a commutative unital Banach algebra is the space of real valued bounded continuous functions on a nonempty topological space equipped with the supremum norm and pointwise multiplication.

(iii) A third example of a unital Banach algebra is the space $\ell^1(\mathbb{Z})$ of bi-infinite summable sequences $(x_i)_{i\in\mathbb{Z}}$ of real numbers with the convolution product defined by $(x * y)_i := \sum_{j\in\mathbb{Z}} x_j y_{i-j}$ for $x, y \in \ell^1(\mathbb{Z})$.

(iv) A fourth example of a Banach algebra is the space $L^1(\mathbb{R}^n)$ of Lebesgue integrable functions on $\mathbb{R}^n$ (modulo equality almost everywhere), where multiplication is given by convolution (see [**75**, Section 7.5]). This Banach algebra does not admit a unit. A candidate for a unit would be the Dirac delta function at the origin which is not actually a function but a measure. The convolution product extends to the space of signed Borel measures on $\mathbb{R}^n$ and they form a commutative unital Banach algebra.

Let $\mathcal{A}$ be a complex Banach algebra and let

$$f(z) = \sum_{n=0}^{\infty} c_n z^n \tag{1.5.5}$$

be a power series with complex coefficients $c_n \in \mathbb{C}$ and convergence radius

$$\rho := \frac{1}{\limsup_{n\to\infty} |c_n|^{1/n}} > 0. \tag{1.5.6}$$

Choose an element $a \in \mathcal{A}$ with $\|a\| < \rho$. Then the sequence $(c_n a^n)_{n\in\mathbb{N}}$ satisfies the inequality $\sum_{n=0}^{\infty} \|c_n a^n\| \leq |c_0| \, \|1\!\!1\| + \sum_{n=1}^{\infty} |c_n| \, \|a\|^n < \infty$, so the sequence $\xi_n := \sum_{i=0}^{n} c_i a^i$ converges by Lemma 1.5.1. Denote the limit by

$$f(a) := \sum_{n=0}^{\infty} c_n a^n \tag{1.5.7}$$

for $a \in \mathcal{A}$ with $\|a\| < \rho$.

Exercise 1.5.4. The map $f : \{a \in \mathcal{A} \,|\, \|a\| < \rho\} \to \mathcal{A}$ defined by (1.5.7) is continuous. **Hint:** For $n \in \mathbb{N}$ define $f_n : X \to X$ by $f_n(a) := \sum_{i=0}^{n} c_i a^i$. Prove that f_n is continuous. Prove that the sequence f_n converges uniformly to f on the set $\{a \in \mathcal{A} \,|\, \|a\| \leq r\}$ for every $r < \rho$.

Theorem 1.5.5 (Inverse). *Let $\mathcal{A}$ be a real unital Banach algebra.*

(i) *For every $a \in \mathcal{A}$ the limit*

$$r_a := \lim_{n\to\infty} \|a^n\|^{1/n} = \inf_{n\in\mathbb{N}} \|a^n\|^{1/n} \le \|a\| \tag{1.5.8}$$

exists. It is called the **spectral radius** *of a.*

(ii) *If $a \in \mathcal{A}$ satisfies $r_a < 1$, then the element $\mathbb{1} - a$ is invertible and*

$$(\mathbb{1} - a)^{-1} = \sum_{n=0}^{\infty} a^n. \tag{1.5.9}$$

(iii) *The group $\mathcal{G} \subset \mathcal{A}$ of invertible elements is an open subset of $\mathcal{A}$ and the map $\mathcal{G} \to \mathcal{G} : a \mapsto a^{-1}$ is continuous. More precisely, if $a \in \mathcal{G}$ and $b \in \mathcal{A}$ satisfy $\|a-b\|\|a^{-1}\| < 1$, then $b \in \mathcal{G}$ and $b^{-1} = \sum_{n=0}^{\infty}(\mathbb{1} - a^{-1}b)^n a^{-1}$ and so*

$$\|b^{-1} - a^{-1}\| \le \frac{\|a-b\|\|a^{-1}\|^2}{1 - \|a-b\|\|a^{-1}\|}, \qquad \|b^{-1}\| \le \frac{\|a^{-1}\|}{1 - \|a-b\|\|a^{-1}\|}. \tag{1.5.10}$$

Proof. We prove part (i). Let $a \in \mathcal{A}$, define

$$r := \inf_{n\in\mathbb{N}} \|a^n\|^{1/n} \ge 0,$$

and fix a real number $\varepsilon > 0$. Choose $m \in \mathbb{N}$ such that

$$\|a^m\|^{1/m} < r + \varepsilon$$

and define

$$M := \max_{\ell=0,1,\dots,m-1} \left(\frac{\|a\|}{r+\varepsilon}\right)^{\ell}.$$

Fix two integers $k \ge 0$ and $0 \le \ell \le m-1$ and let $n := km + \ell$. Then

$$\begin{aligned}
\|a^n\|^{1/n} &= \left\|a^{km}a^{\ell}\right\|^{1/n} \\
&\le \|a\|^{\ell/n} \|a^m\|^{k/n} \\
&\le \|a\|^{\ell/n} (r+\varepsilon)^{km/n} \\
&= \left(\frac{\|a\|}{r+\varepsilon}\right)^{\ell/n} (r+\varepsilon) \\
&\le M^{1/n}(r+\varepsilon).
\end{aligned}$$

Since $\lim_{n\to\infty} M^{1/n} = 1$, there is an integer $n_0 \in \mathbb{N}$ such that

$$\|a^n\|^{1/n} < r + 2\varepsilon$$

for every integer $n \ge n_0$. Hence the limit r_a in (1.5.8) exists and is equal to r. This proves part (i).

We prove part (ii). Let $a \in \mathcal{A}$ and assume $r_a < 1$. Choose a real number α such that

$$r_a < \alpha < 1.$$

Then there exists an $n_0 \in \mathbb{N}$ such that

$$\|a^n\|^{1/n} \le \alpha$$

for every integer $n \ge n_0$. Hence

$$\|a^n\| \le \alpha^n \qquad \text{for every integer } n \ge n_0.$$

This implies $\sum_{n=0}^{\infty} \|a^n\| < \infty$, so the sequence

$$b_n := \sum_{i=0}^{n} a^i$$

converges by Lemma 1.5.1. Denote the limit by b. Since

$$b_n(\mathbb{1} - a) = (\mathbb{1} - a)b_n = \mathbb{1} - a^{n+1}$$

for all $n \in \mathbb{N}$ and $\lim_{n\to\infty} \left\|a^{n+1}\right\| \le \lim_{n\to\infty} \alpha^{n+1} = 0$, it follows that

$$b(\mathbb{1} - a) = (\mathbb{1} - a)b = \mathbb{1}.$$

Hence $\mathbb{1} - a$ is invertible and $(\mathbb{1} - a)^{-1} = b$. This proves part (ii).

We prove part (iii). Fix an element $a \in \mathcal{G}$ and let $b \in \mathcal{A}$ such that

$$\|a - b\| \, \|a^{-1}\| < 1.$$

Then $\left\|\mathbb{1} - a^{-1}b\right\| < 1$ and hence

$$a^{-1}b = \mathbb{1} - (\mathbb{1} - a^{-1}b) \in \mathcal{G}, \qquad (a^{-1}b)^{-1} = \sum_{n=0}^{\infty} (\mathbb{1} - a^{-1}b)^n$$

by part (ii). Hence $b = a(a^{-1}b) \in \mathcal{G}$ and

$$b^{-1} = \sum_{n=0}^{\infty} (\mathbb{1} - a^{-1}b)^n a^{-1}$$

and so

$$\begin{aligned} \|b^{-1} - a^{-1}\| &\le \sum_{n=1}^{\infty} \|a - b\|^n \|a^{-1}\|^{n+1} \\ &= \frac{\|a - b\| \|a^{-1}\|^2}{1 - \|a - b\| \|a^{-1}\|}. \end{aligned}$$

Thus $B_{\|a^{-1}\|^{-1}}(a) \subset \mathcal{G}$ and the map $B_{\|a^{-1}\|^{-1}}(a) \to \mathcal{G} : b \mapsto b^{-1}$ is continuous. This proves part (iii) and Theorem 1.5.5. □

Definition 1.5.6 (Invertible Operator). Let X and Y be Banach spaces. A bounded linear operator $A : X \to Y$ is called **invertible** if there exists a bounded linear operator $B : Y \to X$ such that

$$BA = 1\!\!1_X, \qquad AB = 1\!\!1_Y.$$

The operator B is uniquely determined by A and is denoted by

$$B =: A^{-1}.$$

It is called the **inverse** of A. When $X = Y$, the space of invertible bounded linear operators in $\mathcal{L}(X)$ is denoted by

$$\mathrm{Aut}(X) := \left\{ A \in \mathcal{L}(X) \,\middle|\, \text{there is a } B \in \mathcal{L}(X) \text{ such that } AB = BA = 1\!\!1 \right\}.$$

The **spectral radius** of a bounded linear operator $A \in \mathcal{L}(X)$ is the real number $r_A \geq 0$ defined by

$$r_A := \lim_{n\to\infty} \|A^n\|^{1/n} = \inf_{n\in\mathbb{N}} \|A^n\|^{1/n} \leq \|A\|. \tag{1.5.11}$$

Corollary 1.5.7 (Spectral Radius). *Let X and Y be Banach spaces. Then the following hold.*

(i) *If $A \in \mathcal{L}(X)$ has spectral radius $r_A < 1$, then*

$$1\!\!1_X - A \in \mathrm{Aut}(X), \qquad (1\!\!1_X - A)^{-1} = \sum_{n=0}^{\infty} A^n.$$

(ii) $\mathrm{Aut}(X)$ *is an open subset of $\mathcal{L}(X)$ with respect to the norm topology and the map* $\mathrm{Aut}(X) \to \mathrm{Aut}(X) : A \mapsto A^{-1}$ *is continuous.*

(iii) *Let $A, P \in \mathcal{L}(X, Y)$ be bounded linear operators. Assume A is invertible and $\|P\|\|A^{-1}\| < 1$. Then $A - P$ is invertible,*

$$(A - P)^{-1} = \sum_{n=0}^{\infty} (A^{-1}P)^n A^{-1}, \tag{1.5.12}$$

and

$$\|(A - P)^{-1} - A^{-1}\| \leq \frac{\|P\|\|A^{-1}\|^2}{1 - \|P\|\|A^{-1}\|}. \tag{1.5.13}$$

Proof. Assertions (i) and (ii) follow from Theorem 1.5.5 with $\mathcal{A} = \mathcal{L}(X)$. To prove part (iii), observe that $\|A^{-1}P\| \leq \|A^{-1}\|\|P\| < 1$. Hence it follows from part (i) that the operator $1\!\!1_X - A^{-1}P$ is invertible and that its inverse is given by $(1\!\!1_X - A^{-1}P)^{-1} = \sum_{k=0}^{\infty}(A^{-1}P)^k$. Multiply this identity by A^{-1} on the right to obtain (1.5.12). The inequality (1.5.13) follows directly from (1.5.12) and the limit formula for a geometric series. This proves Corollary 1.5.7. □

1.6. The Baire Category Theorem

The Baire Category Theorem is a powerful tool in functional analysis. It provides conditions under which a subset of a complete metric space is dense. In fact, it describes a class of dense subsets such that every countable intersection of sets in this class belongs again to this class and hence is still a dense subset. Here are the relevant definitions.

Definition 1.6.1 (Baire Category). Let (X, d) be a metric space.

(i) A subset $A \subset X$ is called **nowhere dense** if its closure $\overline{A}$ has an empty interior.

(ii) A subset $A \subset X$ is said to be **meagre** if it is a countable union of nowhere dense subsets of X.

(iii) A subset $A \subset X$ is said to be **nonmeagre** if it is not meagre.

(iv) A subset $A \subset X$ is called **residual** if its complement is meagre.

This definition does not exclude the possibility that X might be the empty set, in which case every subset of X is both meagre and residual. In the literature meagre sets are often called **of the first category (in the sense of Baire)**, nonmeagre sets are called **of the second category**, and residual sets are called **comeagre**. The next lemma summarizes some elementary consequences of these definitions.

Lemma 1.6.2. *Let (X, d) be a metric space. Then the following hold.*

(i) *A subset $A \subset X$ is nowhere dense if and only if its complement $X \setminus A$ contains a dense open subset of X.*

(ii) *If $B \subset X$ is meagre and $A \subset B$, then A is meagre.*

(iii) *If $A \subset X$ is nonmeagre and $A \subset B \subset X$, then B is nonmeagre.*

(iv) *Every countable union of meagre sets is again meagre.*

(v) *Every countable intersection of residual sets is again residual.*

(vi) *A subset of X is residual if and only if it contains a countable intersection of dense open subsets of X.*

Proof. The complement of the closure of a subset of X is the interior of the complement and vice versa. Thus every subset $A \subset X$ satisfies

$$X \setminus \operatorname{int}(\overline{A}) = \overline{X \setminus \overline{A}} = \overline{\operatorname{int}(X \setminus A)}.$$

This shows that a subset $A \subset X$ is nowhere dense if and only if the interior of $X \setminus A$ is dense in X, i.e. $X \setminus A$ contains a dense open subset of X. This proves (i). Parts (ii), (iii), (iv), and (v) follow directly from the definitions.

We prove (vi). Let $R \subset X$ be a residual set and define $A := X \setminus R$. Then there is a sequence of nowhere dense subsets $A_i \subset X$ such that $A = \bigcup_{i=1}^{\infty} A_i$. Define $U_i := X \setminus \overline{A_i} = \mathrm{int}(X \setminus A_i)$. Then U_i is a dense open set by (i) and

$$\bigcap_{i=1}^{\infty} U_i = X \setminus \bigcup_{i=1}^{\infty} \overline{A_i} \subset X \setminus \bigcup_{i=1}^{\infty} A_i = X \setminus A = R.$$

Conversely, suppose that there is a sequence of dense open subsets $U_i \subset X$ such that $\bigcap_{i=1}^{\infty} U_i \subset R$. Define $A_i := X \setminus U_i$ and $A := \bigcup_{i=1}^{\infty} A_i$. Then A_i is nowhere dense by (i) and hence A is meagre by definition. Moreover,

$$X \setminus R \subset X \setminus \bigcap_{i=1}^{\infty} U_i = \bigcup_{i=1}^{\infty} (X \setminus U_i) = \bigcup_{i=1}^{\infty} A_i = A.$$

Hence $X \setminus R$ is meagre by part (ii) and this proves Lemma 1.6.2. □

Lemma 1.6.3. *Let (X, d) be a metric space. The following are equivalent.*

(i) *Every residual subset of X is dense.*

(ii) *If $U \subset X$ is a nonempty open set, then U is nonmeagre.*

(iii) *If $A_i \subset X$ is a sequence of closed sets with empty interior, then their union has empty interior.*

(iv) *If $U_i \subset X$ is a sequence of dense open sets, then their intersection is dense in X.*

Proof. We prove that (i) implies (ii). Assume (i) and let $U \subset X$ be a nonempty open set. Then its complement $X \setminus U$ is not dense and so is not residual by (i). Hence U is not meagre.

We prove that (ii) implies (iii). Assume (ii) and let A_i be a sequence of closed subsets of X with empty interior. Then their union A is meagre. Hence the interior of A is also meagre by part (ii) of Lemma 1.6.2. Hence the interior of A is empty by (ii).

We prove that (iii) implies (iv). Assume (iii) and let U_i be a sequence of dense open subsets of X. For $i \in \mathbb{N}$ define $A_i := X \setminus U_i$. Then A_i is a sequence of closed subsets of X with empty interior. Hence $A := \bigcup_{i=1}^{\infty} A_i$ has empty interior by (iii) and so the set

$$R := \bigcap_{i=1}^{\infty} U_i = \bigcap_{i=1}^{\infty} (X \setminus A_i) = X \setminus A$$

is dense in X.

We prove that (iv) implies (i). Assume (iv) and let $R \subset X$ be residual. Then, by part (vi) of Lemma 1.6.2, there exists a sequence of dense open subsets $U_i \subset X$ such that $\bigcap_{i \in \mathbb{N}} U_i \subset R$. By (iv) the set $\bigcap_{i \in \mathbb{N}} U_i$ is dense in X and hence so is R. This proves Lemma 1.6.3. □

Theorem 1.6.4 (Baire Category Theorem). *Let (X, d) be a nonempty complete metric space. Then the following hold.*

(i) *Every residual subset of X is dense.*

(ii) *If $U \subset X$ is a nonempty open set, then U is nonmeagre.*

(iii) *If $A_i \subset X$ is a sequence of closed sets with empty interior, then their union has empty interior.*

(iv) *If $U_i \subset X$ is a sequence of open dense sets, then their intersection is dense in X.*

(v) *Every residual subset of X is nonmeagre.*

Proof. The first four assertions are equivalent by Lemma 1.6.3.

We prove that (ii) implies (v). Let $R \subset X$ be a residual set. Then $X \setminus R$ is meagre by definition. If the set R were meagre as well, then

$$X = (X \setminus R) \cup R$$

would also be meagre by part (iv) of Lemma 1.6.2, and this would contradict part (ii) because X is nonempty. Thus R is nonmeagre.

We prove part (iv). Thus assume that $U_i \subset X$ is a sequence of dense open sets. Fix an element $x_0 \in X$ and a constant $\varepsilon_0 > 0$. We must prove that $B_{\varepsilon_0}(x_0) \cap \bigcap_{i=1}^{\infty} U_i \neq \emptyset$. We claim that there exist sequences

$$x_k \in U_k, \qquad 0 < \varepsilon_k < 2^{-k}, \qquad k = 1, 2, 3, \dots, \tag{1.6.1}$$

such that

$$\overline{B_{\varepsilon_k}(x_k)} \subset U_k \cap B_{\varepsilon_{k-1}}(x_{k-1}) \tag{1.6.2}$$

for every integer $k \geq 1$. For $k = 1$ observe that $U_1 \cap B_{\varepsilon_0}(x_0)$ is a nonempty open set because U_1 is dense in X. Choose any element $x_1 \in U_1 \cap B_{\varepsilon_0}(x_0)$ and choose $\varepsilon_1 > 0$ such that $\varepsilon_1 < 1/2$ and $\overline{B_{\varepsilon_1}(x_1)} \subset U_1 \cap B_{\varepsilon_0}(x_0)$. If x_{k-1} and ε_{k-1} have been found for some integer $k \geq 2$, use the fact that U_k is dense in X to find x_k and ε_k such that (1.6.1) and (1.6.2) hold.

More precisely, this argument requires the axiom of dependent choice (see page 6). Define the set

$$\mathbf{X} := \Big\{(k, x, \varepsilon) \,\big|\, k \in \mathbb{N},\, x \in X,\, 0 < \varepsilon < 2^{-k},\, \overline{B_\varepsilon(x)} \subset U_k \cap B_{\varepsilon_0}(x_0)\Big\}$$

and define the map $\mathbf{A} : \mathbf{X} \to 2^{\mathbf{X}}$ by

$$\mathbf{A}(k, x, \varepsilon) := \Big\{(k', x', \varepsilon') \in \mathbf{X} \,\big|\, k' = k + 1,\, \overline{B_{\varepsilon'}(x')} \subset B_\varepsilon(x)\Big\}$$

for $(k, x, \varepsilon) \in \mathbf{X}$. Then $\mathbf{X} \neq \emptyset$ and $\mathbf{A}(k, x, \varepsilon) \neq \emptyset$ for all $(k, x, \varepsilon) \in \mathbf{X}$, because U_k is open and dense in X for all k. Hence the existence of the sequences x_k and ε_k follows from the axiom of dependent choice.

Now let $x_k \in U_k$ and $\varepsilon_k > 0$ be sequences that satisfy (1.6.1) and (1.6.2). Then

$$d(x_k, x_{k-1}) < \varepsilon_{k-1} \le 2^{1-k}$$

for all $k \in \mathbb{N}$. Hence

$$d(x_k, x_\ell) \le \sum_{i=k}^{\ell-1} d(x_i, x_{i+1}) < \sum_{i=k}^{\ell-1} 2^{-i} < 2^{1-k}$$

for all $k, \ell \in \mathbb{N}$ with $\ell > k$. Thus $(x_k)_{k\in\mathbb{N}}$ is a Cauchy sequence in X. Since X is complete the sequence $(x_k)_{k\in\mathbb{N}}$ converges. Denote its limit by

$$x^* := \lim_{k\to\infty} x_k.$$

Since $x_\ell \in B_{\varepsilon_k}(x_k)$ for every $\ell \ge k$ it follows that

$$x^* \in \overline{B_{\varepsilon_k}(x_k)} \subset U_k \qquad \text{for all } k \in \mathbb{N}.$$

Moreover,

$$x^* \in \overline{B_{\varepsilon_1}(x_1)} \subset B_{\varepsilon_0}(x_0).$$

This shows that the intersection

$$B_{\varepsilon_0}(x_0) \cap \bigcap_{i=1}^{\infty} U_i$$

is nonempty for all $x_0 \in X$ and all $\varepsilon_0 > 0$. Hence the set $\bigcap_{i=1}^{\infty} U_i$ is dense in X as claimed. This proves part (iv) and Theorem 1.6.4. □

The desired class of dense subsets of our nonempty complete metric space is the collection of residual sets. Every residual set is dense by part (i) of Theorem 1.6.4 and every countable intersection of residual sets is again residual by part (v) of Lemma 1.6.2. It is often convenient to use the characterization of a residual set as one that contains a countable intersection of dense open sets in part (vi) of Lemma 1.6.2. A very useful consequence of the Baire Category Theorem is the assertion that a nonempty complete metric space cannot be expressed as a countable union of nowhere dense subsets (part (ii) of Theorem 1.6.4 with $U = X$).

We emphasize that, while the assumption of the Baire Category Theorem (completeness) depends on the distance function in a crucial way, the conclusion (every countable intersection of dense open subsets is dense) is purely topological. Thus the Baire Category Theorem extends to many metric spaces that are not complete. All that is required is the existence of a complete distance function that induces the same topology as the original distance function.

Example 1.6.5. Let (M,d) be a complete metric space and let $X \subset M$ be a nonempty open set. Then the conclusions of the Baire Category Theorem hold for the metric space (X,d_X) with $d_X := d|_{X\times X} : X\times X \to [0,\infty)$, even though (X,d_X) may not be complete. To see this, let $U_i \subset X$ be a sequence of dense open subsets of X, choose $x_0 \in X$ and $\varepsilon_0 > 0$ such that $\overline{B_{\varepsilon_0}(x_0)} \subset X$, and repeat the argument in the proof of Theorem 1.6.4 to show that $\overline{B_{\varepsilon_0}(x_0)} \cap \bigcap_{i=1}^\infty U_i \neq \emptyset$. All that is needed is the fact that the closure $\overline{B_{\varepsilon_1}(x_1)}$ that contains the sequence x_k is complete with respect to the induced metric.

Example 1.6.6. The conclusions of the Baire Category Theorem hold for the topological vector space $C^\infty([0,1])$ of smooth functions $f:[0,1]\to\mathbb{R}$, equipped with the C^∞ topology. By definition, a sequence $f_n \in C^\infty([0,1])$ converges to $f \in C^\infty([0,1])$ with respect to the C^∞ topology if and only if, for each integer $\ell \geq 0$, the sequence of ℓth derivatives $f_n^{(\ell)} : [0,1]\to\mathbb{R}$ converges uniformly to the ℓth derivative $f^{(\ell)} : [0,1]\to\mathbb{R}$ as n tends to infinity. This topology is induced by the distance function

$$d(f,g) := \sum_{\ell=0}^\infty 2^{-\ell}\frac{\left\|f^{(\ell)}-g^{(\ell)}\right\|_\infty}{1+\left\|f^{(\ell)}-g^{(\ell)}\right\|_\infty},$$

where $\|u\|_\infty := \sup_{0\leq t\leq 1}|u(t)|$ denotes the supremum norm of a continuous function $u:[0,1]\to\mathbb{R}$, and $(C^\infty([0,1]),d)$ is a complete metric space.

Example 1.6.7. A residual subset of $\mathbb{R}^n$ may have Lebesgue measure zero. Namely, choose a bijection $\mathbb{N}\to\mathbb{Q}^n : k\mapsto x_k$ and, for $\varepsilon>0$, define

$$U_\varepsilon := \bigcup_{k=1}^\infty B_{2^{-k}\varepsilon}(x_k).$$

This is a dense open subset of $\mathbb{R}^n$ and its Lebesgue measure is less than $(2\varepsilon)^n$. Hence $R := \bigcap_{i=1}^\infty U_{1/i}$ is a residual set of Lebesgue measure zero and its complement

$$A := \mathbb{R}^n \setminus R = \bigcup_{i=1}^\infty \left(\mathbb{R}^n\setminus U_{1/i}\right)$$

is a meagre set of full Lebesgue measure.

Example 1.6.8. The conclusions of the Baire Category Theorem do not hold for the metric space $X=\mathbb{Q}$ of rational numbers with the standard distance function given by $d(x,y) := |x-y|$ for $x,y\in\mathbb{Q}$. Every one element subset of X is nowhere dense and every subset of X is both meagre and residual.

1.7. Problems

Exercise 1.7.1 (Precompact Sets). Let X and Y be topological spaces such that Y is Hausdorff. Let $f : X \to Y$ be a continuous map and let $A \subset X$ be a precompact subset of X (i.e. its closure $\overline{A}$ is compact). Prove that $B := f(A)$ is a precompact subset of Y. **Hint:** Show that $f(\overline{A}) \subset \overline{B}$. If $\overline{A}$ is compact and Y is Hausdorff show that $f(\overline{A}) = \overline{B}$.

Exercise 1.7.2 (Totally Bounded Sets). Let A be a subset of a metric space. Show that A is totally bounded if and only if $\overline{A}$ is totally bounded.

Exercise 1.7.3 (Complete and Closed Subspaces). Let (X, d_X) be a metric space, let $Y \subset X$ be a subset, and denote by $d_Y := d_X|_{Y\times Y}$ the induced distance function on Y. Prove the following.

(a) If (Y, d_Y) is complete, then Y is a closed subset of X.

(b) If (X, d_X) is complete and $Y \subset X$ is closed, then (Y, d_Y) is complete.

Exercise 1.7.4 (Completion of a Metric Space). Let (X, d) be a metric space. A **completion** of (X, d) is a triple $(\overline{X}, \overline{d}, \iota)$, consisting of a complete metric space $(\overline{X}, \overline{d})$ and an isometric embedding $\iota : X \to \overline{X}$ with a dense image.

(a) Every completion $(\overline{X}, \overline{d}, \iota)$ of (X, d) has the following **universality property**: If (Y, d_Y) is a complete metric space and $\phi : X \to Y$ is a **1-Lipschitz map** (i.e. a Lipschitz continuous map with Lipschitz constant one), then there exists a unique 1-Lipschitz map $\overline{\phi} : \overline{X} \to Y$ such that

$$\phi = \overline{\phi} \circ \iota.$$

(b) If $(\overline{X}_1, \overline{d}_1, \iota_1)$ and $(\overline{X}_2, \overline{d}_2, \iota_2)$ are completions of (X, d), then there exists a unique isometry $\psi : \overline{X}_1 \to \overline{X}_2$ such that $\psi \circ \iota_1 = \iota_2$.

(c) (X, d) admits a completion. **Hint:** The space $C_b(X)$ of bounded continuous functions $f : X \to \mathbb{R}$ is a Banach space with the supremum norm. Let $x_0 \in X$ and, for $x \in X$, define $f_x \in C_b(X)$ by

$$f_x(y) := d(y, x) - d(y, x_0) \qquad \text{for } y \in X.$$

Prove that the map $X \to C_b(X) : x \mapsto f_x$ is an isometric embedding and that the closure of its image is a completion of (X, d).

(d) Let $(\overline{X}, \overline{d})$ be a complete metric space and let $\iota : X \to \overline{X}$ be a 1-Lipschitz map that satisfies the universality property in (a). Prove that $(\overline{X}, \overline{d}, \iota)$ is a completion of (X, d).

Exercise 1.7.5 (Completion of a Normed Vector Space). The completion of a normed vector space is a Banach space.

Exercise 1.7.6 (Operator Norm). This exercise shows that the supremum in the definition of the operator norm need not be a maximum (see Definition 1.2.1). Consider the Banach space $X := C([-1,1])$ of continuous functions $f : [-1,1] \to \mathbb{R}$ equipped with the supremum norm and define the bounded linear functional

$$\Lambda : C([-1,1]) \to \mathbb{R}$$

by

$$\Lambda(f) := \int_0^1 f(t)\,dt - \int_{-1}^0 f(t)\,dt \qquad \text{for } f \in C([-1,1]).$$

Prove that there does not exist a function $f \in C([-1,1])$ such that $\|f\|_\infty = 1$ and $|\Lambda(f)| = \|\Lambda\| = 2$.

Exercise 1.7.7 (Continuously Differentiable Functions). Let $I := [0,1]$ be the unit interval and denote by $C^1(I)$ the space of continuously differentiable functions $f : I \to \mathbb{R}$ (with one-sided derivatives at $t = 0$ and $t = 1$). Define

$$\|f\|_{C^1} := \sup_{0 \le t \le 1} |f(t)| + \sup_{0 \le t \le 1} |f'(t)| \qquad \text{for } f \in C^1(I). \tag{1.7.1}$$

(a) Prove that $C^1(I)$ is a Banach space with the norm (1.7.1).

(b) Show that the inclusion $\iota : C^1(I) \to C(I)$ is a bounded linear operator.

(c) Let $B \subset C^1(I)$ be the unit ball. Show that $\iota(B)$ has compact closure.

(d) Is $\iota(B)$ a closed subset of $C(I)$?

(e) Does the linear operator $\iota : C^1(I) \to C(I)$ have a dense image?

Exercise 1.7.8 (Integration Against a Kernel). Let $I := [0,1]$, let $K : I \times I \to \mathbb{R}$ be a continuous function, and define the linear operator $T_K : C(I) \to C(I)$ by

$$(T_K f)(t) := \int_0^1 K(t,s) f(s)\,ds \qquad \text{for } f \in C(I) \text{ and } 0 \le t \le 1.$$

Prove that T_K is continuous. Let $B \subset C(I)$ be the unit ball and prove that its image $T_K(B)$ has a compact closure in $C(I)$.

Exercise 1.7.9 (Fekete's Lemma). Let $(\alpha_n)_{n\in\mathbb{N}}$ be a sequence of real numbers and suppose that there exists a constant $c \ge 0$ such that

$$\alpha_{n+m} \le \alpha_n + \alpha_m + c \qquad \text{for all } n, m \in \mathbb{N}.$$

Prove that $\lim_{n\to\infty} \alpha_n/n = \inf_{n\in\mathbb{N}} \alpha_n/n$. Here both sides of the equation may be minus infinity. Compare this with part (i) of Theorem 1.5.5 by taking $\alpha_n := \log \|a^n\|$.

Exercise 1.7.10 (The Inverse in a Unital Banach Algebra). Let $\mathcal{A}$ be a unital Banach algebra and let $a, b \in \mathcal{A}$ such that $\mathbb{1} - ab$ is invertible. Prove that $\mathbb{1} - ba$ is invertible. **Hint:** An explicit formula for the inverse of $\mathbb{1} - ba$ in terms of the inverse of $\mathbb{1} - ab$ can be guessed by expanding $(\mathbb{1} - ab)^{-1}$ and $(\mathbb{1} - ba)^{-1}$ formally as geometric series (Theorem 1.5.5).

Exercise 1.7.11 (Cantor's Intersection Theorem). The **diameter** of a nonempty subset A of a metric space (X, d) is defined by

$$\operatorname{diam}(A) := \sup_{x,y \in A} d(x, y). \tag{1.7.2}$$

(a) Prove that a metric space (X, d) is complete if and only if every nested sequence $A_1 \supset A_2 \supset A_3 \supset \cdots$ of nonempty closed subsets $A_n \subset X$ satisfying $\lim_{n\to\infty} \operatorname{diam}(A_n) = 0$ has a nonempty intersection (consisting of a single point).

(b) Find an example of a complete metric space and a nested sequence of nonempty closed bounded sets whose intersection is empty. **Hint:** Consider the unit sphere in an infinite-dimensional Hilbert space.

Exercise 1.7.12 (Convergence Along Arithmetic Sequences). Let $f : [0, \infty) \to \mathbb{R}$ be a continuous function such that

$$\lim_{n\to\infty} f(nt) = 0 \qquad \text{for all } t > 0.$$

Prove that

$$\lim_{x\to\infty} f(x) = 0.$$

Hint: Fix a constant $\varepsilon > 0$ and show that the set

$$A_n := \{t > 0 \,|\, |f(mt)| \le \varepsilon \text{ for every integer } m \ge n\}$$

has a nonempty interior for some $n \in \mathbb{N}$ (using the Baire Category Theorem 1.6.4). Assume without loss of generality that $[a, b] \subset A_n$ for $0 < a < b$ with $n(b - a) \ge a$. Deduce that $|f(x)| \le \varepsilon$ for all $x \ge na$.

Exercise 1.7.13 (Nowhere Differentiable Continuous Functions). Prove that the set

$$\mathcal{R} := \{f : [0, 1] \to \mathbb{R} \,|\, f \text{ is continuous and nowhere differentiable}\}$$

is residual in the Banach space $C([0, 1])$ and hence is dense. (This result is due to Stefan Banach and was proved in 1931.) **Hint:** Prove that the set

$$\mathcal{U}_n := \left\{ f \in C([0, 1]) \,\middle|\, \sup_{\substack{0 \le s \le 1 \\ s \ne t}} \left| \frac{f(s) - f(t)}{s - t} \right| > n \text{ for all } t \in [0, 1] \right\}$$

is open and dense in $C([0, 1])$ for every $n \in \mathbb{N}$ and that $\bigcap_{n=1}^{\infty} \mathcal{U}_n \subset \mathcal{R}$.

The proof of the Baire Category Theorem uses the axiom of dependent choice. A theorem of Blair asserts that the Baire Category Theorem is equivalent to the axiom of dependent choice. That the axiom of dependent choice follows from the Baire Category Theorem is the content of the next exercise.

Exercise 1.7.14 (Baire Category and Dependent Choice). Let $\mathbf{X}$ be a nonempty set and let $\mathbf{A}: \mathbf{X} \to 2^{\mathbf{X}}$ be a map which assigns to each $\mathbf{x} \in \mathbf{X}$ a nonempty subset $\mathbf{A}(\mathbf{x}) \subset \mathbf{X}$. Use Theorem 1.6.4 to prove that there is a sequence $(\mathbf{x}_n)_{n\in\mathbb{N}}$ in $\mathbf{X}$ such that $\mathbf{x}_{n+1} \in \mathbf{A}(\mathbf{x}_n)$ for all $n \in \mathbb{N}$.

Hint: Denote by $\mathcal{X} := \mathbf{X}^{\mathbb{N}}$ the set of all sequences $\xi = (\mathbf{x}_n)_{n\in\mathbb{N}}$ in $\mathbf{X}$ and define the function $d: \mathcal{X} \times \mathcal{X} \to [0, \infty)$ by $d(\xi, \xi) := 0$ and

$$d(\xi, \eta) := 2^{-n}, \qquad n := \min\{k \in \mathbb{N} \,|\, \mathbf{x}_k \neq \mathbf{y}_k\},$$

for every pair of distinct sequences $\xi = (\mathbf{x}_n)_{n\in\mathbb{N}}, \eta = (\mathbf{y}_n)_{n\in\mathbb{N}} \in \mathcal{X}$. Prove that $(\mathcal{X}, d)$ is a complete metric space. For $k \in \mathbb{N}$ define

$$\mathcal{U}_k := \left\{ \xi = (\mathbf{x}_n)_{n\in\mathbb{N}} \in \mathbf{X}^{\mathbb{N}} \,\middle|\, \begin{array}{l} \text{there is an integer } \ell > k \\ \text{such that } \mathbf{x}_\ell \in \mathbf{A}(\mathbf{x}_k) \end{array} \right\}.$$

Prove that $\mathcal{U}_k$ is a dense open subset of $\mathcal{X}$ for every $k \in \mathbb{N}$ and deduce that the set $\mathcal{R} := \bigcap_{k\in\mathbb{N}} \mathcal{U}_k$ is nonempty. Construct the desired sequence as a suitable subsequence of an element $\xi = (\mathbf{x}_n)_{n\in\mathbb{N}} \in \mathcal{R}$.

Exercise 1.7.15 (Borel Measurable Linear Operators).

(a) Sets with the Baire property. A subset B of a topological space is said to have the **Baire property** if there exists an open set U such that the symmetric difference $B\Delta U := (B \setminus U) \cup (U \setminus B)$ is meagre, i.e. B and U differ by a meagre set (see Definition 1.6.1). Prove that the collection of all sets with the Baire property is the smallest σ-algebra containing the Borel sets and the meagre sets.

(b) Pettis' Lemma. Let X be a Banach space and let $B \subset X$ be a non-meagre subset that has the Baire property. Prove that the set $B - B$ is a neighborhood of the origin. In particular, if B is a linear subspace of X, then $B = X$. **Hint:** Let U be an open subset of X such that $B\Delta U$ is meagre. Show that $U \neq \emptyset$, fix an element $x \in U$, and find an open neighborhood V of the origin such that $x + V - V \subset U$. For every $v \in V$ show that $U \cap (v + U) \neq \emptyset$ and deduce that $B \cap (v + B) \neq \emptyset$.

(c) Borel measurable linear operators. Let $f: X \to Y$ be a Borel measurable linear operator from a Banach space X to a separable normed vector space Y. Prove that f is continuous. **Hint:** $B := \{x \in X \,|\, \|f(x)\|_Y < 1/2\}$ is a nonmeagre Borel set.

Chapter 2

Principles of Functional Analysis

This chapter is devoted to the three fundamental *principles of functional analysis*. The first is the *Uniform Boundedness Principle* in Section 2.1. It asserts that every pointwise bounded family of bounded linear operators on a Banach space is bounded. The second is the *Open Mapping Theorem* in Section 2.2. It asserts that every surjective bounded linear operator between two Banach spaces is open. An important corollary is the *Inverse Operator Theorem* which asserts that every bijective bounded linear operator between two Banach spaces has a bounded inverse. An equivalent result is the *Closed Graph Theorem* which asserts that a linear operator between two Banach spaces is bounded if and only if its graph is a closed linear subspace of the product space. The third fundamental principle in functional analysis is the *Hahn–Banach Theorem* in Section 2.3. It asserts that every bounded linear functional on a linear subspace of a normed vector space extends to a bounded linear functional on the entire normed vector space. A slightly stronger version of the Hahn–Banach Theorem, in which the norm is replaced by a quasi-seminorm, can be reformulated as the geometric assertion that two convex subsets of a normed vector space can be separated by a closed hyperplane whenever one of them has nonempty interior. There are in fact many variants of the Hahn–Banach Theorem, including one for positive linear functionals on ordered vector spaces, which is used to establish the separation of convex sets. Another application of the Hahn–Banach Theorem is a criterion for a linear subspace to be dense. The final section of this chapter discusses reflexive Banach spaces and includes an exposition of the James space.

2.1. Uniform Boundedness

Let X be a set. A family $\{f_i\}_{i\in I}$ of functions $f_i : X \to Y_i$, indexed by the elements of a set I and each taking values in a normed vector space Y_i, is called **pointwise bounded** if

$$\sup_{i\in I} \|f_i(x)\|_{Y_i} < \infty \qquad \text{for all } x \in X. \tag{2.1.1}$$

Theorem 2.1.1 (Uniform Boundedness). *Let X be a Banach space, let I be any set, and, for each $i \in I$, let Y_i be a normed vector space and let $A_i : X \to Y_i$ be a bounded linear operator. Assume that the operator family $\{A_i\}_{i\in I}$ is pointwise bounded. Then $\sup_{i\in I} \|A_i\| < \infty$.*

Proof. See page 51. □

Lemma 2.1.2. *Let (X, d) be a nonempty complete metric space, let I be any set, and, for each $i \in I$, let $f_i : X \to \mathbb{R}$ be a continuous function. Assume that the family $\{f_i\}_{i\in I}$ is pointwise bounded. Then there exist a point $x_0 \in X$ and a number $\varepsilon > 0$ such that*

$$\sup_{i\in I} \sup_{x\in B_\varepsilon(x_0)} |f_i(x)| < \infty.$$

Proof. For $n \in \mathbb{N}$ and $i \in I$ define the set

$$F_{n,i} := \left\{ x \in X \,\middle|\, |f_i(x)| \le n \right\}.$$

This set is closed because f_i is continuous. Hence the set

$$F_n := \bigcap_{i\in I} F_{n,i} = \left\{ x \in X \,\middle|\, \sup_{i\in I} |f_i(x)| \le n \right\}$$

is closed for every $n \in \mathbb{N}$. Moreover,

$$X = \bigcup_{n\in\mathbb{N}} F_n,$$

because the family $\{f_i\}_{i\in I}$ is pointwise bounded. Since (X, d) is a nonempty complete metric space, it follows from the Baire Category Theorem 1.6.4 that the sets F_n cannot all be nowhere dense. Since these sets are all closed, there exists an integer $n \in \mathbb{N}$ such that F_n has nonempty interior. Hence there exist an integer $n \in \mathbb{N}$, a point $x_0 \in X$, and a number $\varepsilon > 0$ such that $B_\varepsilon(x_0) \subset F_n$. Hence

$$\sup_{i\in I} \sup_{x\in B_\varepsilon(x_0)} |f_i(x)| \le n$$

and this proves Lemma 2.1.2. □

Proof of Theorem 2.1.1. Define the function $f_i : X \to \mathbb{R}$ by

$$f_i(x) := \|A_i x\|_{Y_i}$$

for $x \in X$ and $i \in I$. Then f_i is continuous for each i and the family $\{f_i\}_{i\in I}$ is pointwise bounded by assumption. Since X is a Banach space, Lemma 2.1.2 asserts that there exist a vector $x_0 \in X$ and a constant $\varepsilon > 0$ such that

$$c := \sup_{i\in I} \sup_{x\in B_\varepsilon(x_0)} \|A_i x\|_{Y_i} < \infty.$$

Hence, for all $x \in X$ and all $i \in I$, we have

$$\|x - x_0\|_X \le \varepsilon \qquad \Longrightarrow \qquad \|A_i x\|_{Y_i} \le c. \tag{2.1.2}$$

Let $i \in I$ and $x \in X$ such that $\|x\|_X = 1$. Then $\|A_i(x_0 \pm \varepsilon x)\|_{Y_i} \le c$ and so

$$\begin{aligned}\|A_i x\|_{Y_i} &= \frac{1}{2\varepsilon}\|A_i(x_0+\varepsilon x) - A_i(x_0-\varepsilon x)\|_{Y_i} \\ &\le \frac{1}{2\varepsilon}\|A_i(x_0+\varepsilon x)\|_{Y_i} + \frac{1}{2\varepsilon}\|A_i(x_0-\varepsilon x)\|_{Y_i} \le \frac{c}{\varepsilon}.\end{aligned}$$

Hence

$$\|A_i\| = \sup_{x\in X\setminus\{0\}} \frac{\|A_i x\|_{Y_i}}{\|x\|_X} = \sup_{\substack{x\in X \\ \|x\|_X=1}} \|A_i x\|_{Y_i} \le \frac{c}{\varepsilon} \tag{2.1.3}$$

for all $i \in I$ and this proves Theorem 2.1.1. □

Remark 2.1.3. The above argument in the proof of Theorem 2.1.1, which asserts that (2.1.2) implies (2.1.3), can be rewritten as the inequality

$$\sup_{\substack{x\in X \\ \|x-x_0\|_X<\varepsilon}} \|Ax\|_Y \ge \varepsilon\|A\| \tag{2.1.4}$$

for all $A \in \mathcal{L}(X,Y)$, all $x_0 \in X$, and all $\varepsilon > 0$. With this understood, one can prove the Uniform Boundedness Theorem as follows (see Sokal [**80**]). Let $\{A_i\}_{i\in I}$ be a sequence of bounded linear operators $A_i : X \to Y_i$ such that $\sup_{i\in I}\|A_i\| = \infty$. Then the axiom of countable choice asserts that there is a sequence $i_n \in I$ such that $\|A_{i_n}\| \ge 4^n$ for all $n \in \mathbb{N}$. Now use the axiom of dependent choice, and the estimate (2.1.4) with $A = A_{i_n}$ and $\varepsilon = 1/3^n$, to find a sequence $x_n \in X$ such that, for all $n \in \mathbb{N}$,

$$\|x_n - x_{n-1}\|_X \le \frac{1}{3^n}, \qquad \|A_{i_n}x_n\|_{Y_{i_n}} \ge \frac{2}{3}\frac{1}{3^n}\|A_{i_n}\|.$$

Then $(x_n)_{n\in\mathbb{N}}$ is a Cauchy sequence and so converges to an element $x^* \in X$ such that $\|x^* - x_n\|_X \le \frac{1}{2}\frac{1}{3^n}$. Thus

$$\|A_{i_n}x^*\|_{Y_{i_n}} \ge \left(\frac{2}{3} - \frac{1}{2}\right)\frac{1}{3^n}\|A_{i_n}\| \ge \frac{1}{6}\left(\frac{4}{3}\right)^n$$

for all $n \in \mathbb{N}$ and so the operator family $\{A_i\}_{i\in I}$ is not pointwise bounded. This argument circumvents the Baire Category Theorem.

The Uniform Boundedness Theorem is also known as the **Banach–Steinhaus Theorem**. A useful consequence is that the limit of a pointwise convergent sequence of bounded linear operators is again a bounded linear operator. This is the content of Theorem 2.1.5 below.

Definition 2.1.4. Let X and Y be normed vector spaces. A sequence of bounded linear operators $A_i : X \to Y$, $i \in \mathbb{N}$, is said to **converge strongly** to a bounded linear operator $A : X \to Y$ if $Ax = \lim_{i\to\infty} A_i x$ for all $x \in X$.

Theorem 2.1.5 (Banach–Steinhaus). *Let X and Y be Banach spaces and let $A_i : X \to Y$, $i \in \mathbb{N}$, be a sequence of bounded linear operators. Then the following are equivalent.*

(i) *The sequence $(A_i x)_{i\in\mathbb{N}}$ converges in Y for every $x \in X$.*

(ii) $\sup_{i\in\mathbb{N}} \|A_i\| < \infty$ *and there is a dense subset $D \subset X$ such that $(A_i x)_{i\in\mathbb{N}}$ is a Cauchy sequence in Y for every $x \in D$.*

(iii) $\sup_{i\in\mathbb{N}} \|A_i\| < \infty$ *and there is a bounded linear operator $A : X \to Y$ such that A_i converges strongly to A and $\|A\| \le \liminf_{i\to\infty} \|A_i\|$.*

The equivalence of (i) and (iii) continues to hold when Y is not complete. The equivalence of (ii) and (iii) continues to hold when X is not complete.

Proof. That (iii) implies both (i) and (ii) is obvious.

We prove that (i) implies (iii). Since convergent sequences are bounded, the sequence $(A_i)_{i\in\mathbb{N}}$ is pointwise bounded. Since X is complete it follows from Theorem 2.1.1 that $\sup_{i\in\mathbb{N}} \|A_i\| < \infty$. Define the map $A : X \to Y$ by $Ax := \lim_{i\to\infty} A_i x$ for $x \in X$. This map is linear and

$$\|Ax\|_Y = \lim_{i\to\infty} \|A_i x\|_Y = \liminf_{i\to\infty} \|A_i x\|_Y \le \liminf_{i\to\infty} \|A_i\| \, \|x\|_X \tag{2.1.5}$$

for all $x \in X$. Hence A is bounded and $\|A\| \le \liminf_{i\to\infty} \|A_i\| < \infty$.

We prove that (ii) implies (iii). Define $c := \sup_{i\in\mathbb{N}} \|A_i\| < \infty$. Let $x \in X$ and $\varepsilon > 0$. Choose $\xi \in D$ such that $c\,\|x - \xi\| < \frac{\varepsilon}{3}$. Since $(A_i\xi)_{i\in\mathbb{N}}$ is a Cauchy sequence, there exists an integer $n_0 \in \mathbb{N}$ such that $\|A_i\xi - A_j\xi\|_Y < \frac{\varepsilon}{3}$ for all $i, j \in \mathbb{N}$ with $i, j \ge n_0$. This implies

$$\begin{aligned}
\|A_i x - A_j x\|_Y &\le \|A_i x - A_i\xi\|_Y + \|A_i\xi - A_j\xi\|_Y + \|A_j\xi - A_j x\|_Y \\
&\le \|A_i\| \, \|x - \xi\|_X + \|A_i\xi - A_j\xi\|_Y + \|A_j\| \, \|\xi - x\|_X \\
&\le 2c\,\|x - \xi\|_X + \|A_i\xi - A_j\xi\|_Y < \tfrac{2\varepsilon}{3} + \tfrac{\varepsilon}{3} = \varepsilon
\end{aligned}$$

for all $i, j \in \mathbb{N}$ with $i, j \ge n_0$. Hence $(A_i x)_{i\in\mathbb{N}}$ is a Cauchy sequence and so it converges because Y is complete. The limit operator A satisfies (2.1.5) and this proves Theorem 2.1.5. □

Example 2.1.6. This example shows that the hypothesis that X is complete cannot be removed in Theorems 2.1.1 and 2.1.5. Consider the space

$$X := \left\{ x = (x_i)_{i \in \mathbb{N}} \in \mathbb{R}^{\mathbb{N}} \,\middle|\, \exists n \in \mathbb{N} \, \forall i \in \mathbb{N} : i \geq n \implies x_i = 0 \right\}$$

with the supremum norm $\|x\| := \sup_{i \in \mathbb{N}} |x_i|$. This is a normed vector space. It is not complete, but is a linear subspace of ℓ^∞ whose closure $\overline{X} = c_0$ is the subspace of sequences of real numbers that converge to zero. Define the linear operators $A_n : X \to X$ and $A : X \to X$ by

$$A_n x := (x_1, 2x_2, \ldots, nx_n, 0, 0, \ldots), \qquad Ax := (ix_i)_{i \in \mathbb{N}}$$

for $n \in \mathbb{N}$ and $x = (x_i)_{i \in \mathbb{N}} \in X$. Then $Ax = \lim_{n \to \infty} A_n x$ for every $x \in X$ and $\|A_n\| = n$ for every $n \in \mathbb{N}$. Thus the sequence $\{A_n x\}_{n \in \mathbb{N}}$ is bounded for every $x \in X$, the linear operator A is not bounded, and the sequence A_n converges strongly to A.

Corollary 2.1.7 (Bilinear Map). *Let X be a Banach space and let Y and Z be normed vector spaces (over $\mathbb{R}$ or $\mathbb{C}$). Let $B : X \times Y \to Z$ be a bilinear map. Then the following are equivalent.*

(i) *B is* **bounded**, *i.e. there is a constant $c \geq 0$ such that*

$$\|B(x, y)\|_Z \leq c \|x\|_X \|y\|_Y$$

for all $x \in X$ and all $y \in Y$.

(ii) *B is continuous.*

(iii) *For every $x \in X$ the linear map $Y \to Z : y \mapsto B(x, y)$ is continuous and, for every $y \in Y$, the linear map $X \to Z : x \mapsto B(x, y)$ is continuous.*

Proof. If (i) holds, then B is locally Lipschitz continuous and hence is continuous. Thus (i) implies (ii). That (ii) implies (iii) is obvious. We prove that (iii) implies (i). Thus assume (iii), define

$$S := \{ y \in Y \mid \|y\|_Y = 1 \},$$

and, for $y \in S$, define the linear operator $A_y : X \to Z$ by $A_y(x) := B(x, y)$. This operator is continuous by (iii) and hence is bounded by Theorem 1.2.2. Now let $x \in X$. Then the linear map $Y \to Z : y \mapsto A_y x = B(x, y)$ is continuous by (iii) and hence $\sup_{y \in S} \|A_y x\|_Z < \infty$ by Theorem 1.2.2. Hence $c := \sup_{y \in S} \|A_y\| < \infty$ by Theorem 2.1.1. Thus

$$\|B(x, y)\|_Z \leq c \|x\|_X \qquad \text{for all } x \in X \text{ and all } y \in S.$$

This implies (i) and completes the proof of Corollary 2.1.7. □

2.2. Open Mappings and Closed Graphs

2.2.1. The Open Mapping Theorem. A map $f : X \to Y$ between topological spaces is called **open** if the image of every open subset of X under f is an open subset of Y.

Theorem 2.2.1 (Open Mapping Theorem). *Let X, Y be Banach spaces and let $A : X \to Y$ be a surjective bounded linear operator. Then A is open.*

Proof. See page 56. $\square$

The key step in the proof of Theorem 2.2.1 is the next lemma, which asserts that the closure $\overline{A(B)}$ of the image of the open unit ball $B \subset X$ under a surjective bounded linear operator $A : X \to Y$ contains an open ball in Y centered at the origin. Its proof relies on the Baire Category Theorem 1.6.4. Lemma 2.2.3 below asserts that if an open ball in Y centered at the origin is contained in $\overline{A(B)}$, then it is contained in $A(B)$.

Lemma 2.2.2. *Let X, Y, and A be as in Theorem 2.2.1. Then there exists a constant $\delta > 0$ such that*

$$\{y \in Y \,|\, \|y\|_Y < \delta\} \subset \overline{\{Ax \,|\, x \in X,\, \|x\|_X < 1\}}. \tag{2.2.1}$$

Proof. For $C \subset Y$ and $\lambda > 0$ define $\lambda C := \{\lambda y \,|\, y \in C\}$. Consider the sets

$$B := \{x \in X \,|\, \|x\|_X < 1\}, \qquad C := A(B) = \{Ax \,|\, x \in X,\, \|x\|_X < 1\}.$$

Then $X = \bigcup_{n\in\mathbb{N}} nB$ and so $Y = \bigcup_{n\in\mathbb{N}} A(nB) = \bigcup_{n\in\mathbb{N}} nC$ because A is surjective. Since Y is complete, at least one of the sets nC is not nowhere dense, by the Baire Category Theorem 1.6.4. Hence the set $\overline{nC}$ has a nonempty interior for some $n \in \mathbb{N}$ and this implies that the set $\overline{2^{-1}C}$ has a nonempty interior. Choose $y_0 \in Y$ and $\delta > 0$ such that

$$B_\delta(y_0) \subset \overline{2^{-1}C}.$$

We claim that (2.2.1) holds with this constant δ. To see this, fix an element $y \in Y$ such that $\|y\|_Y < \delta$. Then $y_0 + y \in \overline{2^{-1}C}$ and $y_0 \in \overline{2^{-1}C}$. Hence there exist sequences $x_i, x_i' \in 2^{-1}B$ such that

$$y_0 + y = \lim_{i\to\infty} Ax_i', \qquad y_0 = \lim_{i\to\infty} Ax_i.$$

Hence $x_i' - x_i \in B$, so $A(x_i' - x_i) \in C$ and

$$y = \lim_{i\to\infty} A(x_i' - x_i) \in \overline{C}.$$

Thus (2.2.1) holds as claimed. This proves Lemma 2.2.2. $\square$

Lemma 2.2.3. *Let X and Y be Banach spaces and let $A : X \to Y$ be a bounded linear operator. If $\delta > 0$ satisfies* (2.2.1), *then*

$$\{y \in Y \,|\, \|y\|_Y < \delta\} \subset \{Ax \,|\, x \in X,\, \|x\|_X < 1\}. \tag{2.2.2}$$

Proof. The proof is based on the following observation.

Claim. *Let $y \in Y$ with $\|y\|_Y < \delta$. Then there is a sequence $(x_k)_{k\in\mathbb{N}_0}$ in X such that*

$$\begin{aligned} &\|x_0\|_X < \frac{\|y\|_Y}{\delta}, \quad \|x_k\|_X < \frac{\delta - \|y\|_Y}{\delta 2^k} \quad \textit{for } k = 1,2,3,\dots, \\ &\|y - Ax_0 - \cdots - Ax_k\|_Y < \frac{\delta - \|y\|_Y}{2^{k+1}} \quad \textit{for } k = 0,1,2,\dots. \end{aligned} \tag{2.2.3}$$

We prove the claim by an induction argument. By (2.2.1) the closed ball of radius δ in Y is contained in the closure of the image under A of the open ball of radius one in X. Hence every nonzero vector $y \in Y$ satisfies

$$y \in \overline{\{Ax \,|\, x \in X,\, \|x\|_X < \delta^{-1}\|y\|_Y\}}. \tag{2.2.4}$$

Fix an element $y \in Y$ such that $\|y\|_Y < \delta$ and define

$$\varepsilon := \delta - \|y\|_Y > 0.$$

Then, by (2.2.4), there exists a vector $x_0 \in X$ such that $\|x_0\|_X < \delta^{-1}\|y\|_Y$ and $\|y - Ax_0\|_Y < \varepsilon 2^{-1}$. Use (2.2.4) with y replaced by $y - Ax_0$ to find a vector $x_1 \in X$ such that $\|x_1\|_X < \varepsilon\delta^{-1}2^{-1}$ and $\|y - Ax_0 - Ax_1\|_Y < \varepsilon 2^{-2}$. Once the vectors $x_0, \dots, x_k$ have been found such that (2.2.3) holds, we have $\|y - \sum_{i=0}^k Ax_i\|_Y < \varepsilon 2^{-k-1}$ and so, by (2.2.4), there exists an $x_{k+1} \in X$ such that $\|x_{k+1}\|_X < \varepsilon\delta^{-1}2^{-k-1}$ and $\|y - \sum_{i=0}^k Ax_i - Ax_{k+1}\|_Y < \varepsilon 2^{-k-2}$. Hence the existence of a sequence $(x_k)_{k\in\mathbb{N}_0}$ in X that satisfies (2.2.3) follows from the axiom of dependent choice (see page 6). This proves the claim.

Now fix an element $y \in Y$ such that $\|y\|_Y < \delta$. By the claim, there is a sequence $(x_k)_{k\in\mathbb{N}_0}$ in X that satisfies (2.2.3) and hence $\sum_{k=0}^\infty \|x_k\|_X < 1$. Since X is complete, it then follows from Lemma 1.5.1 that the limit

$$x := \sum_{k=0}^\infty x_k = \lim_{k\to\infty} \sum_{i=0}^k x_i$$

exists. This limit satisfies

$$\|x\|_X \le \sum_{k=0}^\infty \|x_k\|_X < 1, \qquad Ax = \lim_{k\to\infty} \sum_{i=0}^k Ax_i = y.$$

Here the last equation follows from (2.2.3). This proves the inclusion (2.2.2) and Lemma 2.2.3. □

Proof of Theorem 2.2.1. Let $\delta > 0$ be the constant of Lemma 2.2.2 and let $B \subset X$ be the open unit ball. Then $B_\delta(0;Y) \subset \overline{A(B)}$ by Lemma 2.2.2 and hence $B_\delta(0;Y) \subset A(B)$ by Lemma 2.2.3.

Now fix an open set $U \subset X$. Let $y_0 \in A(U)$ and choose $x_0 \in U$ such that $Ax_0 = y_0$. Since U is open there is an $\varepsilon > 0$ such that $B_\varepsilon(x_0) \subset U$. We prove that $B_{\delta\varepsilon}(y_0) \subset A(U)$. Choose $y \in Y$ such that $\|y - y_0\|_Y < \delta\varepsilon$. Then $\left\|\varepsilon^{-1}(y - y_0)\right\|_Y < \delta$ and hence there exists a $\xi \in X$ such that

$$\|\xi\|_X < 1, \qquad A\xi = \varepsilon^{-1}(y - y_0).$$

This implies $x_0 + \varepsilon\xi \in B_\varepsilon(x_0) \subset U$ and hence

$$y = y_0 + \varepsilon A\xi = A(x_0 + \varepsilon\xi) \in A(U).$$

Thus we have proved that, for every $y_0 \in A(U)$, there exists a number $\varepsilon > 0$ such that $B_{\delta\varepsilon}(y_0) \subset A(U)$. Hence $A(U)$ is an open subset of Y and this proves Theorem 2.2.1. □

If $A : X \to Y$ is a surjective bounded linear operator between Banach spaces, then it descends to a bijective bounded linear operator from the quotient space $X/\ker(A)$ to Y (see Theorem 1.2.14). The next corollary asserts that the induced operator $\overline{A} : X/\ker(A) \to Y$ has a bounded inverse whose norm is bounded above by δ^{-1}, where the constant $\delta > 0$ is as in Lemma 2.2.2.

Corollary 2.2.4. *Let X, Y, and A be as in Theorem 2.2.1 and let $\delta > 0$ be the constant of Lemma 2.2.2. Then*

$$\inf_{\substack{x \in X \\ Ax = y}} \|x\|_X \le \delta^{-1} \|y\|_Y \qquad \text{for all } y \in Y. \tag{2.2.5}$$

Proof. Let $y \in Y$ and choose $c > \delta^{-1}\|y\|_Y$. Then $\left\|c^{-1}y\right\|_Y < \delta$ and so, by Lemma 2.2.2 and Lemma 2.2.3, there exists an element $\xi \in X$ such that $A\xi = c^{-1}y$ and $\|\xi\|_X < 1$. Hence $x := c\xi$ satisfies $\|x\|_X = c\|\xi\|_X < c$ and $Ax = cA\xi = y$. This proves (2.2.5) and Corollary 2.2.4. □

An important consequence of the Open Mapping Theorem is the special case of Corollary 2.2.4 where A is bijective.

Theorem 2.2.5 (Inverse Operator Theorem). *Let X and Y be Banach spaces and let $A : X \to Y$ be a bijective bounded linear operator. Then the inverse operator $A^{-1} : Y \to X$ is bounded.*

Proof. By Theorem 2.2.1 the linear operator $A : X \to Y$ is open. Hence its inverse is continuous and is therefore bounded by Theorem 1.2.2. Alternatively, use Corollary 2.2.4 to deduce that $\|A^{-1}\| \le \delta^{-1}$, where $\delta > 0$ is the constant of Lemma 2.2.2. □

Example 2.2.6. This example shows that the hypothesis that X and Y are complete cannot be removed in Theorems 2.2.1 and 2.2.5. As in Example 2.1.6, let $X \subset \ell^\infty$ be the subspace of sequences $x = (x_k)_{k\in\mathbb{N}}$ of real numbers that vanish for sufficiently large k, equipped with the supremum norm. Thus X is a normed vector space but is not a Banach space. Define the operator $A : X \to X$ by $Ax := (k^{-1}x_k)_{k\in\mathbb{N}}$ for $x = (x_k)_{k\in\mathbb{N}} \in X$. Then A is a bijective bounded linear operator but its inverse is unbounded.

Example 2.2.7. Here is an example where X is complete and Y is not. Let $X = Y = C([0, 1])$ be the space of continuous functions $f : [0, 1] \to \mathbb{R}$ equipped with the norms

$$\|f\|_X := \sup_{0\le t\le 1} |f(t)|, \qquad \|f\|_Y := \sqrt{\int_0^1 |f(t)|^2\, dt}.$$

Then X is a Banach space, Y is a normed vector space, and the identity

$$A = \mathrm{id} : X \to Y$$

is a bijective bounded linear operator with an unbounded inverse.

Example 2.2.8. Here is an example where Y is complete and X is not. This example requires the axiom of choice. Let Y be an infinite-dimensional Banach space and choose an unbounded linear functional $\Phi : Y \to \mathbb{R}$. The existence of such a linear functional is shown in part (iv) of Example 1.2.10 and its kernel is a dense linear subspace of Y by Exercise 1.3.9. Define the normed vector space $(X, \|\cdot\|_X)$ by

$$X := \{(x, t) \in Y \times \mathbb{R} \,|\, \Phi(x) = 0\}, \qquad \|(x, t)\|_X := \|x\|_Y + |t|$$

for $(x, t) \in X$. Then X is not complete. Choose a vector $y_0 \in Y$ such that

$$\Phi(y_0) = 1$$

and define the linear map $A : X \to Y$ by

$$A(x, t) := x + ty_0 \qquad \text{for } (x, t) \in X.$$

Then A is a bijective bounded linear operator. Its inverse is given by

$$A^{-1}y = (y - \Phi(y)y_0, \Phi(y))$$

for $y \in H$ and hence is unbounded.

Example 2.2.8 relies on a decomposition of a Banach space as a direct sum of two linear subspaces where one of them is closed and the other is dense. The next corollary establishes an important estimate for a pair of *closed* subspaces of a Banach space X whose **direct sum** is equal to X.

Corollary 2.2.9. *Let X be a Banach space and let $X_1, X_2 \subset X$ be two closed linear subspaces such that*

$$X = X_1 \oplus X_2,$$

i.e. $X_1 \cap X_2 = \{0\}$ and every vector $x \in X$ can be written as $x = x_1 + x_2$ with $x_1 \in X_1$ and $x_2 \in X_2$. Then there exists a constant $c \geq 0$ such that

$$\|x_1\| + \|x_2\| \leq c\,\|x_1 + x_2\| \tag{2.2.6}$$

for all $x_1 \in X_1$ and all $x_2 \in X_2$.

Proof. The vector space $X_1 \times X_2$ is a Banach space with the norm function

$$X_1 \times X_2 \to [0, \infty) : (x_1, x_2) \mapsto \|(x_1, x_2)\| := \|x_1\| + \|x_2\|$$

(see Exercise 1.2.15) and the linear operator $A : X_1 \times X_2 \to X$, defined by

$$A(x_1, x_2) := x_1 + x_2$$

for $(x_1, x_2) \in X_1 \times X_2$, is bijective by assumption and bounded by the triangle inequality. Hence its inverse is bounded by the Inverse Operator Theorem 2.2.5. This proves Corollary 2.2.9. □

2.2.2. The Closed Graph Theorem. It is often interesting to consider linear operators on a Banach space X whose domains are not the entire Banach space but instead are linear subspaces of X. In most of the interesting cases the domains are dense linear subspaces. Here is a first elementary example.

Example 2.2.10. Let $X := C([0,1])$ be the Banach space of continuous real valued functions $f : [0,1] \to \mathbb{R}$ equipped with the supremum norm. Let

$$\mathrm{dom}(A) := C^1([0,1]) = \{f : [0,1] \to \mathbb{R} \,|\, f \text{ is continuously differentiable}\}$$

and define the linear operator $A : \mathrm{dom}(A) \to X$ by

$$Af := f' \qquad \text{for } f \in C^1([0,1]).$$

The linear subspace $\mathrm{dom}(A) = C^1([0,1])$ is dense in $X = C([0,1])$ by the Weierstraß Approximation Theorem. Moreover, the graph of A, defined by

$$\mathrm{graph}(A) := \{(f, g) \in X \times X \,|\, f \in \mathrm{dom}(A),\, g = Af\},$$

is a closed linear subspace of $X \times X$. Namely, if $f_n \in C^1([0,1])$ is a sequence of continuously differentiable functions such that the pair (f_n, Af_n) converges to (f, g) in $X \times X$, then f_n converges uniformly to f and f_n' converges uniformly to g, and hence f is continuously differentiable with $f' = g$ by the fundamental theorem of calculus.

Here is a general definition of operators with closed graphs.

Definition 2.2.11 (Closed Operator). Let X, Y be Banach spaces, let $\mathrm{dom}(A) \subset X$ be a linear subspace, and let $A : \mathrm{dom}(A) \to Y$ be a linear operator. The operator A is called **closed** if its graph

$$\mathrm{graph}(A) := \left\{(x, y) \in X \times Y \,\middle|\, x \in \mathrm{dom}(A),\ y = Ax\right\} \tag{2.2.7}$$

is a closed linear subspace of $X \times Y$. Explicitly, this means that, if $(x_n)_{n\in\mathbb{N}}$ is a sequence in the domain of A such that x_n converges to a vector $x \in X$ and Ax_n converges to a vector $y \in Y$, then $x \in \mathrm{dom}(A)$ and $y = Ax$. The **graph norm of** A on the linear subspace $\mathrm{dom}(A) \subset X$ is the norm function $\mathrm{dom}(A) \to [0, \infty) : x \mapsto \|x\|_A$ defined by

$$\|x\|_A := \|x\|_X + \|Ax\|_Y \tag{2.2.8}$$

for $x \in \mathrm{dom}(A)$.

Note that a linear operator $A : X \supset \mathrm{dom}(A) \to Y$ is always a bounded linear operator with respect to the graph norm. In Example 2.2.10 the graph norm of A on $\mathrm{dom}(A) = C^1([0,1])$ agrees with the usual C^1 norm

$$\|f\|_{C^1} = \sup_{0\le t\le 1} |f(t)| + \sup_{0\le t\le 1} |f'(t)| \qquad \text{for } f \in C^1([0,1]), \tag{2.2.9}$$

and $C^1([0,1])$ is a Banach space with this norm.

Exercise 2.2.12. Let X, Y be Banach spaces and let $A : \mathrm{dom}(A) \to Y$ be a linear operator, defined on a linear subspace $\mathrm{dom}(A) \subset X$. Prove that the graph of A is a closed subspace of $X \times Y$ if and only if $\mathrm{dom}(A)$ is a Banach space with respect to the graph norm.

The notion of an unbounded linear operator with a dense domain will only become relevant much later in this book when we deal with the spectral theory of linear operators (see Chapter 6). For now it is sufficient to consider linear operators from a Banach space X to a Banach space Y that are defined on the entire space X, rather than just a subspace of X. In this situation it turns out that the closed graph condition is equivalent to boundedness. This is the content of the Closed Graph Theorem, which can be derived as a consequence of the Open Mapping Theorem and vice versa.

Theorem 2.2.13 (Closed Graph Theorem). *Let X and Y be Banach spaces and let $A : X \to Y$ be a linear operator. Then A is bounded if and only if its graph is a closed linear subspace of $X \times Y$.*

Proof. Assume first that A is bounded. Then A is continuous by Theorem 1.2.2. Hence, if $(x_n)_{n\in\mathbb{N}}$ is a sequence in X such that x_n converges to $x \in X$ and Ax_n converges to $y \in Y$, we must have $y = \lim_{n\to\infty} Ax_n = Ax$ and hence $(x, y) \in \mathrm{graph}(A)$.

Conversely, suppose that $\Gamma := \mathrm{graph}(A) = \{(x,y) \in X \times Y \mid y = Ax\}$ is a closed linear subspace of $X \times Y$. Then Γ is a Banach space with the norm

$$\|(x,y)\|_\Gamma := \|x\|_X + \|y\|_Y \qquad \text{for } (x,y) \in \Gamma$$

and the projection

$$\pi : \Gamma \to X, \qquad \pi(x,y) := x \quad \text{for } (x,y) \in \Gamma,$$

is a bijective bounded linear operator. Its inverse is the linear map

$$\pi^{-1} : X \to \Gamma, \qquad \pi^{-1}(x) = (x, Ax) \quad \text{for } x \in X,$$

and is bounded by the Inverse Operator Theorem 2.2.5. Hence there exists a constant $c > 0$ such that $\|x\|_X + \|Ax\|_Y = \left\|\pi^{-1}(x)\right\|_\Gamma \le c\,\|x\|_X$ for all $x \in X$. Thus A is bounded and this proves Theorem 2.2.13. □

Exercise 2.2.14. **(i)** Derive the Inverse Operator Theorem 2.2.5 from the Closed Graph Theorem 2.2.13.

(ii) Derive the Open Mapping Theorem 2.2.1 from the Inverse Operator Theorem 2.2.5. **Hint:** Consider the induced operator $\overline{A} : X/\ker(A) \to Y$ and use Theorem 1.2.14.

Example 2.2.15. **(i)** The hypothesis that X is complete cannot be removed in Theorem 2.2.13. Let $X := C^1([0,1])$ and $Y := C([0,1])$, both equipped with the supremum norm, and define $A : X \to Y$ by $Af := f'$. Then A is unbounded and has a closed graph (see Example 2.2.10).

(ii) The hypothesis that Y is complete cannot be removed in Theorem 2.2.13. Let X be an infinite-dimensional Banach space, let $\Phi : X \to \mathbb{R}$ be an unbounded linear functional, let $Y := \ker(\Phi) \times \mathbb{R}$ with $\|(x,t)\|_Y := \|x\|_X + |t|$ for $(x,t) \in Y$, choose an element $x_0 \in X$ such that $\Phi(x_0) = 1$, and define the linear operator $A : X \to Y$ by $Ax := (x - \Phi(x)x_0, \Phi(x))$ for $x \in X$. Then A is unbounded and has a closed graph (see Example 2.2.8).

Let X and Y be Banach spaces and let $A : X \to Y$ be a linear operator. The Closed Graph Theorem asserts that the following are equivalent.

(i) The operator A is continuous, i.e. for every sequence $(x_n)_{n\in\mathbb{N}}$ in X and all $x \in X$ we have

$$\lim_{n\to\infty} x_n = x \qquad \Longrightarrow \qquad Ax = \lim_{n\to\infty} Ax_n.$$

(ii) The operator A has a closed graph, i.e. for every sequence $(x_n)_{n\in\mathbb{N}}$ in X and all $x, y \in X$ we have

$$\begin{array}{c} \lim_{n\to\infty} x_n = x \\ \lim_{n\to\infty} Ax_n = y \end{array} \qquad \Longrightarrow \qquad Ax = \lim_{n\to\infty} Ax_n.$$

Thus the closed graph condition is much easier to verify for linear operators than boundedness. Examples are the next two corollaries.

Corollary 2.2.16 (Hellinger–Toeplitz Theorem). *Let H be a real Hilbert space and let $A : H \to H$ be a* **symmetric linear operator**, *i.e.*

$$\langle x, Ay\rangle = \langle Ax, y\rangle \qquad \text{for all } x, y \in H. \tag{2.2.10}$$

Then A is bounded.

Proof. By Theorem 2.2.13 it suffices to prove that A has a closed graph. Thus assume that $(x_n)_{n\in\mathbb{N}}$ is a sequence in H and $x, y \in H$ are vectors such that

$$\lim_{n\to\infty} x_n = x, \qquad \lim_{n\to\infty} Ax_n = y.$$

Then

$$\langle y, z\rangle = \lim_{n\to\infty}\langle Ax_n, z\rangle = \lim_{n\to\infty}\langle x_n, Az\rangle = \langle x, Az\rangle = \langle Ax, z\rangle$$

for all $z \in H$ and hence $Ax = y$. This proves Corollary 2.2.16. □

Corollary 2.2.17 (Douglas Factorization [23]). *Let X, Y, Z be Banach spaces and let $A : X \to Y$ and $B : Z \to Y$ be bounded linear operators. Assume A is injective. Then the following are equivalent.*

(i) $\mathrm{im}(B) \subset \mathrm{im}(A)$.

(ii) *There exists a bounded linear operator $T : Z \to X$ such that $AT = B$.*

Proof. If (ii) holds, then $\mathrm{im}(B) = \mathrm{im}(AT) \subset \mathrm{im}(A)$. Conversely, suppose that $\mathrm{im}(B) \subset \mathrm{im}(A)$ and define

$$T := A^{-1} \circ B : Z \to X.$$

Then T is a linear operator and $AT = B$. We prove that T has a closed graph. To see this, let $(z_n)_{n\in\mathbb{N}}$ be a sequence in Z such that the limits

$$z := \lim_{n\to\infty} z_n, \qquad x := \lim_{n\to\infty} Tz_n$$

exist. Then

$$Ax = \lim_{n\to\infty} ATz_n = \lim_{n\to\infty} Bz_n = Bz$$

and hence $x = Tz$. Thus T has a closed graph and hence is bounded by Theorem 2.2.13. This proves Corollary 2.2.17. □

The hypothesis that A is injective cannot be removed in Corollary 2.2.17. For example, take $X = \ell^\infty$, $Y = Z = \ell^\infty/c_0$, and $B = \mathrm{id}$. Then the projection $A : \ell^\infty \to \ell^\infty/c_0$ does not have a bounded right inverse (see Exercise 2.5.1).

2.2.3. Closeable Operators. For a linear operator that is defined on a proper linear subspace it is an interesting question whether it can be extended to a linear operator with a closed graph. Such linear operators are called closeable.

Definition 2.2.18 (Closeable Operator). Let X and Y be Banach spaces, let $\mathrm{dom}(A) \subset X$ be a linear subspace, and let $A : \mathrm{dom}(A) \to Y$ be a linear operator. The operator A is called **closeable** if there exists a closed linear operator $A' : \mathrm{dom}(A') \to Y$ on a subspace $\mathrm{dom}(A') \subset X$ such that

$$\mathrm{dom}(A) \subset \mathrm{dom}(A'), \qquad A'x = Ax \quad \text{for all } x \in \mathrm{dom}(A). \tag{2.2.11}$$

Lemma 2.2.19 (Characterization of Closeable Operators). *Let X and Y be Banach spaces, let $\mathrm{dom}(A) \subset X$ be a linear subspace, and let $A : \mathrm{dom}(A) \to Y$ be a linear operator. Then the following are equivalent.*

(i) *A is closeable.*

(ii) *The projection $\pi_X : \overline{\mathrm{graph}(A)} \to X$ onto the first factor is injective.*

(iii) *If $(x_n)_{n\in\mathbb{N}}$ is a sequence in $\mathrm{dom}(A)$ and $y \in Y$ is a vector such that $\lim_{n\to\infty} x_n = 0$ and $\lim_{n\to\infty} Ax_n = y$, then $y = 0$.*

Proof. That (i) implies (iii) follows from the fact that $y = A'0 = 0$ for every closed extension $A' : \mathrm{dom}(A') \to Y$ of A.

We prove that (iii) implies (ii). The closure of any linear subspace of a normed vector space is again a linear subspace. Hence $\overline{\mathrm{graph}(A)}$ is a linear subspace of $X \times Y$ and the projection $\pi_X : \overline{\mathrm{graph}(A)} \to X$ onto the first factor is a linear map by definition. By (iii) the kernel of this linear map is the zero subspace and hence it is injective.

We prove that (ii) implies (i). Define

$$\mathrm{dom}(A') := \pi_X\left(\overline{\mathrm{graph}(A)}\right) \subset X.$$

This is a linear subspace and the map $\pi_X : \overline{\mathrm{graph}(A)} \to \mathrm{dom}(A')$ is bijective by (ii). Denote its inverse by $\pi_X^{-1} : \mathrm{dom}(A') \to \overline{\mathrm{graph}(A)}$ and denote by

$$\pi_Y : \overline{\mathrm{graph}(A)} \to Y$$

the projection onto the second factor. Then

$$A' := \pi_Y \circ \pi_X^{-1} : \mathrm{dom}(A') \to Y$$

is a linear operator, its graph is the linear subspace

$$\mathrm{graph}(A') = \overline{\mathrm{graph}(A)} \subset X \times Y,$$

and (2.2.11) holds because $\mathrm{graph}(A) \subset \mathrm{graph}(A')$. Thus we have proved Lemma 2.2.19. $\square$

Example 2.2.20. Let $H = L^2(\mathbb{R})$ and define $\Lambda : \mathrm{dom}(\Lambda) \to \mathbb{R}$ by

$$\mathrm{dom}(\Lambda) := \left\{ f \in L^2(\mathbb{R}) \,\middle|\, \begin{array}{l} \text{there exists a constant } c > 0 \text{ such that} \\ f(t) = 0 \text{ for almost all } t \in \mathbb{R} \setminus [-c, c] \end{array} \right\}$$

and

$$\Lambda(f) := \int_{-\infty}^{\infty} f(t)\, dt \qquad \text{for } f \in \mathrm{dom}(\Lambda).$$

This linear functional is not closeable because the sequence $f_n \in \mathrm{dom}(\Lambda)$, given by $f_n(t) := \frac{1}{n}$ for $|t| \le n$ and $f_n(t) := 0$ for $|t| > n$, satisfies

$$\|f_n\|_{L^2} = \frac{2}{n}, \qquad \Lambda(f_n) = 2$$

for all $n \in \mathbb{N}$. (See Lemma 2.2.19.)

Example 2.2.21. Let $H = L^2(\mathbb{R})$ and define $\Lambda : \mathrm{dom}(\Lambda) \to \mathbb{R}$ by

$$\mathrm{dom}(\Lambda) := C_c(\mathbb{R}), \qquad \Lambda(f) := f(0)$$

for $f \in C_c(\mathbb{R})$ (the space of compactly supported continuous real valued functions $f : \mathbb{R} \to \mathbb{R}$). This linear functional is not closeable because there exists a sequence of continuous functions $f_n : \mathbb{R} \to \mathbb{R}$ with compact support such that $f_n(0) = 1$ and $\|f_n\|_{L^2} \le \frac{1}{n}$ for all $n \in \mathbb{N}$. (See Lemma 2.2.19.)

Exercise 2.2.22 (Linear Functionals). Let X be a real Banach space, let $Y \subset X$ be a linear subspace, and let $\Lambda : Y \to \mathbb{R}$ be a linear functional. Show that Λ is closeable if and only if Λ is bounded. **Hint:** Use the Hahn–Banach Theorem (Corollary 2.3.4) in Section 2.3 below.

Example 2.2.23 (Symmetric Operators). Let H be a Hilbert space and let $A : \mathrm{dom}(A) \to H$ be a linear operator, defined on a dense linear subspace $\mathrm{dom}(A) \subset H$. Suppose A is **symmetric**, i.e.

$$\langle x, Ay \rangle = \langle Ax, y \rangle \qquad \text{for all } x, y \in \mathrm{dom}(A). \tag{2.2.12}$$

Then A is closeable. To see this, choose a sequence $x_n \in \mathrm{dom}(A)$ such that $\lim_{n\to\infty} \|x_n\| = 0$ and the sequence Ax_n converges to an element $y \in H$ as n tends to infinity. Then

$$\langle y, z \rangle = \lim_{n\to\infty} \langle Ax_n, z \rangle = \lim_{n\to\infty} \langle x_n, Az \rangle = 0$$

for all $z \in \mathrm{dom}(A)$. Since $\mathrm{dom}(A)$ is a dense subspace of H, there exists a sequence $z_i \in \mathrm{dom}(A)$ that converges to y as i tends to infinity. Hence

$$\|y\|^2 = \langle y, y \rangle = \lim_{i\to\infty} \langle y, z_i \rangle = 0$$

and so $y = 0$. Thus A is closeable by Lemma 2.2.19.

Example 2.2.24 (Differential Operators). This example shows that every differential operator is closeable. Let $\Omega \subset \mathbb{R}^n$ be a nonempty open set, fix a constant $1 < p < \infty$, and consider the Banach space $X := L^p(\Omega)$ (with respect to the Lebesgue measure on Ω). Then the space

$$\operatorname{dom}(A) := C_0^\infty(\Omega)$$

of smooth functions $u : \Omega \to \mathbb{R}$ with compact support is a dense linear subspace of $L^p(\Omega)$ (see [**75**, Thm, 4.15]). Let $m \in \mathbb{N}$ and, for every multi-index $\alpha = (\alpha_1, \dots, \alpha_n) \in \mathbb{N}_0^n$ with $|\alpha| = \alpha_1 + \cdots + \alpha_n \leq m$, let $a_\alpha : \Omega \to \mathbb{R}$ be a smooth function. Define the operator $A : C_0^\infty(\Omega) \to L^p(\Omega)$ by

$$Au := \sum_{|\alpha| \leq m} a_\alpha \partial^\alpha u. \tag{2.2.13}$$

Here the sum runs over all multi-indices $\alpha = (\alpha_1, \dots, \alpha_n) \in \mathbb{N}_0^n$ with $|\alpha| \leq m$ and $\partial^\alpha = \frac{\partial^{|\alpha|}}{\partial x_1^{\alpha_1} \cdots \partial x_n^{\alpha_n}}$. We prove that A is closeable.

To see this, define the constant $1 < q < \infty$ by $1/p + 1/q = 1$ and define the **formal adjoint** of A as the operator $B : C_0^\infty(\Omega) \to L^q(\Omega)$, given by

$$Bv := \sum_{|\alpha| \leq m} (-1)^{|\alpha|} \partial^\alpha (a_\alpha v)$$

for $v \in C_0^\infty(\Omega)$. Then integration by parts shows that

$$\int_\Omega v(Au) = \int_\Omega (Bv)u \tag{2.2.14}$$

for all $u, v \in C_0^\infty(\Omega)$. Now let $u_k \in C_0^\infty(\Omega)$ be a sequence of smooth functions with compact support and let $v \in L^p(\Omega)$ such that

$$\lim_{k\to\infty} \|u_k\|_{L^p} = 0, \qquad \lim_{k\to\infty} \|v - Au_k\|_{L^p} = 0.$$

Then, for every test function $\phi \in C_0^\infty(\Omega)$, we have

$$\int_\Omega \phi v = \lim_{k\to\infty} \int_\Omega \phi(Au_k) = \lim_{k\to\infty} \int_\Omega (B\phi)u_k = 0.$$

Since $C_0^\infty(\Omega)$ is dense in $L^q(\Omega)$, this implies that

$$\int_\Omega \phi v = 0 \qquad \text{for all } \phi \in L^q(\Omega).$$

Now take $\phi := \operatorname{sign}(v)|v|^{p-1} \in L^q(\Omega)$ to obtain $\int_\Omega |v|^p = 0$ and hence v vanishes almost everywhere. Hence it follows from Lemma 2.2.19 that the linear operator $A : C_0^\infty(\Omega) \to L^p(\Omega)$ is closeable as claimed.

2.3. Hahn–Banach and Convexity

2.3.1. The Hahn–Banach Theorem. The Hahn–Banach Theorem deals with bounded linear functionals on a subspace of a Banach space X and asserts that every such functional extends to a bounded linear functional on all of X. This theorem continues to hold in the more general setting where X is any real vector space and boundedness is replaced by a bound relative to a given quasi-seminorm on X.

Definition 2.3.1 (Quasi-Seminorm). Let X be a real vector space. A function $p : X \to \mathbb{R}$ is called a **quasi-seminorm** if it satisfies

$$p(x+y) \le p(x) + p(y), \qquad p(\lambda x) = \lambda p(x) \tag{2.3.1}$$

for all $x, y \in X$ and all $\lambda \ge 0$. It is called a **seminorm** if it is a quasi-seminorm and $p(\lambda x) = |\lambda| p(x)$ for all $x \in X$ and all $\lambda \in \mathbb{R}$. A seminorm has nonnegative values, because $2p(x) = p(x) + p(-x) \ge p(0) = 0$ for all $x \in X$. Thus a seminorm satisfies all the axioms of a norm except nondegeneracy (i.e. there may be nonzero elements $x \in X$ such that $p(x) = 0$).

Theorem 2.3.2 (Hahn–Banach). *Let X be a normed vector space and let $p : X \to \mathbb{R}$ be a quasi-seminorm. Let $Y \subset X$ be a linear subspace and let $\phi : Y \to \mathbb{R}$ be a linear functional such that $\phi(x) \le p(x)$ for all $x \in Y$. Then there exists a linear functional $\Phi : X \to \mathbb{R}$ such that*

$$\Phi|_Y = \phi, \qquad \Phi(x) \le p(x) \quad \text{for all } x \in X.$$

Proof. See page 66. □

Lemma 2.3.3. *Let X, p, Y, and ϕ be as in Theorem 2.3.2. Let $x_0 \in X \setminus Y$ and define $Y' := Y \oplus \mathbb{R}x_0$. Then there exists a linear functional $\phi' : Y' \to \mathbb{R}$ such that $\phi'|_Y = \phi$ and $\phi'(x) \le p(x)$ for all $x \in Y'$.*

Proof. An extension $\phi' : Y' \to \mathbb{R}$ of the linear functional $\phi : Y \to \mathbb{R}$ is uniquely determined by its value $a := \phi'(x_0) \in \mathbb{R}$ on x_0. This extension satisfies the required condition $\phi'(x) \le p(x)$ for all $x \in Y'$ if and only if

$$\phi(y) + \lambda a \le p(y + \lambda x_0) \qquad \text{for all } y \in Y \text{ and all } \lambda \in \mathbb{R}. \tag{2.3.2}$$

If this holds, then

$$\phi(y) \pm a \le p(y \pm x_0) \qquad \text{for all } y \in Y. \tag{2.3.3}$$

Conversely, if (2.3.3) holds and $\lambda > 0$, then

$$\begin{aligned} \phi(y) + \lambda a &= \lambda\left(\phi(\lambda^{-1}y) + a\right) \le \lambda p(\lambda^{-1}y + x_0) = p(y + \lambda x_0), \\ \phi(y) - \lambda a &= \lambda\left(\phi(\lambda^{-1}y) - a\right) \le \lambda p(\lambda^{-1}y - x_0) = p(y - \lambda x_0). \end{aligned}$$

This shows that (2.3.2) is equivalent to (2.3.3). Thus it remains to find a real number $a \in \mathbb{R}$ that satisfies (2.3.3). Equivalently, a must satisfy

$$\phi(y) - p(y - x_0) \le a \le p(y + x_0) - \phi(y) \qquad \text{for all } y \in Y. \tag{2.3.4}$$

To see that such a number exists, fix two vectors $y, y' \in Y$. Then

$$\begin{aligned}\phi(y) + \phi(y') &= \phi(y + y') \\ &\le p(y + y') \\ &= p(y + x_0 + y' - x_0) \\ &\le p(y + x_0) + p(y' - x_0).\end{aligned}$$

Thus

$$\phi(y') - p(y' - x_0) \le p(y + x_0) - \phi(y)$$

for all $y, y' \in Y$ and this implies

$$\sup_{y' \in Y} \big(\phi(y') - p(y' - x_0)\big) \le \inf_{y \in Y} \big(p(y + x_0) - \phi(y)\big).$$

Hence there exists a real number $a \in \mathbb{R}$ that satisfies (2.3.4) and this proves Lemma 2.3.3. □

Proof of Theorem 2.3.2. Define the set

$$\mathscr{P} := \left\{ (Z, \psi) \;\middle|\; \begin{array}{l} Z \text{ is a linear subspace of } X \text{ and} \\ \psi : Z \to \mathbb{R} \text{ is a linear functional such that} \\ Y \subset Z,\ \psi|_Y = \phi, \text{ and } \psi(x) \le p(x) \text{ for all } x \in Z \end{array} \right\}.$$

This set is partially ordered by the relation

$$(Z, \psi) \preccurlyeq (Z', \psi') \qquad \stackrel{\text{def}}{\Longleftrightarrow} \qquad Z \subset Z' \text{ and } \psi'|_Z = \psi$$

for $(Z, \psi), (Z', \psi') \in \mathscr{P}$. A chain in $\mathscr{P}$ is a totally ordered subset $\mathscr{C} \subset \mathscr{P}$. Every nonempty chain $\mathscr{C} \subset \mathscr{P}$ has a supremum (Z_0, ψ_0) given by

$$Z_0 := \bigcup_{(Z,\psi) \in \mathscr{C}} Z, \qquad \psi_0(x) := \psi(x) \quad \text{for all } (Z, \psi) \in \mathscr{C} \text{ and all } x \in Z.$$

Hence it follows from the Lemma of Zorn that $\mathscr{P}$ has a maximal element (Z, ψ). By Lemma 2.3.3 every such maximal element satisfies $Z = X$ and this proves Theorem 2.3.2. □

A special case of the Hahn–Banach Theorem is where the quasi-seminorm is actually a norm. In this situation the Hahn–Banach Theorem is an existence result for bounded linear functionals on real and complex normed vector spaces. It takes the following form.

Corollary 2.3.4 (Real Case). *Let X be a normed vector space over $\mathbb{R}$, let $Y \subset X$ be a linear subspace, let $\phi : Y \to \mathbb{R}$ be a linear functional, and let $c \geq 0$ such that $|\phi(x)| \leq c\,\|x\|$ for all $x \in Y$. Then there exists a bounded linear functional $\Phi : X \to \mathbb{R}$ such that*

$$\Phi|_Y = \phi, \qquad |\Phi(x)| \leq c\,\|x\| \quad \textit{for all } x \in X.$$

Proof. By Theorem 2.3.2 with $p(x) := c\,\|x\|$, there exists a linear functional $\Phi : X \to \mathbb{R}$ such that $\Phi|_Y = \phi$ and $\Phi(x) \leq c\,\|x\|$ for all $x \in X$. Since $\Phi(-x) = -\Phi(x)$ it follows that $|\Phi(x)| \leq c\,\|x\|$ for all $x \in X$ and this proves Corollary 2.3.4. □

Corollary 2.3.5 (Complex Case). *Let X be a normed vector space over $\mathbb{C}$, let $Y \subset X$ be a linear subspace, let $\psi : Y \to \mathbb{C}$ be a complex linear functional, and let $c \geq 0$ such that $|\psi(x)| \leq c\,\|x\|$ for all $x \in Y$. Then there exists a bounded complex linear functional $\Psi : X \to \mathbb{C}$ such that*

$$\Psi|_Y = \psi, \qquad |\Psi(x)| \leq c\,\|x\| \quad \textit{for all } x \in X.$$

Proof. By Corollary 2.3.4 there exists a real linear functional $\Phi : X \to \mathbb{R}$ such that

$$\Phi|_X = \operatorname{Re}\psi$$

and $|\Phi(x)| \leq c\,\|x\|$ for all $x \in X$. Define $\Psi : X \to \mathbb{C}$ by

$$\Psi(x) := \Phi(x) - \mathbf{i}\Phi(\mathbf{i}x) \qquad \text{for } x \in X.$$

Then $\Psi : X \to \mathbb{C}$ is complex linear and, for all $x \in Y$, we have

$$\begin{aligned} \Psi(x) &= \Phi(x) - \mathbf{i}\Phi(\mathbf{i}x) \\ &= \operatorname{Re}(\psi(x)) - \mathbf{i}\operatorname{Re}(\psi(\mathbf{i}x)) \\ &= \operatorname{Re}(\psi(x)) - \mathbf{i}\operatorname{Re}(\mathbf{i}\psi(x)) \\ &= \operatorname{Re}(\psi(x)) + \mathbf{i}\operatorname{Im}(\psi(x)) \\ &= \psi(x). \end{aligned}$$

To prove the estimate, fix a vector $x \in X$ such that $\Psi(x) \neq 0$ and choose a real number $\theta \in \mathbb{R}$ such that

$$e^{\mathbf{i}\theta} = |\Psi(x)|^{-1}\,\Psi(x).$$

Then

$$|\Psi(x)| = e^{-\mathbf{i}\theta}\Psi(x) = \Psi(e^{-\mathbf{i}\theta}x) = \Phi(e^{-\mathbf{i}\theta}x) \leq c\|e^{-\mathbf{i}\theta}x\| = c\|x\|.$$

Here the third equality follows from the fact that $\Psi(e^{-\mathbf{i}\theta}x)$ is real. This proves Corollary 2.3.5. □

2.3.2. Positive Linear Functionals. The Hahn–Banach Theorem has several important applications. The first is an extension theorem for positive linear functionals on ordered vector spaces. Recall that a partial order is a transitive, anti-symmetric, reflexive relation.

Definition 2.3.6 (Ordered Vector Space). An **ordered vector space** is a pair $(X, \preccurlyeq)$, where X is a real vector space and $\preccurlyeq$ is a partial order on X that satisfies the following two axioms for all $x, y, z \in X$ and all $\lambda \in \mathbb{R}$.

(O1) If $0 \preccurlyeq x$ and $0 \le \lambda$, then $0 \preccurlyeq \lambda x$.

(O2) If $x \preccurlyeq y$, then $x + z \preccurlyeq y + z$.

In this situation the set $P := \{x \in X \mid 0 \preccurlyeq x\}$ is called the **positive cone**. A linear functional $\Phi : X \to \mathbb{R}$ is called **positive** if $\Phi(x) \ge 0$ for all $x \in P$.

Theorem 2.3.7 (Hahn–Banach for Positive Linear Functionals). *Let $(X, \preccurlyeq)$ be an ordered vector space and let $P \subset X$ be the positive cone. Let $Y \subset X$ be a linear subspace satisfying the following condition.*

(O3) *For each $x \in X$ there exists a $y \in Y$ such that $x \preccurlyeq y$.*

Let $\phi : Y \to \mathbb{R}$ be a positive linear functional, i.e. $\phi(y) \ge 0$ for all $y \in Y \cap P$. Then there is a positive linear functional $\Phi : X \to \mathbb{R}$ such that $\Phi|_Y = \phi$.

Proof. The proof has three steps.

Step 1. *For every $x \in X$ the set $\{y \in Y \mid x \preccurlyeq y\}$ is nonempty and the restriction of ϕ to this set is bounded below.*

Fix an element $x \in X$. Then the set $\{y \in Y \mid x \preccurlyeq y\}$ is nonempty by (O3). It follows also from (O3) that there exists a $y_0 \in Y$ such that $-x \preccurlyeq -y_0$. Thus we have $y_0 \preccurlyeq x$ by (O2). If $y \in Y$ satisfies $x \preccurlyeq y$, then $y_0 \preccurlyeq y$ and this implies $\phi(y_0) \le \phi(y)$, because ϕ is positive. This proves Step 1.

Step 2. *By Step 1 the formula*

$$p(x) := \inf\{\phi(y) \mid y \in Y,\ x \preccurlyeq y\} \qquad \text{for } x \in X \tag{2.3.5}$$

defines a function $p : X \to \mathbb{R}$. This function is a quasi-seminorm and satisfies $p(y) = \phi(y)$ for all $y \in Y$.

Let $x_1, x_2 \in X$ and $\varepsilon > 0$. For $i = 1, 2$ choose $y_i \in Y$ such that $x_i \preccurlyeq y_i$ and $\phi(y_i) < p(x_i) + \varepsilon/2$. Then $x_1 + x_2 \preccurlyeq x_1 + y_2 \preccurlyeq y_1 + y_2$ by (O2), and so

$$p(x_1 + x_2) \le \phi(y_1 + y_2) = \phi(y_1) + \phi(y_2) < p(x_1) + p(x_2) + \varepsilon.$$

This implies $p(x_1 + x_2) \le p(x_1) + p(x_2)$ for all $x_1, x_2 \in X$.

Now let $x \in X$ and $\lambda > 0$. Then $\{y \in Y \mid \lambda x \preccurlyeq y\} = \{\lambda y \mid y \in Y,\, x \preccurlyeq y\}$ by (O1) and hence

$$p(\lambda x) = \inf_{\substack{y \in Y \\ \lambda x \preccurlyeq y}} \phi(y) = \inf_{\substack{y \in Y \\ x \preccurlyeq y}} \phi(\lambda y) = \inf_{\substack{y \in Y \\ x \preccurlyeq y}} \lambda \phi(y) = \lambda p(x).$$

Moreover $p(0) = 0$ by definition, and hence p is a quasi-seminorm. The formula $p(y) = \phi(y)$ for $y \in Y$ follows directly from the definition of p in (2.3.5) and this proves Step 2.

Step 3. *We prove Theorem* 2.3.7.

By Step 2 and the Hahn–Banach Theorem 2.3.2, there exists a linear functional $\Phi : X \to \mathbb{R}$ such that $\Phi|_Y = \phi$ and $\Phi(x) \le p(x)$ for all $x \in X$. If $x \in P$, then $-x \preccurlyeq 0 \in Y$, hence $\Phi(-x) \le p(-x) \le \phi(0) = 0$, and so $\Phi(x) \ge 0$. This proves Theorem 2.3.7. □

Exercise 2.3.8. Give a direct proof of Theorem 2.3.7 based on the Lemma of Zorn. **Hint:** *If $(X, \preccurlyeq)$ is an ordered vector space, $Y \subset X$ is a linear subspace satisfying (O3), $\phi : Y \to \mathbb{R}$ is a positive linear functional, and $x_0 \in X \setminus Y$, then there is a positive linear functional $\psi : Y \oplus \mathbb{R}x_0 \to \mathbb{R}$ such that $\psi|_Y = \phi$.* To see this, find a real number $a \in \mathbb{R}$ that satisfies the conditions

$$x_0 \preccurlyeq y \qquad \Longrightarrow \qquad a \le \phi(y)$$

and

$$y \preccurlyeq x_0 \qquad \Longrightarrow \qquad \phi(y) \le a$$

for all $y \in Y$.

Exercise 2.3.9. This exercise shows that the assumption (O3) cannot be removed in Theorem 2.3.7. The space $X := BC(\mathbb{R})$ of bounded continuous real valued functions on $\mathbb{R}$ is an ordered vector space with

$$f \preccurlyeq g \qquad \stackrel{\text{def}}{\Longleftrightarrow} \qquad f(t) \le g(t) \quad \text{for all } t \in \mathbb{R}.$$

The subspace $Y := C_c(\mathbb{R})$ of compactly supported continuous functions does not satisfy (O3) and the positive linear functional

$$C_c(\mathbb{R}) \to \mathbb{R} : f \mapsto \int_{-\infty}^{\infty} f(t)\, dt$$

does not extend to a positive linear functional on $BC(\mathbb{R})$. **Hint:** Every positive linear functional on $BC(\mathbb{R})$ is bounded with respect to the sup-norm.

2.3.3. Separation of Convex Sets. The second application of the Hahn–Banach Theorem concerns a pair of disjoint convex sets in a normed vector space. They can be separated by a hyperplane whenever one of them has nonempty interior (see Figure 2.3.1). The result and its proof carry over to general topological vector spaces (see Theorem 3.1.11 below).

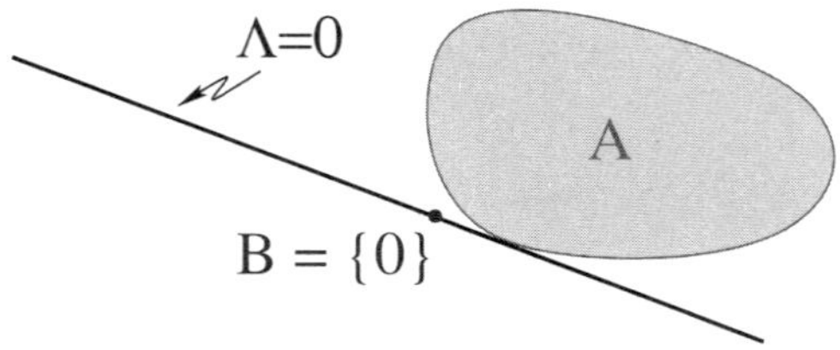

Figure 2.3.1. Two convex sets, separated by a hyperplane.

Theorem 2.3.10 (Separation of Convex Sets). *Let X be a real normed vector space and let $A, B \subset X$ be nonempty disjoint convex sets such that $\mathrm{int}(A) \neq \emptyset$. Then there exist a nonzero bounded linear functional $\Lambda : X \to \mathbb{R}$ and a constant $c \in \mathbb{R}$ such that $\Lambda(x) \geq c$ for all $x \in A$ and $\Lambda(x) \leq c$ for all $x \in B$. Moreover, every such bounded linear functional satisfies $\Lambda(x) > c$ for all $x \in \mathrm{int}(A)$.*

Proof. See page 71. □

Exercise 2.3.11. This exercise shows that the hypothesis that one of the convex sets has nonempty interior cannot be removed in Theorem 2.3.10. Consider the Hilbert space $H = \ell^2$ and define

$$A := \left\{ x \in \ell^2 \,\middle|\, \begin{array}{l} \exists n \in \mathbb{N}\ \forall i \in \mathbb{N} \\ i < n \implies x_i > 0 \\ i \geq n \implies x_i = 0 \end{array} \right\}, \quad B := \left\{ x \in \ell^2 \,\middle|\, \begin{array}{l} \exists n \in \mathbb{N}\ \forall i \in \mathbb{N} \\ i < n \implies x_i = 0 \\ i \geq n \implies x_i > 0 \end{array} \right\}.$$

Show that A, B are nonempty disjoint convex subsets of ℓ^2 with empty interior whose closures agree. If $\Lambda : \ell^2 \to \mathbb{R}$ is a bounded linear functional and c is a real number such that $\Lambda(x) \geq c$ for all $x \in A$ and $\Lambda(x) \leq c$ for all $x \in B$, show that $\Lambda = 0$ and $c = 0$.

Exercise 2.3.12. Define $A := \left\{ x \in \ell^2 \,|\, x_i = 0 \text{ for } i > 1 \right\}$ and

$$B := \left\{ x = (x_i)_{i=1}^\infty \in \mathbb{R}^{\mathbb{N}} \,\middle|\, |ix_i - i^{1/3}| \leq x_1 \text{ for all } i > 1 \right\} \subset \ell^2.$$

Show that A, B are nonempty disjoint closed convex subsets of ℓ^2 and $A - B$ is dense in ℓ^2. Deduce that A, B cannot be separated by an affine hyperplane.

Lemma 2.3.13. *Let X be a normed vector space and let $A \subset X$ be a convex set. Then* $\mathrm{int}(A)$ *and* $\overline{A}$ *are convex sets. Moreover, if* $\mathrm{int}(A) \neq \emptyset$, *then* $A \subset \overline{\mathrm{int}(A)}$.

Proof. The proof of convexity of $\mathrm{int}(A)$ and $\overline{A}$ is left as an exercise (see also Lemma 3.1.10). Let $x_0 \in \mathrm{int}(A)$ and choose $\delta > 0$ such that $B_\delta(x_0) \subset A$. If $x \in A$, then the set $U_x := \{tx + (1-t)y \,|\, y \in B_\delta(x_0),\, 0 < t < 1\} \subset A$ is open and hence $x \in \overline{U}_x \subset \overline{\mathrm{int}(A)}$. □

Lemma 2.3.14. *Let X be a normed vector space, let $A \subset X$ be a convex set with nonempty interior, let $\Lambda : X \to \mathbb{R}$ be a nonzero bounded linear functional, and let $c \in \mathbb{R}$ such that $\Lambda(x) \geq c$ for all $x \in$* $\mathrm{int}(A)$. *Then $\Lambda(x) \geq c$ for all $x \in A$ and $\Lambda(x) > c$ for all $x \in$* $\mathrm{int}(A)$.

Proof. Since A is convex and has nonempty interior, we have $A \subset \overline{\mathrm{int}(A)}$ by Lemma 2.3.13, and so $\Lambda(x) \geq c$ for all $x \in A$ because Λ is continuous. Now let $x \in \mathrm{int}(A)$, choose $x_0 \in X$ such that $\Lambda(x_0) = 1$, and choose $t > 0$ such that $x - tx_0 \in A$. Then $\Lambda(x) = t + \Lambda(x - tx_0) \geq t + c > c$. □

Proof of Theorem 2.3.10. The proof has three steps.

Step 1. *Let X be a real normed vector space, let $U \subset X$ be a nonempty open convex set such that $0 \notin U$, and define $P := \{tx \,|\, x \in U,\, t \in \mathbb{R},\, t \geq 0\}$. Then P is a convex subset of X and satisfies the following.*

(P1) *If $x \in P$ and $\lambda \geq 0$, then $\lambda x \in P$.*

(P2) *If $x, y \in P$, then $x + y \in P$.*

(P3) *If $x \in P$ and $-x \in P$, then $x = 0$.*

If $x, y \in P \setminus \{0\}$, choose $x_0, x_1 \in U$ and $t_0, t_1 > 0$ such that $x = t_0 x_0$ and $y = t_1 x_1$; then $z := \frac{t_0}{t_0+t_1} x_0 + \frac{t_1}{t_0+t_1} x_1 \in U$ and hence $x + y = (t_0 + t_1) z \in P$. This proves (P2). That P satisfies (P1) is obvious and that it satisfies (P3) follows from the fact that $0 \notin U$. By (P1) and (P2) the set P is convex.

Step 2. *Let X and U be as in Step 1. Then there exists a bounded linear functional $\Lambda : X \to \mathbb{R}$ such that $\Lambda(x) > 0$ for all $x \in U$.*

Let P be as in Step 1. Then it follows from (P1), (P2), (P3) that the relation

$$x \preccurlyeq y \quad \overset{\text{def}}{\Longleftrightarrow} \quad y - x \in P$$

defines a partial order $\preccurlyeq$ on X that satisfies (O1) and (O2).

Let $x_0 \in U$. Then the linear subspace $Y := \mathbb{R}x_0$ satisfies (O3). Namely, if $x \in X$, then $x_0 - tx \in U \subset P$ for $t > 0$ sufficiently small and so $x \preccurlyeq t^{-1}x_0$. Moreover, the linear functional $Y \to \mathbb{R} : tx_0 \mapsto t$ is positive by (P3). Hence, by Theorem 2.3.7, there is a linear functional $\Lambda : X \to \mathbb{R}$ such that $\Lambda(tx_0) = t$

for all $t \in \mathbb{R}$ and $\Lambda(x) \geq 0$ for all $x \in P$. We prove that this functional is bounded. Choose $\delta > 0$ such that $\overline{B}_\delta(x_0) \subset P$, and let $x \in X$ with $\|x\| \leq 1$. Then $x_0 - \delta x \in P$, hence $\Lambda(x_0 - \delta x) \geq 0$, and so $\Lambda(x) \leq \delta^{-1}\Lambda(x_0) = \delta^{-1}$. Thus $|\Lambda(x)| \leq \delta^{-1}\|x\|$ for all $x \in X$. Since $U \subset P$, we have $\Lambda(x) \geq 0$ for all $x \in U$, and so $\Lambda(x) > 0$ for all $x \in U$ by Lemma 2.3.14.

Step 3. *We prove Theorem* 2.3.10.

Let X, A, B be as in Theorem 2.3.10. Then $U := \mathrm{int}(A) - B$ is a nonempty open convex set and $0 \notin U$. Hence, by Step 2, there exists a bounded linear functional $\Lambda : X \to \mathbb{R}$ such that $\Lambda(x) > 0$ for all $x \in U$. Thus $\Lambda(x) > \Lambda(y)$ for all $x \in \mathrm{int}(A)$ and all $y \in B$. This implies $\Lambda(x) \geq c := \sup_{y \in B} \Lambda(y)$ for all $x \in \mathrm{int}(A)$. Hence $\Lambda(x) \geq c$ for all $x \in A$ and $\Lambda(x) > c$ for all $x \in \mathrm{int}(A)$ by Lemma 2.3.14. This proves Theorem 2.3.10. □

Definition 2.3.15 (Hyperplane). Let X be a real normed vector space. A **hyperplane** in X is a closed linear subspace of codimension one. An **affine hyperplane** is a translate of a hyperplane. An **open half-space** is a set of the form $\{x \in X \mid \Lambda(x) > c\}$ where $\Lambda : X \to \mathbb{R}$ is a nonzero bounded linear functional and $c \in \mathbb{R}$.

Exercise 2.3.16. Show that $H \subset X$ is an affine hyperplane if and only if there exist a nonzero bounded linear functional $\Lambda : X \to \mathbb{R}$ and a real number $c \in \mathbb{R}$ such that $H = \Lambda^{-1}(c)$.

Let X, A, B, Λ, c be as in Theorem 2.3.10. Then $H := \Lambda^{-1}(c)$ is an affine hyperplane that separates the convex sets A and B. It divides X into two connected components such that the interior of A is contained in one of them and B is contained in the closure of the other.

Corollary 2.3.17. *Let X be a real Banach space and let $A \subset X$ be an open convex set such that $0 \notin A$. Let $Y \subset X$ be a linear subspace such that $Y \cap A = \emptyset$. Then there is a hyperplane $H \subset X$ such that*

$$Y \subset H, \qquad H \cap A = \emptyset.$$

Proof. Assume without loss of generality that Y is closed, consider the quotient $X' := X/Y$, and denote by $\pi : X \to X'$ the obvious projection. Then π is open by Theorem 2.2.1, so $A' := \pi(A) \subset X'$ is an open convex set that does not contain the origin. Hence Theorem 2.3.10 asserts that there is a bounded linear functional $\Lambda' : X' \to \mathbb{R}$ such that $\Lambda'(x') > 0$ for all $x' \in A'$. Hence $\Lambda := \Lambda' \circ \pi : X \to \mathbb{R}$ is a bounded linear functional such that $Y \subset \ker(\Lambda)$ and $\Lambda(x) > 0$ for all $x \in A$. So $H := \ker(\Lambda)$ is the required hyperplane. □

Corollary 2.3.18. *Let X be a real normed vector space and let $A \subset X$ be a nonempty open convex set. Then A is the intersection of all open half-spaces containing A.*

Proof. Let $y \in X \setminus A$. Then, by Theorem 2.3.10 with $B = \{y\}$, there is a $\Lambda \in X^*$ and a $c \in \mathbb{R}$ such that $\Lambda(x) > c$ for all $x \in A$ and $\Lambda(y) \le c$. Hence there is an open half-space containing A but not y. □

Corollary 2.3.19. *Let X be a real normed vector space and let $A, B \subset X$ be nonempty disjoint convex sets such that A is closed and B is compact. Then there exists a bounded linear functional $\Lambda : X \to \mathbb{R}$ such that*

$$\inf_{x\in A} \Lambda(x) > \sup_{y\in B} \Lambda(y).$$

Proof. We prove first that

$$\delta := \inf_{x\in A,\, y\in B} \|x - y\| > 0.$$

Choose sequences $x_n \in A$ and $y_n \in B$ such that

$$\lim_{n\to\infty} \|x_n - y_n\| = \delta.$$

Since B is compact, we may assume, by passing to a subsequence if necessary, that the sequence $(y_n)_{n\in\mathbb{N}}$ converges to an element $y \in B$. If $\delta = 0$ it would follow that the sequence $(x_n - y_n)_{n\in\mathbb{N}}$ converges to zero, so the sequence $x_n = y_n + (x_n - y_n)$ converges to y, and so $y \in A$, because A is closed, contradicting the fact that $A \cap B = \emptyset$. Thus $\delta > 0$ as claimed. Hence

$$U := \bigcup_{x\in A} B_\delta(x)$$

is an open convex set that contains A and is disjoint from B. Thus, by Theorem 2.3.10, there is a bounded linear functional $\Lambda : X \to \mathbb{R}$ such that

$$\Lambda(x) > c := \sup_{y\in B} \Lambda(y) \qquad \text{for all } x \in U.$$

Choose $\xi \in X$ such that $\|\xi\| < \delta$ and $\varepsilon := \Lambda(\xi) > 0$. Then every $x \in A$ satisfies $x - \xi \in U$ and hence

$$\Lambda(x) - \varepsilon = \Lambda(x - \xi) > c.$$

This proves Corollary 2.3.19. □

Exercise 2.3.20. Let X be a real normed vector space and let $A \subset X$ be a nonempty convex set. Prove that $\overline{A}$ is the intersection of all closed half-spaces of X containing A.

2.3.4. The Closure of a Linear Subspace. The third application of the Hahn–Banach Theorem is a characterization of the closure of a linear subspace of a real normed vector space X. Recall that the dual space of X is the space

$$X^* := \mathcal{L}(X, \mathbb{R})$$

of real valued bounded linear functionals on X. At this point it is convenient to introduce an alternative notation for the elements of the dual space. Denote a bounded linear functional on X by $x^* : X \to \mathbb{R}$ and denote the value of this linear functional on an element $x \in X$ by

$$\langle x^*, x \rangle := x^*(x).$$

This notation is reminiscent of the inner product on a Hilbert space and there are in fact many parallels between the pairing

$$X^* \times X \to \mathbb{R} : (x^*, x) \mapsto \langle x^*, x \rangle \tag{2.3.6}$$

and inner products on Hilbert spaces. Recall that X^* is a Banach space with respect to the norm

$$\|x^*\| := \sup_{x \in X \setminus \{0\}} \frac{|\langle x^*, x \rangle|}{\|x\|} \qquad \text{for } x^* \in X^* \tag{2.3.7}$$

(see Theorem 1.3.1). It follows directly from (2.3.7) that

$$|\langle x^*, x \rangle| \le \|x^*\| \, \|x\| \tag{2.3.8}$$

for all $x^* \in X^*$ and all $x \in X$, in analogy to the Cauchy–Schwarz inequality. Hence the pairing (2.3.6) is continuous by Corollary 2.1.7.

Definition 2.3.21 (Annihilator). Let X be a real normed vector space. For any subset $S \subset X$ define the **annihilator** of S as the space of bounded linear functionals on X that vanish on S and denote it by

$$S^\perp := \left\{ x^* \in X^* \,\middle|\, \langle x^*, x \rangle = 0 \text{ for all } x \in S \right\}. \tag{2.3.9}$$

Since the pairing (2.3.6) is continuous, the annihilator $S^\perp$ is a closed linear subspace of X^* for every subset $S \subset X$. As before, the closure of a subset $Y \subset X$ is denoted by $\overline{Y}$.

Theorem 2.3.22. *Let X be a real normed vector space, let $Y \subset X$ be a linear subspace, and let $x_0 \in X \setminus \overline{Y}$. Then*

$$\delta := d(x_0, Y) := \inf_{y \in Y} \|x_0 - y\| > 0 \tag{2.3.10}$$

and there exists a bounded linear functional $x^* \in Y^\perp$ such that

$$\|x^*\| = 1, \qquad \langle x^*, x_0 \rangle = \delta.$$

Proof. We prove first that the number δ in (2.3.10) is positive. Suppose by contradiction that $\delta = 0$. Then, by the axiom of countable choice, there exists a sequence $(y_n)_{n\in\mathbb{N}}$ in Y such that $\|x_0 - y_n\| < 1/n$ for all $n \in \mathbb{N}$. This implies that y_n converges to x_0 and hence $x_0 \in \overline{Y}$, in contradiction to our assumption. This shows that $\delta > 0$ as claimed.

Now define the subspace $Z \subset X$ by

$$Z := Y \oplus \mathbb{R}x_0 = \{y + tx_0 \,|\, y \in Y,\, t \in \mathbb{R}\}$$

and define the linear functional $\psi : Z \to \mathbb{R}$ by

$$\psi(y + tx_0) := \delta t \qquad \text{for } y \in Y \text{ and } t \in \mathbb{R}.$$

This functional is well defined because $x_0 \notin Y$. It satisfies $\psi(y) = 0$ for all $y \in Y$ and $\psi(x_0) = \delta$. Moreover, if $y \in Y$ and $t \in \mathbb{R} \setminus \{0\}$, then

$$\frac{|\psi(y + tx_0)|}{\|y + tx_0\|} = \frac{|t|\delta}{\|y + tx_0\|} = \frac{\delta}{\|t^{-1}y + x_0\|} \leq 1.$$

Here the last inequality follows from the definition of δ. With this understood, it follows from Corollary 2.3.4 that there exists a bounded linear functional $x^* \in X^*$ such that

$$\|x^*\| \leq 1$$

and

$$\langle x^*, x\rangle = \psi(x) \qquad \text{for all } x \in Z.$$

The norm of x^* is actually equal to one because

$$\|x^*\| \geq \sup_{y\in Y} \frac{|\psi(x_0 + y)|}{\|x_0 + y\|} = \sup_{y\in Y} \frac{|\delta|}{\|x_0 + y\|} = 1$$

by definition of δ. Moreover,

$$\langle x^*, x_0\rangle = \psi(x_0) = \delta$$

and

$$\langle x^*, y\rangle = \psi(y) = 0 \quad \text{for all } y \in Y.$$

This proves Theorem 2.3.22. □

Corollary 2.3.23. *Let X be a real normed vector space and let $x_0 \in X$ be a nonzero vector. Then there exists a bounded linear functional $x^* \in X^*$ such that*

$$\|x^*\| = 1, \qquad \langle x^*, x_0\rangle = \|x_0\|.$$

Proof. This follows directly from Theorem 2.3.22 with $Y := \{0\}$. □

The next corollary characterizes the closure of a linear subspace and gives rise to a criterion for a linear subspace to be dense.

Corollary 2.3.24 (Closure of a Subspace). *Let X be a real normed vector space, let $Y \subset X$ be a linear subspace, and let $x \in X$. Then*

$$x \in \overline{Y} \qquad \Longleftrightarrow \qquad \langle x^*, x\rangle = 0 \text{ for all } x^* \in Y^\perp.$$

Proof. If $x \in \overline{Y}$ and $x^* \in Y^\perp$, then there is a sequence $(y_n)_{n\in\mathbb{N}}$ in Y that converges to x and so $\langle x^*, x\rangle = \lim_{n\to\infty}\langle x^*, y_n\rangle = 0$. If $x \notin \overline{Y}$, then there is an element $x^* \in Y^\perp$ such that $\langle x^*, x\rangle > 0$ by Theorem 2.3.22. □

Corollary 2.3.25 (Dense Subspaces). *Let X be a real normed vector space and let $Y \subset X$ be a linear subspace. Then Y is dense in X if and only if $Y^\perp = \{0\}$.*

Proof. By Corollary 2.3.24 we have $\overline{Y} = X$ if and only if $\langle x^*, x\rangle = 0$ for all $x^* \in Y^\perp$ and all $x \in X$, and this is equivalent to $Y^\perp = \{0\}$. □

The next corollary asserts that the dual space of a quotient is a subspace of the dual space and vice versa.

Corollary 2.3.26 (Dual Spaces of Subspaces and Quotients). *Let X be a real normed vector space and let $Y \subset X$ be a linear subspace. Then the following hold.*

(i) *The linear map*

$$X^*/Y^\perp \to Y^* : [x^*] \mapsto x^*|_Y \tag{2.3.11}$$

is an isometric isomorphism.

(ii) *Assume Y is closed and let $\pi : X \to X/Y$ be the canonical projection, given by $\pi(x) := x + Y$ for $x \in X$. Then the linear map*

$$(X/Y)^* \to Y^\perp : \Lambda \mapsto \Lambda \circ \pi \tag{2.3.12}$$

is an isometric isomorphism.

Proof. We prove part (i). The linear map

$$X^* \to Y^* : x^* \mapsto x^*|_Y$$

vanishes on $Y^\perp$ and hence descends to the quotient $X^*/Y^\perp$. The resulting map (2.3.11) is injective by definition. Now fix any bounded linear functional $y^* \in Y^*$. Then Corollary 2.3.4 asserts that there is a bounded linear functional $x^* \in X^*$ such that

$$x^*|_Y = y^*, \qquad \|x^*\| = \|y^*\|.$$

Moreover, if $\xi^* \in X^*$ satisfies $\xi^*|_Y = y^*$, then $\|\xi^*\| \geq \|y^*\| = \|x^*\|$. Hence x^* minimizes the norm among all bounded linear functionals on X that restrict to y^* on Y. Thus $\|x^* + Y^\perp\|_{X^*/Y^\perp} = \|x^*\| = \|y^*\|$, and this shows that the map (2.3.11) is an isometric isomorphism.

We prove part (ii). Fix a bounded linear functional

$$\Lambda : X/Y \to \mathbb{R}$$

and define

$$x^* := \Lambda \circ \pi : X \to \mathbb{R}.$$

Then x^* is a bounded linear functional on X and $x^*|_Y = 0$. Thus

$$x^* \in Y^\perp.$$

Conversely, fix an element $x^* \in Y^\perp$. Then x^* vanishes on Y and hence descends to a unique linear map $\Lambda : X/Y \to \mathbb{R}$ such that

$$\Lambda \circ \pi = x^*.$$

To prove that Λ is bounded, observe that

$$\Lambda(x+Y) = \langle x^*, x\rangle = \langle x^*, x+y\rangle \le \|x^*\| \, \|x+y\|$$

for all $x \in X$ and all $y \in Y$, hence

$$|\Lambda(x+Y)| \le \|x^*\| \inf_{y\in Y} \|x+y\| = \|x^*\| \, \|x+Y\|_{X/Y}$$

for all $x \in X$, and hence

$$\|\Lambda\| \le \|x^*\|.$$

Conversely

$$\begin{aligned}\langle x^*, x\rangle &= \Lambda(x+Y) \\ &\le \|\Lambda\| \, \|x+Y\|_{X/Y} \\ &\le \|\Lambda\| \, \|x\|\end{aligned}$$

for all $x \in X$ and so

$$\|x^*\| \le \|\Lambda\|.$$

Hence the linear map (2.3.12) is an isometric isomorphism. This proves Corollary 2.3.26. □

Corollary 2.3.27. *Let X be a real normed vector space and let $Y \subset X$ be a closed linear subspace. Then*

$$\inf_{\xi^*\in Y^\perp} \|x^* + \xi^*\| = \sup_{y\in Y\setminus\{0\}} \frac{\langle x^*, y\rangle}{\|y\|} \qquad \text{for all } x^* \in X^* \tag{2.3.13}$$

and

$$\|x^*\| = \sup_{x\in X\setminus Y} \frac{\langle x^*, x\rangle}{\inf_{y\in Y}\|x+y\|} \qquad \text{for all } x^* \in Y^\perp. \tag{2.3.14}$$

Proof. This follows directly from Corollary 2.3.26. □

2.3.5. Complemented Subspaces. A familiar observation in linear algebra is that, for every subspace $Y \subset X$ of a finite-dimensional vector space X, there exists another subspace $Z \subset X$ such that $X = Y \oplus Z$. This continues to hold for infinite-dimensional vector spaces. However, it does not hold, in general, for closed subspaces of normed vector spaces. Here is the relevant definition.

Definition 2.3.28 (Complemented Subspace). Let X be a normed vector space. A closed subspace $Y \subset X$ is called **complemented** if there exists a closed subspace $Z \subset X$ such that $Y \cap Z = \{0\}$ and $X = Y \oplus Z$. A bounded linear operator $P : X \to X$ is called a **projection** if $P^2 = P$.

Exercise 2.3.29. Let X be a Banach space, let $Y \subset X$ be a closed linear subspace, and let $\pi : X \to X/Y$ be the canonical projection. (**Warning:** The term *projection* is used here with two different meanings.) Prove that the following are equivalent.

(i) Y is complemented.

(ii) There is a projection $P : X \to X$ such that $\mathrm{im}(P) = Y$.

(iii) There is a bounded linear operator $T : X/Y \to X$ such that $\pi \circ T = \mathrm{id}$. (The operator T, if it exists, is called a **right inverse** of π.)

Hint: For (i) $\Longrightarrow$ (ii) use Corollary 2.2.9. For (ii) $\Longrightarrow$ (i) define $Z := \ker(P)$. For (ii) $\Longrightarrow$ (iii) let $T[x] := x - Px$. For (iii) $\Longrightarrow$ (ii) let $P := 1\!\!1 - T \circ \pi$.

Lemma 2.3.30. *Let X be a normed vector space and let $Y \subset X$ be a closed linear subspace such that $\dim(Y) < \infty$ or $\dim(X/Y) < \infty$. Then Y is complemented.*

Proof. Assume $n := \dim(X/Y) < \infty$ and choose vectors $x_1, \dots, x_n \in X$ whose equivalence classes $[x_i] := x_i + Y$ form a basis of X/Y. Then the linear subspace $Z := \mathrm{span}\{x_1, \dots, x_n\}$ is closed by Corollary 1.2.7 and satisfies $X = Y \oplus Z$.

Now assume $n := \dim Y < \infty$ and choose a basis $x_1, \dots, x_n$ of Y. By the Hahn–Banach Theorem (Corollary 2.3.4) there exist bounded linear functionals $x_1^*, \dots, x_n^* \in X^*$ that satisfy $\langle x_i^*, x_j \rangle = \delta_{ij}$. Then the subspace

$$Z := \{x \in X \mid \langle x_i^*, x \rangle = 0 \text{ for } i = 1, \dots, n\}$$

is closed by Theorem 1.2.2. Moreover, $x - \sum_{i=1}^n \langle x_i^*, x \rangle x_i \in Z$ for all $x \in X$ and hence $X = Y \oplus Z$. This proves Lemma 2.3.30. □

There are examples of closed subspaces of infinite-dimensional Banach spaces that are not complemented. The simplest such example is the subspace $c_0 \subset \ell^\infty$. Phillips' Lemma asserts that it is not complemented. The proof is outlined in Exercise 2.5.1 below.

2.3.6. Orthonormal Bases.

Definition 2.3.31. Let H be an infinite-dimensional real Hilbert space. A sequence $(e_i)_{i\in\mathbb{N}}$ in H is called a **(countable) orthonormal basis** if

$$\langle e_i, e_j\rangle = \delta_{ij} := \begin{cases} 1, & \text{if } i = j, \\ 0, & \text{if } i \neq j, \end{cases} \qquad \text{for all } i, j \in \mathbb{N}, \tag{2.3.15}$$

$$x \in H, \quad \langle e_i, x\rangle = 0 \text{ for all } i \in \mathbb{N} \qquad \Longrightarrow \qquad x = 0. \tag{2.3.16}$$

If $(e_i)_{i\in\mathbb{N}}$ is an orthonormal basis, then (2.3.15) implies that the e_i are linearly independent and (2.3.16) asserts that the set $E := \operatorname{span}(\{e_i \,|\, i \in \mathbb{N}\})$ is a dense linear subspace of H (Corollary 2.3.25).

Exercise 2.3.32. Show that an infinite-dimensional Hilbert space H admits a countable orthonormal basis if and only if it is separable. **Hint:** Assume H is separable. Choose a dense sequence, construct a linearly independent subsequence spanning a dense subspace, and use *Gram–Schmidt.*

Exercise 2.3.33. Let H be a separable Hilbert space and let $\{e_i\}_{i\in\mathbb{N}}$ be an orthonormal basis. Show that the map $\ell^2 \to H : x = (x_i)_{i\in\mathbb{N}} \mapsto \sum_{i=1}^\infty x_i e_i$ is well defined (i.e. $\xi_n := \sum_{i=1}^n x_i e_i$ is a Cauchy sequence in H for all $x \in \ell^2$) and defines a Hilbert space isometry. Deduce that

$$x = \sum_{i=1}^\infty \langle e_i, x\rangle e_i, \qquad \|x\|^2 = \sum_{i=1}^\infty \langle e_i, x\rangle^2 \qquad \text{for all } x \in H. \tag{2.3.17}$$

Example 2.3.34. The sequences $e_i := (\delta_{ij})_{j\in\mathbb{N}}$ for $i \in \mathbb{N}$ form an orthonormal basis of ℓ^2.

Example 2.3.35 (Fourier Series). The functions $e_k(t) := e^{2\pi \mathbf{i} kt}$, $k \in \mathbb{Z}$, form an orthonormal basis of the complex Hilbert space $L^2(\mathbb{R}/\mathbb{Z}, \mathbb{C})$. It is equipped with the complex valued **Hermitian inner product**

$$\langle f, g\rangle := \int_0^1 \overline{f(t)} g(t)\, dt \qquad \text{for } f, g \in L^2(\mathbb{R}/\mathbb{Z}, \mathbb{C}), \tag{2.3.18}$$

that is complex anti-linear in the first variable and complex linear in the second variable. To verify completeness, one can fix a continuous function $f : \mathbb{R}/\mathbb{Z} \to \mathbb{C}$, define $f_n := \sum_{k=-n}^n \langle e_k, f\rangle e_k$ for $n \in \mathbb{N}_0$, and prove that the sequence $n^{-1}(f_0 + f_1 + \cdots + f_{n-1})$ converges uniformly to f (Fejér's Theorem).

Example 2.3.36. The functions $s_n(t) := \sqrt{2}\sin(\pi nt)$ for $n \in \mathbb{N}$ form an orthonormal basis of the Hilbert space $L^2([0,1])$ and so do the functions $c_0(t) := 1$ and $c_n(t) := \sqrt{2}\cos(\pi nt)$ for $n \in \mathbb{N}$. **Exercise:** Use completeness in Example 2.3.35 to verify the completeness axiom (2.3.16) for these two orthonormal bases.

2.4. Reflexive Banach Spaces

2.4.1. The Bidual Space. Let X be a real normed vector space. The **bidual space** of X is the dual space of the dual space and is denoted by

$$X^{**} := (X^*)^* = \mathcal{L}(X^*, \mathbb{R}).$$

There is a natural map $\iota = \iota_X : X \to X^{**}$ which assigns to every element $x \in X$ the linear functional $\iota(x) : X^* \to \mathbb{R}$ whose value at x^* is obtained by evaluating the bounded linear functional $x^* : X \to \mathbb{R}$ at the point $x \in X$. Thus the map $\iota : X \to X^{**}$ is defined by

$$\iota(x)(x^*) := \langle x^*, x\rangle \tag{2.4.1}$$

for $x \in X$ and $x^* \in X^*$. It is a consequence of the Hahn–Banach Theorem that the linear map $\iota : X \to X^{**}$ is an isometric embedding.

Lemma 2.4.1. *Let X be a real normed vector space. Then the linear map $\iota : X \to X^{**}$ is an isometric embedding. In particular,*

$$\|x\| = \sup_{x^* \in X^* \setminus \{0\}} \frac{|\langle x^*, x\rangle|}{\|x^*\|} \tag{2.4.2}$$

for all $x \in X$.

Proof. That the map $\iota : X \to X^{**}$ is linear follows directly from the definition. To prove that it preserves the norm, fix a nonzero vector $x_0 \in X$. Then, by Corollary 2.3.23, there exists a bounded linear functional $x_0^* \in X^*$ such that $\|x_0^*\| = 1$ and $\langle x_0^*, x_0\rangle = \|x_0\|$. Hence

$$\|x_0\| = \frac{|\langle x_0^*, x_0\rangle|}{\|x_0^*\|} \le \|\iota(x_0)\| = \sup_{x^* \in X^* \setminus \{0\}} \frac{|\langle x^*, x_0\rangle|}{\|x^*\|} \le \|x_0\|.$$

Here the last inequality follows from (2.3.8). This proves Lemma 2.4.1. □

Corollary 2.4.2. *Let X be a real normed vector space and let $Y \subset X$ be a closed linear subspace. Then, for every $x \in X$,*

$$\inf_{y \in Y} \|x + y\| = \sup_{x^* \in Y^\perp \setminus \{0\}} \frac{|\langle x^*, x\rangle|}{\|x^*\|}. \tag{2.4.3}$$

Proof. The left-hand side of equation (2.4.3) is the norm of the equivalence class $[x] = x + Y$ in the quotient space X/Y. The right-hand side is the norm of the bounded linear functional

$$\iota_{X/Y}(x + Y) : (X/Y)^* \cong Y^\perp \to \mathbb{R}$$

(see Corollary 2.3.26). Hence equation (2.4.3) follows from Lemma 2.4.1 with X replaced by X/Y. This proves Corollary 2.4.2. □

2.4.2. Reflexive Banach Spaces.

Definition 2.4.3 (Reflexive Banach Space). A real normed vector space X is called **reflexive** if the isometric embedding $\iota : X \to X^{**}$ in (2.4.1) is bijective. A reflexive normed vector space is necessarily complete by Theorem 1.3.1.

Theorem 2.4.4. *Let X be a Banach space. Then the following hold.*

(i) *X is reflexive if and only if X^* is reflexive.*

(ii) *If X is reflexive and $Y \subset X$ is a closed linear subspace, then the subspace Y and the quotient space X/Y are reflexive.*

Proof. We prove part (i). Assume X is reflexive and let $\Lambda : X^{**} \to \mathbb{R}$ be a bounded linear functional. Define

$$x^* := \Lambda \circ \iota : X \to \mathbb{R},$$

where $\iota = \iota_X : X \to X^{**}$ is the isometric embedding in (2.4.1). Since X is reflexive, this map ι is bijective. Fix an element $x^{**} \in X^{**}$ and define

$$x := \iota^{-1}(x^{**}) \in X.$$

Then

$$\Lambda(x^{**}) = \Lambda \circ \iota(x) = \langle x^*, x\rangle = \langle \iota(x), x^*\rangle = \langle x^{**}, x^*\rangle.$$

Here the first and last equation follow from the fact that $x^{**} = \iota(x)$, the second equation follows from the definition of $x^* = \Lambda \circ \iota$, and the third equation follows from the definition of the map ι in (2.4.1). This shows that

$$\Lambda = \iota_{X^*}(x^*),$$

where $\iota_{X^*} : X^* \to X^{***}$ is the isometric embedding in (2.4.1) with X replaced by X^*. This shows that the dual space X^* is reflexive.

Conversely, assume X^* is reflexive. The subspace $\iota(X)$ of X^{**} is complete by Lemma 2.4.1 and is therefore closed. We prove that $\iota(X)$ is a dense subspace of X^{**}. To see this, let $\Lambda : X^{**} \to \mathbb{R}$ be any bounded linear functional on X^{**} that vanishes on the image of ι, so that $\Lambda \circ \iota = 0$. Since X^* is reflexive, there exists an element $x^* \in X^*$ such that

$$\Lambda(x^{**}) = \langle x^{**}, x^*\rangle$$

for every $x^{**} \in X^{**}$. Since $\Lambda \circ \iota = 0$, this implies

$$\langle x^*, x\rangle = \langle \iota(x), x^*\rangle = \Lambda(\iota(x)) = 0$$

for all $x \in X$, hence $x^* = 0$, and hence $\Lambda = 0$. Thus the annihilator of the linear subspace $\iota(X) \subset X^{**}$ is zero, and so $\iota(X)$ is dense in X^{**} by Corollary 2.3.25. Hence $\iota(X) = X^{**}$ and this proves part (i).

We prove part (ii). Assume X is reflexive and let

$$Y \subset X$$

be a closed linear subspace. We prove first that Y is a reflexive Banach space. Define the linear operator

$$\pi : X^* \to Y^*$$

by

$$\pi(x^*) := x^*|_Y$$

for $x^* \in X^*$. Fix an element $y^{**} \in Y^{**}$ and define $x^{**} \in X^{**}$ by

$$x^{**} := y^{**} \circ \pi : X^* \to \mathbb{R}.$$

Since X is reflexive, there exists a unique element $y \in X$ such that

$$\iota_X(y) = x^{**}.$$

Every element $x^* \in Y^\perp$ satisfies $\pi(x^*) = 0$ and hence

$$\begin{aligned} \langle x^*, y \rangle &= \langle \iota_X(y), x^* \rangle \\ &= \langle x^{**}, x^* \rangle \\ &= \langle y^{**} \circ \pi, x^* \rangle \\ &= \langle y^{**}, \pi(x^*) \rangle \\ &= 0. \end{aligned}$$

In other words, $\langle x^*, y \rangle = 0$ for all $x^* \in Y^\perp$ and so

$$y \in \overline{Y} = Y$$

by Corollary 2.3.24. Now fix any element $y^* \in Y^*$. Then Corollary 2.3.4 asserts that there exists an element $x^* \in X^*$ such that

$$y^* = x^*|_Y = \pi(x^*)$$

and so

$$\begin{aligned} \langle y^{**}, y^* \rangle &= \langle y^{**}, \pi(x^*) \rangle \\ &= \langle x^{**}, x^* \rangle \\ &= \langle \iota(y), x^* \rangle \\ &= \langle x^*, y \rangle \\ &= \langle y^*, y \rangle. \end{aligned}$$

This shows that

$$\iota_Y(y) = y^{**}.$$

Since $y^{**} \in Y^{**}$ was chosen arbitrarily, this proves that the subspace Y is a reflexive Banach space.

Next we prove that the quotient

$$Z := X/Y$$

is reflexive. Let

$$\pi : X \to X/Y$$

be the canonical projection given by

$$\pi(x) := [x] = x + Y \qquad \text{for } x \in X$$

and define the linear operator $T : Z^* \to Y^\perp$ by

$$Tz^* := z^* \circ \pi : X \to \mathbb{R} \qquad \text{for } z^* \in Z^*.$$

Note that $Tz^* \in Y^\perp$ because $(Tz^*)(y) = z^*(\pi(y)) = 0$ for all $y \in Y$. Moreover, T is an isometric isomorphism by Corollary 2.3.26.

Now fix an element $z^{**} \in Z^{**}$. Then the map

$$z^{**} \circ T^{-1} : Y^\perp \to \mathbb{R}$$

is a bounded linear functional on a linear subspace of X^*. Hence, by Corollary 2.3.4, there exists a bounded linear functional $x^{**} : X^* \to \mathbb{R}$ such that

$$\langle x^{**}, x^* \rangle = \langle z^{**}, T^{-1}x^* \rangle \qquad \text{for all } x^* \in Y^\perp.$$

This condition on x^{**} can be expressed in the form

$$\langle x^{**}, z^* \circ \pi \rangle = \langle z^{**}, z^* \rangle \qquad \text{for all } z^* \in Z^*.$$

Since X is reflexive, there exists an element $x \in X$ such that

$$\iota_X(x) = x^{**}.$$

Define

$$z := [x] = \pi(x) \in Z.$$

Then, for all $z^* \in Z^*$, we have

$$\begin{aligned} \langle z^{**}, z^* \rangle &= \langle x^{**}, z^* \circ \pi \rangle \\ &= \langle \iota(x), z^* \circ \pi \rangle \\ &= \langle z^* \circ \pi, x \rangle \\ &= \langle z^*, \pi(x) \rangle \\ &= \langle z^*, z \rangle. \end{aligned}$$

This shows that

$$\iota_Z(z) = z^{**}.$$

Since $z^{**} \in Z^{**}$ was chosen arbitrarily, it follows that Z is reflexive. This proves Theorem 2.4.4. □

Example 2.4.5. **(i)** Every finite-dimensional normed vector space X is reflexive, because $\dim X = \dim X^* = \dim X^{**}$ (see Corollary 1.2.9).

(ii) Every Hilbert space H is reflexive by Theorem 1.4.4. **Exercise:** The composition of the isomorphisms $H \cong H^* \cong H^{**}$ is the map in (2.4.1).

(iii) Let $(M, \mathcal{A}, \mu)$ be a measure space and let $1 < p, q < \infty$ such that

$$\frac{1}{p} + \frac{1}{q} = 1.$$

Then $L^p(\mu)^* \cong L^q(\mu)$ (Example 1.3.3) and this implies that the Banach space $L^p(\mu)$ is reflexive. **Exercise:** Prove that the composition of the isomorphisms $L^p(\mu) \cong L^q(\mu)^* \cong L^p(\mu)^{**}$ is the map in (2.4.1).

(iv) Let $c_0 \subset \ell^\infty$ be the subspace of sequences $x = (x_i)_{i \in \mathbb{N}}$ of real numbers that converge to zero, equipped with the supremum norm. Then the map $\ell^1 \to c_0^* : y \mapsto \Lambda_y$, which assigns to every sequence $y = (y_i)_{i \in \mathbb{N}} \in \ell^1$ the bounded linear functional $\Lambda_y : c_0 \to \mathbb{R}$ defined by $\Lambda_y(x) := \sum_{i=1}^\infty x_i y_i$ for $x = (x_i)_{i \in \mathbb{N}} \in c_0$, is a Banach space isometry (see Example 1.3.7). This implies $c_0^{**} \cong (\ell^1)^* \cong \ell^\infty$ (see Example 1.3.6), and so c_0 is not reflexive. **Exercise:** The composition of the isometric embedding $\iota : c_0 \to c_0^{**}$ in (2.4.1) with the Banach space isometry $c_0^{**} \cong \ell^\infty$ is the canonical inclusion.

(v) The Banach space ℓ^1 is not reflexive. To see this, denote by $c \subset \ell^\infty$ the space of Cauchy sequences of real numbers and consider the bounded linear functional that assigns to each Cauchy sequence $x = (x_i)_{i \in \mathbb{N}} \in c$ its limit $\lim_{i \to \infty} x_i$. By the Hahn–Banach Theorem this functional extends to a bounded linear functional $\Lambda : \ell^\infty \to \mathbb{R}$ (see Corollary 2.3.4), which does not belong to the image of the inclusion $\iota : \ell^1 \to (\ell^1)^{**} \cong (\ell^\infty)^*$.

(vi) Let (M, d) be a compact metric space and let $X = C(M)$ be the Banach space of continuous real valued functions on M with the supremum norm (see part (v) of Example 1.1.3). Suppose M is an infinite set. Then $C(M)$ is not reflexive. To see this, let $A = \{a_1, a_2, \dots\} \subset M$ be a countably infinite subset such that $(a_i)_{i \in \mathbb{N}}$ is a Cauchy sequence and $a_i \neq a_j$ for $i \neq j$. Then $C_A(M) := \{f \in C(M) \,|\, f|_A = 0\}$ is a closed linear subspace of $C(M)$ and the quotient $C(M)/C_A(M)$ is isometrically isomorphic to the space c of Cauchy sequences of real numbers via $C(M)/C_A(M) \to c : [f] \mapsto (f(a_i))_{i=1}^\infty$. By Theorem 2.4.4 the Banach space c is not reflexive, because the closed subspace $c_0 \subset c$ is not reflexive by (iv) above. Hence $C(M)/C_A(M)$ is not reflexive, and so $C(M)$ is not reflexive by Theorem 2.4.4.

(vii) The dual space of the Banach space $C(M)$ in (vi) is isomorphic to the Banach space $\mathcal{M}(M)$ of signed Borel measures on M (see Example 1.3.8). Since $C(M)$ is not reflexive, neither is the space $\mathcal{M}(M)$ by Theorem 2.4.4.

2.4.3. Separable Banach Spaces. Recall that a normed vector space is called **separable** if it contains a countable dense subset (see Definition 1.1.6). Thus a Banach space X is separable if and only if there exists a sequence $e_1, e_2, e_3, \dots$ in X such that the linear subspace of all (finite) linear combinations of the e_i is dense in X. If such a sequence exists, the required countable dense subset can be constructed as the set of all rational linear combinations of the e_i.

Theorem 2.4.6. *Let X be a normed vector space. The following hold.*

(i) *If X^* is separable, then X is separable.*

(ii) *If X is reflexive and separable, then X^* is separable.*

Proof. We prove part (i). Thus assume X^* is separable and choose a dense sequence $(x_i^*)_{i\in\mathbb{N}}$ in X^*. Choose a sequence $x_i \in X$ such that

$$\|x_i\| = 1, \qquad \langle x_i^*, x_i\rangle \geq \frac{1}{2}\|x_i^*\| \qquad \text{for all } i \in \mathbb{N}.$$

Let $Y \subset X$ be the linear subspace of all finite linear combinations of the x_i. We prove that Y is dense in X. To see this, fix any element $x^* \in Y^\perp$. Then there is a sequence $i_k \in \mathbb{N}$ such that $\lim_{k\to\infty}\|x^* - x_{i_k}^*\| = 0$. This implies

$$\begin{aligned}\|x_{i_k}^*\| &\leq 2|\langle x_{i_k}^*, x_{i_k}\rangle| = 2|\langle x_{i_k}^* - x^*, x_{i_k}\rangle| \\ &\leq 2\|x_{i_k}^* - x^*\|\|x_{i_k}\| = 2\|x_{i_k}^* - x^*\|.\end{aligned}$$

The last term on the right converges to zero as k tends to infinity, and hence $x^* = \lim_{k\to\infty} x_{i_k}^* = 0$. This shows that $Y^\perp = \{0\}$. Hence Y is dense in X by Corollary 2.3.25 and this proves part (i). If X is reflexive and separable, then X^{**} is separable, and so X^* is separable by (i). This proves part (ii) and Theorem 2.4.6. □

Example 2.4.7. (i) Finite-dimensional Banach spaces are separable.

(ii) The space ℓ^p is separable for $1 \leq p < \infty$, and $(\ell^1)^* \cong \ell^\infty$ is not separable. The subspace $c_0 \subset \ell^\infty$ of all sequences that converge to zero is separable.

(iii) Let M be a second countable locally compact Hausdorff space, denote by $\mathcal{B} \subset 2^M$ its Borel σ-algebra, and let $\mu : \mathcal{B} \to [0, \infty]$ be a locally finite Borel measure. Then the space $L^p(\mu)$ is separable for $1 \leq p < \infty$. (See for example [**75**, Thm. 4.13].)

(iv) Let (M, d) be a compact metric space. Then the Banach space $C(M)$ of continuous functions with the supremum norm is separable. Its dual space $\mathcal{M}(M)$ of signed Borel measures is in general not separable.

2.4.4. The James Space. In 1950 Robert C. James [**37, 38**] discovered a remarkable example of a nonreflexive Banach space J that is isometrically isomorphic to its bidual space J^{**}. In this example the image of the canonical isometric embedding

$$\iota : J \to J^{**}$$

in (2.4.1) is a closed subspace of codimension one. Our exposition follows Megginson [**59**].

Recall that $c_0 \subset \ell^\infty$ is the Banach space of all sequences $(x_i)_{i\in\mathbb{N}} \in \mathbb{R}^{\mathbb{N}}$ that converge to zero, equipped with the supremum norm

$$\|x\|_\infty := \sup_{i\in\mathbb{N}} |x_i| \qquad \text{for } x = (x_i)_{i\in\mathbb{N}} \in c_0.$$

By Example 1.3.7 the dual space of c_0 is isomorphic to the space ℓ^1 of absolutely summable sequences of real numbers with the norm

$$\|x\|_1 := \sum_{i=1}^\infty |x_i| \qquad \text{for } x = (x_i)_{i\in\mathbb{N}} \in \ell^1.$$

Recall also that ℓ^2 is the Hilbert space of all square summable sequences of real numbers with the norm

$$\|x\|_2 := \left(\sum_{i=1}^\infty |x_i|^2\right)^{1/2} \qquad \text{for } x = (x_i)_{i\in\mathbb{N}} \in \ell^2.$$

Definition 2.4.8 (The James Space). Let $\mathcal{P} \subset 2^{\mathbb{N}}$ be the collection of all nonempty finite subsets of $\mathbb{N}$ and write the elements of $\mathcal{P}$ in the form $\mathbf{p} = (p_1, p_2, \dots, p_k)$ with $1 \le p_1 < p_2 < \cdots < p_k$. For each $\mathbf{p} = (p_1, p_2, \dots, p_k) \in \mathcal{P}$ and each sequence $x = (x_i)_{i\in\mathbb{N}}$ of real numbers define the number $\|x\|_{\mathbf{p}} \in [0, \infty)$ by $\|x\|_{\mathbf{p}} := 0$ when $k = 1$ and by

$$\|x\|_{\mathbf{p}} := \sqrt{\frac{1}{2}\left(\sum_{j=1}^{k-1} \left|x_{p_j} - x_{p_{j+1}}\right|^2 + \left|x_{p_k} - x_{p_1}\right|^2\right)} \tag{2.4.4}$$

when $k \ge 2$. The **James space** is the normed vector space defined by

$$J := \left\{ x \in c_0 \,\middle|\, \sup_{\mathbf{p}\in\mathcal{P}} \|x\|_{\mathbf{p}} < \infty \right\} \tag{2.4.5}$$

and

$$\|x\|_J := \sup_{\mathbf{p}\in\mathcal{P}} \|x\|_{\mathbf{p}} \tag{2.4.6}$$

for $x \in J$.

Before moving on to the main result of this section (Theorem 2.4.14) we explore some of the basic properties of the James space. This is the content of the next five lemmas.

Lemma 2.4.9. *The set J in (2.4.5) is a linear subspace of c_0 and $\|\cdot\|_J$ is a norm on J. With this norm J is a Banach space. Moreover,*

$$\|x\|_\infty \le \|x\|_J \le \sqrt{2}\,\|x\|_2 \qquad \text{for all } x \in c_0, \tag{2.4.7}$$

and thus $\ell^2 \subset J \subset c_0$.

Proof. By definition, $\|x+y\|_J \le \|x\|_J + \|y\|_J$ and $\|\lambda x\|_J = |\lambda|\|x\|_J$ for all $x, y \in c_0$ and all $\lambda \in \mathbb{R}$. Hence J is a linear subspace of c_0.

To prove the first inequality in (2.4.7), fix an element $\mathbf{p} = (i,j) \in \mathcal{P}$. Then $|x_i - x_j| = \|x\|_{\mathbf{p}} \le \|x\|_J$ for all $x \in c_0$ and all $i, j \in \mathbb{N}$ with $i < j$. Hence $|x_i| = \lim_{j\to\infty}|x_i - x_j| \le \|x\|_J$ for all $x \in c_0$ and all $i \in \mathbb{N}$. Now fix any element $\mathbf{p} = (p_1, p_2, \dots, p_k) \in \mathcal{P}$. Then

$$\begin{aligned}
\|x\|_{\mathbf{p}}^2 &= \frac{1}{2}\sum_{j=1}^{k-1} \left|x_{p_j} - x_{p_{j+1}}\right|^2 + \frac{1}{2}\left|x_{p_k} - x_{p_1}\right|^2 \\
&\le \sum_{j=1}^{k-1} \left|x_{p_j}\right|^2 + \sum_{j=1}^{k-1} \left|x_{p_{j+1}}\right|^2 + \left|x_{p_k}\right|^2 + \left|x_{p_1}\right|^2 \\
&= 2\sum_{j=1}^{k} \left|x_{p_j}\right|^2 \le 2\,\|x\|_2^2
\end{aligned}$$

for all $x \in c_0$. Take the supremum over all $\mathbf{p} \in \mathcal{P}$ to obtain $\|x\|_J \le \sqrt{2}\|x\|_2$. This proves (2.4.7). By (2.4.7) there are natural inclusions $\ell^2 \subset J \subset c_0$. Moreover, it follows from (2.4.7) that $\|x\|_J \ne 0$ for every $x \in J \setminus \{0\}$ and so $(J, \|\cdot\|_J)$ is a normed vector space.

We prove that J is complete. Let $(x_n)_{n\in\mathbb{N}}$ be a Cauchy sequence in J. Then $(\|x_n\|_J)_{n\in\mathbb{N}}$ is a Cauchy sequence in $\mathbb{R}$, so the limit $C := \lim_{n\to\infty}\|x_n\|_J$ exists. Moreover, $(x_n)_{n\in\mathbb{N}}$ is a Cauchy sequence in c_0 by (2.4.7) and hence converges in the supremum norm to an element $x \in c_0$. Thus

$$\|x\|_{\mathbf{p}} = \lim_{n\to\infty}\|x_n\|_{\mathbf{p}} \le \lim_{n\to\infty}\|x_n\|_J = C$$

for all $\mathbf{p} \in \mathcal{P}$. Take the supremum over all $\mathbf{p} \in \mathcal{P}$ to obtain $x \in J$. We must prove that $\lim_{n\to\infty}\|x_n - x\|_J = 0$. To see this, fix a number $\varepsilon > 0$ and choose an integer $n_0 \in \mathbb{N}$ such that $\|x_n - x_m\|_J < \varepsilon/2$ for all integers $m, n \ge n_0$. Then $\|x_n - x\|_{\mathbf{p}} = \lim_{m\to\infty}\|x_n - x_m\|_{\mathbf{p}} \le \sup_{m\ge n_0}\|x_n - x_m\|_J \le \varepsilon/2$ for all $\mathbf{p} \in \mathcal{P}$ and all $n \ge n_0$. Thus $\|x_n - x\|_J = \sup_{\mathbf{p}\in\mathcal{P}}\|x_n - x\|_{\mathbf{p}} \le \varepsilon/2 < \varepsilon$ for every integer $n \ge n_0$. This shows that $\lim_{n\to\infty}\|x_n - x\|_J = 0$ and completes the proof of Lemma 2.4.9. □

The next goal is to prove that ℓ^2 is dense in J. For this it is convenient to introduce another norm on J. For every sequence $x = (x_i)_{i \in \mathbb{N}}$ of real numbers and every $\mathbf{p} = (p_1, p_2, \dots, p_k) \in \mathcal{P}$ define $|||x|||_{\mathbf{p}} := |x_{p_1}|$ in the case $k = 1$ and

$$|||x|||_{\mathbf{p}} := \sqrt{\frac{1}{2}\left(|x_{p_1}|^2 + \sum_{j=1}^{k-1} \left|x_{p_j} - x_{p_{j+1}}\right|^2 + |x_{p_k}|^2\right)} \tag{2.4.8}$$

in the case $k \geq 2$. Denote the supremum of these numbers over all $\mathbf{p} \in \mathcal{P}$ by

$$|||x|||_J := \sup_{\mathbf{p} \in \mathcal{P}} |||x|||_{\mathbf{p}}. \tag{2.4.9}$$

This is a norm on J that is equivalent to $\|\cdot\|_J$. Care must be taken. The second estimate in (2.4.10) below holds for $x \in c_0$ but not for all $x \in \ell^\infty$.

Lemma 2.4.10. *Every $x \in c_0$ satisfies the inequalities*

$$\frac{1}{\sqrt{2}} \|x\|_J \leq |||x|||_J \leq \|x\|_J\,. \tag{2.4.10}$$

Moreover, the function $J \to [0, \infty) : x \mapsto |||x|||_J$ is a norm.

Proof. Let $x \in c_0$ and $\mathbf{p} = (p_1, \dots, p_k) \in \mathcal{P}$. Then

$$\begin{aligned}
\|x\|_{\mathbf{p}}^2 &= \frac{1}{2}\left(\sum_{j=1}^{k} \left|x_{p_j} - x_{p_{j+1}}\right|^2 + |x_{p_k} - x_{p_1}|^2\right) \\
&\leq \sum_{j=1}^{k} \left|x_{p_j} - x_{p_{j+1}}\right|^2 + |x_{p_k}|^2 + \|x_{p_1}\|^2 \\
&= 2|||x|||_{\mathbf{p}}^2 \leq 2|||x|||_J^2.
\end{aligned}$$

Take the supremum over all $\mathbf{p} \in \mathcal{P}$ to obtain the inequality $\|x\|_J \leq \sqrt{2}|||x|||_J$. Now define $\mathbf{q}_n := (p_1, \dots, p_k, n)$ for every integer $n > p_k$. Then

$$\begin{aligned}
|||x|||_{\mathbf{p}}^2 &= \frac{1}{2}\left(\sum_{j=1}^{k} \left|x_{p_j} - x_{p_{j+1}}\right|^2 + |x_{p_k}|^2 + |x_{p_1}|^2\right) \\
&= \lim_{n \to \infty} \frac{1}{2}\left(\sum_{j=1}^{k} \left|x_{p_j} - x_{p_{j+1}}\right|^2 + |x_{p_k} - p_n|^2 + |p_n - x_{p_1}|^2\right) \\
&= \lim_{n \to \infty} \|x\|_{\mathbf{q}_n}^2 \leq \|x\|_J^2\,.
\end{aligned}$$

Take the supremum over all $\mathbf{p} \in \mathcal{P}$ to obtain the inequality $|||x|||_J \leq \|x\|_J$. This proves Lemma 2.4.10. □

Lemma 2.4.11. *The subspace ℓ^2 is dense in J.*

Proof. Fix a nonzero element $x \in J$ and a real number $\varepsilon > 0$, and choose a constant $0 < \delta < |||x|||_J$ such that

$$2\delta|||x|||_J < \varepsilon^2. \tag{2.4.11}$$

We claim that there are elements $n \in \mathbb{N}$ and $\mathbf{p} = (p_1, \dots, p_k) \in \mathcal{P}$ such that

$$\sup_{i \geq n} |x_i| < \delta, \qquad |||x|||_{\mathbf{p}} > |||x|||_J - \delta, \qquad p_k = n. \tag{2.4.12}$$

Namely, choose $n \in \mathbb{N}$ such that $\sup_{i \geq n} |x_i| < \delta$ and $\mathbf{p}_0 = (p_1, \dots, p_{k-1}) \in \mathcal{P}$ such that $|||x|||_{\mathbf{p}_0} > |||x|||_J - \delta$. Next choose $p_k > p_{k-1}$ so large that $p_k \geq n$ and the tuple $\mathbf{p} := (p_1, \dots, p_k)$ satisfies

$$|||x|||_{\mathbf{p}} = \sqrt{|||x|||_{\mathbf{p}_0}^2 - \tfrac{1}{2}|x_{p_{k-1}}|^2 + \tfrac{1}{2}|x_{p_{k-1}} - x_{p_k}|^2 + \tfrac{1}{2}|x_{p_k}|^2} > |||x|||_J - \delta.$$

Then increase n, if necessary, to obtain $p_k = n$.

Define $\xi := (x_1, \dots, x_n, 0, \dots)$. We prove that

$$|||x - \xi|||_J < \varepsilon. \tag{2.4.13}$$

To see this, let $\mathbf{q} = (q_1, \dots, q_\ell) \in \mathcal{P}$. If $q_\ell \leq n$, then $|||x - \xi|||_{\mathbf{q}} = 0$. Thus assume $q_\ell > n$, let $j \in \{1, \dots, \ell\}$ be the smallest element such that $q_j > n$, and define $\mathbf{q}' := (q_j, q_{j+1}, \dots, q_\ell) \in \mathcal{P}$. Then

$$|||x - \xi|||_{\mathbf{q}} = |||x|||_{\mathbf{q}'}. \tag{2.4.14}$$

Now consider the tuple $\mathbf{p}' := (p_1, \dots, p_k, q_j, q_{j+1}, \dots, q_\ell) \in \mathcal{P}$. By (2.4.12), it satisfies the inequality

$$\begin{aligned}
|||x|||_J^2 &\geq |||x|||_{\mathbf{p}'}^2 \\
&= |||x|||_{\mathbf{p}}^2 + |||x|||_{\mathbf{q}'}^2 + \tfrac{1}{2}\left|x_{p_k} - x_{q_j}\right|^2 - \tfrac{1}{2}\left|x_{p_k}\right|^2 - \tfrac{1}{2}\left|x_{q_j}\right|^2 \\
&> \left(|||x|||_J - \delta\right)^2 - \delta^2 + |||x|||_{\mathbf{q}'}^2 \\
&= |||x|||_J^2 - 2\delta|||x|||_J + |||x|||_{\mathbf{q}'}^2.
\end{aligned}$$

This implies $|||x|||_{\mathbf{q}'}^2 < 2\delta|||x|||_J$ and hence

$$|||x - \xi|||_{\mathbf{q}} = |||x|||_{\mathbf{q}'} < \sqrt{2\delta|||x|||_J} < \varepsilon$$

by (2.4.11) and (2.4.14). Take the supremum over all elements $\mathbf{q} \in \mathcal{P}$ to obtain the inequality (2.4.13). By (2.4.13) the set c_{00} of all finite sequences is dense in J and so is the subspace ℓ^2. This proves Lemma 2.4.11. □

The following lemma shows that the standard basis vectors $e_i := (\delta_{ij})_{j \in \mathbb{N}}$ form a Schauder basis of J (see Exercise 2.5.12).

Lemma 2.4.12. *For each $n \in \mathbb{N}$ define the projection $\Pi_n : J \to J$ by*

$$\Pi_n(x) := \sum_{i=1}^n x_i e_i \qquad \text{for } x = (x_i)_{i\in\mathbb{N}} \in J. \tag{2.4.15}$$

Then

$$\|\Pi_n(x)\|_J \le \|x\|_J, \qquad \|x - \Pi_n(x)\|_J \le \|x\|_J \tag{2.4.16}$$

for all $n \in \mathbb{N}$ and all $x \in J$, and

$$\lim_{n\to\infty} \|x - \Pi_n(x)\|_J = 0 \tag{2.4.17}$$

for all $x \in J$.

Proof. We prove (2.4.16). Fix an element $x \in J$, a positive integer n, and an element $\mathbf{p} = (p_1, \dots, p_k) \in \mathcal{P}$. If $p_k \le n$, then $\|\Pi_n(x)\|_{\mathbf{p}} = \|x\|_{\mathbf{p}}$ and, if $p_1 > n$, then $\|\Pi_n(x)\|_{\mathbf{p}} = 0$. Thus assume

$$p_1 \le n < p_k,$$

let $\ell \in \{1, \dots, k-1\}$ be the largest element such that $p_\ell \le n$, and define

$$\mathbf{q} := (p_1, \dots, p_\ell).$$

Then

$$\begin{aligned}\|\Pi_n(x)\|_{\mathbf{p}}^2 &= \|x\|_{\mathbf{q}}^2 - \tfrac{1}{2}|x_{p_\ell} - x_{p_1}|^2 + \tfrac{1}{2}|x_{p_\ell}|^2 + \tfrac{1}{2}|x_{p_1}|^2 \\ &= |||x|||_{\mathbf{q}}^2 \\ &\le \|x\|_J^2\end{aligned}$$

by Lemma 2.4.10. Thus $\|\Pi_n(x)\|_{\mathbf{p}} \le \|x\|_J$ for all $\mathbf{p} \in \mathcal{P}$ and this proves the first inequality in (2.4.16). To prove the second inequality in (2.4.16), observe that $\|x - \Pi_n(x)\|_{\mathbf{p}} = \|x\|_{\mathbf{p}}$ whenever $p_1 > n$ and $\|x - \Pi_n(x)\|_{\mathbf{p}} = 0$ whenever $p_k \le n$. Thus assume $p_1 \le n < p_k$, let $\ell \in \{2, \dots, k\}$ be the smallest element such that $p_\ell > n$, and define $\mathbf{q} := (p_\ell, \dots, p_k)$. Then

$$\begin{aligned}\|x - \Pi_n(x)\|_{\mathbf{p}}^2 &= \|x\|_{\mathbf{q}}^2 - \tfrac{1}{2}|x_{p_\ell} - x_{p_k}|^2 + \tfrac{1}{2}|x_{p_\ell}|^2 + \tfrac{1}{2}|x_{p_k}|^2 \\ &= |||x|||_{\mathbf{q}}^2 \\ &\le \|x\|_J^2\end{aligned}$$

by Lemma 2.4.10. Thus $\|x - \Pi_n(x)\|_{\mathbf{p}} \le \|x\|_J$ for all $\mathbf{p} \in \mathcal{P}$ and this proves the second inequality in (2.4.16).

We prove (2.4.17). When $x \in \ell^2$ this follows from (2.4.7). Since ℓ^2 is dense in J by Lemma 2.4.11, it follows from the estimate (2.4.16) and the Banach–Steinhaus Theorem 2.1.5 that (2.4.17) holds for all $x \in J$. This proves Lemma 2.4.12. □

With this preparation we are in a position to examine the dual space of the James space J. Fix a bounded linear functional $\Lambda : J \to \mathbb{R}$. By (2.4.7), the inclusion $\ell^2 \hookrightarrow J$ is a bounded linear operator and, by Lemma 2.4.11, it has a dense image. Thus the composition of Λ with this inclusion is a bounded linear functional $\Lambda|_{\ell^2} : \ell^2 \to \mathbb{R}$. Hence, by the Riesz Representation Theorem 1.4.4, there exists a unique sequence $y = (y_i)_{i\in\mathbb{N}} \in \ell^2$ such that

$$\Lambda(x) = \sum_{i=1}^{\infty} y_i x_i = \langle y, x\rangle \qquad \text{for all } x \in \ell^2 \subset J, \tag{2.4.18}$$

and, conversely, Λ is uniquely determined by this sequence $y \in \ell^2$. Thus the dual space of J can be identified with the space of all $y \in \ell^2$ such that

$$\|y\|_{J^*} := \sup_{0\neq x\in\ell^2} \frac{|\langle y, x\rangle|}{\|x\|_J} < \infty. \tag{2.4.19}$$

By (2.4.7) and (2.4.19), every $y \in J^*$ satisfies the inequalities

$$\frac{1}{\sqrt{2}} \|y\|_2 \le \|y\|_{J^*} \le \|y\|_1 . \tag{2.4.20}$$

Thus there are canonical inclusions

$$\ell^1 \subset J^* \subset \ell^2 \subset J \subset c_0.$$

At this point it is convenient to make use of two concepts that will only be introduced in Chapters 3 and 4. These are the dual operator $A^* : Y^* \to X^*$ of a bounded linear operator $A : X \to Y$ (Definition 4.1.1) and the weak* topology on the dual space of a Banach space (Example 3.1.9). A useful fact is that the dual operator has the same operator norm as the original operator (Lemma 4.1.2). Under our identification of J^* with a subspace of ℓ^2, the dual operator of the projection $\Pi_n : J \to J$ in (2.4.15) is the operator

$$\Pi_n : J^* \to J^*, \qquad \Pi_n(y) := \sum_{i=1}^{n} y_i e_i \quad \text{for } y = (y_i)_{i\in\mathbb{N}} \in J^*. \tag{2.4.21}$$

Thus it follows from the estimates in (2.4.16) that

$$\|\Pi_n(y)\|_{J^*} \le \|y\|_{J^*}, \qquad \|y - \Pi_n(y)\|_{J^*} \le \|y\|_{J^*} \tag{2.4.22}$$

for all $y \in J^*$ and all $n \in \mathbb{N}$. Moreover, the dual space of c_0 can be identified with ℓ^1 (Example 1.3.7) and the dual operator of the inclusion $J \hookrightarrow c_0$ is then the inclusion $\ell^1 \hookrightarrow J^*$. Hence it follows from general considerations that ℓ^1 is dense in J^* with respect to the weak* topology (Theorem 4.1.8). The next lemma shows that ℓ^1 is dense in J^* with respect to the norm topology.

Lemma 2.4.13. *Every $y \in J^*$ satisfies*

$$\lim_{n\to\infty} \|y - \Pi_n(y)\|_{J^*} = 0. \tag{2.4.23}$$

Proof. Fix an element $y \in J^*$. We prove that

$$\varepsilon_n := \|y - \Pi_n(y)\|_{J^*} = \sup_{\substack{0\neq x\in J\\ \Pi_n(x)=0}} \frac{|\langle y, x\rangle|}{\|x\|_J} \geq \varepsilon_{n+1} \tag{2.4.24}$$

for all $n \in \mathbb{N}$. To see this, fix an integer $n \in \mathbb{N}$ and recall from Lemma 2.4.12 that $\|x - \Pi_n(x)\|_J \leq \|x\|_J$ for all $x \in J$. Hence

$$\begin{aligned}
\|y - \Pi_n(y)\|_{J^*} &= \sup_{0\neq x\in J} \frac{|\langle y - \Pi_n(y), x\rangle|}{\|x\|_J} \\
&\leq \sup_{\substack{x\in J\\ \Pi_n(x)\neq x}} \frac{|\langle y, x - \Pi_n(x)\rangle|}{\|x - \Pi_n(x)\|_J} \\
&= \sup_{\substack{0\neq x\in J\\ \Pi_n(x)=0}} \frac{|\langle y, x\rangle|}{\|x\|_J} \\
&\leq \sup_{0\neq x\in J} \frac{|\langle y - \Pi_n(y), x\rangle|}{\|x\|_J} \\
&= \|y - \Pi_n(y)\|_{J^*}.
\end{aligned}$$

This proves the second equality in (2.4.24). This equality also shows that the sequence $(\varepsilon_n)_{n\in\mathbb{N}}$ is nonincreasing. Thus we have proved (2.4.24).

Now suppose, by contradiction, that $\lim_{n\to\infty} \varepsilon_n = \inf_{n\in\mathbb{N}} \varepsilon_n > 0$. Choose a constant $0 < \varepsilon < \inf_{n\in\mathbb{N}} \varepsilon_n$. Then, by (2.4.24) and the axiom of countable choice, there exists a sequence of sequences $x_n = (x_{n,i})_{i\in\mathbb{N}} \in J$ such that

$$\Pi_n(x_n) = 0, \qquad \|x_n\|_J = 1, \qquad \langle y, x_n\rangle > \varepsilon \tag{2.4.25}$$

for all $n \in \mathbb{N}$. Since c_{00} is dense in J by Lemma 2.4.12, the sequence can be chosen such that $x_n \in c_{00}$ for all $n \in \mathbb{N}$. By Lemma 2.4.9 each element x_n satisfies $\|x_n\|_\infty \leq \|x_n\|_J = 1$. Define the map $\kappa : \mathbb{N} \to \mathbb{N}$ by

$$\kappa(n) := \max\{i \in \mathbb{N} \,|\, x_{n,i} \neq 0\} \qquad \text{for } n \in \mathbb{N}. \tag{2.4.26}$$

Then $\kappa(n) > n$ for all $n \in \mathbb{N}$. Next define the sequence $n_j \in \mathbb{N}$ by $n_1 := 1$ and $n_{j+1} := \kappa(n_j) > n_j$ for $j \in \mathbb{N}$, and define the sequence $\xi = (\xi_i)_{i\in\mathbb{N}} \in c_0$ by $\xi_1 := 0$ and

$$\xi_i := \frac{x_{n_j,i}}{j} \qquad \text{for } j \in \mathbb{N} \text{ and } n_j + 1 \leq i \leq n_{j+1} = \kappa(n_j). \tag{2.4.27}$$

This sequence converges to zero because $|x_{n_j,i}| \le 1$ for all i and j. Moreover, it follows from (2.4.25), (2.4.26), and (2.4.27) that

$$\langle y, \Pi_{n_k}(\xi)\rangle = \sum_{i=1}^{n_k} y_i\xi_i = \sum_{j=1}^{k-1} \frac{\langle y, x_{n_j}\rangle}{j} \ge \sum_{j=1}^{k-1} \frac{\varepsilon}{j} \qquad \text{for all } k \in \mathbb{N}. \tag{2.4.28}$$

Now let $\mathbf{p} = (p_1, \dots, p_\ell) \in \mathcal{P}$. If $p_1 = 1$, then $|||\xi|||_{\mathbf{p}} = |||\xi|||_{(p_2,\dots,p_\ell)}$ in the case $\ell \ge 2$ and $|||\xi|||_{\mathbf{p}} = 0$ in the case $\ell = 1$. Thus assume $p_1 \ge 2$ and define

$$\mathcal{J} := \{j \in \mathbb{N} \mid \text{ there exists an } i \in \{1,\dots,\ell\} \text{ such that } n_j < p_i \le n_{j+1}\}.$$

Then $\mathcal{J} \ne \emptyset$. Let $m := \max \mathcal{J}$ and define

$$\begin{aligned} k_j &:= \min\{i \in \{1,\dots,\ell\} \mid n_j < p_i \le n_{j+1}\}, \\ \ell_j &:= \max\{i \in \{1,\dots,\ell\} \mid n_j < p_i \le n_{j+1}\}, \\ \mathbf{p}_j &:= (p_{k_j}, \dots, p_{\ell_j}) \end{aligned}$$

for each $j \in \mathcal{J}$. Then $\{1,\dots,\ell\} = \bigcup_{j\in\mathcal{J}}\{k_j,\dots,\ell_j\}$ because $p_1 \ge 2$, and

$$|||\xi|||_{\mathbf{p}_j} = j^{-1}|||x_{n_j}|||_{\mathbf{p}_j} \le j^{-1}|||x_{n_j}|||_J \le j^{-1}\left\|x_{n_j}\right\|_J = j^{-1}$$

for all $j \in \mathcal{J}$ by (2.4.10) and (2.4.25). Hence

$$\begin{aligned} 2|||\xi|||_{\mathbf{p}}^2 &= \left|\xi_{p_1}\right|^2 + \sum_{m\ne j\in\mathcal{J}} \sum_{i=k_j}^{\ell_j} \left|\xi_{p_i} - \xi_{p_{i+1}}\right|^2 + \sum_{i=k_m}^{\ell_m - 1} \left|\xi_{p_i} - \xi_{p_{i+1}}\right|^2 + \left|\xi_{p_\ell}\right|^2 \\ &\le 2\sum_{j\in\mathcal{J}} \left(\left|\xi_{p_{k_j}}\right|^2 + \sum_{i=k_j}^{\ell_j - 1} \left|\xi_{p_i} - \xi_{p_{i+1}}\right|^2 + \left|\xi_{p_{\ell_j}}\right|^2 \right) \\ &= 4\sum_{j\in\mathcal{J}} |||\xi|||_{\mathbf{p}_j}^2 \\ &\le 4\sum_{j\in\mathcal{J}} \frac{1}{j^2} \\ &\le \frac{2}{3}\pi^2. \end{aligned}$$

Take the supremum over all $\mathbf{p} \in \mathcal{P}$ and use Lemma 2.4.10 to obtain

$$\|\xi\|_J \le \sqrt{2}|||\xi|||_J = \sup_{\mathbf{p}\in\mathcal{P}} \sqrt{2}|||\xi|||_{\mathbf{p}} \le \sqrt{\frac{2}{3}}\pi < \infty$$

and so $\xi \in J$. It then follows from Lemma 2.4.12 that

$$\|\Pi_{n_k}(\xi)\|_J \le \|\xi\|_J \le \sqrt{\frac{2}{3}}\pi$$

for all $k \in \mathbb{N}$, in contradiction to the fact that the sequence $\langle y, \Pi_{n_k}(\xi)\rangle$ is unbounded by (2.4.28). This proves Lemma 2.4.13. □

We are now in a position to prove the main result of this subsection.

Theorem 2.4.14 (James). *The James space J is isometrically isomorphic to its bidual space J^{**} and the image of the canonical inclusion*

$$\iota : J \to J^{**}$$

has codimension one in J^{**}.

Proof. The proof has seven steps.

Step 1. *Let $\Lambda : J^* \to \mathbb{R}$ be a bounded linear functional and define*

$$z_i := \Lambda(e_i) \qquad \textit{for } i \in \mathbb{N}.$$

Then $z := (z_i)_{i\in\mathbb{N}} \in \ell^\infty$,

$$\Lambda(\Pi_n(y)) = \langle y, \Pi_n(z)\rangle \tag{2.4.29}$$

for all $n \in \mathbb{N}$ and all $y \in J^*$, and

$$\Lambda(y) = \lim_{n\to\infty} \langle y, \Pi_n(z)\rangle \tag{2.4.30}$$

for all $y \in J^*$.

For every $i \in \mathbb{N}$ we have

$$|\langle e_i, e_i\rangle| = 1 = \|e_i\|_J$$

and thus

$$1 \le \|e_i\|_{J^*} = \sup_{0\ne x\in J} \frac{|\langle x, e_i\rangle|}{\|x\|_J} = \sup_{0\ne x\in J} \frac{|x_i|}{\|x\|_J} \le \sup_{0\ne x\in J} \frac{\|x\|_\infty}{\|x\|_J} \le 1$$

by Lemma 2.4.9. Hence

$$\|e_i\|_{J^*} = 1 \qquad \text{for all } i \in \mathbb{N}.$$

This implies

$$|z_i| = |\Lambda(e_i)| \le \|\Lambda\| \, \|e_i\|_{J^*} = \|\Lambda\|$$

for all $i \in \mathbb{N}$ and so $z \in \ell^\infty$. Now let $y = (y_i)_{i\in\mathbb{N}} \in J^*$. Then

$$\Lambda(\Pi_n(y)) = \sum_{i=1}^n y_i \Lambda(e_i) = \sum_{i=1}^n y_i z_i = \langle y, \Pi_n(z)\rangle \qquad \text{for all } n \in \mathbb{N}$$

and this proves (2.4.29). It follows from (2.4.23) and (2.4.29) that

$$\begin{aligned}
\lim_{n\to\infty} |\Lambda(y) - \langle y, \Pi_n(z)\rangle| &= \lim_{n\to\infty} |\Lambda(y - \Pi_n(y))| \\
&\le \lim_{n\to\infty} \|\Lambda\| \, \|y - \Pi_n(y)\|_{J^*} \\
&= 0.
\end{aligned}$$

This proves (2.4.30) and Step 1.

Step 2. *Let $\Lambda : J^* \to \mathbb{R}$ and $z \in \ell^\infty$ be as in Step 1. Then*

$$\sup_{p\in\mathcal{P}} \max\{\|z\|_{\mathbf{p}}, |||z|||_{\mathbf{p}}\} \le \|\Lambda\|. \tag{2.4.31}$$

Fix an element $\mathbf{p} = (p_1, \dots, p_k) \in \mathcal{P}$ and choose an integer $n \ge p_k$. Then

$$\begin{aligned}
\max\{\|z\|_{\mathbf{p}}, |||z|||_{\mathbf{p}}\} &= \max\{\|\Pi_n(z)\|_{\mathbf{p}}, |||\Pi_n(z)|||_{\mathbf{p}}\} \\
&\le \|\Pi_n(z)\|_J \\
&= \sup_{0\ne y\in J^*} \frac{|\langle y, \Pi_n(z)\rangle|}{\|y\|_{J^*}} \\
&= \sup_{0\ne y\in J^*} \frac{|\Lambda(\Pi_n(y))|}{\|y\|_{J^*}} \\
&\le \sup_{0\ne y\in J^*} \frac{\|\Lambda\| \, \|\Pi_n(y)\|_{J^*}}{\|y\|_{J^*}} \\
&\le \|\Lambda\|.
\end{aligned}$$

Here the second step follows from Lemma 2.4.10, the third step follows from Lemma 2.4.1, the fourth step follows from (2.4.29), and the last step follows from (2.4.22). This proves (2.4.31) and Step 2.

Step 3. *Let $z = (z_i)_{i\in\mathbb{N}} \in \ell^\infty$ be a bounded sequence such that*

$$\sup_{p\in\mathcal{P}} \max\{\|z\|_{\mathbf{p}}, |||z|||_{\mathbf{p}}\} < \infty. \tag{2.4.32}$$

Then z is a Cauchy sequence and the sequence $x := (x_i)_{i\in\mathbb{N}}$, defined by

$$\lambda := \lim_{j\to\infty} z_j, \qquad x_i := z_i - \lambda \quad \text{for } i \in \mathbb{N}, \tag{2.4.33}$$

is an element of J.

Suppose, by contradiction, that z is not a Cauchy sequence. Then there exist two subsequences $(z_{p_i})_{i\in\mathbb{N}}$ and $(z_{q_i})_{i\in\mathbb{N}}$ converging to different limits. Passing to further subsequences we may assume that $p_i < q_i < p_{i+1}$ for all $i \in \mathbb{N}$ and that there exists a constant $\varepsilon > 0$ such that $|z_{p_i} - z_{q_j}| > \varepsilon$ for all $i, j \in \mathbb{N}$. For $n \in \mathbb{N}$ consider the tuple

$$\mathbf{p}_n := (p_1, q_1, p_2, q_2, \dots, p_n, q_n).$$

Then $\|z\|_{\mathbf{p}_n} > \sqrt{n}\varepsilon$ for all $n \in \mathbb{N}$, in contradiction to (2.4.32). This shows that z is a Cauchy sequence. Now the sequence x in (2.4.33) converges to zero, by definition, and satisfies

$$\|x\|_J = \sup_{\mathbf{p}\in\mathcal{P}} \|x\|_{\mathbf{p}} = \sup_{\mathbf{p}\in\mathcal{P}} \|z\|_{\mathbf{p}} < \infty$$

by (2.4.32). Hence $x \in J$ and this proves Step 3.

Step 4. *Let $z = (z_i)_{i\in\mathbb{N}} \in \ell^\infty$ be a bounded sequence that satisfies* (2.4.32) *and let $\lambda \in \mathbb{R}$ and $x \in J$ be given by* (2.4.33). *Then the limit*

$$\begin{aligned}\Lambda(y) &:= \lim_{n\to\infty} \langle y, \Pi_n(z)\rangle \\ &= \lim_{n\to\infty} \langle y, \Pi_n(x)\rangle + \lambda \lim_{n\to\infty} \sum_{i=1}^n y_i\end{aligned} \tag{2.4.34}$$

exists for every $y \in J^*$ and defines a linear functional $\Lambda : J^* \to \mathbb{R}$.

That the sequence $(\sum_{i=1}^n y_i)_{n\in\mathbb{N}}$ converges for $y \in \ell^1$ is obvious. Moreover, the subspace ℓ^1 is dense in J^* by Lemma 2.4.13, and

$$\begin{aligned}\left|\sum_{i=1}^n y_i\right| &= |\langle \mathbb{1}_n, y\rangle| \\ &\le \|\mathbb{1}_n\|_J \|y\|_{J^*} \\ &= \|y\|_{J^*}\end{aligned}$$

for all $n \in \mathbb{N}$. Here

$$\mathbb{1}_n := (1, \dots, 1, 0, \dots)$$

denotes the sequence whose first n entries are equal to one, followed by zeros. Hence the sequence of functionals

$$J^* \to \mathbb{R} : y \mapsto y_1 + \cdots + y_n$$

is uniformly bounded and converges for all y belonging to the dense subspace $\ell^1 \subset J^*$. Thus it follows from the Banach–Steinhaus Theorem 2.1.5 that the sequence $(\sum_{i=1}^n y_i)_{n\in\mathbb{N}}$ converges for all $y \in J^*$. Hence it follows from Step 3 and Lemma 2.4.12 that the limit in (2.4.34) exists for all $y \in J^*$ and this proves Step 4.

Step 5. *Let $z \in \ell^\infty$ be a sequence that satisfies* (2.4.32) *and let $\Lambda : J^* \to \mathbb{R}$ be the linear map defined by* (2.4.34) *in Step 4. Then Λ is a bounded linear functional on J^* and its norm is*

$$\|\Lambda\| = \sup_{p\in\mathcal{P}} \max\{\|z\|_{\mathbf{p}}, |||z|||_{\mathbf{p}}\}. \tag{2.4.35}$$

We prove that

$$\frac{|\Lambda(y)|}{\|y\|_{J^*}} \le \sup_{p\in\mathcal{P}} \max\{\|z\|_{\mathbf{p}}, |||z|||_{\mathbf{p}}\} \qquad \text{for all } y \in J^* \setminus \{0\}. \tag{2.4.36}$$

To see this, note first that

$$\|x\|_J = \sup_{p\in\mathcal{P}} \|x\|_{\mathbf{p}} = \sup_{p\in\mathcal{P}} \max\{\|x\|_{\mathbf{p}}, |||x|||_{\mathbf{p}}\} \tag{2.4.37}$$

for all $x \in J$ by Lemma 2.4.10.

Next we prove the inequality

$$\sup_{p\in\mathcal{P}} \max\{\|\Pi_n(z)\|_{\mathbf{p}}, |||\Pi_n(z)|||_{\mathbf{p}}\} \le \sup_{p\in\mathcal{P}} \max\{\|z\|_{\mathbf{p}}, |||z|||_{\mathbf{p}}\} \tag{2.4.38}$$

for all $n \in \mathbb{N}$. To see this, fix two elements

$$\mathbf{p} = (p_1, \dots, p_k) \in \mathcal{P}, \qquad n \in \mathbb{N}.$$

Then $\|\Pi_n(z)\|_{\mathbf{p}} = |||\Pi_n(z)|||_{\mathbf{p}} = 0$ whenever $p_1 > n$, and $|||\Pi_n(z)|||_{\mathbf{p}} = |||z|||_{\mathbf{p}}$ and $\|\Pi_n(z)\|_{\mathbf{p}} = \|z\|_{\mathbf{p}}$ whenever $p_k \le n$. Thus assume

$$p_1 \le n < p_k$$

and denote by $\ell \in \{1, \dots, k-1\}$ the largest number such that $p_\ell \le n$. Consider the element

$$\mathbf{q} := (p_1, \dots, p_\ell) \in \mathcal{P}.$$

It satisfies

$$\begin{aligned}
2\,\|\Pi_n(z)\|_{\mathbf{p}}^2 &= 2|||\Pi_n(z)|||_{\mathbf{p}}^2 \\
&= |z_{p_1}|^2 + \sum_{j=1}^{\ell-1} |z_{p_j} - z_{p_{j+1}}|^2 + |z_{p_\ell}|^2 \\
&= 2|||z|||_{\mathbf{q}}^2.
\end{aligned}$$

This proves (2.4.38).

Now take $x = \Pi_n(z)$. Then, by (2.4.37) and (2.4.38),

$$\begin{aligned}
\frac{|\langle y, \Pi_n(z)\rangle|}{\|y\|_{J^*}} &\le \|\Pi_n(z)\|_J \\
&= \sup_{p\in\mathcal{P}} \max\{\|\Pi_n(z)\|_{\mathbf{p}}, |||\Pi_n(z)|||_{\mathbf{p}}\} \\
&\le \sup_{p\in\mathcal{P}} \max\{\|z\|_{\mathbf{p}}, |||z|||_{\mathbf{p}}\}
\end{aligned}$$

for all $y \in J^* \setminus \{0\}$ and all $n \in \mathbb{N}$. Take the limit $n \to \infty$. Then it follows from the definition of Λ in Step 4 via equation (2.4.34) that

$$\begin{aligned}
\frac{|\Lambda(y)|}{\|y\|_{J^*}} &= \lim_{n\to\infty} \frac{|\langle y, \Pi_n(z)\rangle|}{\|y\|_{J^*}} \\
&\le \sup_{p\in\mathcal{P}} \max\{\|z\|_{\mathbf{p}}, |||z|||_{\mathbf{p}}\}
\end{aligned}$$

for all $y \in J^* \setminus \{0\}$. This proves (2.4.36). Thus $\Lambda : J^* \to \mathbb{R}$ is a bounded linear functional. Now take the supremum over all $y \in J^* \setminus \{0\}$ to obtain

$$\begin{aligned}
\|\Lambda\| &= \sup_{0\ne y\in J^*} \frac{|\Lambda(y)|}{\|y\|_{J^*}} \\
&\le \sup_{p\in\mathcal{P}} \max\{\|z\|_{\mathbf{p}}, |||z|||_{\mathbf{p}}\}.
\end{aligned}$$

The converse inequality was established in Step 2 and this proves Step 5.

Step 6. *The canonical inclusion $\iota : J \to J^{**}$ has a codimension-one image.*

By Step 1, Step 2, Step 4, and Step 5, the bidual space of J is naturally isomorphic to the space

$$J^{**} := \left\{ z \in \ell^\infty \,\middle|\, \sup_{\mathbf{p}\in\mathcal{P}} \max\{\|z\|_{\mathbf{p}}, |||z|||_{\mathbf{p}}\} < \infty \right\}.$$

The correspondence assigns to a sequence $z \in J^{**}$ the bounded linear functional $\Lambda : J^* \to \mathbb{R}$ given by (2.4.34). That it is well defined for every $z \in J^{**}$ was proved in Step 4, that it is bounded was proved in Step 5, and that every bounded linear functional on J^* is of this form was proved in Steps 1 and 2. It was also proved in Step 5 that the identification of J^{**} with the dual space of J^* is an isometry with respect to the norm on J^{**}, defined by

$$\|z\|_{J^{**}} := \sup_{\mathbf{p}\in\mathcal{P}} \max\{\|z\|_{\mathbf{p}}, |||z|||_{\mathbf{p}}\} \qquad \text{for } x \in J^{**}.$$

Under this identification, the canonical inclusion $\iota : J \to J^{**}$ is the obvious inclusion of J into J^{**} as a subset. It is an isometric embedding by the general observation in Lemma 2.4.1 (see also Lemma 2.4.10 and equation (2.4.37)). Moreover, the constant sequence $\mathbb{1} := (1, 1, 1, \dots)$ is a unit vector in J^{**} and

$$J^{**} = J \oplus \mathbb{R}\mathbb{1}$$

by Step 3. This proves Step 6.

Step 7. *The map*

$$J \to J^{**} : x = (x_i)_{i\in\mathbb{N}} \mapsto (x_{i+1} - x_1)_{i\in\mathbb{N}}$$

is an isometric isomorphism.

The map is bijective by Step 3. If $x = (x_i)_{i\in\mathbb{N}} \in J$ and $z = (z_i)_{i\in\mathbb{N}} \in J^{**}$ are related by the conditions

$$x_1 = -\lim_{j\to\infty} z_j, \qquad x_{i+1} - x_1 = z_i \qquad \text{for } i \in \mathbb{N},$$

then

$$\|z\|_{(p_1,\dots,p_k)} = \|x\|_{(p_1+1,\dots,p_k+1)}, \qquad |||z|||_{(p_1,\dots,p_k)} = \|x\|_{(1,p_1+1,\dots,p_k+1)}$$

for all $(p_1, \dots, p_k) \in \mathcal{P}$, and hence

$$\|x\|_J = \sup_{\mathbf{p}\in\mathcal{P}} \|x\|_{\mathbf{p}} = \sup_{\mathbf{p}\in\mathcal{P}} \max\{\|z\|_{\mathbf{p}}, |||z|||_{\mathbf{p}}\} = \|z\|_{J^{**}}.$$

This proves Step 7 and Theorem 2.4.14. □

Remark 2.4.15. **(i)** Let X be a real Banach space. A **Schauder basis** of X is a sequence $(e_i)_{i\in\mathbb{N}}$ in X such that, for every $x \in X$, there exists a unique sequence $(x_i)_{i\in\mathbb{N}} \in \mathbb{R}^{\mathbb{N}}$ such that

$$\lim_{n\to\infty}\Big\|x - \sum_{i=1}^{n} x_i e_i\Big\| = 0. \tag{2.4.39}$$

Associated to every Schauder basis $(e_i)_{i\in\mathbb{N}}$ of X is a unique sequence of bounded linear functionals $e_i^* \in X^*$ such that $\langle e_i^*, e_j\rangle = \delta_{ij}$ for all $i, j \in \mathbb{N}$ (see Exercise 2.5.12). Thus the sequence $x_i = \langle e_i^*, x\rangle$ is characterized by the condition (2.4.39). A Schauder basis $(e_i)_{i\in\mathbb{N}}$ is called **normalized** if $\|e_i\| = 1$ for all $i \in \mathbb{N}$. Associated to every Schauder basis $(e_i)_{i\in\mathbb{N}}$ and every $n \in \mathbb{N}$ is a projection $\Pi_n : X \to X$ via

$$\Pi_n(x) := \sum_{i=1}^{n} \langle e_i^*, x\rangle e_i \qquad \text{for } x \in X. \tag{2.4.40}$$

The operator sequence $\Pi_n \in \mathcal{L}(X)$ is bounded by Exercise 2.5.12. A Schauder basis $(e_i)_{i\in\mathbb{N}}$ is called **monotone** if $\|\Pi_n\| \le 1$ for all $n \in \mathbb{N}$. It is called **shrinking** if $\lim_{n\to\infty}\|\Pi_n^*(x^*) - x^*\|_{X^*} = 0$ for every $x^* \in X^*$ and so the sequence $(e_i^*)_{i\in\mathbb{N}}$ is a Schauder basis of X^*. It is called **boundedly complete** if, for every sequence $(x_i)_{i\in\mathbb{N}} \in \mathbb{R}^{\mathbb{N}}$ such that $\sup_{n\in\mathbb{N}}\|\sum_{i=1}^{n} x_i e_i\| < \infty$, the sequence $\sum_{i=1}^{n} x_i e_i$ converges in X.

(ii) By Lemma 2.4.12 the standard basis $(e_i)_{i\in\mathbb{N}}$ of the James space J is a normalized monotone Schauder basis and, by Lemma 2.4.13, it is shrinking. It is not boundedly complete, because the constant sequence $x_i = 1$ satisfies $\|\sum_{i=1}^{n} e_i\|_J = 1$, however, the sequence $\sum_{i=1}^{n} e_i$ does not converge in J.

(iii) The standard basis $(e_i)_{i\in\mathbb{N}}$ of the dual space J^* is again normalized and monotone. One can deduce from Lemma 2.4.13 that this basis is boundedly complete. However, it is not shrinking, because the closure of the span of the dual sequence in J^{**} is the proper subspace $J \subset J^{**}$ by Theorem 2.4.14.

(iv) A theorem of Robert C. James asserts that a Banach space X with a Schauder basis $(e_i)_{i\in\mathbb{N}}$ is reflexive if and only if the basis is both shrinking and boundedly complete.

(v) A Schauder basis $(e_i)_{i\in\mathbb{N}}$ of a Banach space X is called **unconditional** if the sequence $(e_{\sigma(i)})_{i\in\mathbb{N}}$ is a Schauder basis for every bijection $\sigma : \mathbb{N} \to \mathbb{N}$. The James space J does not admit an unconditional Schauder basis.

(vi) There are many examples of Schauder bases, such as any orthonormal basis of a separable Hilbert space, which is always normalized, monotone, unconditional, shrinking, and boundedly complete.

(vii) The reader may verify that the standard basis of ℓ^p for $1 < p < \infty$ is normalized, monotone, unconditional, boundedly complete, and shrinking. For $p = 1$ it is still normalized, monotone, unconditional, and boundedly complete, but no longer shrinking. The Banach space ℓ^∞ does not admit a Schauder basis, because it is not separable.

(viii) There exist separable Banach spaces that do not admit Schauder bases. Examples are Banach spaces that do not have the approximation property (see Exercises 4.2.11 and 4.2.12).

Remark 2.4.16. (i) A **complex structure** on a real Banach space X is a bounded linear operator $I : X \to X$ such that

$$I^2 = -\mathbb{1}.$$

Such a complex structure induces a complex structure $I^{**} : X^{**} \to X^{**}$ on the bidual space such that the canonical inclusion $\iota : X \to X^{**}$ satisfies

$$\iota \circ I = I^{**} \circ \iota.$$

Thus the complex structure descends to the quotient space $X^{**}/\iota(X)$. In the case of the James space $X = J$, this quotient has one real dimension. Hence it does not admit a complex structure, and neither does the James space J.

(ii) Consider the product

$$X := J \times J^*$$

of the James space J with its dual, equipped with the norm

$$\|(x, y)\|_X := \sqrt{\|x\|_J^2 + \|y\|_{J^*}^2} \qquad \text{for } (x, y) \in J \times J^*.$$

By Theorem 2.4.14 the space X is isometrically isomorphic to its dual space. However, it is not reflexive.

(iii) The James space J is an example of a nonreflexive Banach space whose bidual space is separable.

(iv) Another question answered in the negative by the James space is whether a separable Banach space that is isometrically isomorphic to its bidual space must be reflexive. The James space satisfies both conditions, but is not reflexive.

(v) The James space J is an example of an infinite-dimensional Banach space that is not isomorphic to the product space

$$X := J \times J$$

(equipped with any product norm as in Subsection 1.2.3). This is because the canonical inclusion $\iota : X \to X^{**}$ has codimension two by Theorem 2.4.14. Moreover, X admits a complex structure and J does not.

2.5. Problems

Exercise 2.5.1 (Phillips' Lemma). Prove that the subspace

$$c_0 \subset \ell^\infty$$

of all sequences of real numbers that converge to zero is not complemented. This result is due to Phillips [**65**]. The hints are based on [**3**, p. 45].

Hint 1: *There exists an uncountable collection $\{A_i\}_{i\in I}$ of infinite subsets $A_i \subset \mathbb{N}$ such that $A_i \cap A_{i'}$ is a finite set for all $i, i' \in I$ such that $i \neq i'$.*

For example, take

$$I := \mathbb{R} \setminus \mathbb{Q},$$

choose a bijection $\mathbb{N} \to \mathbb{Q} : n \mapsto a_n$, choose sequences $(n_{i,k})_{k\in\mathbb{N}}$ in $\mathbb{N}$, one for each $i \in I$, such that $\lim_{k\to\infty} a_{n_{i,k}} = i$ for all $i \in I = \mathbb{R} \setminus \mathbb{Q}$, and define

$$A_i := \{n_{i,k} \,|\, k \in \mathbb{N}\} \subset \mathbb{N} \qquad \text{for } i \in I.$$

Hint 2: *Let $Q : \ell^\infty \to \ell^\infty$ be a bounded linear operator with $c_0 \subset \ker(Q)$. Then there exists an infinite subset $A \subset \mathbb{N}$ such that $Q(x) = 0$ for every sequence $x = (x_j)_{j\in\mathbb{N}} \in \ell^\infty$ that satisfies $x_j = 0$ for all $j \in \mathbb{N} \setminus A$.*

The set A can be taken as one of the sets A_i in Hint 1. Argue by contradiction and suppose that, for each $i \in I$, there exists a sequence

$$x_i = (x_{ij})_{j\in\mathbb{N}} \in \ell^\infty$$

such that

$$Q(x_i) \neq 0, \qquad \|x_i\|_\infty = 1, \qquad x_{ij} = 0 \text{ for all } j \in \mathbb{N} \setminus A_i.$$

Define the maps $Q_n : \ell^\infty \to \mathbb{R}$ by $Q(x) =: (Q_n(x))_{n\in\mathbb{N}}$ for $x \in \ell^\infty$. For each pair of integers $n, k \in \mathbb{N}$ define the set

$$I_{n,k} := \{i \in I \,|\, |Q_n(x_i)| \geq 1/k\}.$$

Fix a finite set $I' \subset I_{n,k}$ and consider the value of the operator Q on the element

$$x := \sum_{i\in I'} \varepsilon_i x_i, \qquad \varepsilon_i := \mathrm{sign}(Q_n(x_i)).$$

Use the fact that the set

$$B := \{j \in \mathbb{N} \,|\, \exists\, i, i' \in I' \text{ such that } i \neq i' \text{ and } x_{ij} \neq 0 \neq x_{i'j}\}$$

is finite to deduce that $|Q_n(x)| \leq \|Q(x)\| \leq \|Q\|$ and so

$$\#I_{n,k} \leq k\,\|Q\| \qquad \text{for all } n, k \in \mathbb{N}.$$

This contradicts the fact that the set $I = \bigcup_{n,k\in\mathbb{N}} I_{n,k}$ is uncountable.

Hint 3: *There is no bounded linear operator $Q : \ell^\infty \to \ell^\infty$ with* $\ker(Q) = c_0$.

Exercise 2.5.2 (Uniform Boundedness and Open Mappings). The Uniform Boundedness Principle, the Open Mapping Theorem, and the Closed Graph Theorem do not extend to normed vector spaces that are not complete. Let $X = \mathbb{R}^\infty$ be the vector space of sequences $x = (x_i)_{i\in\mathbb{N}}$ of real numbers with only finitely many nonzero terms. For $x \in X$ define

$$\|x\|_1 := \sum_{i=1}^{\infty} |x_i| , \qquad \|x\|_\infty := \sup_{i\in\mathbb{N}} |x_i|_\infty .$$

Prove the following.

(a) For $n \in \mathbb{N}$ define the linear functional $\Lambda_n : X \to \mathbb{R}$ by $\Lambda_n(x) := nx_n$. Then Λ_n is bounded for all $n \in \mathbb{N}$ and $\sup_{n\in\mathbb{N}} |\Lambda_n(x)| < \infty$ for all $x \in X$, however, $\sup_{n\in\mathbb{N}} \|\Lambda_n\|_{X^*} = \infty$ (for either norm on X).

(b) The identity operator $\mathrm{id} : (X, \|\cdot\|_1) \to (X, \|\cdot\|_\infty)$ is bounded but does not have a bounded inverse.

(c) The identity operator $\mathrm{id} : (X, \|\cdot\|_\infty) \to (X, \|\cdot\|_1)$ has a closed graph but is not bounded.

Exercise 2.5.3 (Zabreĭko's Lemma).

(a) Prove **Zabreĭko's Lemma**. *Let X be a Banach space and let $p : X \to \mathbb{R}$ be a seminorm. Then the following are equivalent.*

(i) *p is continuous.*

(ii) *There exists a constant $c > 0$ such that $p(x) \le c\,\|x\|$ for all $x \in X$.*

(iii) *The seminorm p is countably subadditive, i.e.*

$$p\left(\sum_{i=1}^{\infty} x_i\right) \le \sum_{i=1}^{\infty} p(x_i)$$

for every absolutely convergent series $x = \sum_{i=1}^{\infty} x_i$ in X.

Hint: See Definition 2.3.1 for seminorms and Lemma 1.5.1 for absolutely convergent series. To prove that (iii) implies (ii), define the sets

$$A_n := \{x \in X \,|\, p(x) \le n\} , \qquad F_n := \overline{\{x \in X \,|\, p(x) \le n\}}$$

for $n \in \mathbb{N}$. Show that F_n is convex and symmetric for each n. Use the Baire Category Theorem 1.6.4 to prove that there exists an $n \in \mathbb{N}$ such that F_n contains the open unit ball $B := \{x \in X \mid \|x\| < 1\}$. Prove that $B \subset A_n$ by mimicking the proof of the open mapping theorem (Lemma 2.2.3).

(b) Deduce the Uniform Boundedness Principle, the Open Mapping Theorem, and the Closed Graph Theorem from Zabreĭko's Lemma.

Exercise 2.5.4 (Complex Hahn–Banach). The dual space of a complex normed vector space X is the space of bounded complex linear functionals $x^* : X \to \mathbb{C}$. Adapt Corollaries 2.3.23–2.3.26 of the Hahn–Banach Theorem and their proofs to complex normed vector spaces.

Exercise 2.5.5 (Fourier Series of Continuous Functions). This exercise shows that there exist continuous functions whose Fourier series do not converge uniformly. Denote by $C(\mathbb{R}/2\pi\mathbb{Z}, \mathbb{C})$ the space of continuous 2π-periodic complex valued functions $f : \mathbb{R} \to \mathbb{C}$ equipped with the supremum norm.

(a) For $n \in \mathbb{N}$ the **Dirichlet kernel** $D_n \in C(\mathbb{R}/2\pi\mathbb{Z}, \mathbb{C})$ is defined by

$$D_n(t) := \sum_{k=-n}^{n} e^{\mathbf{i}kt} = \frac{\sin((n+\frac{1}{2})t)}{\sin(\frac{1}{2}t)} \qquad \text{for } t \in \mathbb{R}. \tag{2.5.1}$$

Prove that $\|D_n\|_{L^1} \geq \frac{8}{\pi}\sum_{k=1}^{n}\frac{1}{k}$.

(b) The nth Fourier expansion of a function $f \in C(\mathbb{R}/2\pi\mathbb{Z}, \mathbb{C})$ is defined by

$$(\mathcal{F}_n(f))(x) := (D_n * f)(x) = \sum_{k=-n}^{n} \int_0^{2\pi} f(t) e^{\mathbf{i}k(x-t)}\, dt \tag{2.5.2}$$

for $x \in \mathbb{R}$. Prove that the operator $\mathcal{F}_n : C(\mathbb{R}/2\pi\mathbb{Z}, \mathbb{C}) \to C(\mathbb{R}/2\pi\mathbb{Z}, \mathbb{C})$ has the operator norm $\|\mathcal{F}_n\| = \|D_n\|_{L^1}$.

(c) Deduce from the Uniform Boundedness Principle (Theorem 2.1.1) that there exists a function $f \in C(\mathbb{R}/2\pi\mathbb{Z}, \mathbb{C})$ such that the sequence $\mathcal{F}_n(f)$ does not converge uniformly.

Exercise 2.5.6 (Fourier Series of Integrable Functions). The **Fourier coefficients** of a function $f \in L^1([0, 2\pi], \mathbb{C})$ are given by

$$\widehat{f}(k) := \int_0^{2\pi} e^{-\mathbf{i}t} f(t)\, dt \qquad \text{for } k \in \mathbb{Z}, \tag{2.5.3}$$

and the **Fourier series of** f is $\mathscr{F}(f) := \big(\widehat{f}(k)\big)_{k\in\mathbb{Z}}$.

(a) Prove the **Riemann–Lebesgue Lemma**, which asserts that

$$\lim_{|k|\to\infty} \left|\widehat{f}(k)\right| = 0$$

for all $f \in L^1([0, 2\pi], \mathbb{C})$.

(b) Denote by $c_0(\mathbb{Z}, \mathbb{C}) \subset \ell^\infty(\mathbb{Z}, \mathbb{C})$ the closed subspace of all bi-infinite sequences of complex numbers that converge to zero as $|k|$ tends to infinity. Prove that the bounded linear operator $\mathscr{F} : L^1([0, 2\pi], \mathbb{C}) \to c_0(\mathbb{Z}, \mathbb{C})$ has a dense image but is not surjective. **Hint:** Investigate the Fourier coefficients of the Dirichlet kernels in Exercise 2.5.5.

Exercise 2.5.7 (Banach Limits). Let ℓ^∞ be the Banach space of bounded sequences of real numbers with the supremum norm as in part (ii) of Example 1.1.3 and define the shift operator $T : \ell^\infty \to \ell^\infty$ by

$$Tx := (x_{n+1})_{n\in\mathbb{N}} \qquad \text{for } x = (x_n)_{n\in\mathbb{N}} \in \ell^\infty.$$

Consider the subspace

$$Y := \operatorname{im}(\mathrm{id} - T) = \{x - Tx \,|\, x \in \ell^\infty\}.$$

Prove the following.

(a) The subspace $c_0 \subset \ell^\infty$ of all sequences that converge to zero is contained in the closure of Y.

(b) Let $\mathbb{1} = (1, 1, 1, \dots) \in \ell^\infty$ be the constant sequence with entries 1. Prove that $\sup_{n\in\mathbb{N}} |1 + x_{n+1} - x_n| \geq 1$ for all $x \in \ell^\infty$ and deduce that

$$d(\mathbb{1}, Y) = \inf_{y\in Y} \|\mathbb{1} - y\|_\infty = 1.$$

(c) By the Hahn–Banach Theorem 2.3.22 there exists a bounded linear functional $\Lambda : \ell^\infty \to \mathbb{R}$ such that

$$\Lambda(\mathbb{1}) = 1, \qquad \|\Lambda\| = 1, \qquad \Lambda(x - Tx) = 0 \quad \text{for all } x \in \ell^\infty. \tag{2.5.4}$$

Prove that any such functional has the following properties.

(i) $\Lambda(Tx) = \Lambda(x)$ for all $x \in \ell^\infty$.

(ii) If $x \in \ell^\infty$ satisfies $x_n \geq 0$ for all $n \in \mathbb{N}$, then $\Lambda(x) \geq 0$.

(iii) $\liminf_{n\to\infty} x_n \leq \Lambda(x) \leq \limsup_{n\to\infty} x_n$ for all $x \in \ell^\infty$.

(iv) If $x \in \ell^\infty$ converges, then $\Lambda(x) = \lim_{n\to\infty} x_n$.

(d) Let Λ be as in (c). Find $x, y \in \ell^\infty$ such that $\Lambda(xy) \neq \Lambda(x)\Lambda(y)$. **Hint:** Consider the sequence $x_n := (-1)^n$ and show that $\Lambda(x) = 0$.

(e) Let Λ be as in (c). Prove that there does not exist a sequence $y \in \ell^1$ such that $\Lambda(x) = \sum_{n=1}^\infty x_n y_n$ for all $x \in \ell^\infty$. **Hint:** Any such sequence would have nonnegative entries $y_n \geq 0$ by part (ii) in (c) and satisfy $\sum_{n=1}^\infty y_n = 1$. Hence $\sum_{n=1}^N y_n > 0$ for some $N \in \mathbb{N}$ in contradiction to part (iv) in (c).

Exercise 2.5.8 (Minkowski Functionals). Let X be a normed vector space and let $C \subset X$ be a convex subset such that $0 \in C$. The **Minkowski functional** of C is the function

$$p : X \to [0, \infty]$$

defined by

$$p(x) := \inf\left\{\lambda > 0 \,|\, \lambda^{-1}x \in C\right\} \qquad \text{for } x \in X. \tag{2.5.5}$$

The convex set C is called **absorbing** if, for every $x \in X$, there is a $\lambda > 0$ such that $\lambda^{-1}x \in C$. Let p be the Minkowski functional of C.

(a) Prove that $p(x+y) \leq p(x) + p(y)$ and $p(\lambda x) = \lambda p(x)$ for all $x, y \in X$ and all $\lambda > 0$.

(b) Prove that C is absorbing if and only if p takes values in $[0, \infty)$ and hence is a quasi-seminorm (see Definition 2.3.1).

(c) Suppose C is absorbing. Find conditions on C which ensure that p is a seminorm or a norm. Do this both for real scalars and complex scalars.

(d) Prove that p is continuous if and only if zero is an interior point of C. In this case, show that $\mathrm{int}(C) = p^{-1}([0,1))$ and $\overline{C} = p^{-1}([0,1])$.

Exercise 2.5.9 (Reflexive Banach Spaces). Let X be a normed vector space and let $Y \subset X$ be a closed subspace. Assume Y and X/Y are reflexive. Prove that X is reflexive.

Exercise 2.5.10 (Schatten's Projective Tensor Product). Let X and Y be real normed vector spaces.

(a) For every normed vector space Z, the space $\mathcal{B}(X, Y; Z)$ of bounded bilinear maps $B : X \times Y \to Z$ is a normed vector space with the norm

$$\|B\| := \sup_{\substack{x \in X \setminus \{0\} \\ y \in Y \setminus \{0\}}} \frac{\|B(x,y)\|_Z}{\|x\|_X \, \|y\|_Y} \qquad \text{for } B \in \mathcal{B}(X, Y; Z).$$

(b) The map

$$\mathcal{B}(X, Y; Z) \to \mathcal{L}(X, \mathcal{L}(Y, Z)) : B \mapsto (x \mapsto B(x, \cdot))$$

is an isometric isomorphism.

(c) Associated to every pair $(x, y) \in X \times Y$ is a linear functional

$$x \otimes y \in \mathcal{B}(X, Y; \mathbb{R})^*$$

defined by $\langle x \otimes y, B \rangle := B(x, y)$ for $B \in \mathcal{B}(X, Y; \mathbb{R})$. It satisfies

$$\|x \otimes y\| = \|x\|_X \, \|y\|_Y \, .$$

Hint: Use the Hahn–Banach Theorem to prove the inequality $\|x \otimes y\| \geq \|x\|_X \, \|y\|_Y$. Namely, consider the bilinear functional $B : X \times Y \to \mathbb{R}$, defined by $B(x, y) := \langle x^*, x \rangle \langle y^*, y \rangle$ for suitable elements $x^* \in X^*$ and $y^* \in Y^*$ of the dual spaces.

(d) Let $X \otimes Y \subset \mathcal{B}(X, Y; \mathbb{R})^*$ be the smallest closed subspace containing the image of the bilinear map $X \times Y \to \mathcal{B}(X, Y; \mathbb{R})^* : (x, y) \mapsto x \otimes y$ in (c). Then, for every normed vector space Z, the map

$$\mathcal{L}(X \otimes Y, Z) \to \mathcal{B}(X, Y; Z) : A \mapsto B_A$$

defined by $B_A(x, y) := A(x \otimes y)$ for $x, y \in X$ and $A \in \mathcal{L}(X \otimes Y, Z)$ is an isometric isomorphism.

Exercise 2.5.11 (Strict Convexity and Hahn–Banach).

(a) Prove **Ruston's Theorem:** *The following properties of a normed vector space X are equivalent.*

(i) *If $x, y \in X$ satisfy $x \neq y$ and $\|x\| = \|y\| = 1$, then $\|x + y\| < 2$.*

(ii) *If $x, y \in X$ satisfy $x \neq 0 \neq y$ and $\|x + y\| = \|x\| + \|y\|$, then $x = \lambda y$ for some $\lambda > 0$.*

(iii) *If $x^* \in X^*$ is a nonzero bounded linear functional then there exists at most one element $x \in X$ such that $\|x\| = 1$ and $\langle x^*, x\rangle = \|x^*\|$.*

The normed vector space X is called **strictly convex** *if it satisfies these equivalent conditions.*

Condition (i) says that the unit sphere contains no nontrivial line segment. Condition (ii) says that equality in the triangle inequality occurs only in the trivial situation. Condition (iii) says that the support hyperplane $H_{x^*} := \{x \in X \mid \langle x^*, x\rangle = \|x^*\|\}$ meets the unit sphere in at most one point. (Note that $\inf_{x \in H_{x^*}} \|x\| = 1$.)

(b) For which p is $L^p([0,1])$ strictly convex? Is $C([0,1])$ strictly convex?

(c) If X is a normed vector space such that X^* is strictly convex, $Y \subset X$ is a linear subspace, and $y^* : Y \to \mathbb{R}$ is a bounded linear functional, then there is a unique $x^* \in X^*$ such that $x^*|_Y = y^*$ and $\|x^*\| = \|y^*\|$.

Exercise 2.5.12 (Schauder Bases). Let X be a separable real Banach space and let $(e_i)_{i\in\mathbb{N}}$ be a **Schauder basis** of X. This means that, for each element $x \in X$, there exists a unique sequence $(x_i)_{i\in\mathbb{N}}$ of real numbers such that

$$\lim_{n\to\infty} \left\| x - \sum_{i=1}^{n} x_i e_i \right\| = 0. \tag{2.5.6}$$

Let $n \in \mathbb{N}$ and define the map $\Pi_n : X \to X$ by

$$\Pi_n(x) := \sum_{i=1}^{n} x_i e_i \tag{2.5.7}$$

for $x \in X$, where $(x_i)_{i\in\mathbb{N}}$ is the unique sequence that satisfies (2.5.6).

(a) Prove that the operators $\Pi_n : X \to X$ are linear and satisfy

$$\Pi_n \circ \Pi_m = \Pi_m \circ \Pi_n = \Pi_m \tag{2.5.8}$$

for all integers $n \geq m \geq 1$. In particular, they are projections.

(b) Define a map $X \to [0,\infty) : x \mapsto |||x|||$ by the formula

$$|||x||| := \sup_{n\in\mathbb{N}} \|\Pi_n(x)\| \qquad \text{for } x \in X. \tag{2.5.9}$$

Prove that this is a norm and that $\|x\| \le |||x|||$ for all $x \in X$.

(c) Prove that $(X, |||\cdot|||)$ is a Banach space. **Hint:** Let $(x_k)_{k\in\mathbb{N}}$ be a Cauchy sequence in $(X, |||\cdot|||)$. Then $(x_k)_{k\in\mathbb{N}}$ is a Cauchy sequence in $(X, \|\cdot\|)$ by (b). Hence there is an $x \in X$ such that $\lim_{k\to\infty} \|x - x_k\| = 0$. Also, $(\Pi_n(x_k))_{k\in\mathbb{N}}$ is a Cauchy sequence in $(X, \|\cdot\|)$ for all n. Thus there is a sequence $(\xi_n)_{n\in\mathbb{N}}$ in X such that $\lim_{k\to\infty} \|\xi_n - \Pi_n(x_k)\| = 0$ for all $n \in \mathbb{N}$. Prove that

$$\Pi_m(\xi_n) = \xi_m \qquad \text{for all integers } n \ge m \ge 1. \tag{2.5.10}$$

(The restriction of Π_m to every finite-dimensional subspace is continuous.) Let $\varepsilon > 0$. Choose $k_0 \in \mathbb{N}$ such that $|||x_k - x_\ell||| < \varepsilon/3$ for all $k, \ell \ge k_0$. Then choose $n_0 \in \mathbb{N}$ such that $\|x_{k_0} - \Pi_n(x_{k_0})\| < \varepsilon/3$ for all $n \ge n_0$. Then

$$\begin{aligned}\|x - \xi_n\| &= \lim_{k\to\infty} \|x_k - \Pi_n(x_k)\| \\ &\le \lim_{k\to\infty} \Big(2|||x_k - x_{k_0}||| + \|x_{k_0} - \Pi_n(x_{k_0})\|\Big) \\ &< \varepsilon\end{aligned}$$

for $n \ge n_0$. Deduce that $\xi_n = \Pi_n(x)$ for all n and $\lim_{k\to\infty} |||x - x_k||| = 0$.

(d) Prove that there exists a constant $c > 0$ such that

$$\sup_{n\in\mathbb{N}} \|\Pi_n(x)\| \le c\,\|x\| \qquad \text{for all } x \in X. \tag{2.5.11}$$

Hint: Use parts (b) and (c) and the Open Mapping Theorem 2.2.1.

Exercise 2.5.13 (The Canonical Inclusion). Let X be a normed vector space and let $\iota_X : X \to X^{**}$ be the canonical inclusion defined by (2.4.1).

(a) Show that $(\iota_X)^*\iota_{X^*} = \mathrm{id}_{X^*}$ and determine the kernel of the projection

$$P := \iota_{X^*}(\iota_X)^* : X^{***} \to X^{***}.$$

(b) Assume X is complete. Show that X is reflexive if and only if

$$\iota_{X^*}(\iota_X)^* = \mathrm{id}_{X^{***}}.$$

(c) Linton's Pullback. *Let $Y \subset X$ be a closed subspace and let $j : Y \to X$ be the obvious inclusion. Then $\iota_X \circ j = j^{**} \circ \iota_Y : Y \to X^{**}$. This map is an isometric embedding of Y into X^{**} whose image is*

$$\iota_X \circ j(Y) = j^{**} \circ \iota_Y(Y) = \iota_X(X) \cap j^{**}(Y^{**}) \subset X^{**}.$$

(d) Deduce from Linton's Pullback that Y is reflexive whenever X is reflexive.

(e) Show that X is reflexive if and only if $\iota_{X^{**}} = (\iota_X)^{**}$.

Note. This exercise requires the notion of the dual operator, introduced in Definition 4.1.1 below.

Chapter 3

The Weak and Weak* Topologies

This chapter is devoted to the study of the weak topology on a Banach space X and the weak* topology on its dual space X^*. With these topologies X and X^* are locally convex Hausdorff topological vector spaces and the elementary properties of such spaces are discussed in Section 3.1. In particular, it is shown that the closed unit ball in X^* is the weak* closure of the unit sphere, and that a linear functional on X^* is continuous with respect to the weak* topology if and only if it belongs to the image of the canonical embedding $\iota : X \to X^{**}$. The central result of this chapter is the Banach–Alaoglu Theorem in Section 3.2 which asserts that the unit ball in the dual space X^* is compact with respect to the weak* topology. This theorem has important consequences in many fields of mathematics. Further properties of the weak* topology on the dual space are established in Section 3.3. It is shown that a linear subspace of X^* is weak* closed if and only if its intersection with the closed unit ball is weak* closed. A consequence of the Banach–Alaoglu Theorem is that the unit ball in a reflexive Banach space is weakly compact. A theorem of Eberlein–Šmulyan asserts that this property characterizes reflexive Banach spaces (Section 3.4). The Kreĭn–Milman Theorem in Section 3.5 asserts that every nonempty compact convex subset of a locally convex Hausdorff topological vector space is the convex hull of its extremal points. Combining the Kreĭn–Milman Theorem with the Banach–Alaoglu Theorem, one can prove that every homeomorphism of a compact metric space admits an invariant ergodic Borel probability measure. Some properties of such ergodic measures are explored in Section 3.6.

3.1. Topological Vector Spaces

3.1.1. Definition and Examples. Recall that the **product topology** on a product $X \times Y$ of two topological spaces X and Y is defined as the weakest topology on $X \times Y$ that contains all subsets of the form $U \times V$ where $U \subset X$ and $V \subset Y$ are open. Equivalently, it is the weakest topology on $X \times Y$ such that the projections $\pi_X : X \times Y \to X$ and $\pi_Y : X \times Y \to Y$ are continuous.

Definition 3.1.1 (Topological Vector Space). A **topological vector space** is a pair $(X, \mathscr{U})$ where X is a real vector space and $\mathscr{U} \subset 2^X$ is a topology such that the structure maps

$$X \times X \to X : (x, y) \mapsto x + y, \qquad \mathbb{R} \times X \to X : (\lambda, x) \mapsto \lambda x$$

are continuous with respect to the product topologies on $X \times X$ and $\mathbb{R} \times X$. A topological vector space $(X, \mathscr{U})$ is called **locally convex** if, for every open set $U \subset X$ and every $x \in U$, there is an open set $V \subset X$ such that

$$x \in V \subset U, \qquad V \text{ is convex.}$$

Example 3.1.2 (Strong Topology). A normed vector space $(X, \|\cdot\|)$ is a topological vector space with the topology $\mathscr{U}^{\mathrm{s}} := \mathscr{U}(X, \|\cdot\|)$ induced by the norm as in Definition 1.1.2. This is sometimes called the **strong topology** or **norm topology** to distinguish it from other weaker topologies discussed below.

Example 3.1.3 (Smooth Functions). The space $X := C^\infty(\Omega)$ of smooth functions on an open subset $\Omega \subset \mathbb{R}^n$ is a locally convex Hausdorff topological vector space. The topology is given by uniform convergence with all derivatives on compact sets and is induced by the complete metric

$$d(f, g) := \sum_{\ell=1}^{\infty} 2^{-\ell} \frac{\|f - g\|_{C^\ell(K_\ell)}}{1 + \|f - g\|_{C^\ell(K_\ell)}}.$$

Here $K_\ell \subset \Omega$ is an exhausting sequence of compact sets.

Example 3.1.4. Let X be a real vector space. Then $(X, \mathscr{U})$ is a topological vector space with $\mathscr{U} := \{\emptyset, X\}$, but not with the discrete topology.

Example 3.1.5 (Convergence in Measure). Let $(M, \mathcal{A}, \mu)$ be a measure space such that $\mu(M) < \infty$, denote by $\mathcal{L}^0(\mu)$ the vector space of all real valued measurable functions on M, and define

$$L^0(\mu) := \mathcal{L}^0(\mu)/\sim,$$

where the equivalence relation is given by equality almost everywhere. Define a metric on $L^0(\mu)$ by

$$d(f, g) := \int_M \frac{|f - g|}{1 + |f - g|} \, d\mu \qquad \text{for } f, g \in \mathcal{L}^0(\mu).$$

Then $L^0(\mu)$ is a topological vector space with the topology induced by d. A sequence $f_n \in L^0(\mu)$ converges to $f \in L^0(\mu)$ in this topology if and only if it **converges in measure**, i.e.

$$\lim_{n\to\infty} \mu(\{x \in M \,|\, |f_n(x) - f(x)| > \varepsilon\}) = 0 \qquad \text{for all } \varepsilon > 0.$$

The topological vector space $L^0(\mu)$ with the topology of convergence in measure is not locally convex, in general. **Exercise:** Prove that every nonempty open convex subset of $L^0([0,1])$ is the whole space. Deduce that every continuous linear functional $\Lambda : L^0([0,1]) \to \mathbb{R}$ vanishes.

An important class of topological vector spaces is determined by sets of linear functionals as follows. Fix a real vector space X and let

$$\mathcal{F} \subset \{f : X \to \mathbb{R} \,|\, f \text{ is linear}\}$$

be any set of linear functionals on X. Define $\mathscr{U}_{\mathcal{F}} \subset 2^X$ to be the weakest topology on X such that every linear functional $f \in \mathcal{F}$ is continuous. Then the pre-image of an open interval under any of the linear functionals $f \in \mathcal{F}$ is an open subset of X. Hence so is the set

$$V := \{x \in X \,|\, a_i < f_i(x) < b_i \text{ for } i = 1, \dots, m\}$$

for all integers $m \in \mathbb{N}$, all $f_1, \dots, f_m \in \mathcal{F}$, and all $2m$-tuples of real numbers $a_1, \dots, a_m, b_1, \dots, b_m$ such that $a_i < b_i$ for $i = 1, \dots, m$. Denote the collection of all subsets of X of this form by

$$\mathscr{V}_{\mathcal{F}} := \left\{ \bigcap_{i=1}^m f_i^{-1}((a_i, b_i)) \;\middle|\; \begin{array}{l} m \in \mathbb{N}, f_1, \dots, f_m \in \mathcal{F}, \\ a_1, \dots, a_m, b_1, \dots, b_m \in \mathbb{R}, \\ a_i < b_i \text{ for } i = 1, \dots, m \end{array} \right\}. \tag{3.1.1}$$

Lemma 3.1.6. *Let X be a real vector space, let $\mathcal{F} \subset \mathbb{R}^X$ be a set of real valued linear functionals on X, and let $\mathscr{U}_{\mathcal{F}} \subset 2^X$ be the weakest topology on X such that all elements of $\mathcal{F}$ are continuous. Then the following hold.*

(i) *The collection $\mathscr{V}_{\mathcal{F}}$ in* (3.1.1) *is a basis for the topology $\mathscr{U}_{\mathcal{F}}$, i.e.*

$$\mathscr{U}_{\mathcal{F}} = \{U \subset X \,|\, \forall\, x \in U\; \exists\, V \in \mathscr{V}_{\mathcal{F}} \text{ such that } x \in V \subset U\}. \tag{3.1.2}$$

(ii) *$(X, \mathscr{U}_{\mathcal{F}})$ is a locally convex topological vector space.*

(iii) *A sequence $x_n \in X$ converges to an element $x_0 \in X$ with respect to the topology $\mathscr{U}_{\mathcal{F}}$ if and only if $f(x_0) = \lim_{n\to\infty} f(x_n)$ for all $f \in \mathcal{F}$.*

(iv) *The topological space $(X, \mathscr{U}_{\mathcal{F}})$ is Hausdorff if and only if $\mathcal{F}$* **separates points**, *i.e. for every nonzero vector $x \in X$ there exists a linear functional $f \in \mathcal{F}$ such that $f(x) \neq 0$.*

Proof. Part (i) is an exercise with hints. Define the set $\mathscr{U}_{\mathcal{F}} \subset 2^X$ by the right-hand side of equation (3.1.2). Then it follows directly from the definitions that $\mathscr{U}_{\mathcal{F}} \subset 2^X$ is a topology, that every linear function $f : X \to \mathbb{R}$ in $\mathcal{F}$ is continuous with respect to this topology, and that every other topology $\mathscr{U} \subset 2^X$ with respect to which each element of $\mathcal{F}$ is continuous must contain $\mathscr{V}_{\mathcal{F}}$ and hence also $\mathscr{U}_{\mathcal{F}}$. This proves part (i).

We prove part (ii). We prove first that scalar multiplication is continuous with respect to $\mathscr{U}_{\mathcal{F}}$. Fix a set $V \in \mathscr{V}_{\mathcal{F}}$ and let $\lambda_0 \in \mathbb{R}$ and $x_0 \in X$ such that $\lambda_0 x_0 \in V$. Then it follows from the definition of $\mathscr{V}_{\mathcal{F}}$ in (3.1.1) that there exists a constant $\delta > 0$ such that

$$\delta \neq |\lambda_0|, \qquad (\lambda_0 - \delta)x_0 \in V, \qquad (\lambda_0 + \delta)x_0 \in V.$$

Define

$$U := \frac{1}{\lambda_0 - \delta} V \cap \frac{1}{\lambda_0 + \delta} V.$$

Then $U \in \mathscr{V}_{\mathcal{F}}$ and $x_0 \in U$. Moreover, if $x \in U$ and $\lambda \in \mathbb{R}$ satisfies

$$|\lambda - \lambda_0| < \delta,$$

then $(\lambda_0 - \delta)x \in V$ and $(\lambda_0 + \delta)x \in V$ and hence $\lambda x \in V$, because V is convex. This shows that scalar multiplication is continuous.

We prove that addition is continuous. Fix an element $W \in \mathscr{V}_{\mathcal{F}}$ and let $x_0, y_0 \in X$ such that $x_0 + y_0 \in W$. Define the sets

$$U := \frac{1}{2}(x_0 - y_0) + \frac{1}{2}W, \qquad V := \frac{1}{2}(y_0 - x_0) + \frac{1}{2}W.$$

Then $U, V \in \mathscr{V}_{\mathcal{F}}$ by (3.1.1). Moreover, $x_0 \in U$, $y_0 \in V$, and, for all $x \in U$ and all $y \in V$, we have $x + y \in W$ because W is convex. This shows that addition is continuous. Hence $(X, \mathscr{U}_{\mathcal{F}})$ is a topological vector space. That $(X, \mathscr{U}_{\mathcal{F}})$ is locally convex follows from the fact that the elements of $\mathscr{V}_{\mathcal{F}}$ are all convex sets. This proves part (ii).

We prove part (iii). Fix a sequence $(x_n)_{n\in\mathbb{N}}$ in X and an element $x_0 \in X$. Assume x_n converges to x_0 with respect to the topology $\mathscr{U}_{\mathcal{F}}$. Let $f \in \mathcal{F}$ and fix a constant $\varepsilon > 0$. Then the set

$$U := \{x \in X \mid |f(x) - f(x_0)| < \varepsilon\}$$

is an element of $\mathscr{V}_{\mathcal{F}}$ and hence of $\mathscr{U}_{\mathcal{F}}$. Since $x_0 \in U$, there exists a positive integer n_0 such that $x_n \in U$ for every integer $n \geq n_0$. Thus we have proved

$$\forall\, f \in \mathcal{F}\ \forall\, \varepsilon > 0\ \exists\, n_0 \in \mathbb{N}\ \forall\, n \in \mathbb{N} : \big(n \geq n_0 \implies |f(x_n) - f(x_0)| < \varepsilon\big).$$

This means that

$$\lim_{n\to\infty} f(x_n) = f(x_0)$$

for all $f \in \mathcal{F}$.

Conversely suppose that

$$\lim_{n\to\infty} f(x_n) = f(x_0)$$

for all $f \in \mathcal{F}$ and fix a set $U \in \mathscr{U}_{\mathcal{F}}$ such that

$$x_0 \in U.$$

Then there exists a set

$$V = \bigcap_{i=1}^{m} f_i^{-1}((a_i, b_i)) \in \mathscr{V}_{\mathcal{F}}$$

such that $x_0 \in V \subset U$. This means that $a_i < f_i(x_0) < b_i$ for $i = 1, \dots, m$. Since $\lim_{n\to\infty} f_i(x_n) = f_i(x_0)$ for $i = 1, \dots, m$, there is a positive integer n_0 such that $a_i < f_i(x_n) < b_i$ for every integer $n \geq n_0$ and every $i \in \{1, \dots, m\}$. Thus $x_n \in V \subset U$ for every integer $n \geq n_0$ and this proves part (iii).

We prove part (iv). Assume first that X is Hausdorff and let $x \in X \setminus \{0\}$. Then there exists an open set $U \subset X$ such that

$$0 \in U, \qquad x \notin U.$$

Choose a set $V = \bigcap_{i=1}^{m} f_i^{-1}((a_i, b_i)) \in \mathscr{V}_{\mathcal{F}}$ such that

$$0 \in V \subset U.$$

Since $0 \in V$ it follows that $a_i < 0 < b_i$ for all $i \in \{1, \dots, m\}$. Since $x \notin V$, there exists index $i \in \{1, \dots, m\}$ such that $f_i(x) \notin (a_i, b_i)$ and so $f_i(x) \neq 0$.

Conversely suppose that, for every $x \in X$, there exists an element $f \in \mathcal{F}$ such that $f(x) \neq 0$. Let $x_0, x_1 \in X$ such that $x_0 \neq x_1$ and choose $f \in \mathcal{F}$ such that $f(x_1 - x_0) \neq 0$. Choose $\varepsilon > 0$ such that $2\varepsilon < |f(x_1 - x_0)|$ and consider the sets

$$U_i := \{x \in X \,|\, |f(x - x_i)| < \varepsilon\}$$

for $i = 0, 1$. Then $U_0, U_1 \in \mathscr{V}_{\mathcal{F}} \subset \mathscr{U}_{\mathcal{F}}$, $x_0 \in U_0$, $x_1 \in U_1$, and $U_0 \cap U_1 = \emptyset$. This proves part (iv) and Lemma 3.1.6. □

Example 3.1.7 (Product Topology). Let I be any set and consider the space $X := \mathbb{R}^I$ of all functions $x : I \to \mathbb{R}$. This is a real vector space. For $i \in I$ denote the evaluation map at i by $\pi_i : \mathbb{R}^I \to \mathbb{R}$, i.e. $\pi_i(x) := x(i)$ for $x \in \mathbb{R}^I$. Then $\pi_i : X \to \mathbb{R}$ is a linear functional for every $i \in I$. Let

$$\pi := \{\pi_i \,|\, i \in I\}$$

be the collection of all these evaluation maps and denote by $\mathscr{U}_\pi$ the weakest topology on X such that the projection π_i is continuous for every $i \in I$. By Lemma 3.1.6 this topology is given by (3.1.1) and (3.1.2). It is called the **product topology** on $\mathbb{R}^I$. Thus $\mathbb{R}^I$ is a locally convex Hausdorff topological vector space with the product topology.

Example 3.1.8 (Weak Topology). Let X be a real normed vector space.

(i) *The* **weak topology** *on X is the weakest topology $\mathscr{U}^{\mathrm{w}} \subset 2^X$ with respect to which every bounded linear functional $\Lambda : X \to \mathbb{R}$ is continuous.* It is the special case of the topology $\mathscr{U}_{\mathcal{F}} \subset 2^X$ in Lemma 3.1.6, where $\mathcal{F} := X^*$ is the dual space. By Corollary 2.3.23 the dual space separates points, i.e. for every $x \in X \setminus \{0\}$ there is an $x^* \in X^*$ such that $\langle x^*, x\rangle \neq 0$. Hence Lemma 3.1.6 asserts that $(X, \mathscr{U}^{\mathrm{w}})$ is a locally convex Hausdorff topological vector space.

(ii) By Theorem 1.2.2 every bounded linear functional is continuous with respect to the **strong topology** $\mathscr{U}^{\mathrm{s}} := \mathscr{U}(X, \|\cdot\|)$ in Definition 1.1.2. Hence

$$\mathscr{U}^{\mathrm{w}} \subset \mathscr{U}^{\mathrm{s}}.$$

The weak and strong topologies agree when X is finite-dimensional.

(iii) Let $(x_n)_{n\in\mathbb{N}}$ be a sequence in X and let $x \in X$. Then Lemma 3.1.6 asserts that x_n **converges weakly** to x (i.e. in the weak topology) if and only if

$$\langle x^*, x\rangle = \lim_{n\to\infty} \langle x^*, x_n\rangle \qquad \text{for all } x^* \in X^*.$$

In this case we write $x_n \stackrel{\mathrm{w}}{\rightharpoonup} x$ or $x = \mathrm{w}\text{-}\lim_{n\to\infty} x_n$.

Example 3.1.9 (Weak* Topology). Let X be a real normed vector space and let $X^* = \mathcal{L}(X, \mathbb{R})$ be its dual space.

(i) *The* **weak* topology** *on X^* is the weakest topology $\mathscr{U}^{\mathrm{w}^*} \subset 2^{X^*}$ with respect to which the linear functional $\iota(x) : X^* \to \mathbb{R}$ in (2.4.1) is continuous for all $x \in X$.* It is the special case of the topology $\mathscr{U}_{\mathcal{F}} \subset 2^{X^*}$ in Lemma 3.1.6, where $\mathcal{F} := \iota(X) \subset X^{**}$. This collection of linear functionals separates points, i.e. for every $x^* \in X^* \setminus \{0\}$ there is an element $x \in X$ such that $\langle x^*, x\rangle \neq 0$. Hence Lemma 3.1.6 asserts that $(X^*, \mathscr{U}^{\mathrm{w}^*})$ is a locally convex Hausdorff topological vector space.

(ii) Denote by $\mathscr{U}^{\mathrm{s}} \subset 2^{X^*}$ the strong topology induced by the norm, and denote by $\mathscr{U}^{\mathrm{w}} \subset 2^{X^*}$ the weak topology in Example 3.1.8. Then

$$\mathscr{U}^{\mathrm{w}^*} \subset \mathscr{U}^{\mathrm{w}} \subset \mathscr{U}^{\mathrm{s}}.$$

These weak and weak* topologies agree when X is a reflexive Banach space.

(iii) Let $(x_n^*)_{n\in\mathbb{N}}$ be a sequence in X^* and let $x^* \in X^*$. Then Lemma 3.1.6 asserts that x_n^* **converges to x^* in the weak* topology** if and only if

$$\langle x^*, x\rangle = \lim_{n\to\infty} \langle x_n^*, x\rangle \qquad \text{for all } x \in X.$$

In this case we write $x_n^* \stackrel{\mathrm{w}^*}{\rightharpoonup} x^*$ or $x^* = \mathrm{w}^*\text{-}\lim_{n\to\infty} x_n^*$.

3.1.2. Convex Sets. This subsection picks up the topic of separating a pair of nonempty disjoint convex sets by a hyperplane. For normed vector spaces this problem was examined in Subsection 2.3.3. The main result (Theorem 2.3.10) and its proof carry over almost verbatim to topological vector spaces (see Theorem 3.1.11). The next lemma shows that the closure and interior of a convex subset of a topological vector space are again convex.

Lemma 3.1.10. *Let X be a topological vector space and let $K \subset X$ be a convex subset. Then the closure $\overline{K}$ and the interior $\mathrm{int}(K)$ are convex subsets of X. Moreover, if $\mathrm{int}(K) \neq \emptyset$, then $K \subset \overline{\mathrm{int}(K)}$.*

Proof. We prove that $\mathrm{int}(K)$ is convex. Let $x_0, x_1 \in \mathrm{int}(K)$, choose a real number $0 < \lambda < 1$, and define

$$x_\lambda := (1-\lambda)x_0 + \lambda x_1.$$

Choose open sets $U_0, U_1 \subset X$ such that $x_0 \subset U_0 \subset K$ and $x_1 \subset U_1 \subset K$ and define

$$U := (U_0 - x_0) \cap (U_1 - x_1) = \{x \in X \,|\, x_0 + x \in U_0,\ x_1 + x \in U_1\}.$$

Then $U \subset X$ is an open set containing the origin such that

$$x_0 + U \subset K, \qquad x_1 + U \subset K.$$

Since K is convex, this implies that $x_\lambda + U$ is an open subset of K containing x_λ. Hence $x_\lambda \in \mathrm{int}(K)$.

We prove that the closure $\overline{K}$ is convex. Let $x_0, x_1 \in \overline{K}$, choose a real number $0 < \lambda < 1$, and define $x_\lambda := (1-\lambda)x_0 + \lambda x_1$. Let U be an open neighborhood of x_λ. Then the set

$$W := \{(y_0, y_1) \in X \times X \,|\, (1-\lambda)y_0 + \lambda y_1 \in U\}$$

is an open neighborhood of the pair (x_0, x_1), by continuity of addition and scalar multiplication. Hence there exist open sets $U_0, U_1 \subset X$ such that

$$x_0 \in U_0, \qquad x_1 \in U_1, \qquad U_0 \times U_1 \subset W.$$

Since $x_0, x_1 \in \overline{K}$, the sets $U_0 \cap K$ and $U_1 \cap K$ are nonempty. Choose elements

$$y_0 \in U_0 \cap K, \qquad y_1 \in U_1 \cap K.$$

Then $(y_0, y_1) \in U_0 \times U_1 \subset W$ and hence $y_\lambda := (1-\lambda)y_0 + \lambda y_1 \in U \cap K$. Thus $U \cap K \neq \emptyset$ for every open neighborhood U of x_λ and so $x_\lambda \in \overline{K}$.

We prove the last assertion. Assume $\mathrm{int}(K) \neq \emptyset$ and let $x \in K$. Then the set $U_x := \{tx + (1-t)y \,|\, y \in \mathrm{int}(K),\ 0 < t < 1\}$ is open and contained in K. Hence $U_x \subset \mathrm{int}(K)$ and so $x \in \overline{U}_x \subset \overline{\mathrm{int}(K)}$. This proves Lemma 3.1.10. $\square$

Theorem 3.1.11 (Separation of Convex Sets). *Let X be a topological vector space and let $A, B \subset X$ be nonempty disjoint convex sets such that A is open. Then there is a continuous linear functional $\Lambda : X \to \mathbb{R}$ such that*

$$\Lambda(x) > \sup_{y \in B} \Lambda(y) \qquad \text{for all } x \in A.$$

Proof. Assume first that $B = \{0\}$. Then the set

$$P := \{tx \,|\, x \in A,\, t \geq 0\}$$

satisfies the conditions (P1), (P2), (P3) on page 71. Hence $(X, \preccurlyeq)$ is an ordered vector space with the partial order defined by $x \preccurlyeq y$ iff $y - x \in P$. Let $x_0 \in A$. Then the linear subspace

$$Y := \mathbb{R}x_0$$

satisfies (O3) on page 68. Hence Theorem 2.3.7 asserts that there exists a positive linear functional $\Lambda : X \to \mathbb{R}$ such that

$$\Lambda(x_0) = 1.$$

If $x \in A$, then $x - tx_0 \in A$ for $t > 0$ sufficiently small because A is open and hence $\Lambda(x) \geq t > 0$.

We prove that Λ is continuous. To see this, define

$$U := \{x \in X \,|\, \Lambda(x) > 0\}$$

and fix an element $x \in U$. Then

$$V := \left\{y \in X \,|\, x_0 + \Lambda(x)^{-1}(y - x) \in A\right\}$$

is an open set such that $x \in V \subset U$. This shows that U is an open set and hence so is the set

$$\Lambda^{-1}((a, b)) = (ax_0 + U) \cap (bx_0 - U)$$

for every pair of real numbers $a < b$. Hence Λ is continuous and this proves the result for $B = \{0\}$.

To prove the result in general, observe that

$$U := A - B \subset X$$

is a nonempty open convex set such that $0 \notin U$. Hence, by the special case, there is a continuous linear functional $\Lambda : X \to \mathbb{R}$ such that $\Lambda(x) > 0$ for all $x \in U$. Thus $\Lambda(x) > \Lambda(y)$ for all $x \in A$ and all $y \in B$. Define $c := \sup_{y \in B} \Lambda(y)$ and choose $\xi \in X$ such that $\Lambda(\xi) = 1$. If $x \in A$, then, since A is open, there exists a number $\delta > 0$ such that $x - \delta\xi \in A$, and so $\Lambda(x) = \Lambda(x - \delta\xi) + \delta \geq c + \delta$. Hence $\Lambda(x) > c$ for all $x \in A$. This proves Theorem 3.1.11. □

Theorem 3.1.12 (The Topology $\mathscr{U}_\mathcal{F}$). *Let X be a real vector space and let $\mathcal{F} \subset \{\Lambda : X \to \mathbb{R} \,|\, \Lambda \text{ is linear}\}$ be a linear subspace of the space of all linear functionals on X. Let $\mathscr{U}_\mathcal{F} \subset 2^X$ be the weakest topology on X such that each $\Lambda \in \mathcal{F}$ is continuous. This topology has the following properties.*

(i) *A linear functional $\Lambda : X \to \mathbb{R}$ is continuous if and only if it has a closed kernel if and only if $\Lambda \in \mathcal{F}$.*

(ii) *The closure of a linear subspace $E \subset X$ is $\overline{E} = \bigcap_{\Lambda \in \mathcal{F},\, E \subset \ker(\Lambda)} \ker(\Lambda)$.*

(iii) *A linear subspace $E \subset X$ is closed if and only if, for all $x \in X$,*

$$x \in E \qquad \iff \qquad \Lambda(x) = 0 \text{ for all } \Lambda \in \mathcal{F} \text{ such that } E \subset \ker(\Lambda).$$

(iv) *A linear subspace $E \subset X$ is dense if and only if, for all $\Lambda \in \mathcal{F}$,*

$$E \subset \ker(\Lambda) \qquad \Longrightarrow \qquad \Lambda = 0.$$

Proof. See page 118. □

Lemma 3.1.13. *Let X be a real vector space and let $n \in \mathbb{N}$. Then the following hold for every n-tuple $\Lambda_1, \dots, \Lambda_n : X \to \mathbb{R}$ of linear independent linear functionals on X.*

(i) *There exist vectors $x_1, \dots, x_n \in X$ such that*

$$\Lambda_i(x_j) = \delta_{ij} := \begin{cases} 1, & \text{if } i = j, \\ 0, & \text{if } i \neq j \end{cases} \qquad \text{for } i, j = 1, \dots, n. \tag{3.1.3}$$

(ii) *If $\Lambda : X \to \mathbb{R}$ is a linear functional such that $\bigcap_{i=1}^n \ker(\Lambda_i) \subset \ker(\Lambda)$, then $\Lambda \in \operatorname{span}\{\Lambda_1, \dots, \Lambda_n\}$.*

Proof. The proof is by induction on n. Part (i) holds for $n = 1$ by definition. We prove that (i)$_n$ implies (ii)$_n$ and (ii)$_n$ implies (i)$_{n+1}$ for all $n \in \mathbb{N}$.

Fix an integer $n \in \mathbb{N}$ and assume (i)$_n$. Let $\Lambda : X \to \mathbb{R}$ be a linear functional such that $\bigcap_{i=1}^n \ker(\Lambda_i) \subset \ker(\Lambda)$. Since (i) holds for n, there exist vectors $x_1, \dots, x_n \in \mathbb{N}$ such that $\Lambda_i(x_j) = \delta_{ij}$ for $i, j = 1, \dots, n$. Fix an element $x \in X$. Then $x - \sum_{i=1}^n \Lambda_i(x) x_i \in \bigcap_{j=1}^n \ker(\Lambda_j) \subset \ker(\Lambda)$ and this implies $\Lambda(x) = \sum_{i=1}^n \Lambda_i(x)\Lambda(x_i)$. Thus $\Lambda = \sum_{i=1}^n \Lambda(x_i)\Lambda_i$, so (ii) holds for n.

Now assume (ii)$_n$. Let $\Lambda_1, \dots, \Lambda_{n+1} : X \to \mathbb{R}$ be linearly independent linear functionals and define $Z_i := \bigcap_{j \neq i} \ker(\Lambda_j)$ for $i = 1, \dots, n+1$. Then $\Lambda_i \notin \operatorname{span}\{\Lambda_j \,|\, j \neq i\}$ for $i = 1, \dots, n+1$. Since (ii) holds for n, this implies that, for each $i \in \{1, \dots, n+1\}$, there exists a vector $x_i \in Z_i$ such that $\Lambda_i(x_i) = 1$. Thus (i) holds with n replaced by $n+1$. This completes the induction argument and the proof of Lemma 3.1.13. □

Lemma 3.1.14. *Let X be a real vector space and let $\Lambda_1, \dots, \Lambda_n : X \to \mathbb{R}$ and $\Lambda : X \to \mathbb{R}$ be linear functionals. Then the following are equivalent.*

(i) $\bigcap_{i=1}^n \ker(\Lambda_i) \subset \ker(\Lambda)$.

(ii) $\Lambda \in \operatorname{span}\{\Lambda_1, \dots, \Lambda_n\}$.

(iii) *There exists a constant $c \geq 0$ such that*

$$|\Lambda(x)| \leq c \max_{i=1,\dots,n} |\Lambda_i(x)| \qquad \text{for all } x \in X. \tag{3.1.4}$$

Proof. We prove that (i) implies (ii). Thus assume (i) and choose a maximal subset $J \subset \{1, \dots, n\}$ such that the Λ_j with $j \in J$ are linearly independent. Then $\bigcap_{j \in J} \ker(\Lambda_j) = \bigcap_{i=1}^n \ker(\Lambda_i) \subset \ker(\Lambda)$ by (i) and so it follows from Lemma 3.1.13 that $\Lambda \in \operatorname{span}\{\Lambda_j \,|\, j \in J\}$. Thus (ii) holds.

We prove that (ii) implies (iii). Thus assume (ii) and choose real numbers $c_1, \dots, c_n$ such that $\Lambda = \sum_{i=1}^n c_i \Lambda_i$. Define $c := \sum_{i=1}^n |c_i|$. Then

$$|\Lambda(x)| = \left| \sum_{i=1}^n c_i \Lambda_i(x) \right| \leq \sum_{i=1}^n |c_i| \, |\Lambda_i(x)| \leq c \max_{i=1,\dots,n} |\Lambda_i(x)|$$

for all $x \in X$ and so (iii) holds. That (iii) implies (i) is obvious and this proves Lemma 3.1.14. □

Proof of Theorem 3.1.12. We prove (i). If $\Lambda \in \mathcal{F}$, then Λ is continuous by definition of the topology $\mathscr{U}_{\mathcal{F}}$. If Λ is continuous, then Λ has a closed kernel by definition of continuity. Thus it remains to prove that, if Λ has a closed kernel, then $\Lambda \in \mathcal{F}$. Thus assume Λ has a closed kernel and, without loss of generality, that $\Lambda \neq 0$. Choose $x \in X$ such that $\Lambda(x) = 1$. Then $x \in X \setminus \ker(\Lambda)$ and the set $X \setminus \ker(\Lambda)$ is open. Hence there is an integer $n > 0$, a constant $\varepsilon > 0$, and elements $\Lambda_1, \dots, \Lambda_n \in \mathcal{F} \setminus \{0\}$ such that

$$V := \bigcap_{i=1}^n \{ y \in X \,|\, |\Lambda_i(y) - \Lambda_i(x)| < \varepsilon \} \subset X \setminus \ker(\Lambda).$$

We prove that

$$\bigcap_{i=1}^n \ker(\Lambda_i) \subset \ker(\Lambda). \tag{3.1.5}$$

Namely, choose $y \in X$ such that $\Lambda_i(y) = 0$ for $i = 1, \dots, n$. Then $x + ty \in V$ and hence $x + ty \notin \ker(\Lambda)$ for all $t \in \mathbb{R}$. Thus $1 + t\Lambda(y) = \Lambda(x + ty) \neq 0$ for all $t \in \mathbb{R}$ and this implies $\Lambda(y) = 0$. This proves (3.1.5). It follows from (3.1.5) and Lemma 3.1.14 that

$$\Lambda \in \operatorname{span}\{\Lambda_1, \dots, \Lambda_n\} \subset \mathcal{F}$$

and this proves part (i).

We prove (ii). Let $E \subset X$ be a linear subspace. If $\Lambda \in \mathcal{F}$ vanishes on E, then $\overline{E} \subset \ker(\Lambda)$ because $\ker(\Lambda)$ is a closed subset of X that contains E. Conversely, let $x \in X \setminus \overline{E}$. Since $(X, \mathscr{U}_{\mathcal{F}})$ is locally convex by part (ii) of Lemma 3.1.6, and $X \setminus \overline{E}$ is open, there exists a convex open set $U \in \mathscr{U}_{\mathcal{F}}$ such that $x \in U$ and $U \cap E = \emptyset$. Since U and E are convex, Theorem 3.1.11 asserts that there exists a continuous linear functional $\Lambda : X \to \mathbb{R}$ such that $\Lambda(x) > \sup_{y \in E} \Lambda(y)$. Since E is a linear subspace, this implies $E \subset \ker(\Lambda)$. Since $\Lambda \in \mathcal{F}$ by part (i), it follows that $x \notin \bigcap_{\Lambda \in \mathcal{F},\, E \subset \ker(\Lambda)} \ker(\Lambda)$. This proves part (ii). Parts (iii) and (iv) follow directly from (ii) and this proves Theorem 3.1.12. □

Theorem 3.1.12 has several important consequences for the weak and weak* topologies. These are summarized in the next two subsections.

3.1.3. Elementary Properties of the Weak Topology. There are more strongly closed sets in an infinite-dimensional Banach space than there are weakly closed sets. However, for convex sets both notions agree. Thus a linear subspace of a Banach space is closed if and only if it is weakly closed.

Lemma 3.1.15 (Closed Convex Sets Are Weakly Closed). *Let X be a real normed vector space and let $K \subset X$ be a convex subset. Then K is closed if and only if it is weakly closed.*

Proof. Let $K \subset X$ be a closed convex set. We prove it is weakly closed. To see this, fix an element $x_0 \in X \setminus K$. Then there is a constant $\delta > 0$ such that $B_\delta(x_0) \cap K = \emptyset$. By Theorem 2.3.10 with $A := B_\delta(x_0)$ and $B := K$, there exist an $x^* \in X^*$ and a $c \in \mathbb{R}$ such that $\langle x^*, x \rangle > c$ for all $x \in B_\delta(x_0)$ and $\langle x^*, x \rangle \leq c$ for all $x \in K$. Thus

$$U := \{x \in X \mid \langle x^*, x \rangle > c\}$$

is a weakly open set that contains x_0 and is disjoint from K. This shows that $X \setminus K$ is weakly open and hence K is weakly closed. Conversely, every weakly closed subset of X is closed and this proves Lemma 3.1.15. □

Lemma 3.1.16 (Bounded Linear Functionals Are Weakly Continuous). *Let X be a real normed vector space and let $\Lambda : X \to \mathbb{R}$ be a linear functional. Then Λ is continuous with respect to the norm topology on X if and only if it is continuous with respect to the weak topology.*

Proof. This follows from part (i) of Theorem 3.1.12. □

At this point it is useful to introduce the concept of the pre-annihilator.

Definition 3.1.17 (Pre-Annihilator). Let X be a real normed vector space and let $T \subset X^*$ be a subset of the dual space $X^* = \mathcal{L}(X, \mathbb{R})$. The set

$$^{\perp}T := \{x \in X \mid \langle x^*, x\rangle = 0 \text{ for all } x^* \in T\} \tag{3.1.6}$$

is called the **pre-annihilator** or **left annihilator** or **joint kernel** of T. It is a closed linear subspace of X.

Corollary 3.1.18 (Weak Closure of a Subspace). *Let X be a real normed vector space and let $E \subset X$ be a linear subspace. Then the following hold.*

(i) *The closure of E is the subspace $\overline{E} = {}^{\perp}(E^{\perp})$ and agrees with the weak closure of E.*

(ii) *E is closed if and only if E is weakly closed if and only if $E = {}^{\perp}(E^{\perp})$.*

(iii) *E is dense if and only if E is weakly dense if and only if $E^{\perp} = \{0\}$.*

Proof. The formula $\overline{E} = {}^{\perp}(E^{\perp})$ for the closure of E is a restatement of Corollary 2.3.24. That this subspace is also the weak closure of E follows from part (ii) of Theorem 3.1.12 and also from Lemma 3.1.15. This proves (i). Parts (ii) and (iii) follow directly from (i) and this proves Corollary 3.1.18. □

The next lemma shows that the limit of a weakly convergent sequence in a Banach space is contained in the closed convex hull of the sequence.

Definition 3.1.19. Let X be a real vector space and let $S \subset X$. The set

$$\operatorname{conv}(S) := \left\{ \sum_{i=1}^{n} \lambda_i x_i \,\middle|\, n \in \mathbb{N},\, x_i \in S,\, \lambda_i \geq 0,\, \sum_{i=1}^{n} \lambda_i = 1 \right\} \tag{3.1.7}$$

is convex and is called the **convex hull of** S. If X is a topological vector space, then the closure of the convex hull of a set $S \subset X$ is called the **closed convex hull of** S and is denoted by $\overline{\operatorname{conv}}(S)$.

Lemma 3.1.20 (Mazur). *Let X be a real normed vector space and let $x_i \in X$ be a sequence that converges weakly to x. Then*

$$x \in \overline{\operatorname{conv}}(\{x_i \mid i \in \mathbb{N}\}),$$

i.e. for every $\varepsilon > 0$ there exist an $n \in \mathbb{N}$ and real numbers $\lambda_1, \dots, \lambda_n$ such that $\lambda_i \geq 0$ for all i, $\sum_{i=1}^n \lambda_i = 1$, and $\|x - \sum_{i=1}^n \lambda_i x_i\| < \varepsilon$.

Proof. The set $K := \operatorname{conv}(\{x_i \mid i \in \mathbb{N}\})$ is convex and so is its strong closure $\overline{K}$ by Lemma 3.1.10. Hence $\overline{K}$ is weakly closed by Lemma 3.1.15. Since $x_i \in K$ converges weakly to x it follows that $x \in \overline{K}$. □

It follows from Lemma 3.1.20 that the weak limit of every weakly convergent sequence in the unit sphere $S \subset X$ in a Banach space X is contained in the closed unit ball $B = \mathrm{conv}(S) = \overline{\mathrm{conv}}(S)$. In fact, it turns out that B is the weak closure of S whenever X is infinite-dimensional, and so $\mathscr{U}^{\mathrm{w}} \subsetneq \mathscr{U}^{\mathrm{s}}$.

Lemma 3.1.21 (Weak Closure of the Unit Sphere). *Let X be an infinite-dimensional real normed vector space and define*

$$S := \{x \in X \mid \|x\| = 1\}, \qquad B := \{x \in X \mid \|x\| \le 1\}. \tag{3.1.8}$$

Then B is the weak closure of S.

Proof. The set B is weakly closed by Lemma 3.1.15 and hence contains the weak closure of S. We prove that B is contained in the weak closure of S. To see this, let $x_0 \in B$ and let $U \subset X$ be a weakly open set containing x_0. Then there exist elements $x_1^*, \dots, x_n^* \in X^*$ and a constant $\varepsilon > 0$ such that

$$V := \{x \in X \mid |\langle x_i^*, x - x_0\rangle| < \varepsilon \text{ for } i = 1, \dots, n\} \subset U.$$

Since X is infinite-dimensional, there is a nonzero vector $\xi \in X$ such that

$$\langle x_i^*, \xi\rangle = 0 \qquad \text{for } i = 1, \dots, n.$$

Since $\|x_0\| \le 1$ there exists a real number t such that

$$\|x_0 + t\xi\| = 1.$$

Hence $x_0 + t\xi \in V \cap S$ and so $U \cap S \ne \emptyset$. Thus x_0 belongs to the weak closure of S and this completes the proof of Lemma 3.1.21. □

In view of Lemma 3.1.21 one might ask whether every element of B is the limit of a weakly convergent sequence in S. The answer is negative in general. For example, the next exercise shows that a sequence in ℓ^1 converges weakly if and only if it converges strongly. Thus the limit of every weakly convergent sequence of norm one in ℓ^1 has again norm one. The upshot is that the weak closure of a subset of a Banach space is in general much bigger than the set of all limits of weakly convergent sequences in that subset.

Exercise 3.1.22 (Schur's Theorem). Let $x_n = (x_{n,i})_{i\in\mathbb{N}}$ for $n \in \mathbb{N}$ be a sequence in ℓ^1 that converges weakly to an element $x = (x_i)_{i\in\mathbb{N}} \in \ell^1$. Prove that $\lim_{n\to\infty} \|x_n - x\|_{\ell^1} = 0$. (See also Exercise 3.7.3.)

Exercise 3.1.23. Let X be a Banach space and suppose X^* is separable. Let $S \subset X$ be a bounded set and let $x \in X$ be an element in the weak closure of S. Prove that there is a sequence $(x_n)_{n\in\mathbb{N}}$ in S that converges weakly to x.

Exercise 3.1.24. Let X be a normed vector space. Prove that the canonical inclusion $\iota : X \to X^{**}$ is continuous with respect to the weak topology on X and the weak* topology on X^{**}.

3.1.4. Elementary Properties of the Weak* Topology. When X is a Banach space and Y is a dense subspace, the dual spaces X^* and Y^* are canonically isomorphic because every bounded linear functional on Y extends uniquely to a bounded linear functional on X. The extension has the same norm as the original linear functional on Y and hence the canonical isomorphism $X^* \to Y^* : x^* \mapsto x^*|_Y$ is an isometry. However, the weak* topologies of X^* and Y^* may differ dramatically. Namely, by part (i) of Theorem 3.1.12 the space of weak* continuous linear functionals on Y^* can be identified with the original normed vector space Y and so may be much smaller than the space of weak* continuous linear functionals on X^*. In other words, the completion of a normed vector space is a Banach space and both spaces have the same dual space, however, their weak* topologies differ. Thus great care must be taken when dealing with the weak* topology of the dual space of a normed vector space versus that of the dual space of a Banach space.

Corollary 3.1.25 (Weak* Continuous Linear Functionals). *Let X be a real normed vector space and let*

$$\Lambda : X^* \to \mathbb{R}$$

be a linear functional on its dual space. Then the following are equivalent.

(i) *Λ is continuous with respect to the weak* topology on X^*.*

(ii) *The kernel of Λ is a weak* closed linear subspace of X^*.*

(iii) *Λ belongs to the image of the inclusion $\iota : X \to X^{**}$ in (2.4.1), i.e. there exists an element $x \in X$ such that $\Lambda(x^*) = \langle x^*, x\rangle$ for all $x^* \in X^*$.*

Proof. This follows directly from part (i) of Theorem 3.1.12 and the definition of the weak* topology in Example 3.1.9. □

Corollary 3.1.26 (Weak* Closure of a Subspace). *Let X be a real normed vector space and let $E \subset X^*$ be a linear subspace of its dual space. Then the following hold.*

(i) *The linear subspace $({}^{\perp}E)^{\perp}$ is the weak* closure of E.*

(ii) *E is weak* closed if and only if $E = ({}^{\perp}E)^{\perp}$.*

(iii) *E is weak* dense in X^* if and only if ${}^{\perp}E = \{0\}$.*

Proof. By Corollary 3.1.25 the pre-annihilator of E is the space of weak* continuous linear functionals on X^* that vanish on E. Hence part (i) follows from part (ii) of Theorem 3.1.12. Part (ii) follows directly from (i). Part (iii) follows from (i) and the fact that any subset $S \subset X$ satisfies $S^{\perp} = X^*$ if and only if $S \subset \{0\}$ by Corollary 2.3.4. This proves Corollary 3.1.26. □

Corollary 3.1.27 (Separation of Convex Sets). *Let X be a real normed vector space and let $A, B \subset X^*$ be nonempty disjoint convex sets such that A is weak* open. Then there exists an element $x \in X$ such that*

$$\langle x^*, x\rangle > \sup_{y^*\in B} \langle y^*, x\rangle \qquad \text{for all } x^* \in A.$$

Proof. Theorem 3.1.11 and Corollary 3.1.25. □

Corollary 3.1.28 (Weak* Closure of the Unit Sphere). *Let X be an infinite-dimensional real normed vector space and define*

$$S^* := \{x^* \in X^* \mid \|x^*\| = 1\}, \qquad B^* := \{x^* \in X^* \mid \|x^*\| \le 1\}.$$

Then B^* is the weak closure of S^*.*

Proof. Let $F_x := \{x^* \in X^* \,|\, \langle x^*, x\rangle \le 1\}$ for $x \in S$ (the unit sphere in X). Then F_x is weak* closed for all $x \in S$, and hence so is $B^* = \bigcap_{x\in S} F_x$. Now let $K \subset X^*$ be the weak* closure of S^*. Then $K \subset B^*$ because B^* is a weak* closed set containing S^*, and $B^* \subset K$ because K is a weakly closed set containing S^* and B^* is the weak closure of S^* by Lemma 3.1.21. □

Corollary 3.1.29 (Goldstine's Theorem). *Let X be a real normed vector space and let $\iota : X \to X^{**}$ be the inclusion* (2.4.1). *The following hold.*

(i) *$\iota(X)$ is weak* dense in X^{**}.*

(ii) *Assume X is infinite-dimensional and denote by $S \subset X$ the closed unit sphere. Then the weak* closure of $\iota(S)$ is the closed unit ball $B^{**} \subset X^{**}$.*

Proof. By definition ${}^\perp\iota(X) = \{0\}$, so (i) holds by Corollary 3.1.26. To prove (ii), let $K \subset X^{**}$ be the weak* closure of $\iota(S)$. Then $K \subset B^{**}$ because B^{**} is weak* closed by Corollary 3.1.28. Moreover, the set $\iota^{-1}(K) \subset X$ is weakly closed by Exercise 3.1.24 and $S \subset \iota^{-1}(K)$. Hence $B \subset \iota^{-1}(K)$ by Lemma 3.1.21, hence $\iota(B) \subset K$, and so K is the weak* closure of $\iota(B)$. Thus K is convex by Lemma 3.1.10. Now let $x_0^{**} \in X^{**} \setminus K$ and choose a convex weak* open neighborhood $U \subset X^{**}$ of x_0^{**} such that $U \cap K = \emptyset$. Then, by Corollary 3.1.27, there exists an element $x_0^* \in X^*$ such that

$$\langle x_0^{**}, x_0^*\rangle > \sup_{x^{**}\in K} \langle x^{**}, x_0^*\rangle \ge \sup_{x\in S}\langle \iota(x), x_0^*\rangle = \sup_{x\in S}\langle x_0^*, x\rangle = \|x_0^*\|.$$

Hence $\|x_0^{**}\| > 1$ and so $x_0^{**} \notin B^{**}$. Thus $B^{**} \subset K \subset B^{**}$ and so $B^{**} = K$. This proves Corollary 3.1.29. □

Corollary 3.1.29 shows that, in contrast to the weak topology, a closed linear subspace of X^* is not necessarily weak* closed. For example, the space c_0 (Example 1.3.7) is a closed linear subspace of $\ell^\infty \cong (\ell^1)^*$ but is dense with respect to the weak* topology and so is not weak* closed. The study of the weak* closed subspaces will be taken up again in Section 3.3.

3.2. The Banach–Alaoglu Theorem

3.2.1. The Separable Case. We prove two versions of the Banach–Alaoglu Theorem. The first version holds for separable normed vector spaces and asserts that every bounded sequence in the dual space has a weak* convergent subsequence.

Theorem 3.2.1 (Banach–Alaoglu: The Separable Case). *Let X be a separable real normed vector space. Then every bounded sequence in the dual space X^* has a weak* convergent subsequence.*

Proof. Let $D = \{x_1, x_2, x_3, \dots\} \subset X$ be a countable dense subset and let $(x_n^*)_{n\in\mathbb{N}}$ be a bounded sequence in X^*. Then the standard diagonal sequence argument shows that there is a subsequence $(x_{n_i}^*)_{i\in\mathbb{N}}$ such that the sequence of real numbers $(\langle x_{n_i}^*, x_k\rangle)_{i\in\mathbb{N}}$ converges for every $k \in \mathbb{N}$. More precisely, the sequence $(\langle x_n^*, x_1\rangle)_{n\in\mathbb{N}}$ is bounded and hence has a convergent subsequence $(\langle x_{n_{i,1}}^*, x_1\rangle)_{i\in\mathbb{N}}$. Since the sequence $(\langle x_{n_{i,1}}^*, x_2\rangle)_{i\in\mathbb{N}}$ is bounded it has a convergent subsequence $(\langle x_{n_{i,2}}^*, x_2\rangle)_{i\in\mathbb{N}}$. Continue by induction and use the axiom of dependent choice (see page 6) to construct a sequence of subsequences $(x_{n_{i,k}})_{i\in\mathbb{N}}$ such that, for every $k \in \mathbb{N}$, $(x_{n_{i,k+1}})_{i\in\mathbb{N}}$ is a subsequence of $(x_{n_{i,k}})_{i\in\mathbb{N}}$ and the sequence $(\langle x_{n_{i,k}}^*, x_k\rangle)_{i\in\mathbb{N}}$ converges. Now consider the diagonal subsequence $x_{n_i}^* := x_{n_{i,i}}^*$. Then the sequence $(\langle x_{n_i}^*, x_k\rangle)_{i\in\mathbb{N}}$ converges for every $k \in \mathbb{N}$ as claimed.

With this understood, it follows from the equivalence of (ii) and (iii) in Theorem 2.1.5, with $Y = \mathbb{R}$ and A_i replaced by the bounded linear functional $x_{n_i}^* : X \to \mathbb{R}$, that there exists an element $x^* \in X^*$ such that

$$\langle x^*, x\rangle = \lim_{i\to\infty} \langle x_{n_i}^*, x\rangle$$

for all $x \in X$. Hence the sequence $(x_{n_i}^*)_{i\in\mathbb{N}}$ converges to x^* in the weak* topology as claimed. This proves Theorem 3.2.1. □

Example 3.2.2. This example shows that the hypothesis that X is separable cannot be removed in Theorem 3.2.1. The Banach space $X = \ell^\infty$ with the supremum norm is not separable. For $n \in \mathbb{N}$ define the bounded linear functional $\Lambda_n : \ell^\infty \to \mathbb{R}$ by $\Lambda_n(x) := x_n$ for $x = (x_i)_{i\in\mathbb{N}} \in \ell^\infty$. Then the sequence $(\Lambda_n)_{n\in\mathbb{N}}$ in X^* does not have a weak* convergent subsequence. To see this, let $n_1 < n_2 < n_3 < \cdots$ be any sequence of positive integers and define the sequence $x = (x_i)_{i\in\mathbb{N}} \in \ell^\infty$ by $x_i := 1$ for $i = n_{2k}$ with $k \in \mathbb{N}$ and by $x_i := -1$ otherwise. Then $\Lambda_{n_k}(x) = x_{n_k} = (-1)^k$ and hence the sequence of real numbers $(\Lambda_{n_k}(x))_{k\in\mathbb{N}}$ does not converge. Thus the subsequence $(\Lambda_{n_k})_{k\in\mathbb{N}}$ in X^* does not converge in the weak* topology.

3.2.2. Invariant Measures. Let (M,d) be a compact metric space and let $\phi : M \to M$ be a homeomorphism. Denote by $\mathcal{B} \subset 2^M$ the Borel σ-algebra. The space $C(M)$ of all continuous functions $f : M \to \mathbb{R}$ with the supremum norm is a separable Banach space (Example 1.1.3) and its dual space is isomorphic to the space $\mathcal{M}(M)$ of signed Borel measures $\mu : \mathcal{B} \to \mathbb{R}$ (Example 1.3.8), equipped with the norm function

$$\|\mu\| := \sup_{B\in\mathcal{B}} \big(\mu(B) - \mu(M \setminus B)\big)$$

for $\mu \in \mathcal{M}(M)$. A Borel measure $\mu : \mathcal{B} \to [0,\infty)$ is called a **probability measure** if $\|\mu\| = \mu(M) = 1$. A probability measure $\mu : \mathcal{B} \to [0,1]$ is called **ϕ-invariant** if

$$\int_M (f\circ\phi)\,d\mu = \int_M f\,d\mu \qquad \text{for all } f \in C(M). \tag{3.2.1}$$

The set

$$\mathcal{M}(\phi) := \left\{\mu \in \mathcal{M}(M) \,\middle|\, \begin{array}{l}\mu(B) \geq 0 \text{ for all } B \in \mathcal{B},\\ \mu(M) = 1, \text{ and } \mu \text{ satisfies (3.2.1)}\end{array}\right\} \tag{3.2.2}$$

of ϕ-invariant Borel probability measures is a weak* closed convex subset of the unit sphere in $\mathcal{M}(M)$. The next lemma shows that it is nonempty.

Lemma 3.2.3. *Every homeomorphism of a compact metric space admits an invariant Borel probability measure.*

Proof. Let $\phi : M \to M$ be a homeomorphism of a compact metric space. Fix an element $x_0 \in X$ and, for every integer $n \geq 1$, define the Borel probability measure $\mu_n : \mathcal{B} \to [0,1]$ by

$$\int_M f\,d\mu_n := \frac{1}{n}\sum_{k=0}^{n-1} f(\phi^k(x_0)) \qquad \text{for } f \in C(M).$$

Here $\phi^0 := \mathrm{id} : M \to M$ and $\phi^k := \phi\circ\cdots\circ\phi$ denotes the kth iterate of ϕ for $k \in \mathbb{N}$. By Theorem 3.2.1, the sequence μ_n has a weak* convergent subsequence $(\mu_{n_i})_{i\in\mathbb{N}}$. Its weak* limit is a Borel measure $\mu : \mathcal{B} \to [0,\infty)$ such that

$$\|\mu\| = \int_M 1\,d\mu = \lim_{i\to\infty}\int_M 1\,d\mu_{n_i} = 1$$

and

$$\int_M (f\circ\phi)\,d\mu = \lim_{i\to\infty}\frac{1}{n_i}\sum_{k=1}^{n_i} f(\phi^k(x_0)) = \lim_{i\to\infty}\frac{1}{n_i}\sum_{k=0}^{n_i-1} f(\phi^k(x_0)) = \int_M f\,d\mu$$

for all $f \in C(M)$. Hence $\mu \in \mathcal{M}(\phi)$. □

3.2.3. The General Case. The second version of the Banach–Alaoglu Theorem applies to all real normed vector spaces and asserts that the closed unit ball in the dual space is weak* compact.

Theorem 3.2.4 (Banach–Alaoglu: The General Case). *Let X be a real normed vector space. Then the closed unit ball*

$$B^* := \{x^* \in X^* \mid \|x^*\| \le 1\} \tag{3.2.3}$$

in the dual space X^* is weak compact.*

Proof. This is an application of Tychonoff's Theorem A.2.1. The parameter space is $I = X$. Associated to each $x \in X$ is the compact interval

$$K_x := [-\|x\|, \|x\|] \subset \mathbb{R}.$$

The product of these compact intervals is the space

$$K := \prod_{x \in X} K_x = \{f : X \to \mathbb{R} \mid |f(x)| \le \|x\| \text{ for all } x \in X\} \subset \mathbb{R}^X.$$

Define

$$L := \{f : X \to \mathbb{R} \mid f \text{ is linear}\} \subset \mathbb{R}^X.$$

The intersection of K and L is the closed unit ball

$$B^* := \{x^* \in X^* \mid \|x^*\| \le 1\} = L \cap K.$$

By definition, the weak* topology on $B^* = L \cap K$ is induced by the product topology on $\mathbb{R}^X$ (see Example 3.1.7). Moreover L is a closed subset of $\mathbb{R}^X$ with respect to the product topology. To see this, fix elements $x, y \in X$ and $\lambda \in \mathbb{R}$ and define the maps $\phi_{x,y} : \mathbb{R}^X \to \mathbb{R}$ and $\psi_{x,\lambda} : \mathbb{R}^X \to \mathbb{R}$ by

$$\phi_{x,y}(f) := f(x+y) - f(x) - f(y), \qquad \psi_{x,\lambda}(f) := f(\lambda x) - \lambda f(x).$$

By definition of the product topology, these maps are continuous and this implies that the set

$$L = \bigcap_{x,y \in X} \phi_{x,y}^{-1}(0) \cap \bigcap_{x \in X, \lambda \in \mathbb{R}} \psi_{x,\lambda}^{-1}(0)$$

is closed with respect to the product topology. Since K is a compact subset of $\mathbb{R}^X$ by Tychonoff's Theorem A.2.1 and $\mathbb{R}^X$ is a Hausdorff space by Example 3.1.7, it follows that $B^* = L \cap K$ is a closed subset of a compact set and hence is compact. This proves Theorem 3.2.4. □

The next theorem characterizes the weak* compact subsets of the dual space of a separable Banach space.

Theorem 3.2.5 (Weak* Compact Subsets). *Let X be a separable Banach space and let $K \subset X^*$. Then the following are equivalent.*

(i) *K is weak* compact.*

(ii) *K is bounded and weak* closed.*

(iii) *K is* **sequentially weak* compact**, *i.e. every sequence in K has a weak* convergent subsequence with limit in K.*

(iv) *K is bounded and* **sequentially weak* closed**, *i.e. if $x^* \in X^*$ is the weak* limit of a sequence in K, then $x^* \in K$.*

The implications (i) $\iff$ (ii) *and* (ii) $\implies$ (iv) *and* (iii) $\implies$ (iv) *continue to hold when X is not separable.*

Proof. We prove that (i) implies (ii). Assume K is weak* compact. Then K is weak* closed, because the weak* topology on X^* is Hausdorff. To prove that K is bounded, fix an element $x \in X$. Then the function

$$K \to \mathbb{R} : x^* \mapsto \langle x^*, x \rangle$$

is continuous with respect to the weak* topology and hence is bounded. Thus

$$\sup_{x^* \in K} |\langle x^*, x \rangle| < \infty \qquad \text{for all } x \in X.$$

Hence it follows from the Uniform Boundedness Theorem 2.1.1 that

$$\sup_{x^* \in K} \|x^*\| < \infty$$

and so K is bounded.

We prove that (ii) implies (i). Assume K is bounded and weak* closed. Choose $c > 0$ such that

$$\|x^*\| \le c \qquad \text{for all } x^* \in K.$$

Since the set

$$cB^* = \{x^* \in X^* \mid \|x^*\| \le c\}$$

is weak* compact by Theorem 3.2.4 and $K \subset cB^*$ is weak* closed, it follows that K is weak* compact.

We prove that (ii) implies (iii). Assume K is bounded and weak* closed. Let $(x_n^*)_{n\in\mathbb{N}}$ be a sequence in K. This sequence is bounded by assumption and hence, by Theorem 3.2.1, has a weak* convergent subsequence because X is separable. Let $x^* \in X^*$ be the weak* limit of that subsequence. Since K is weak* closed it follows that $x^* \in K$. Thus K is sequentially weak* compact.

We prove that (iii) implies (iv). Assume K is sequentially weak* compact. Then K is bounded because every weak* convergent sequence is bounded by the Uniform Boundedness Theorem 2.1.1. Moreover K is sequentially weak* closed by uniqueness of the weak* limit. (If $x_n^* \in K$ converges to $x^* \in X^*$ in the weak* topology, then it has a subsequence that weak* converges to an element $y^* \in K$ and so $x^* = y^* \in K$.)

We prove that (iv) implies (ii). Assume K is bounded and sequentially weak* closed. We must prove that K is weak* closed. Let $x_0^* \in X^*$ be an element of the weak* closure of K. Choose a countable dense subset $\{x_k \,|\, k \in \mathbb{N}\}$ of X. Then the set

$$U_n := \left\{ x^* \in X^* \,\middle|\, |\langle x^* - x_0^*, x_k \rangle| < \frac{1}{n} \text{ for } k = 1, \ldots, n \right\}$$

is a weak* open neighborhood of x_0^* for every $n \in \mathbb{N}$. Hence $U_n \cap K \neq \emptyset$ for all $n \in \mathbb{N}$ and so it follows from the axiom of countable choice that there exists a sequence $(x_n^*)_{n\in\mathbb{N}}$ in X^* such that, for all $n \in \mathbb{N}$, we have $x_n^* \in U_n \cap K$. This sequence satisfies $|\langle x_n^* - x_0^*, x_k \rangle| \leq 1/n$ for all $k, n \in \mathbb{N}$ such that $n \geq k$. Thus

$$\lim_{n\to\infty} \langle x_n^*, x_k \rangle = \langle x_0^*, x_k \rangle \qquad \text{for all } k \in \mathbb{N}.$$

Since the sequence $(x_n^*)_{n\in\mathbb{N}}$ in X^* is bounded, and the sequence $(x_k)_{k\in\mathbb{N}}$ is dense in X, it follows from Theorem 2.1.5 that

$$\lim_{n\to\infty} \langle x_n^*, x \rangle = \langle x_0^*, x \rangle \qquad \text{for all } x \in X.$$

Hence $(x_n^*)_{n\in\mathbb{N}}$ is a sequence in K that weak* converges to x_0^* and so $x_0^* \in K$. This proves Theorem 3.2.5. □

Corollary 3.2.6. *Let (M, d) be a compact metric space and let $\phi : M \to M$ be a homeomorphism. Then the set $\mathcal{M}(\phi)$ of ϕ-invariant Borel probability measures on M is a weak* compact convex subset of $\mathcal{M}(M) = C(M)^*$.*

Proof. The set $\mathcal{M}(\phi)$ is convex, bounded, and weak* closed by definition (see Subsection 3.2.2). Hence it is weak* compact by Theorem 3.2.5. □

Example 3.2.7. The hypothesis that X is complete cannot be removed in Theorem 3.2.5. Let c_{00} be the set of all sequences $x = (x_i)_{i\in\mathbb{N}} \in \ell^\infty$ with only finitely many nonzero entries, equipped with the supremum norm. Its closure is the space $c_0 \subset \ell^\infty$ in Example 1.3.7 and so its dual space is isomorphic to ℓ^1. A sequence of bounded linear functionals $\Lambda_n : c_{00} \to \mathbb{R}$ converges to the bounded linear functional $\Lambda : c_{00} \to \mathbb{R}$ in the weak* topology if and only if $\lim_{n\to\infty} \Lambda_n(e_i) = \Lambda(e_i)$ for all $i \in \mathbb{N}$, where $e_i := (\delta_{ij})_{j\in\mathbb{N}}$. For $n \in \mathbb{N}$ define $\Lambda_n : X \to \mathbb{R}$ by $\Lambda_n(x) := x_n$ for $x = (x_i)_{i\in\mathbb{N}} \in X$. Then $n\Lambda_n$ converges to zero in the weak* topology, and hence $K := \{n\Lambda_n \,|\, n \in \mathbb{N}\} \cup \{0\}$ is an unbounded weak* compact subset of $c_{00}^* \cong \ell^1$.

Example 3.2.8. The Banach space $X = \ell^\infty$ is not separable. We prove that (i) does not imply (iii) and (iv) does not imply any of the other assertions in Theorem 3.2.5 for $X = \ell^\infty$. The closed unit ball in $(\ell^\infty)^*$ is weak* compact by Theorem 3.2.4 but is not sequentially weak* compact. Namely, for each $n \in \mathbb{N}$ the bounded linear functional $\Lambda_n : \ell^\infty \to \mathbb{R}$, defined by

$$\Lambda_n(x) := x_n$$

for $x = (x_i)_{i\in\mathbb{N}} \in \ell^\infty$, has the norm $\|\Lambda_n\| = 1$ and the sequence $(\Lambda_n)_{n\in\mathbb{N}}$ in $(\ell^\infty)^*$ does not have a weak* convergent subsequence by Example 3.2.2. Moreover, the bounded set

$$K := \{\Lambda_n \,|\, n \in \mathbb{N}\} \subset (\ell^\infty)^*$$

is sequentially weak* closed, but is neither sequentially weak* compact nor weak* compact. (**Exercise:** Find a sequence of weak* open subsets $U_n \subset (\ell^\infty)^*$ such that $\Lambda_n \in U_n \setminus U_m$ for all $m, n \in \mathbb{N}$ with $m \neq n$.)

Example 3.2.9. Let M be a locally compact Hausdorff space which is sequentially compact but not compact. (An example is an uncountable well-ordered set M such that every element of M has only countably many predecessors, equipped with the order topology, as in [**75**, Example 3.6].) Let $\delta : M \to C_0(M)^*$ be the embedding defined in Exercise 3.2.10 below. Then $K := \delta(M)$ is a sequentially weak* compact set in $C_0(M)^*$ and is not weak* compact. So (iii) does not imply (i) in Theorem 3.2.5 for $X = C_0(M)$.

Exercise 3.2.10. Let M be a locally compact Hausdorff space. A continuous function $f : M \to \mathbb{R}$ is said to **vanish at infinity** if, for every $\varepsilon > 0$, there is a compact set $K \subset M$ such that $\sup_{x\in M\setminus K}|f(x)| < \varepsilon$. Denote by $C_0(M)$ the space of all continuous functions $f : M \to \mathbb{R}$ that vanish at infinity.

(i) Prove that $C_0(M)$ is a Banach space with the supremum norm.

(ii) Prove that the map $\delta : M \to C_0(M)^*$, which assigns to each $x \in M$ the bounded linear functional

$$\delta_x : C_0(M) \to \mathbb{R}$$

given by

$$\delta_x(f) := f(x) \qquad \text{for } f \in C_0(M),$$

is a homeomorphism onto its image

$$\delta(M) \subset C_0(M)^*,$$

equipped with the weak* topology.

3.3. The Banach–Dieudonné Theorem

This section is devoted to a theorem of Banach–Dieudonné which implies that a linear subspace of the dual space of a Banach space X is weak* closed if and only if its intersection with the closed unit ball in X^* is weak* closed. This result will play a central role in the proof of the Eberlein–Šmulyan Theorem 3.4.1, which characterizes reflexive Banach spaces in terms of weak compactness of the closed unit ball.

Theorem 3.3.1 (Banach–Dieudonné). *Let X be a real Banach space, let $E \subset X^*$ be a linear subspace of the dual space $X^* = \mathcal{L}(X, \mathbb{R})$, and let $B^* := \{x^* \in X^* \mid \|x^*\| \le 1\}$. Assume that the intersection*

$$E \cap B^* = \{x^* \in E \mid \|x^*\| \le 1\}$$

is weak closed and let $x_0^* \in X^* \setminus E$. Then*

$$\inf_{x^* \in E} \|x^* - x_0^*\| > 0 \tag{3.3.1}$$

and, if $0 < \delta < \inf_{x^* \in E} \|x^* - x_0^*\|$, then there is a vector $x_0 \in X$ such that

$$\langle x_0^*, x_0 \rangle = 1, \qquad \|x_0\| \le \delta^{-1}, \qquad \langle x^*, x_0 \rangle = 0 \text{ for all } x^* \in E. \tag{3.3.2}$$

Proof. See page 131. □

The last condition in (3.3.2) asserts that x_0 is an element of the pre-annihilator ${}^\perp E$ (see Definition 3.1.17).

Corollary 3.3.2 (Weak* Closed Linear Subspaces). *Let X be a real Banach space and let $E \subset X^*$ be a linear subspace of its dual space. Then the following are equivalent.*

(i) *E is weak* closed.*

(ii) *$E \cap B^*$ is weak* closed.*

(iii) *$({}^\perp E)^\perp = E$.*

Proof. That (i) implies (ii) follows from the fact that the closed unit ball $B^* \subset X^*$ is weak* closed by Corollary 3.1.28.

We prove that (ii) implies (iii). The inclusion $E \subset ({}^\perp E)^\perp$ follows directly from the definitions. To prove the converse, fix an element $x_0^* \in X^* \setminus E$. Then Theorem 3.3.1 asserts that there exists a vector $x_0 \in {}^\perp E$ such that $\langle x_0^*, x_0 \rangle \ne 0$, and this implies $x_0^* \notin ({}^\perp E)^\perp$.

That (iii) implies (i) follows from the fact that, for every $x \in X$, the linear functional $\iota(x) : X^* \to \mathbb{R}$ in (2.4.1) is continuous with respect to the weak* topology by definition, and so the set $S^\perp = \bigcap_{x \in S} \ker(\iota(x))$ is a weak* closed linear subspace of X^* for every subset $S \subset X$ (see also Corollary 3.1.26). This proves Corollary 3.3.2. □

Proof of Theorem 3.3.1. The proof has five steps.

Step 1. $\inf_{x^* \in E} \|x^* - x_0^*\| > 0$.

By assumption, the intersection $E \cap B^*$ is weak* closed and hence is a closed subset of X^*. Let $(x_i^*)_{i \in \mathbb{N}}$ be a sequence in E that converges to an element $x^* \in X^*$. Then the sequence $(x_i^*)_{i \in \mathbb{N}}$ is bounded. Choose a constant $c > 0$ such that $\|x_i^*\| \le c$ for all $i \in \mathbb{N}$. Then $c^{-1}x_i^* \in E \cap B^*$ for all i and so $c^{-1}x^* = \lim_{i \to \infty} c^{-1}x_i^* \in E \cap B^*$. Hence $x^* \in E$. This shows that E is a closed linear subspace of X^*. Since $x_0^* \notin E$, this proves Step 1.

Step 2. *Choose a real number*

$$0 < \delta < \inf_{x^* \in E} \|x^* - x_0^*\| . \tag{3.3.3}$$

Then there exists a sequence of finite subsets $S_1, S_2, S_3, \dots$ of the closed unit ball $B \subset X$ such that, for all $n \in \mathbb{N}$ and all $x^* \in X^*$, we have

$$\left.\begin{array}{l} \|x^* - x_0^*\| \le \delta n \;\; \textit{and} \\ \max_{x \in S_k} |\langle x^* - x_0^*, x\rangle| \le k\delta \\ \textit{for all } k \in \mathbb{N} \textit{ with } 1 \le k < n \end{array}\right. \quad \Longrightarrow \quad x^* \notin E. \tag{3.3.4}$$

For $n = 1$ condition (3.3.4) holds by (3.3.3). Now fix an integer $n \ge 1$ and suppose, by induction, that the finite sets $S_k \subset B$ have been constructed for $k \in \mathbb{N}$ with $k < n$ such that (3.3.4) holds. For every finite set $S \subset B$ define

$$E(S) := \left\{ x^* \in E \;\middle|\; \begin{array}{l} \|x^* - x_0^*\| \le \delta(n+1), \\ \max_{x \in S_k} |\langle x^* - x_0^*, x\rangle| \le \delta k \text{ for } 1 \le k < n, \\ \max_{x \in S} |\langle x^* - x_0^*, x\rangle| \le \delta n \end{array} \right\}.$$

Define

$$R := \|x_0^*\| + \delta(n+1).$$

Since $E \cap B^*$ is weak* closed so is the set

$$K := R(E \cap B^*) = \{x^* \in E \,|\, \|x^*\| \le R = \|x_0^*\| + \delta(n+1)\}.$$

Hence K is weak* compact by Theorem 3.2.5. Moreover, for every finite set

$$S \subset B,$$

the set $E(S)$ is the intersection of K with the weak* closed sets

$$\left\{ x^* \in X^* \;\middle|\; \|x^* - x_0^*\| \le \delta(n+1) \right\},$$

$$\left\{ x^* \in X^* \;\middle|\; \max_{x \in S} \langle x^* - x_0^*, x\rangle \le \delta n \right\},$$

$$\left\{ x^* \in X^* \;\middle|\; \max_{x \in S_k} \langle x^* - x_0^*, x\rangle \le \delta k \right\}, \quad k \in \mathbb{N}, \quad k < n.$$

Hence $E(S) \subset K$ is a weak* closed set for every finite set $S \subset B$.

Now assume, by contradiction, that $E(S) \neq \emptyset$ for every finite set $S \subset B$. Then every finite collection $\mathscr{S} \subset 2^B$ of finite subsets of B satisfies

$$\bigcap_{S\in\mathscr{S}} E(S) = E\left(\bigcup_{S\in\mathscr{S}} S\right) \neq \emptyset,$$

and hence the collection

$$\{E(S)\,|\,S \text{ is a finite subset of } B\}$$

of weak* closed subsets of K has the finite intersection property. Since K is weak* compact, this implies that there exists an element $x^* \in X^*$ such that $x^* \in E(S)$ for every finite set $S \subset B$. This element x^* belongs to the subspace E and satisfies

$$\max_{x\in S_k}\langle x^* - x_0^*, x\rangle \leq \delta k$$

for all $k \in \mathbb{N}$ with $k < n$ as well as

$$\|x^* - x_0^*\| = \sup_{x\in B}|\langle x^* - x_0^*, x\rangle| \leq \delta n$$

in contradiction to (3.3.4). This contradiction shows that there exists a finite set $S \subset B$ such that $E(S) = \emptyset$. With this understood, Step 2 follows from the axiom of dependent choice (see page 6).

Step 3. *Let the constant $\delta > 0$ and the sequence of finite subsets $S_n \subset B$ for $n \in \mathbb{N}$ be as in Step 2. Choose a sequence $(x_i)_{i\in\mathbb{N}}$ in B such that*

$$\bigcup_{n\in\mathbb{N}} \frac{1}{n} S_n = \{x_1, x_2, x_3, \dots\}.$$

Then

$$\sup_{i\in\mathbb{N}} |\langle x^* - x_0^*, x_i\rangle| > \delta$$

for all $x^* \in E$.

Let $x^* \in E$ and choose an integer

$$n \geq \delta^{-1}\|x^* - x_0^*\|.$$

Then $\|x^* - x_0^*\| \leq \delta n$ and therefore $n \geq 2$ by (3.3.3). Hence, by Step 2, there exist an integer $k \in \{1, \dots, n-1\}$ and an element $x \in S_k$ such that

$$|\langle x^* - x_0^*, x\rangle| > \delta k.$$

Choose $i \in \mathbb{N}$ such that $k^{-1}x = x_i$. Then

$$|\langle x^* - x_0^*, x_i\rangle| > \delta$$

and this proves Step 3.

Step 4. *Let $(x_i)_{i\in\mathbb{N}}$ be as in Step 3. Then $\lim_{i\to\infty}\|x_i\| = 0$. Moreover, there exists a summable sequence $\alpha = (\alpha_i)_{i\in\mathbb{N}} \in \ell^1$ such that*

$$\sum_{i=1}^{\infty}\alpha_i\langle x_0^*, x_i\rangle = 1, \quad \sum_{i=1}^{\infty}\alpha_i\langle x^*, x_i\rangle = 0 \text{ for all } x^* \in E, \quad \sum_{i=1}^{\infty}|\alpha_i| \le \delta^{-1}.$$

It follows from the definition that $\lim_{i\to\infty}\|x_i\| = 0$. Define the bounded linear operator $T : X^* \to c_0$ (with values in the Banach space $c_0 \subset \ell^\infty$ of sequences of real numbers that converge to zero) by

$$Tx^* := (\langle x^*, x_i\rangle)_{i\in\mathbb{N}} \qquad \text{for } x^* \in X^*.$$

Then, by Step 3,

$$\|Tx^* - Tx_0^*\|_\infty > \delta \qquad \text{for all } x^* \in E.$$

Hence it follows from the Hahn–Banach Theorem 2.3.22 with $Y = T(E)$ and Example 1.3.7 that there exists an element $\beta = (\beta_i)_{i\in\mathbb{N}} \in \ell^1 \cong c_0^*$ such that

$$\langle \beta, Tx_0^*\rangle \ge \delta, \qquad \langle \beta, Tx^*\rangle = 0 \text{ for all } x^* \in E^*, \qquad \|\beta\|_1 = 1.$$

Hence the sequence $\alpha = (\alpha_i)_{i\in\mathbb{N}} \in \ell^1$ with the entries $\alpha_i := \langle \beta, Tx_0^*\rangle^{-1}\beta_i$ for $i \in \mathbb{N}$ satisfies the requirements of Step 4.

Step 5. *Let $(x_i)_{i\in\mathbb{N}}$ be the sequence in Step 3 and let $(\alpha_i)_{i\in\mathbb{N}}$ be the summable sequence of real numbers in Step 4. Then the limit*

$$x_0 := \sum_{i=1}^{\infty}\alpha_i x_i = \lim_{n\to\infty}\sum_{i=1}^{n}\alpha_i x_i \tag{3.3.5}$$

exists in X and satisfies the requirements of Theorem 3.3.1.

Since $\|x_i\| \le 1$ for all $i \in \mathbb{N}$, we have

$$\sum_{i=1}^{\infty}\|\alpha_i x_i\| \le \sum_{i=1}^{\infty}|\alpha_i| \le \delta^{-1}.$$

Since X is a Banach space, this implies that the limit (3.3.5) exists and satisfies $\|x_0\| \le \delta^{-1}$ (see Lemma 1.5.1). Moreover, by Step 4,

$$\langle x_0^*, x_0\rangle = \sum_{i=1}^{\infty}\alpha_i\langle x_0^*, x_i\rangle = 1, \qquad \langle x^*, x_0\rangle = \sum_{i=1}^{\infty}\alpha_i\langle x^*, x_i\rangle = 0$$

for all $x^* \in E$. This proves Theorem 3.3.1. □

3.4. The Eberlein–Šmulyan Theorem

If X is a reflexive Banach space, then the weak and weak* topologies agree on its dual space $X^* = \mathcal{L}(X, \mathbb{R})$, hence the closed unit ball in X^* is weakly compact by the Banach–Alaoglu Theorem 3.2.4, and so the closed unit ball in X is also weakly compact. The Eberlein–Šmulyan Theorem asserts that this property characterizes reflexivity. It also asserts that weak compactness of the closed unit ball is equivalent to sequential weak compactness.

Theorem 3.4.1 (Eberlein–Šmulyan). *Let X be a real Banach space and let $B := \{x \in X \mid \|x\| \le 1\}$ be the closed unit ball. Then the following are equivalent.*

(i) *X is reflexive.*

(ii) *B is weakly compact.*

(iii) *B is sequentially weakly compact.*

(iv) *Every bounded sequence in X has a weakly convergent subsequence.*

Proof. See page 136. □

Remark 3.4.2 (James' Theorem). A theorem of Robert C. James [**39**] asserts the following.

Let $C \subset X$ be a nonempty bounded weakly closed subset of a Banach space over the reals. Then C is weakly compact if and only if every bounded linear functional on X attains its maximum over C.

That the condition is necessary for weak compactness follows from the fact that every bounded linear functional on X is continuous with respect to the weak topology (Lemma 3.1.16). The converse is highly nontrivial and requires the construction of a bounded linear functional on X that fails to attain its maximum over C whenever C is not weakly compact. This goes beyond the scope of this book and we refer to the original paper by James [**39**] as well as the work of Holmes [**36**] and Pryce [**70**].

Combining James' Theorem with Theorem 3.4.1, one obtains the following result [**40**]. *A Banach space X is reflexive if and only if, for every bounded linear functional $x^* \in X^*$, there exists an element $x \in X$ such that*

$$\|x\| = 1, \qquad \langle x^*, x \rangle = \|x^*\| .$$

If X is reflexive, the existence of such an element x can be deduced from the Hahn–Banach Theorem (Corollary 2.3.23).

The proof of Theorem 3.4.1 relies on Helly's Theorem, a precursor to the Hahn–Banach Theorem proved in 1921, which shows when a finite system of linear equations has a solution.

Lemma 3.4.3 (Helly's Theorem). *Let X be a real normed vector space and let $x_1^*, \dots, x_n^* \in X^*$ and $c_1, \dots, c_n \in \mathbb{R}$. Fix a real number $M \geq 0$. Then the following are equivalent.*

(i) *For every $\varepsilon > 0$ there exists an $x \in X$ such that*

$$\|x\| < M + \varepsilon, \qquad \langle x_i^*, x\rangle = c_i \text{ for } i = 1, \dots, n. \tag{3.4.1}$$

(ii) *Every vector $\lambda = (\lambda_1, \dots, \lambda_n) \in \mathbb{R}^n$ satisfies the inequality*

$$\left|\sum_{i=1}^n \lambda_i c_i\right| \leq M \left\|\sum_{i=1}^n \lambda_i x_i^*\right\|. \tag{3.4.2}$$

Proof. We prove that (i) implies (ii). Fix a constant $\varepsilon > 0$. By (i) there exists a vector $x \in X$ such that (3.4.1) holds. Hence

$$\begin{aligned}\left|\sum_{i=1}^n \lambda_i c_i\right| &= \left|\left\langle\sum_{i=1}^n \lambda_i x_i^*, x\right\rangle\right| \\ &\leq \|x\| \left\|\sum_{i=1}^n \lambda_i x_i^*\right\| \\ &\leq (M+\varepsilon)\left\|\sum_{i=1}^n \lambda_i x_i^*\right\|.\end{aligned}$$

Since $\varepsilon > 0$ was chosen arbitrarily, this proves (ii).

We prove that (ii) implies (i). Thus assume (ii) holds and suppose first that $x_1^*, \dots, x_n^*$ are linearly independent. Then, by Lemma 3.1.13, there exist vectors $x_1, \dots, x_n \in X$ such that $\langle x_i^*, x_j\rangle = \delta_{ij}$ for $i, j = 1, \dots, n$. Define

$$Z := {}^\perp\{x_1^*, \dots, x_n^*\}.$$

We prove that $Z^\perp = \mathrm{span}\{x_1^*, \dots, x_n^*\}$. Let $x^* \in Z^\perp$. Then, for all $x \in X$,

$$x - \sum_{i=1}^n \langle x_i^*, x\rangle x_i \in Z$$

and hence

$$0 = \left\langle x^*, x - \sum_{i=1}^n \langle x_i^*, x\rangle x_i\right\rangle = \left\langle x^* - \sum_{i=1}^n \langle x^*, x_i\rangle x_i^*, x\right\rangle.$$

This shows that $x^* = \sum_{i=1}^n \langle x^*, x_i\rangle x_i^* \in \mathrm{span}\{x_1^*, \dots, x_n^*\}$ for all $x^* \in Z^\perp$. The converse inclusion is obvious.

Now define

$$x := \sum_{j=1}^{n} c_j x_j.$$

Then $\langle x_i^*, x\rangle = c_i$ for $i = 1, \dots, n$ and every other solution of this equation has the form $x + z$ with $z \in Z$. Hence it follows from Corollary 2.4.2 that

$$\begin{aligned}
\inf_{z\in Z} \|x + z\| &= \sup_{x^*\in Z^\perp} \frac{|\langle x^*, x\rangle|}{\|x^*\|} \\
&= \sup_{\lambda\in\mathbb{R}^n} \frac{|\langle \sum_i \lambda_i x_i^*, x\rangle|}{\|\sum_i \lambda_i x_i^*\|} \\
&= \sup_{\lambda\in\mathbb{R}^n} \frac{|\sum_i \lambda_i c_i|}{\|\sum_i \lambda_i x_i^*\|} \\
&\le M.
\end{aligned}$$

This proves (i) for linearly independent n-tuples $x_1^*, \dots, x_n^* \in X^*$.

To prove the result in general, choose a subset $J \subset \{1, \dots, n\}$ such that the x_j^* for $j \in J$ are linearly independent and span the same subspace as $x_1^*, \dots, x_n^*$. Fix a constant $\varepsilon > 0$. Then, by what we have just proved, there exists an $x \in X$ such that $\|x\| < M + \varepsilon$ and $\langle x_j^*, x\rangle = c_j$ for $j \in J$. Let $i \in \{1, \dots, n\} \setminus J$. Then there exist real numbers λ_j for $j \in J$ such that $\sum_{j\in J} \lambda_j x_j^* = x_i^*$. Hence $\sum_{j\in J} \lambda_j c_j = c_i$ by (3.4.2) and so $\langle x_i^*, x\rangle = c_i$. Thus x satisfies (3.4.1) and this proves Lemma 3.4.3. □

Proof of Theorem 3.4.1. Assume X is reflexive. Then $\iota : X \to X^{**}$ is a Banach space isometry and hence is a homeomorphism with respect to the weak topology on both spaces. Since X^* is reflexive by Theorem 2.4.4, the weak topology on X^{**} agrees with the weak* topology. Hence it follows from Theorem 3.2.4 that the closed unit ball $B^{**} \subset X^{**}$ is weakly compact, and hence so is the closed unit ball $B \subset X$. This shows that (i) implies (ii).

We prove that (ii) implies (i). Thus assume that the closed unit ball in X is weakly compact and fix a nonzero element $x^{**} \in X^{**}$.

Claim. *For every finite set $S \subset X^*$ there is an element $x^* \in X^*$ such that*

$$\|x\| \le 2\,\|x^{**}\|, \qquad \langle x^*, x\rangle = \langle x^{**}, x^*\rangle \quad \textit{for all } x \in S.$$

To see this, write $S = \{x_1^*, \dots, x_n^*\}$ and define $c_i := \langle x^{**}, x_i^*\rangle$ for $i = 1, \dots, n$. Then every vector $\lambda = (\lambda_1, \dots, \lambda_n) \in \mathbb{R}^n$ satisfies the inequality

$$\left|\sum_{i=1}^{n} \lambda_i c_i\right| = \left|\left\langle x^{**}, \sum_{i=1}^{n} \lambda_i x_i^*\right\rangle\right| \le \|x^{**}\| \left\|\sum_{i=1}^{n} \lambda_i x_i^*\right\|.$$

Thus the claim follows from Lemma 3.4.3 with $\varepsilon := M := \|x^{**}\| > 0$.

We prove that x^{**} belongs to the image of the inclusion $\iota : X \to X^{**}$. Denote by $\mathscr{S} \subset 2^{X^*}$ the set of all finite subsets $S \subset X^*$. For $S \in \mathscr{S}$ define

$$K(S) := \left\{ x \in X \,\middle|\, \|x\| \le 2\,\|x^{**}\| \text{ and } \langle x^*, x\rangle = \langle x^{**}, x^*\rangle \text{ for all } x^* \in S \right\}.$$

Then, for every finite set $S \subset X^*$, the set $K(S)$ is nonempty by the claim, is weakly closed by Lemma 3.1.15, and is contained in cB, where $c := 2\,\|x^{**}\|$. The set cB is weakly compact by (ii) and the collection $\{K(S) \,|\, S \in \mathscr{S}\}$ has the finite intersection property because

$$\bigcap_{i=1}^m K(S_i) = K\left(\bigcup_{i=1}^m S_i\right) \ne \emptyset \qquad \text{for all } S_1, \dots, S_m \in \mathscr{S}.$$

Hence

$$\bigcap_{S \in \mathscr{S}} K(S) \ne \emptyset$$

and so there exists an $x \in X$ such that $x \in K(S)$ for all $S \subset \mathscr{S}$. This shows that $\langle x^*, x\rangle = \langle x^{**}, x^*\rangle$ for all $x^* \in X^*$, and thus $x^{**} = \iota(x)$. Thus we have proved that (ii) implies (i).

We prove that (i) implies (iii). Assume first that X is separable as well as reflexive. Then X^* is separable by Theorem 2.4.6 and is reflexive by Theorem 2.4.4. Let $(x_n)_{n\in\mathbb{N}}$ be a sequence in the closed unit ball $B \subset X$. Then $(\iota(x_n))_{n\in\mathbb{N}}$ is a bounded sequence in X^{**} and hence has a weak* convergent subsequence $(\iota(x_{n_i}))_{i\in\mathbb{N}}$ by Theorem 3.2.1. Since $\iota : X \to X^{**}$ is a homeomorphism with respect to the weak topologies of X and X^*, it follows that the sequence $(x_{n_i})_{i\in\mathbb{N}}$ converges weakly to an element $x \in X$. Since $x_{n_i} \in B$ for all $i \in \mathbb{N}$, it then follows from Lemma 3.1.20 that $x \in B$. This shows that the closed unit ball $B \subset X$ is sequentially weakly compact whenever X is reflexive and separable.

Now assume X is reflexive and let $(x_n)_{n\in\mathbb{N}}$ be a sequence in the closed unit ball $B \subset X$. Let $Y := \overline{\operatorname{span}\{x_n \,|\, n \in \mathbb{N}\}}$ be the smallest closed subspace of X that contains the sequence $(x_n)_{n\in\mathbb{N}}$. Then Y is reflexive by Theorem 2.4.4 and Y is separable by definition. Hence the sequence $(x_n)_{n\in\mathbb{N}}$ has a subsequence that converges weakly to an element of B. Thus B is sequentially weakly compact. This shows that (i) implies (iii).

We prove that (iii) implies (iv). If $(x_n)_{n\in\mathbb{N}}$ is a bounded sequence, then there exists a constant $c > 0$ such that $\|x_n\| \le c$ for all $n \in \mathbb{N}$, hence the sequence $(c^{-1}x_n)_{n\in\mathbb{N}}$ in B has a weakly convergent subsequence by (iii), and hence so does the original sequence $(x_n)_{n\in\mathbb{N}}$. This shows that (iii) implies (iv).

We prove that (iv) implies (i). Thus assume (iv) and choose an element $x_0^{**} \in X^{**}$ such that $\|x_0^{**}\| \le 1$. We prove in three steps that x_0^{**} belongs to the image of the inclusion $\iota : X \to X^{**}$ in (2.4.1).

Step 1. *Let $n \in \mathbb{N}$ and $x_1^*, \dots, x_n^* \in X^*$. Then there is an $x \in X$ such that*

$$\|x\| \le 1, \qquad \langle x_i^*, x \rangle = \langle x_0^{**}, x_i^* \rangle \qquad \text{for } i = 1, \dots, n. \tag{3.4.3}$$

Denote by $S \subset X$ the unit sphere and recall from Corollary 3.1.29 that the weak* closure of $\iota(S)$ is the closed unit ball $B^{**} \subset X^{**}$. For $m \in \mathbb{N}$ the set

$$U_m := \left\{ x^{**} \in X^{**} \,\middle|\, |\langle x^{**} - x_0^{**}, x_i^* \rangle| < \frac{1}{m} \text{ for } i = 1, \dots, n \right\}$$

is a weak* open neighborhood of $x_0^{**} \in B^{**}$ and so $U_m \cap \iota(S) \ne \emptyset$. Hence, by the axiom of countable choice, there exists a sequence $(x_m)_{m\in\mathbb{N}}$ in X such that

$$\|x_m\| = 1, \qquad \iota(x_m) \in U_m \qquad \text{for all } m \in \mathbb{N}.$$

This sequence satisfies

$$|\langle x_i^*, x_m \rangle - \langle x_0^{**}, x_i^* \rangle| < \frac{1}{m} \qquad \text{for all } m \in \mathbb{N} \text{ and } i = 1, \dots, n.$$

By (iv), there exists a weakly convergent subsequence $(x_{m_k})_{k\in\mathbb{N}}$. Denote the weak limit by x. It satisfies $\|x\| \le 1$ by Lemma 3.1.20 and

$$\langle x_i^*, x \rangle = \lim_{k\to\infty} \langle x_i^*, x_{m_k} \rangle = \langle x_0^{**}, x_i^* \rangle \qquad \text{for } i = 1, \dots, n.$$

This proves Step 1.

Step 2. *Define*

$$E := \left\{ x^* \in X^* \,\middle|\, \langle x_0^{**}, x^* \rangle = 0 \right\}$$

and let $B^* \subset X^*$ be the closed unit ball. Then $E \cap B^*$ is weak closed.*

Fix an element x_0^* in the weak* closure of $E \cap B^*$. Then $x_0^* \in B^*$ by Corollary 3.1.28. We must prove that $x_0^* \in E$. Fix a constant $\varepsilon > 0$. We claim that there are sequences $x_n \in B$ and $x_n^* \in E \cap B^*$ such that, for all $n \in \mathbb{N}$,

$$\langle x_i^*, x_n \rangle = \langle x_0^{**}, x_i^* \rangle = \begin{cases} \langle x_0^{**}, x_0^* \rangle, & \text{if } i = 0, \\ 0, & \text{if } i \ge 1, \end{cases} \qquad \text{for } i = 0, \dots, n-1, \tag{3.4.4}$$

and

$$|\langle x_n^* - x_0^*, x_i \rangle| < \varepsilon \qquad \text{for } i = 1, \dots, n. \tag{3.4.5}$$

By Step 1 there exists an element $x_1 \in B$ such that $\langle x_0^*, x_1 \rangle = \langle x_0^{**}, x_0^* \rangle$. Thus x_1 satisfies (3.4.4) for $n = 1$. Moreover, since x_0^* belongs to the weak* closure of $E \cap B^*$, there exists an element $x_1^* \in E \cap B^*$ such that

$$|\langle x_1^* - x_0^*, x_1 \rangle| < \varepsilon.$$

Thus x_1^* satisfies (3.4.5) for $n = 1$.

Now let $n \in \mathbb{N}$ and suppose that $x_i \in B$ and $x_i^* \in E \cap B^*$ have been found for $i = 1, \dots, n$ such that (3.4.4) and (3.4.5) are satisfied. Then, by Step 1, there is an element $x_{n+1} \in B$ such that

$$\langle x_i^*, x_{n+1} \rangle = \langle x_0^{**}, x_i^* \rangle \qquad \text{for } i = 0, \dots, n.$$

Furthermore, since x_0^* belongs to the weak* closure of $E \cap B^*$, there exists an element $x_{n+1}^* \in E \cap B^*$ such that

$$\left| \langle x_{n+1}^* - x_0^*, x_i \rangle \right| < \varepsilon \qquad \text{for } i = 1, \dots, n+1.$$

By the axiom of dependent choice (page 6), this shows that there exist sequences $x_n \in B$ and $x_n^* \in E \cap B^*$ that satisfy (3.4.4) and (3.4.5).

Since $\|x_n\| \le 1$ for all $n \in \mathbb{N}$, it follows from (iv) that there exists a weakly convergent subsequence $(x_{n_k})_{k \in \mathbb{N}}$. Denote the limit by x_0. Then

$$\langle x_m^*, x_0 \rangle = \lim_{k \to \infty} \langle x_m^*, x_{n_k} \rangle = \langle x_0^{**}, x_m^* \rangle = 0 \qquad \text{for all } m \in \mathbb{N}. \tag{3.4.6}$$

Here the second equation follows from (3.4.4) and the last equation follows from the fact that $x_m^* \in E \cap B^*$ for $m \ge 1$. Moreover, Lemma 3.1.20 asserts that $x_0 \in B$ and that there exist an $m \in \mathbb{N}$ and $\lambda_1, \dots, \lambda_m \in \mathbb{R}$ such that

$$\lambda_i \ge 0, \qquad \sum_{i=1}^m \lambda_i = 1, \qquad \left\| x_0 - \sum_{i=1}^m \lambda_i x_i \right\| < \varepsilon. \tag{3.4.7}$$

Hence

$$\begin{aligned}
\left| \langle x_0^{**}, x_0^* \rangle \right| &\le \left| \langle x_0^{**}, x_0^* \rangle - \sum_{i=1}^m \lambda_i \langle x_m^*, x_i \rangle \right| + \left| \left\langle x_m^*, \sum_{i=1}^m \lambda_i x_i - x_0 \right\rangle \right| \\
&\le \sum_{i=1}^m \lambda_i \left| \langle x_0^{**}, x_0^* \rangle - \langle x_m^*, x_i \rangle \right| + \left\| \sum_{i=1}^m \lambda_i x_i - x_0 \right\| \\
&= \sum_{i=1}^m \lambda_i \left| \langle x_0^* - x_m^*, x_i \rangle \right| + \left\| \sum_{i=1}^m \lambda_i x_i - x_0 \right\| \\
&< 2\varepsilon.
\end{aligned}$$

Here the first step uses equation (3.4.6), the second step uses (3.4.7), the third step uses the equation $\langle x_0^{**}, x_0^* \rangle = \langle x_0^*, x_i \rangle$ in (3.4.4), and the last step follows from (3.4.5), (3.4.6), and (3.4.7). Thus $|\langle x_0^{**}, x_0^* \rangle| < 2\varepsilon$ for all $\varepsilon > 0$, therefore $\langle x_0^{**}, x_0^* \rangle = 0$, and so $x_0^* \in E \cap B^*$. This proves Step 2.

Step 3. *There exists an element $x_0 \in X$ such that $\iota(x_0) = x_0^{**}$.*

By Corollary 3.3.2, the linear subspace $E \subset X^*$ in Step 2 is weak* closed. (This is the only place in the proof where we use the fact that X is complete.) Hence it follows from Corollary 3.1.25 that there exists an element $x_0 \in X$ such that $\langle x^*, x_0 \rangle = \langle x_0^{**}, x^* \rangle$ for all $x^* \in X^*$. This proves Step 3 and Theorem 3.4.1. □

3.5. The Kreĭn–Milman Theorem

The Kreĭn–Milman Theorem [**47, 60**] is a general result about compact convex subsets of a locally convex Hausdorff topological vector space. It asserts that every such convex subset is the closed convex hull of its set of extremal points. In particular, the result applies to the dual space of a Banach space, equipped with the weak* topology. Here are the relevant definitions.

Definition 3.5.1 (Extremal Point and Face). Let X be a real vector space and let $K \subset X$ be a nonempty convex subset. A subset

$$F \subset K$$

is called a **face of** K if F is a nonempty convex subset of K and

$$\begin{array}{l} x_0, x_1 \in K,\ 0 < \lambda < 1, \\ (1-\lambda)x_0 + \lambda x_1 \in F \end{array} \quad \Longrightarrow \quad x_0, x_1 \in F. \tag{3.5.1}$$

An element $x \in K$ is called an **extremal point of** K if

$$\begin{array}{l} x_0, x_1 \in K,\ 0 < \lambda < 1, \\ (1-\lambda)x_0 + \lambda x_1 = x \end{array} \quad \Longrightarrow \quad x_0 = x_1 = x. \tag{3.5.2}$$

This means that the singleton $F := \{x\}$ is a face of K or, equivalently, that there is no open line segment in K that contains x (see Figure 3.5.1). Denote the set of extremal points of K by

$$\mathcal{E}(K) := \{x \in K \,|\, x \text{ satisfies (3.5.2)}\}.$$

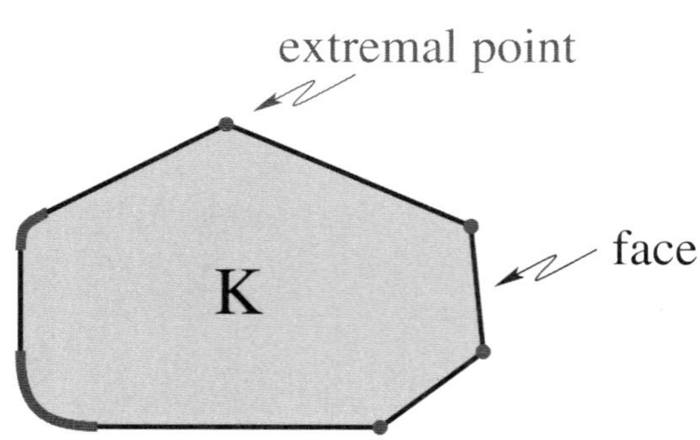

Figure 3.5.1. Extremal points and faces.

Recall that the convex hull of a set $E \subset X$ is denoted by $\mathrm{conv}(E)$ and that its closure, the closed convex hull of E, is denoted by $\overline{\mathrm{conv}}(E)$ whenever X is a topological vector space (see Definition 3.1.19).

Theorem 3.5.2 (Kreĭn–Milman). *Let X be a locally convex Hausdorff topological vector space and let $K \subset X$ be a nonempty compact convex set. Then K is the closed convex hull of its extremal points, i.e. $K = \overline{\mathrm{conv}}(\mathcal{E}(K))$. In particular, K admits an extremal point, i.e. $\mathcal{E}(K) \neq \emptyset$.*

Proof. The proof has five steps.

Step 1. *Let*

$$\mathscr{K} := \{K \subset X \mid K \text{ is a nonempty compact convex set}\}$$

and define the relation $\preccurlyeq$ *on* $\mathscr{K}$ *by*

$$F \preccurlyeq K \quad \stackrel{\text{def}}{\Longleftrightarrow} \quad F \text{ is a face of } K \tag{3.5.3}$$

for $F, K \in \mathscr{K}$. *Then* $(\mathscr{K}, \preccurlyeq)$ *is a partially ordered set and every nonempty chain* $\mathscr{C} \subset \mathscr{K}$ *has an infimum.*

That the relation (3.5.3) is a partial order follows directly from the definition. Moreover, every element $K \in \mathscr{K}$ is a closed set because X is Hausdorff. This implies that every nonempty chain $\mathscr{C} \subset \mathscr{K}$ has an infimum

$$C_0 := \bigcap_{C \in \mathscr{C}} C.$$

This proves Step 1.

Step 2. *If* $K \in \mathscr{K}$ *and* $\Lambda : X \to \mathbb{R}$ *is a continuous linear functional, then*

$$F := K \cap \Lambda^{-1}(\sup_K \Lambda) \in \mathscr{K}$$

and $F \preccurlyeq K$.

Abbreviate $c := \sup_K \Lambda$. Since K is compact and Λ is continuous, the set $F = K \cap \Lambda^{-1}(c)$ is nonempty. Since K is closed and Λ is continuous, the set F is a closed subset of K and hence is compact. Since K is convex and Λ is linear, F is convex. Thus $F \in \mathscr{K}$.

To prove that F is a face of K, fix two elements $x_0, x_1 \in K$ and a real number $0 < \lambda < 1$ such that

$$x := (1 - \lambda)x_0 + \lambda x_1 \in F.$$

Then $(1 - \lambda)\Lambda(x_0) + \lambda\Lambda(x_1) = \Lambda(x) = c$ and hence

$$(1 - \lambda)(c - \Lambda(x_0)) + \lambda(c - \Lambda(x_1)) = 0.$$

Since $c - \Lambda(x_0) \geq 0$ and $c - \Lambda(x_1) \geq 0$, this implies

$$\Lambda(x_0) = \Lambda(x_1) = c$$

and hence $x_0, x_1 \in F$. Thus F is a face of K. This proves Step 2.

Step 3. *Every minimal element of $\mathscr{K}$ is a singleton.*

Fix an element $K \in \mathscr{K}$ which is not a singleton and choose two elements $x_0, x_1 \in K$ such that $x_0 \neq x_1$. Since X is a locally convex Hausdorff space, there exists a convex open set $U_1 \subset X$ such that $x_1 \in U_1$ and $x_0 \notin U_1$. Hence it follows from Theorem 3.1.11 that there exists a continuous linear functional $\Lambda : X \to \mathbb{R}$ such that $\Lambda(x_0) < \Lambda(x)$ for all $x \in U_1$ and so

$$\Lambda(x_0) < \Lambda(x_1).$$

By Step 2, the set $F := K \cap \Lambda^{-1}(\sup_K \Lambda)$ is a face of K and $x_0 \in K \setminus F$. Thus K is not a minimal element of $\mathscr{K}$.

Step 4. *Let $K \in \mathscr{K}$. Then $\mathcal{E}(K) \neq \emptyset$.*

By Step 1 and the Lemma of Zorn, there exists a minimal element $E \in \mathscr{K}$ such that $E \preccurlyeq K$. By Step 3,

$$E = \{x\}$$

is a singleton. Hence $x \in \mathcal{E}(K)$.

Step 5. *Let $K \in \mathscr{K}$. Then $K = \overline{\mathrm{conv}}(\mathcal{E}(K))$.*

It follows directly from the definitions that

$$\overline{\mathrm{conv}}(\mathcal{E}(K)) \subset K.$$

To prove the converse inclusion, assume, by contradiction, that there exists an element

$$x \in K \setminus \overline{\mathrm{conv}}(\mathcal{E}(K)).$$

Since X is a locally convex Hausdorff space, there exists an open convex set $U \subset X$ such that

$$x \in U, \qquad U \cap \overline{\mathrm{conv}}(\mathcal{E}(K)) = \emptyset.$$

Since $\mathcal{E}(K)$ is nonempty by Step 4, it follows from Theorem 3.1.11 that there exists a continuous linear functional $\Lambda : X \to \mathbb{R}$ such that

$$\Lambda(x) > \sup_{\overline{\mathrm{conv}}(\mathcal{E}(K))} \Lambda. \tag{3.5.4}$$

By Step 2, the set

$$F := K \cap \Lambda^{-1}(\sup_K \Lambda)$$

is a face of K and

$$F \cap \mathcal{E}(K) = \emptyset.$$

by (3.5.4). By Step 3, the set F has an extremal point x_0. Then x_0 is also an extremal point of K in contradiction to the fact that $F \cap \mathcal{E}(K) = \emptyset$. This proves Theorem 3.5.2. □

Example 3.5.3. This example shows that the extremal set of a compact convex set need not be compact. Let X be an infinite-dimensional reflexive Banach space. Assume X is **strictly convex**, i.e. for all $x, y \in X$,

$$\|x+y\| = 2\|x\| = 2\|y\|. \quad \Longrightarrow \quad x = y \tag{3.5.5}$$

Then the closed unit ball $B \subset X$ is weakly compact by Theorem 3.4.1 and its extremal set is the unit sphere $\mathcal{E}(B) = S$ (see Exercises 2.5.11 and 3.7.14). Thus the extremal set is not weakly compact and B is the weak closure of its extremal set by Lemma 3.1.21. **Exercise:** Prove that $B = \operatorname{conv}(S)$.

Example 3.5.4 (Infinite-Dimensional Simplex). The infinite product $\mathbb{R}^{\mathbb{N}}$ is a locally convex Hausdorff space with the product topology, induced by the metric

$$d(x,y) := \sum_{i=1}^{\infty} 2^{-i} \frac{|x_i - y_i|}{1 + |x_i - y_i|}$$

for $x = (x_i)_{i\in\mathbb{N}}$ and $y = (y_i)_{i\in\mathbb{N}}$ in $\mathbb{R}^{\mathbb{N}}$. The **infinite-dimensional simplex**

$$\Delta := \left\{ x = (x_i)_{i\in\mathbb{N}} \in \mathbb{R}^{\mathbb{N}} \,\middle|\, x_i \geq 0,\, \sum_{i=1}^{\infty} x_i \leq 1 \right\}$$

is a compact convex subset of $\mathbb{R}^{\mathbb{N}}$ by Tychonoff's Theorem A.2.1. Its set of extremal points is the compact set

$$\mathcal{E}(\Delta) = \{e_i \,|\, i \in \mathbb{N}\} \cup \{0\}, \qquad e_i := (\delta_{ij})_{j\in\mathbb{N}}.$$

The convex hull of $\mathcal{E}(\Delta)$ is strictly contained in Δ and hence is not compact. **Exercise:** The product topology on the infinite-dimensional simplex agrees with the weak* topology it inherits as a subset of $\ell^1 = c_0^*$ (see Example 1.3.7).

Example 3.5.5 (Hilbert Cube). The **Hilbert cube** is the set

$$Q := \left\{ x = (x_i)_{i\in\mathbb{N}} \in \mathbb{R}^{\mathbb{N}} \,|\, 0 \leq x_i \leq 1/i \right\}.$$

This is a compact convex subset of $\mathbb{R}^{\mathbb{N}}$ with respect to the product topology. Its set of extremal points is the compact set

$$\mathcal{E}(Q) = \left\{ x = (x_i)_{i\in\mathbb{N}} \in \mathbb{R}^{\mathbb{N}} \,|\, x_i \in \{0, 1/i\} \right\}.$$

The convex hull of any finite subset of $\mathcal{E}(Q)$ is nowhere dense in Q. Hence

$$\operatorname{conv}(\mathcal{E}(Q)) \subsetneq Q$$

by the Baire Category Theorem 1.6.4. **Exercise:** The product topology on the Hilbert cube agrees with the topology induced by the ℓ^2 norm.

3.6. Ergodic Theory

This section establishes the existence of an ergodic measure for any homeomorphism of a compact metric space. The proof is a fairly straightforward consequence of the Banach–Alaoglu Theorem 3.2.1 and the Kreĭn–Milman Theorem 3.5.2. We also show that the ergodic measures are precisely the extremal points of the convex set of all invariant measures (Theorem 3.6.3). The proof that every ergodic measure is extremal requires von Neumann's Mean Ergodic Theorem 3.6.5, the proof of which will in turn be based on an abstract ergodic theorem for operators on Banach spaces (Theorem 3.6.9).

3.6.1. Ergodic Measures. Let (M, d) be a compact metric space and let $\phi : M \to M$ be a homeomorphism. Denote by $\mathcal{B} \subset 2^M$ the Borel σ-algebra. Recall that the set $\mathcal{M}(\phi)$ of all ϕ-invariant Borel probability measures on M is a nonempty weak* compact convex subset of the space $\mathcal{M}(M) = C(M)^*$ of all signed Borel measures on M (see Subsection 3.2.2 and Corollary 3.2.6).

Definition 3.6.1 (Ergodic Measure). A ϕ-invariant Borel probability measure $\mu : \mathcal{B} \to [0, 1]$ is called **ϕ-ergodic** if, for every Borel set $B \subset M$,

$$\phi(B) = B \qquad \Longrightarrow \qquad \mu(B) \in \{0, 1\}. \tag{3.6.1}$$

The homeomorphism ϕ is called **μ-ergodic** if μ is an ergodic measure for ϕ.

Example 3.6.2. If $x \in M$ is a fixed point of ϕ, then the Dirac measure $\mu = \delta_x$ is ergodic for ϕ. If $\phi = \mathrm{id}$, then the Dirac measure at each point of M is ergodic for ϕ and there are no other ergodic measures.

Theorem 3.6.3 (Ergodic Measures are Extremal). *Let $\mu : \mathcal{B} \to [0, 1]$ be a ϕ-invariant Borel probability measure. Then the following are equivalent.*

(i) *μ is an ergodic measure for ϕ.*

(ii) *μ is an extremal point of $\mathcal{M}(\phi)$.*

Proof. We prove that (ii) implies (i) by an indirect argument. Assume that μ is not ergodic for ϕ. Then there exists a Borel set $\Lambda \subset M$ such that

$$\phi(\Lambda) = \Lambda, \qquad 0 < \mu(\Lambda) < 1.$$

Define $\mu_0, \mu_1 : \mathcal{B} \to [0, 1]$ by

$$\mu_0(B) := \frac{\mu(B \setminus \Lambda)}{1 - \mu(\Lambda)}, \qquad \mu_1(B) := \frac{\mu(B \cap \Lambda)}{\mu(\Lambda)}$$

for $B \in \mathcal{B}$. These are ϕ-invariant Borel probability measures and they are not equal because $\mu_0(\Lambda) = 0$ and $\mu_1(\Lambda) = 1$. Moreover, $\mu = (1 - \lambda)\mu_0 + \lambda\mu_1$ where $\lambda := \mu(\Lambda)$. Hence μ is not an extremal point of $\mathcal{M}(\phi)$. This shows that (ii) implies (i). The converse is proved on page 146. □

Corollary 3.6.4 (Existence of Ergodic Measures). *Every homeomorphism of a compact metric space admits an ergodic measure.*

Proof. The set $\mathcal{M}(\phi)$ of ϕ-invariant Borel probability measures on M is nonempty by Lemma 3.2.3 and is a weak* compact convex subset of $\mathcal{M}(M)$ by Corollary 3.2.6. Hence $\mathcal{M}(\phi)$ has an extremal point μ by Theorem 3.5.2. Thus μ is an ergodic measure by (ii) $\implies$ (i) in Theorem 3.6.3. $\square$

3.6.2. Space and Time Averages. Given a homeomorphism

$$\phi : M \to M$$

of a compact metric space M, a ϕ-ergodic measure

$$\mu : \mathcal{B} \to [0, 1]$$

on the Borel σ-algebra $\mathcal{B} \subset 2^M$, a continuous function $f : M \to \mathbb{R}$, and an element $x \in M$, one can ask the question of whether the sequence of averages $\frac{1}{n}\sum_{k=0}^{n-1} f(\phi^k(x))$ converges. A theorem of Birkhoff [**13**] answers this question in the affirmative for almost every $x \in M$. This is **Birkhoff's Ergodic Theorem**. It asserts that, if μ is a ϕ-ergodic measure, then for every continuous function $f : M \to \mathbb{R}$, there exists a Borel set $\Lambda \subset M$ such that

$$\phi(\Lambda) = \Lambda, \qquad \mu(\Lambda) = 1, \tag{3.6.2}$$

and

$$\int_M f \, d\mu = \lim_{n\to\infty} \frac{1}{n} \sum_{k=0}^{n-1} f(\phi^k(x)) \qquad \text{for all } x \in \Lambda. \tag{3.6.3}$$

In other words, the *time average* of f agrees with the *space average* for almost every orbit of the dynamical system. If ϕ is **uniquely ergodic**, i.e. ϕ admits only one ergodic measure or, equivalently, only one ϕ-invariant Borel probability measure, then equation (3.6.3) actually holds for all $x \in M$. Birkhoff's Ergodic Theorem extends to μ-integrable functions and asserts that the sequence of measurable functions $\frac{1}{n}\sum_{k=0}^{n-1} f \circ \phi^k$ converges **pointwise** almost everywhere to the mean value of f. A particularly interesting case is where f is the characteristic function of a Borel set $B \subset M$. Then the integral of f is the measure of B and it follows from Birkhoff's Ergodic Theorem that

$$\mu(B) = \lim_{n\to\infty} \frac{\#\left\{k \in \{0, \ldots, n-1\} \,\middle|\, \phi^k(x) \in B\right\}}{n} \tag{3.6.4}$$

for μ-almost all $x \in M$. A weaker result is von Neumann's Mean Ergodic Theorem [**62**]. It asserts that the sequence $\frac{1}{n}\sum_{k=0}^{n-1} f \circ \phi^k$ converges to the mean value of f in $L^p(\mu)$ for $1 < p < \infty$. This implies pointwise almost everywhere convergence for a suitable subsequence (see [**75**, Cor. 4.10]).

Theorem 3.6.5 (Von Neumann's Mean Ergodic Theorem). *Let (M,d) be a compact metric space, let $\phi : M \to M$ be a homeomorphism, let $\mu \in \mathcal{M}(\phi)$ be a ϕ-ergodic measure, let $1 < p < \infty$, and let $f \in L^p(\mu)$. Then*

$$\lim_{n\to\infty} \left\| \frac{1}{n} \sum_{k=0}^{n-1} f \circ \phi^k - \int_M f \, d\mu \right\|_{L^p} = 0. \tag{3.6.5}$$

Proof. See page 149. □

Theorem 3.6.5 implies Theorem 3.6.3. The proof has two steps.

Step 1. *Let $\mu_0, \mu_1 \in \mathcal{M}(\phi)$ be ergodic measures such that $\mu_0(\Lambda) = \mu_1(\Lambda)$ for every ϕ-invariant Borel set $\Lambda \subset M$. Then $\mu_0 = \mu_1$.*

Fix a continuous function $f : M \to \mathbb{R}$. Then it follows from Theorem 3.6.5 and [**75**, Cor. 4.10] that there exist Borel sets $B_0, B_1 \subset M$ and a sequence of integers $1 \le n_1 < n_2 < n_3 < \cdots$ such that $\mu_i(B_i) = 1$ and

$$\int_M f \, d\mu_i = \lim_{j\to\infty} \frac{1}{n_j} \sum_{k=0}^{n_j-1} f(\phi^k(x)) \qquad \text{for } x \in B_i \text{ and } i = 0, 1. \tag{3.6.6}$$

For $i = 0, 1$ define $\Lambda_i := \bigcap_{n\in\mathbb{Z}} \phi^n(B_i)$. So Λ_i is a ϕ-invariant Borel set such that $\mu_i(\Lambda_i) = 1$. Thus $\mu_1(\Lambda_0) = \mu_0(\Lambda_0) = 1$ and $\mu_0(\Lambda_1) = \mu_1(\Lambda_1) = 1$ by assumption. This implies that the ϕ-invariant Borel set $\Lambda := \Lambda_0 \cap \Lambda_1$ is nonempty. Since $\Lambda \subset B_0 \cap B_1$, it follows from (3.6.6) that

$$\int_M f \, d\mu_0 = \lim_{j\to\infty} \frac{1}{n_j} \sum_{k=0}^{n_j-1} f(\phi^k(x)) = \int_M f \, d\mu_1 \qquad \text{for all } x \in \Lambda.$$

Thus the integrals of f with respect to μ_0 and μ_1 agree for every continuous function $f : M \to \mathbb{R}$. Hence $\mu_0 = \mu_1$ by uniqueness in the Riesz Representation Theorem (see [**75**, Cor. 3.19]). This proves Step 1.

Step 2. *Let $\mu \in \mathcal{M}(\phi)$ be ergodic. Then μ is an extremal point of $\mathcal{M}(\phi)$.*

Let $\mu_0, \mu_1 \in \mathcal{M}(\phi)$ and $0 < \lambda < 1$ such that $\mu = (1-\lambda)\mu_0 + \lambda\mu_1$. If $B \subset M$ is a Borel set such that $\mu(B) = 0$, then $(1-\lambda)\mu_0(B) + \lambda\mu_1(B) = 0$, and hence $\mu_0(B) = \mu_1(B) = 0$ because $0 < \lambda < 1$. If $B \subset M$ is a Borel set such that $\mu(B) = 1$, then $\mu(M \setminus B) = 0$, hence $\mu_0(M \setminus B) = \mu_1(M \setminus B) = 0$, and therefore $\mu_0(B) = \mu_1(B) = 1$. Now let $\Lambda \subset M$ be a ϕ-invariant Borel set. Then $\mu(\Lambda) \in \{0, 1\}$ because μ is ϕ-ergodic, and hence $\mu_0(\Lambda) = \mu_1(\Lambda) = \mu(\Lambda)$. Thus μ_0 and μ_1 are ϕ-ergodic measures that agree on all ϕ-invariant Borel sets. Hence $\mu_0 = \mu_1 = \mu$ by Step 1 and this proves Step 2.

Step 2 shows that (i) implies (ii) in Theorem 3.6.3. The converse was proved on page 144. □

3.6.3. An Abstract Ergodic Theorem.

Theorem 3.6.5 translates into a theorem about the iterates of a bounded linear operator from a Banach space to itself provided that these iterates are uniformly bounded. For an endomorphism

$$T : X \to X$$

of a vector space X and a positive integer n denote the nth iterate of T by

$$T^n := T \circ \cdots \circ T.$$

For $n = 0$ define

$$T^0 := \mathrm{id}.$$

The ergodic theorem in functional analysis asserts that, if $T : X \to X$ is a bounded linear operator on a reflexive Banach space whose iterates T^n form a bounded sequence of bounded linear operators, then its averages

$$S_n := \frac{1}{n}\sum_{k=1}^{n-1} T^k$$

form a sequence of bounded linear operators that converge strongly to a projection onto the kernel of the operator $\mathbb{1} - T$. Here is the relevant definition.

Definition 3.6.6 (Projection). Let X be a real normed vector space. A bounded linear operator $P : X \to X$ is called a **projection** if

$$P^2 = P.$$

Lemma 3.6.7. *Let X be a real normed vector space and let $P : X \to X$ be a bounded linear operator. Then the following are equivalent.*

(i) *P is a projection.*

(ii) *There exist closed linear subspaces $X_0, X_1 \subset X$ such that*

$$X_0 \cap X_1 = \{0\}, \qquad X_0 \oplus X_1 = X,$$

and

$$P(x_0 + x_1) = x_1$$

for all $x_0 \in X_0$ and all $x_1 \in X_1$.

Proof. If P is a projection, then $P^2 = P$ and hence the linear subspaces

$$X_0 := \ker(P), \qquad X_1 := \mathrm{im}(P) = \ker(\mathbb{1} - P)$$

satisfy the requirements of part (ii). If P is as in (ii), then $P^2 = P$ by definition and $P : X \to X$ is a bounded linear operator by Corollary 2.2.9. This proves Lemma 3.6.7. □

Example 3.6.8. The direct sum of two closed linear subspaces of a Banach space need not be closed. For example, let $X := C([0,1],\mathbb{R})$ be the Banach space of continuous functions $f : [0,1] \to \mathbb{R}$, equipped with the supremum norm. Then the linear subspaces

$$Y := \{(f,g) \in X \times X \mid f = 0\},$$
$$Z := \{(f,g) \in X \times X \mid f \in C^1([0,1]),\ f' = g\}$$

of $X \times X$ are closed, their intersection $Y \cap Z$ is trivial, and their direct sum $Y \oplus Z = \{(f,g) \in X \times X \mid f \in C^1([0,1])\}$ is not closed.

Theorem 3.6.9 (Ergodic Theorem). *Let $T : X \to X$ be a bounded linear operator on a Banach space X. Assume that there is a constant $c \geq 1$ such that*

$$\|T^n\| \leq c \qquad \text{for all } n \in \mathbb{N}. \tag{3.6.7}$$

For $n \in \mathbb{N}$ define the bounded linear operator $S_n : X \to X$ by

$$S_n := \frac{1}{n}\sum_{k=0}^{n-1} T^k. \tag{3.6.8}$$

Then the following hold.

(i) *Let $x \in X$. Then the sequence $(S_n x)_{n\in\mathbb{N}}$ converges if and only if it has a weakly convergent subsequence.*

(ii) *The set*

$$Z := \big\{x \in X \,\big|\, \text{the sequence } (S_n x)_{n\in\mathbb{N}} \text{ converges}\big\} \tag{3.6.9}$$

is a closed T-invariant linear subspace of X and

$$Z = \ker(\mathbb{1} - T) \oplus \overline{\mathrm{im}(\mathbb{1} - T)}. \tag{3.6.10}$$

Moreover, if X is reflexive, then $Z = X$.

(iii) *Define the bounded linear operator*

$$S : Z \to Z$$

by

$$S(x+y) := x \qquad \text{for } x \in \ker(\mathbb{1} - T) \text{ and } y \in \overline{\mathrm{im}(\mathbb{1} - T)}. \tag{3.6.11}$$

Then

$$\lim_{n\to\infty} S_n z = Sz \tag{3.6.12}$$

for all $z \in Z$ and

$$ST = TS = S^2 = S, \qquad \|S\| \leq c. \tag{3.6.13}$$

Proof. See page 150. □

Theorem 3.6.9 implies Theorem 3.6.5. Let $\phi : M \to M$ be a homeomorphism of a compact metric space M and let $\mu \in \mathcal{M}(\phi)$ be an ergodic ϕ-invariant Borel probability measure on M. Define the bounded linear operator $T : L^p(\mu) \to L^p(\mu)$ by

$$Tf := f \circ \phi \qquad \text{for } f \in L^p(\mu).$$

Then $\|Tf\|_p = \|f\|_p$ for all $f \in L^p(\mu)$, by the ϕ-invariance of μ, and so

$$\|T\| = 1.$$

Thus T satisfies the requirement of Theorem 3.6.9. Let $f \in L^p(\mu)$. Since $L^p(\mu)$ is reflexive (Example 1.3.3), Theorem 3.6.9 asserts that the sequence

$$S_n f := \frac{1}{n}\sum_{k=0}^{n-1} T^k f = \frac{1}{n}\sum_{k=0}^{n-1} f \circ \phi^k$$

converges in $L^p(\mu)$ to a function $Sf \in \ker(\mathbb{1}-T)$. It remains to prove that Sf is equal to the constant $c := \int_M f\,d\mu$ almost everywhere. The key to the proof is the fact that every function in the kernel of the operator $\mathbb{1} - T$ is constant (almost everywhere). Once this is understood, it follows that there exists a constant $c \in \mathbb{R}$ such that $Sf = c$ almost everywhere, and hence

$$c = \int_M Sf\,d\mu = \lim_{n\to\infty}\int_M S_n f\,d\mu = \lim_{n\to\infty}\frac{1}{n}\sum_{k=0}^{n-1}\int_M (f\circ\phi^k)\,d\mu = \int_M f\,d\mu.$$

Thus it remains to prove that every function in the kernel of $\mathbb{1} - T$ is constant. Let $g \in L^p(\mu)$ and suppose that $Tg = g$. Choose a representative of the equivalence class of g, still denoted by $g \in \mathcal{L}^p(\mu)$. Then $g(x) = g(\phi(x))$ for almost all $x \in M$. Define

$$E_0 := \{x \in M \,|\, g(x) \neq g(\phi(x))\}, \qquad E := \bigcup_{k\in\mathbb{Z}} \phi^k(E_0).$$

Then $E \subset M$ is a Borel set with $\phi(E) = E$, $\mu(E) = 0$, and $g(\phi(x)) = g(x)$ for every $x \in M \setminus E$. Let $c := \int_M g\,d\mu$ and define $B_-, B_0, B_+ \subset M$ by

$$B_0 := \{x \in M \setminus E \,|\, g(x) = c\}, \qquad B_\pm := \{x \in M \setminus E \,|\, \pm g(x) > c\}.$$

Each of these three Borel sets is invariant under ϕ and hence has measure either zero or one. Moreover, $B_- \cup B_0 \cup B_+ = M \setminus E$ and this implies

$$\mu(B_-) + \mu(B_0) + \mu(B_+) = 1.$$

Hence one of the three sets has measure one and the other two have measure zero. This implies that $\mu(B_0) = 1$, because otherwise either $\int_M g\,d\mu < c$ or $\int_M g\,d\mu > c$. Thus g is equal to its mean value almost everywhere. This proves Theorem 3.6.5. □

Proof of Theorem 3.6.9. The proof has eight steps.

Step 1. *Let $n \in \mathbb{N}$. Then $\|S_n\| \le c$ and $\|S_n(\mathbb{1} - T)\| \le \frac{1+c}{n}$.*

By assumption, we have $\|S_n\| \le \frac{1}{n}\sum_{k=0}^{n-1}\|T^k\| \le c$ for all $n \in \mathbb{N}$. Moreover,

$$S_n(\mathbb{1} - T) = \frac{1}{n}\sum_{k=0}^{n-1} T^k - \frac{1}{n}\sum_{k=1}^{n} T^k = \frac{1}{n}(\mathbb{1} - T^n)$$

and so

$$\|S_n(\mathbb{1} - T)\| \le \frac{1}{n}\big(\|\mathbb{1}\| + \|T^n\|\big) \le \frac{1+c}{n}$$

for all $n \in \mathbb{N}$. This proves Step 1.

Step 2. *Let $x \in X$ such that $Tx = x$. Then $S_n x = x$ for all $n \in \mathbb{N}$ and*

$$\|x\| \le c\,\|x + \xi - T\xi\| \qquad \textit{for all } \xi \in X.$$

Since $Tx = x$ it follows by induction that $T^k x = x$ for all $k \in \mathbb{N}$ and hence

$$x = \frac{1}{n}\sum_{k=0}^{n-1} T^k x = S_n x \qquad \text{for all } n \in \mathbb{N}.$$

Moreover, $\lim_{n\to\infty}\|S_n(\xi - T\xi)\| = 0$ by Step 1 and hence

$$\|x\| = \lim_{n\to\infty}\|x + S_n(\xi - T\xi)\| = \lim_{n\to\infty}\|S_n(x + \xi - T\xi)\| \le c\,\|x + \xi - T\xi\|.$$

Here the inequality holds because $\|S_n\| \le c$ by Step 1. This proves Step 2.

Step 3. *If $x \in \ker(\mathbb{1} - T)$ and $y \in \overline{\mathrm{im}(\mathbb{1} - T)}$, then $\|x\| \le c\,\|x + y\|$.*

Choose a sequence $\xi_n \in X$ such that $y = \lim_{n\to\infty}(\xi_n - T\xi_n)$. Then, by Step 2, we have $\|x\| \le c\|x + \xi_n - T\xi_n\|$ for all $n \in \mathbb{N}$. Take the limit $n \to \infty$ to obtain $\|x\| \le c\|x + y\|$. This proves Step 3.

Step 4. $\ker(\mathbb{1} - T) \cap \overline{\mathrm{im}(\mathbb{1} - K)} = \{0\}$ *and the direct sum*

$$Z := \ker(\mathbb{1} - T) \oplus \overline{\mathrm{im}(\mathbb{1} - T)} \tag{3.6.14}$$

is a closed linear subspace of X.

Let $x \in \ker(\mathbb{1}-T)\cap\overline{\mathrm{im}(\mathbb{1} - T)}$ and define $y := -x$. Then $\|x\| \le c\|x + y\| = 0$ by Step 3 and hence $x = 0$. This shows that $\ker(\mathbb{1} - T) \cap \overline{\mathrm{im}(\mathbb{1} - T)} = \{0\}$. We prove that the subspace Z in (3.6.14) is closed. Let $x_n \in \ker(\mathbb{1} - T)$ and $y_n \in \overline{\mathrm{im}(\mathbb{1} - T)}$ be sequences whose sum $z_n := x_n + y_n$ converges to some element $z \in X$. Then $(z_n)_{n\in\mathbb{N}}$ is a Cauchy sequence and hence $(x_n)_{n\in\mathbb{N}}$ is a Cauchy sequence by Step 3. This implies that $y_n = z_n - x_n$ is a Cauchy sequence and hence $z = x + y$, where $x := \lim_{n\to\infty} x_n \in \ker(\mathbb{1} - T)$ and $y := \lim_{n\to\infty} y_n \in \overline{\mathrm{im}(\mathbb{1} - T)}$. This proves Step 4.

Step 5. *If $z \in Z$, then $Tz \in Z$.*

Let $z \in Z$. Then

$$z = x + y, \qquad x \in \ker(\mathbb{1} - T), \qquad y \in \overline{\mathrm{im}(\mathbb{1} - T)}.$$

Choose a sequence $\xi_i \in X$ such that $y = \lim_{i\to\infty}(\xi_i - T\xi_i)$. Then

$$Ty = \lim_{i\to\infty} T(\xi_i - T\xi_i) = \lim_{i\to\infty} (\mathbb{1} - T)T\xi_i \in \overline{\mathrm{im}(\mathbb{1} - T)}.$$

Hence

$$Tz = Tx + Ty = x + Ty \in Z$$

and this proves Step 5.

Step 6. *Let $x \in \ker(\mathbb{1} - T)$ and $y \in \overline{\mathrm{im}(\mathbb{1} - T)}$. Then*

$$x = \lim_{n\to\infty} S_n(x + y).$$

By Step 1, the sequence

$$\|S_n(\mathbb{1} - T)\xi\| \le \frac{1 + c}{n} \|\xi\|$$

converges to zero as n tends to infinity for every $\xi \in X$. Hence it follows from the estimate $\|S_n\| \le c$ in Step 1 and the Banach–Steinhaus Theorem 2.1.5 that

$$\lim_{n\to\infty} S_n y = 0 \qquad \text{for all } y \in \overline{\mathrm{im}(\mathbb{1} - T)}.$$

Moreover,

$$S_n x = x \qquad \text{for all } n \in \mathbb{N}$$

by Step 2. Hence

$$x = \lim_{n\to\infty} S_n x = \lim_{n\to\infty} S_n(x + y).$$

This proves Step 6.

Step 7. *Let $x, z \in X$. Then the following are equivalent.*

(a) *$Tx = x$ and $z - x \in \overline{\mathrm{im}(\mathbb{1} - T)}$.*

(b) $\lim_{n\to\infty} \|S_n z - x\| = 0$.

(c) *There is a sequence of integers $1 \le n_1 < n_2 < n_3 < \cdots$ such that*

$$\lim_{i\to\infty} \langle x^*, S_{n_i} z\rangle = \langle x^*, x\rangle \qquad \textit{for all } x^* \in X^*.$$

That (a) implies (b) follows immediately from Step 6 and that (b) implies (c) is obvious. We prove that (c) implies (a). Thus assume (c) and fix a bounded linear functional $x^* \in X^*$. Then

$$T^*x^* := x^* \circ T : X \to \mathbb{R}$$

is a bounded linear functional and

$$\begin{aligned}\langle x^*, x - Tx\rangle &= \langle x^* - T^*x^*, x\rangle \\ &= \lim_{i\to\infty} \langle x^* - T^*x^*, S_{n_i}z\rangle \\ &= \lim_{i\to\infty} \langle x^*, (1\!\!1 - T)S_{n_i}z\rangle \\ &= 0.\end{aligned}$$

Here the last equation follows from Step 1. Hence

$$Tx = x$$

by the Hahn–Banach Theorem (Corollary 2.3.23). Next we prove that

$$z - x \in \overline{\mathrm{im}(1\!\!1 - T)}.$$

Assume, by contradiction, that $z - x \in X \setminus \overline{\mathrm{im}(1\!\!1 - T)}$. Then, by the Hahn–Banach Theorem 2.3.22, there exists an element $x^* \in X^*$ such that

$$\text{(3.6.15)} \qquad \langle x^*, z - x\rangle = 1, \qquad \langle x^*, \xi - T\xi\rangle = 0 \quad \text{for all } \xi \in X.$$

This implies $\langle x^*, T^k\xi - T^{k+1}\xi\rangle = 0$ for all $k \in \mathbb{N}$ and all $\xi \in X$. Hence, by induction, $\langle x^*, \xi\rangle = \langle x^*, T^k\xi\rangle$ for every $\xi \in X$ and every integer $k \geq 0$. Thus

$$\langle x^*, S_n z\rangle = \frac{1}{n}\sum_{k=0}^{n-1}\langle x^*, T^k z\rangle = \langle x^*, z\rangle$$

for all $n \in \mathbb{N}$. Hence it follows from (c) that

$$\langle x^*, z - x\rangle = \lim_{i\to\infty}\langle x^*, S_{n_i}z - x\rangle = 0.$$

This contradicts (3.6.15). Thus $z - x \in \overline{\mathrm{im}(1\!\!1 - T)}$ and this proves Step 7.

Step 8. *We prove Theorem* 3.6.9.

The subspace Z in (3.6.14) is closed by Step 4 and is T-invariant by Step 5. Moreover, Step 7 asserts that an element $z \in X$ belongs to Z if and only if the sequence $(S_n z)_{n\in\mathbb{N}}$ converges in the norm topology if and only if $(S_n z)_{n\in\mathbb{N}}$ has a weakly convergent subsequence. If X is reflexive, this holds for all $z \in X$ by Step 1 and Theorem 2.4.4. This proves (i) and (ii).

Define the operator $S : Z \to Z$ by (3.6.11). Then $\|S\| \leq c$ by Step 3, the equation $\lim_{n\to\infty} S_n z = Sz$ for $z \in Z$ follows from Step 6, and $S^2 = S$ by definition. The equation $ST = TS = S$ follows from the fact that S commutes with $T|_Z$ and vanishes on the image of the operator $1\!\!1 - T$. This proves Theorem 3.6.9. □

3.7. Problems

Exercise 3.7.1 (Weak and Strong Convergence). Let H be a real Hilbert space and let $(x_i)_{i\in\mathbb{N}}$ be a sequence in H that converges weakly to $x \in H$. Assume also that

$$\|x\| = \lim_{i\to\infty} \|x_i\|.$$

Prove that $(x_i)_{i\in\mathbb{N}}$ converges strongly to x, i.e.

$$\lim_{i\to\infty} \|x_i - x\| = 0.$$

Exercise 3.7.2 (Weak Convergence and Weak Closure). Let H be an infinite-dimensional separable real Hilbert space and let $(e_n)_{n\in\mathbb{N}}$ be an orthonormal basis of H. Prove the following.

(a) The sequence $(e_n)_{n\in\mathbb{N}}$ converges weakly to zero.

(b) The set

$$A := \left\{\sqrt{n}e_n \,|\, n \in \mathbb{N}\right\}$$

is sequentially weakly closed, but the weak closure of A contains zero. **Hint:** Let $U \subset H$ be a weakly open neighborhood of the origin. Show that there are vectors $y_1, \dots, y_m \in H$ and a number $\varepsilon > 0$ such that

$$V := \{x \in H \,|\, \max_{i=1,\dots,m} |\langle x, y_i\rangle| < \varepsilon\} \subset U.$$

Show that the sequence

$$z_n := \max_{i=1,\dots,m} |\langle e_n, y_i\rangle|$$

is square summable and deduce that $V \cap A \neq \emptyset$.

Exercise 3.7.3 (The Weak Topology of ℓ^1). Prove the following.

(a) The standard basis e_n of ℓ^1 does not converge weakly to zero.

(b) View ℓ^1 as the dual space of c_0 (see Example 1.3.7). Then the standard basis converges to zero in the weak* topology.

(c) Schur's Theorem. *A sequence in ℓ^1 converges (to zero) in the weak topology if and only if it converges (to zero) in the norm topology.*

Exercise 3.7.4 (Weak* Topology and Distance Function). Let X be a separable normed vector space and let $(x_n)_{n\in\mathbb{N}}$ be a dense sequence in the unit ball of X. Prove that the map

$$d(x^*, y^*) := \sum_{n=1}^{\infty} 2^{-n} |\langle x^* - y^*, x_n\rangle| \qquad \text{for } x^*, y^* \in B^* \tag{3.7.1}$$

defines a distance function on the closed unit ball $B^* \subset X^*$. Prove that the topology induced by this distance function is the weak* topology on B^*.

Exercise 3.7.5 (Compact-Open Topology). Let X be a topological space, let Y be a metric space, and let $C(X,Y)$ be the space of continuous functions $f : X \to Y$. The **compact-open topology** on $C(X,Y)$ is the smallest topology such that the set

$$\mathscr{S}(K,V) := \{f \in C(X,Y) \mid f(K) \subset V\}$$

is open for every compact set $K \subset X$ and every open set $V \subset Y$. Thus a set $\mathcal{U} \subset C(X,Y)$ is open with respect to the compact-open topology if and only if, for each $f \in \mathcal{U}$, there are finitely many compact sets $K_1, \dots, K_m \subset X$ and open sets $V_1, \dots, V_m \subset Y$ such that $f \in \bigcap_{i=1}^m \mathscr{S}(K_i, V_i) \subset \mathcal{U}$.

(a) If X is compact, prove that the compact-open topology on $C(X,Y)$ agrees with the topology induced by the metric

$$d(f,g) := \sup_{x \in X} d_Y(f(x), g(x)) \qquad \text{for } f, g \in C(X,Y). \tag{3.7.2}$$

Hint 1: Let $f \in C(X,Y)$ and suppose that $K_1, \dots, K_m \subset X$ are compact sets and $V_1, \dots, V_m \subset Y$ are open sets such that $f(K_i) \subset V_i$ for $i = 1, \dots, m$. Prove that there is a constant $\varepsilon > 0$ such that $B_\varepsilon(f_i(x)) \subset V_i$ for all $x \in K_i$ and all $i \in \{1, \dots, m\}$. Deduce that every $g \in C(X,Y)$ with $d(f,g) < \varepsilon$ satisfies $g(K_i) \subset V_i$ for $i = 1, \dots, m$.

Hint 2: Let $f \in C(X,Y)$ and $\varepsilon > 0$. Find elements $x_1, \dots, x_m \in X$ such that $X = \bigcup_{i=1}^m K_i$, where $K_i := \{x \in X \mid d_Y(f(x_i), f(x)) \le \varepsilon/4\}$. Define

$$\mathcal{U} := \{g \in C(X,Y) \mid g(K_i) \subset V_i \text{ for } i = 1, \dots, m\}, \qquad V_i := B_{\varepsilon/2}(f(x_i)).$$

Show that $f \in \mathcal{U}$ and $d(f,g) < \varepsilon$ for all $g \in \mathcal{U}$.

(b) For each compact subset $K \subset X$ define the seminorm $p_K : C(X,\mathbb{R}) \to \mathbb{R}$ by

$$p_K(f) := \sup_K |f| \qquad \text{for } f \in C(X,\mathbb{R}).$$

Prove that these seminorms generate the compact-open topology, i.e. the compact-open topology on $C(X,\mathbb{R})$ is the smallest topology such that p_K is continuous for every compact set $K \subset X$.

(c) Prove that $C(X,\mathbb{R})$ is a locally convex topological vector space with the compact-open topology.

(d) Prove that a subset $\mathscr{F} \subset C(X,Y)$ is precompact with respect to the compact-open topology if and only if, for every compact set $K \subset X$, the set

$$\mathscr{F}_K := \{f|_K \mid f \in \mathscr{F}\} \subset C(K,Y) \tag{3.7.3}$$

is precompact. **Hint:** Let $\mathscr{K} \subset 2^X$ be the collection of compact subsets. Prove that the map $C(X,Y) \to \prod_{K \in \mathscr{K}} C(K,Y) : f \mapsto (f|_K)_{K \in \mathscr{K}}$ is a homeomorphism onto its image and use Tychonoff's Theorem A.2.1.

(e) Prove the following variant of the Arzelà–Ascoli Theorem.

Arzelà–Ascoli. *Let X be a topological space and let Y be a metric space. A subset $\mathscr{F} \subset C(X,Y)$ is precompact with respect to the compact-open topology if and only if it is pointwise precompact and the set $\mathscr{F}_K \subset C(K,Y)$ in* (3.7.3) *is equi-continuous for every compact set $K \subset X$.*

Hint: Use part (d) and Exercise 1.1.15.

Exercise 3.7.6 (Banach–Alaoglu). Let X be a normed vector space. Deduce the Banach–Alaoglu Theorem 3.2.4 from the Arzelà–Ascoli Theorem in part (e) of Exercise 3.7.5. **Hint:** The closed unit ball in X^* is equi-continuous as a subset of $C(X,\mathbb{R})$. Prove that the compact-open topology on X^* is finer than the weak* topology, i.e. every weak* open subset of X^* is also open with respect to the compact-open topology.

Exercise 3.7.7 (Functions Vanishing at Infinity). Let M be a locally compact Hausdorff space. A continuous real valued function $f : M \to \mathbb{R}$ is said to **vanish at infinity** if, for every $\varepsilon > 0$, there exists a compact set $K \subset M$ such that

$$\sup_{x \in M \setminus K} |f(x)| < \varepsilon.$$

Denote by $C_0(M)$ the space of all continuous functions $f : M \to \mathbb{R}$ that vanish at infinity (see Exercise 3.2.10).

(a) Prove that $C_0(M)$ is a Banach space with the supremum norm.

(b) The dual space $C_0(M)^*$ can be identified with the space $\mathcal{M}(M)$ of signed Radon measures on M with the norm (1.1.4), by the Riesz Representation Theorem (see [**75**, Thm. 3.15 & Ex. 3.35]). Here a **signed Radon measure** on M is a signed Borel measure μ with the property that, for each Borel set $B \subset M$ and each $\varepsilon > 0$, there exists a compact set $K \subset B$ such that $|\mu(A) - \mu(A \cap K)| < \varepsilon$ for every Borel set $A \subset B$.

(c) Prove that the map $\delta : M \to C_0(M)^*$, which assigns to each $x \in M$ the bounded linear functional $\delta_x : C_0(M) \to \mathbb{R}$ given by

$$\delta_x(f) := f(x) \qquad \text{for } f \in C_0(M),$$

is a homeomorphism onto its image $\delta(M) \subset C_0(M)^*$, equipped with the weak* topology. Under the identification in (b) this image is contained in the set

$$P(M) := \{\mu \in \mathcal{M}(M) \,|\, \mu \geq 0,\ \|\mu\| = \mu(M) = 1\}$$

of Radon probability measures. Determine the weak* closure of the set

$$\delta(M) = \{\delta_x \,|\, x \in M\} \subset P(M).$$

Exercise 3.7.8 (Alaoglu–Bourbaki Theorem). Let X and Y be real vector spaces and let

$$Y \times X \to \mathbb{R} : (y, x) \mapsto \langle y, x \rangle \tag{3.7.4}$$

be a nondegenerate pairing. For two subsets $A \subset X$ and $B \subset Y$ define the **polar sets** $A^\circ \subset Y$ and $B_\circ \subset X$ by

$$\begin{aligned} A^\circ &:= \{y \in Y \mid \langle y, a \rangle \le 1 \text{ for all } a \in A\}, \\ B_\circ &:= \{x \in X \mid \langle b, x \rangle \le 1 \text{ for all } b \in B\}. \end{aligned} \tag{3.7.5}$$

Thus A° and $B_\circ$ are intersections of half-spaces.

(a) Suppose X is a real normed vector space, $Y = X^*$ is its dual space, and (3.7.4) is the standard pairing. Let $S \subset X$ and $S^* \subset X^*$ denote the unit spheres and $B \subset X$ and $B^* \subset X^*$ the closed unit balls. Verify that

$$S^0 = B^*, \qquad (S^*)_0 = B.$$

(b) Bipolar Theorem. *Equip X with the topology induced by the linear maps $X \to \mathbb{R} : x \mapsto \langle y, x \rangle$ for $y \in Y$. Then*

$$(A^0)_0 = \overline{\mathrm{conv}}(A \cup \{0\}).$$

(c) Goldstine's Theorem. *If X is a normed vector space and B is the closed unit ball, then the weak* closure of $\iota(B)$ is the closed unit ball in X^{**}.* (See also Corollary 3.1.29.)

(d) Alaoglu–Bourbaki Theorem. *Suppose $(X, \mathscr{U})$ is a locally convex topological vector space over the reals, Y is the space of $\mathscr{U}$-continuous linear functionals $\Lambda : X \to \mathbb{R}$, and (3.7.4) is the standard pairing. Equip Y with the topology $\mathscr{V} \subset 2^Y$ induced by the linear maps $Y \to \mathbb{R} : y \mapsto \langle y, x \rangle$ for $x \in X$. If $A \subset X$ is a $\mathscr{U}$-neighborhood of the origin, then $A^\circ \subset Y$ is $\mathscr{V}$-compact.*

Exercise 3.7.9 (Milman–Pettis Theorem). A normed vector space X over the reals is called **uniformly convex** if, for every $\varepsilon > 0$, there exists a constant $\delta > 0$ such that, for all $x, y \in X$,

$$\|x\| = \|y\| = 1, \quad \|x + y\| > 2 - \delta \qquad \Longrightarrow \qquad \|x - y\| < \varepsilon.$$

The Milman–Pettis Theorem asserts that every uniformly convex Banach space is reflexive. This can be proved as follows.

The proof requires the concept of a net, which generalizes the concept of a sequence. A **directed set** is a nonempty set A, equipped with a reflexive and transitive relation $\preccurlyeq$, such that, for all $\alpha, \beta \in A$, there exists a $\gamma \in A$ with $\alpha \preccurlyeq \gamma$ and $\beta \preccurlyeq \gamma$. Anti-symmetry is not required, so a directed set need not be partially ordered. An example of a directed set is the collection of open neighborhoods of a point x_0 in a topological space X, equipped with the relation $U \preccurlyeq V \iff V \subset U$.

A **net** in a space X is a map

$$A \to X : \alpha \mapsto x_\alpha,$$

defined on a directed set A. A net $(x_\alpha)_{\alpha \in A}$ in a topological space X is said to **converge to** $x \in X$ if, for every open neighborhood $U \subset X$ of x, there exists an element $\alpha_0 \in A$, such that $x_\alpha \in U$ for all $\alpha \in A$ with $\alpha_0 \preccurlyeq \alpha$.

If X and Y are topological spaces, then a map $f : X \to Y$ is continuous if and only if, for every net $(x_\alpha)_{\alpha \in A}$ in X that converges to $x \in X$, the net $(f(x_\alpha))_{\alpha \in A}$ in Y converges to $f(x)$.

Let $(x_\alpha)_{\alpha \in A}$ be a net in X. A **subnet** of $(x_\alpha)_{\alpha \in A}$ is a net of the form $(x_{h(\beta)})_{\beta \in B}$ where $h : B \to A$ is a monotone final map between directed sets. Here the map $h : B \to A$ is called **monotone** if

$$\beta_1 \preccurlyeq \beta_2 \qquad \Longrightarrow \qquad h(\beta_1) \preccurlyeq h(\beta_2)$$

for all $\beta_1, \beta_2 \in B$, and it is called **final** if, for every $\alpha \in A$, there exists an element $\beta \in B$ such that

$$h(\alpha) \preccurlyeq \beta.$$

With this understood, a topological space X is compact if and only if every net in X has a convergent subnet.

A net $(x_\alpha)_{\alpha \in A}$ in a normed vector space X is called a **Cauchy net** if the net $(\|x_\alpha - x_\beta\|)_{(\alpha,\beta) \in A \times A}$ (product order on $A \times A$) converges to zero. If X is a Banach space, then every Cauchy net in X converges.

(a) Let X be a uniformly convex normed vector space. Let $(x_\alpha)_{\alpha \in A}$ be a net in the unit sphere of X such that the net $(\|x_\alpha + x_\beta\|)_{(\alpha,\beta) \in A \times A}$ converges to 2. Prove that $(x_\alpha)_{\alpha \in A}$ is a Cauchy net.

(b) Let X be a normed vector space and let $x^{**} \in X^{**}$ with $\|x^{**}\| = 1$. Prove that there exists a net $(x_\alpha)_{\alpha \in A}$ in the unit sphere of X such that the net $(\iota(x_\alpha))_{\alpha \in A}$ in X^{**} converges to x^{**} with respect to the weak* topology.

(c) Let X be a normed vector space and let $(x_\alpha)_{\alpha \in A}$ be a net in the unit sphere of X such that the net $(\iota(x_\alpha))_{\alpha \in A}$ in X^{**} converges to x^{**} with respect to the weak* topology, where $\|x^{**}\| = 1$. Prove that the net

$$(\iota(x_\alpha + x_\beta))_{(\alpha,\beta) \in A \times A}$$

converges to $2x^{**}$ in the weak* topology. If X is uniformly convex, use (a) to prove that $(x_\alpha)_{\alpha \in A}$ is a Cauchy net.

(d) Assume X is a uniformly convex Banach space, let $x^{**} \in X^{**}$ such that $\|x^{**}\| = 1$, and choose a net $(x_\alpha)_{\alpha \in A}$ as in (b). Use (c) to prove that the net $(x_\alpha)_{\alpha \in A}$ converges to some element $x \in X$. Deduce that $\iota(x) = x^{**}$.

Exercise 3.7.10 (Banach–Mazur Theorem). Let X be a Banach space and let $B^* \subset X^*$ be the closed unit ball in the dual space, equipped with the weak* topology.

(a) Prove that the map $X \to C(B^*) : x \mapsto f_x$, defined by $f_x(x^*) := \langle x^*, x\rangle$ for $x \in X$ and $x^* \in B^*$, is a linear isometric embedding.

(b) If K is a compact metric space, then there is a continuous surjective map $\pi : F \to K$, defined on a closed subset $F \subset \{0,1\}^{\mathbb{N}}$ of the Cantor set. Deduce that there exists a linear isometric embedding $\pi^* : C(K) \to C(F)$. **Hint:** The **Cantor function** is a continuous surjection $\{0,1\}^{\mathbb{N}} \to [0,1]$. Use it to construct a continuous surjection $\{0,1\}^{\mathbb{N}} \to [0,1]^{\mathbb{N}}$ and then find an embedding $K \hookrightarrow [0,1]^{\mathbb{N}}$.

(c) For every closed subset $F \subset [0,1]$ of the unit interval find a linear isometric embedding $\iota_F : C(F) \to C([0,1])$. **Hint:** The complement of F is a countable union of intervals.

(d) Banach–Mazur Theorem. *Every separable Banach space is isometrically isomorphic to a closed subspace of* $C([0,1])$.

Exercise 3.7.11 (Helly's Theorem). (Another proof of Lemma 3.4.3.)

(a) Let X be a normed vector space, let $x_1^*, \dots, x_n^* \in X^*$, and let $c_1, \dots, c_n$ be scalars. Prove that there exists an element $x \in X$ such that

$$\langle x_i^*, x\rangle = c_i \qquad \text{for } i = 1, \dots, n \tag{3.7.6}$$

if and only if there is a constant $M > 0$ such that, for all scalars $\lambda_1, \dots, \lambda_n$,

$$\left|\sum_{i=1}^n \lambda_i c_i\right| \le M \left\|\sum_{i=1}^n \lambda_i x_i^*\right\|. \tag{3.7.7}$$

Hint: Assume $x_1^*, \dots, x_m^*$ are linearly independent and span the same space as $x_1^*, \dots, x_n^*$. Define the map $T : X \to \mathbb{R}^m$ by $Tx := (\langle x_1^*, x\rangle, \dots, \langle x_m^*, x\rangle)$ for $x \in X$. Then T is surjective by Lemma 3.1.13. Use the inequality (3.7.7) to show that every element $x \in T^{-1}(c_1, \dots, c_m)$ satisfies (3.7.6).

(b) Assume (3.7.7) and let $\varepsilon > 0$. Prove that there exists an element $x \in X$ that satisfies (3.7.6) and $\|x\| < M + \varepsilon$. **Hint:** By (a) there exists some element $y \in X$ such that $\langle x_i^*, y\rangle = c_i$ for $i = 1, \dots, n$. Define $Z := \bigcap_{i=1}^n \ker(x_i^*)$. If $y \notin Z$, then, by Theorem 2.3.22, there is an element $x^* \in X^*$ such that

$$\|x^*\| = 1, \qquad x^*|_Z = 0, \qquad \langle x^*, y\rangle = d(y, Z) = \inf_{z \in Z} \|y - z\|.$$

By Lemma 3.1.14 the element x^* is a linear combination of the x_i^*. Use this to deduce from (3.7.7) that $d(y, Z) \le M$. Find $z \in Z$ with $\|y + z\| < M + \varepsilon$. (If $\dim X = \infty$, then $Z \ne \{0\}$, so the norm of $y + z$ can be chosen equal to any number bigger than M.)

Exercise 3.7.12 (Šmulyan–James Theorem). *Let X be a normed vector space. Then the following are equivalent.*

(i) *X is reflexive.*

(ii) *Every bounded sequence in X has a weakly convergent subsequence.*

(iii) *If $C_1 \supset C_2 \supset C_3 \supset \cdots$ is a nested sequence of nonempty bounded closed convex subsets of X, then their intersection is nonempty.*

The implication (*iii*) $\implies$ (*i*) of the Šmulyan–James Theorem strengthens the Eberlein–Šmulyan Theorem 3.4.1.

(a) Prove that (i) implies (ii) and (ii) implies (iii).

(b) Prove that (iii) implies that X is complete.

(c) Let X be a nonreflexive Banach space. Prove that there exists a constant $0 < \alpha < 1$ and an element $x^{**} \in X^{**}$ such that

$$\alpha < d(x^{**}, \iota(X)) \leq \|x^{**}\| < 1. \tag{3.7.8}$$

Hint: Use the Riesz Lemma 1.2.12.

(d) Let $0 < \alpha < 1$ and $x^{**} \in X^{**}$ be as in (c). Find sequences of unit vectors $(x_n^*)_{n\in\mathbb{N}}$ in X^* and $(x_k)_{k\in\mathbb{N}}$ in X such that

$$\langle x_n^*, x_k\rangle = \begin{cases} 0, & \text{if } k < n, \\ \alpha, & \text{if } k \geq n, \end{cases} \qquad \langle x^{**}, x_n^*\rangle = \alpha \quad \text{for all } k, n \in \mathbb{N}. \tag{3.7.9}$$

Hint: Argue by induction. First find unit vectors $x_1^* \in X^*$ and $x_1 \in X$ such that $\langle x^{**}, x_1^*\rangle = \alpha$ and $\langle x_1^*, x_1\rangle = \alpha$. Now let $N > 1$ and assume by induction that unit vectors $x_1, \dots, x_{N-1} \in X$ and $x_1^*, \dots, x_{N-1}^* \in X^*$ have been found that satisfy (3.7.9) for $k, n = 1, \dots, N-1$. With $M := \alpha d(x^{**}, \iota(X)) < 1$ we have $|\lambda_0 \alpha| \leq M \|\lambda_0 x^{**} + \sum_{k=1}^{N-1} \lambda_k \iota(x_k)\|$ for all $\lambda_0, \dots, \lambda_{N-1} \in \mathbb{R}$. Hence, by Helly's Theorem, there exists a unit vector $x_N^* \in X^*$ such that

$$\langle x^{**}, x_N^*\rangle = \alpha, \qquad \langle x_N^*, x_k\rangle = \langle \iota(x_k), x_N^*\rangle = 0 \quad \text{for } k = 1, \dots, N-1.$$

Moreover,

$$\alpha \left|\sum_{n=1}^N \lambda_n\right| = \left|\left\langle x^{**}, \sum_{n=1}^N \lambda_n x_n^*\right\rangle\right| \leq \|x^{**}\| \left\|\sum_{n=1}^N \lambda_n x_n^*\right\|$$

for all $\lambda_1, \dots, \lambda_N \in \mathbb{R}$. Since $\|x^{**}\| < 1$ it follows again from Helly's Theorem that there is a unit vector $x_N \in X$ such that $\langle x_n^*, x_N\rangle = \alpha$ for $n = 1, \dots, N$. This completes the induction step for the proof of (3.7.9).

(e) Let x_k, x_n^* be as in (d) and define $C_N := \overline{\mathrm{conv}}(\{x_k \,|\, k \geq N\})$ for $N \in \mathbb{N}$. Prove that $\langle x_N^*, x\rangle = \alpha$ and $\lim_{n\to\infty}\langle x_n^*, x\rangle = 0$ for all $x \in C_N$. Deduce that the C_N have an empty intersection.

Exercise 3.7.13 (Birkhoff–von Neumann Theorem). An $n \times n$-matrix $M = (m_{ij})$ with nonnegative coefficients $m_{ij} \geq 0$ is called **doubly stochastic** if its row sums and column sums are all equal to one. The Birkhoff–von Neumann Theorem asserts the following.

Every doubly stochastic matrix
is a convex combination of permutation matrices.

Thus the doubly stochastic matrices form a convex set whose extremal points are the permutation matrices. This can be proved as follows.

Let M be a doubly stochastic matrix and denote by $\nu(M)$ the number of positive entries. If $\nu(M) > n$ find a permutation matrix P and a constant $0 < \lambda < 1$ such that the matrix $N := M - \lambda P_1$ has nonnegative entries and strictly fewer positive entries than M. In the case $N \neq 0$ the matrix $M_1 := (1-\lambda)^{-1}N$ is again doubly stochastic with $\nu(M_1) < \nu(M)$, and $M = \lambda P_1 + (1-\lambda)M_1$. Continue by induction until $\nu(M_k) = n$ and so M_k is a permutation matrix. Here is a method to find P_1 and λ.

Hall's Marriage Theorem. *Let X and Y be finite sets and let $\Gamma \subset X \times Y$. Then the following are equivalent.*

(i) *There is an injective map $f : X \to Y$ whose graph is contained in Γ.*

(ii) *For every $A \subset X$ the set*

$$\Gamma(A) := \{y \in Y \mid \textit{there is an } x \in A \textit{ such that } (x, y) \in \Gamma\}$$

satisfies $\#\Gamma(A) \geq \#A$.

Take $X = Y = \{1, \dots, n\}$ and $\Gamma := \{(i, j) \mid m_{ij} > 0\}$. Use the fact that M is doubly stochastic to verify that Γ satisfies (ii). Use the injective map f in (i) to determine the permutation matrix P_1 and take $\lambda := \min_{j=f(i)} m_{ij}$.

Exercise 3.7.14 (Strict Convexity and Extremal Points). A normed vector space is strictly convex (see Example 3.5.3 and Exercise 2.5.11) if and only if the unit sphere is equal to the set of extremal points of the closed unit ball.

Exercise 3.7.15 (A Noncompact Set of Extremal Points). Let $C \subset \mathbb{R}^3$ be the closed convex hull of the set

$$S := \big\{(1 + \cos(\theta), \sin(\theta), 0) \,\big|\, \theta \in \mathbb{R}\big\} \cup \big\{(0, 0, 1), (0, 0, -1)\big\}.$$

Determine the extremal points of C.

Exercise 3.7.16 (Extremal Points of Unit Balls). Determine the extremal points of the closed unit balls in the Banach spaces

$$c_0,\ c,\ C([0,1]),\ \ell^1,\ \ell^p,\ \ell^\infty,\ L^1([0,1]),\ L^p([0,1]),\ L^\infty([0,1])$$

for $1 < p < \infty$.

Exercise 3.7.17 (Hilbert Cube). (See Example 3.5.5.)

(a) Show that the Hilbert cube $Q := \{x = (x_i)_{i\in\mathbb{N}} \in \ell^2 \,|\, 0 \le x_i \le 1/i\}$ is a compact subset of ℓ^2 with respect to the norm topology.

(b) Is the set $R := \{x = (x_i)_{i\in\mathbb{N}} \in \ell^2 \,|\, 0 \le x_i \le 1/\sqrt{i}\}$ compact in ℓ^2 with respect to either the norm topology or the weak topology?

Exercise 3.7.18. Let X be a real normed vector space, let $B^* \subset X^*$ be the closed unit ball in the dual space, and let $\Lambda : X^* \to \mathbb{R}$ be a linear functional such that the restriction $\Lambda|_{B^*} : B^* \to \mathbb{R}$ is weak* continuous. Then there exists an element $x \in X$ such that $\Lambda = \iota(x)$.

Exercise 3.7.19 (Markov–Kakutani Fixed Point Theorem). Let X be a locally convex Hausdorff topological vector space and let $\mathcal{A}$ be a collection of pairwise commuting continuous linear operators $A : X \to X$. Let $C \subset X$ be a nonempty $\mathcal{A}$-invariant compact convex subset of X, so that

$$A(C) \subset C \qquad \text{for all } A \in \mathcal{A}.$$

Then there exists an element $x \in C$ such that $Ax = x$ for all $A \in \mathcal{A}$.

(a) For $A \in \mathcal{A}$ and $k \in \mathbb{N}$ define

$$A_k := \frac{1}{k}\big(\mathbb{1} + A + A^2 + \cdots + A^{k-1}\big).$$

Then $A_k(C)$ is a nonempty compact convex subset of C.

(b) If $A, B \in \mathcal{A}$ and $k, \ell \in \mathbb{N}$, then

$$A_k(B_\ell(C)) \subset A_k(C) \cap B_\ell(C).$$

Use this to prove that the set

$$F := \bigcap_{k\in\mathbb{N}} \bigcap_{A\in\mathcal{A}} A_k(C)$$

is nonempty.

(c) Prove that every element $x \in F$ is a fixed point of $\mathcal{A}$. **Hint:** Fix an element $A \in \mathcal{A}$. If $Ax \ne x$ find a continuous linear functional $\Lambda : X \to \mathbb{R}$ such that $\Lambda(x - Ax) = 1$. Prove that, for every $k \in \mathbb{N}$, there exists an element $y \in C$ such that

$$A_k y = x.$$

Now observe that

$$y - A^k y = k(x - Ax)$$

and deduce that the functional Λ is unbounded on the compact set $C - C$, contradicting continuity.

Exercise 3.7.20 (Bell–Fremlin Theorem). *The axiom of choice is equivalent to the assertion that the closed unit ball in the dual space of every nonzero Banach space has an extremal point.*

(a) Let X be any nonzero Banach space. Use the Banach–Alaoglu Theorem 3.2.4, the Hahn–Banach Theorem 2.3.2, and the Kreĭn–Milman Theorem 3.5.2 to prove that the closed unit ball in X^* has an extremal point.

(b) Let I be any index set and, for each $i \in I$, let X_i be a nonzero Banach space. Define the Banach spaces

$$\bigoplus_{i \in I} X_i := \left\{ x = (x_i)_{i \in I} \,\middle|\, x_i \in X_i \text{ and } \|x\|_1 := \sum_{i \in I} \|x_i\|_{X_i} < \infty \right\} \tag{3.7.10}$$

and

$$\prod_{i \in I} X_i := \left\{ x = (x_i)_{i \in I} \,\middle|\, x_i \in X_i \text{ and } \|x\|_\infty := \sup_{i \in I} \|x_i\|_{X_i} < \infty \right\}. \tag{3.7.11}$$

Prove that $\prod_{i \in I} X_i^*$ is isomorphic to the dual space of $\bigoplus_{i \in I} X_i$.

(c) Let S be a nonempty set. Define $c_0(S)$ to be the space of all functions $f : S \to \mathbb{R}$ that satisfy $\#\{s \in S \,|\, |f(s)| > \varepsilon\} < \infty$ for all $\varepsilon > 0$, equipped with the supremum norm $\|f\|_\infty := \sup_{s \in S} |f(s)|$. Define $\ell^1(S)$ to be the space of all functions $g : S \to \mathbb{R}$ such that $\|g\|_1 := \sum_{s \in S} |g(s)| < \infty$. Prove that $\ell^1(S)$ is isomorphic to the dual space of $c_0(S)$.

(d) Let $(S_i)_{i \in I}$ be a family of pairwise disjoint nonempty sets. Then the Banach space $\prod_{i \in I} \ell^1(S_i)$ is isomorphic to the dual space of $\bigoplus_{i \in I} c_0(S_i)$ by (b) and (c). Suppose the closed unit ball in $\prod_{i \in I} \ell^1(S_i)$ has an extremal point $g = (g_i)_{i \in I}$. Prove that $g_i \neq 0$ for all $i \in I$. Show that, for each $i \in I$, there is a unique element $s_i \in S_i$ such that $g_i(s_i) \neq 0$.

Chapter 4

Fredholm Theory

The purpose of the present chapter is to give a basic introduction to Fredholm operators and their indices including the stability theorem. A Fredholm operator is a bounded linear operator between Banach spaces that has a finite-dimensional kernel, a closed image, and a finite-dimensional cokernel. Its Fredholm index is the difference of the dimensions of kernel and cokernel. The stability theorem asserts that the Fredholm operators of any given index form an open subset of the space of all bounded linear operators between two Banach spaces, with respect to the topology induced by the operator norm. It also asserts that the sum of a Fredholm operator and a compact operator is again Fredholm and has the same index as the original operator. Fredholm operators play an important role in many fields of mathematics, including topology and geometry. There are many important topics that go beyond the scope of the present book. For example, the space of Fredholm operators on an infinite-dimensional Hilbert space is a classifying space for K-theory in that each continuous map from a topological space into the space of Fredholm operators gives rise to a pair of vector bundles (roughly speaking, the kernel and cokernel bundles) whose K-theory class is a homotopy invariant [**5**–**7**, **42**]. Another topic not covered here is Quillen's determinant line bundle over the space of Fredholm operators [**71**, **77**].

The chapter starts with an introduction to the dual of a bounded linear operator. It includes a proof of the Closed Image Theorem which asserts that an operator has a closed image if and only if its dual does. It then moves on to compact operators which map the unit ball to pre-compact subsets of the target space, characterizes Fredholm operators in terms of invertibility modulo compact operators, and establishes the stability theorem.

4.1. The Dual Operator

4.1.1. Definition and Examples. The dual operator of a bounded linear operator between Banach spaces is the induced operator between the dual spaces. Such a dual operator has been implicitly used in the proof of Theorem 3.6.9. Here is the formal definition.

Definition 4.1.1 (Dual Operator). Let X and Y be real normed vector spaces, denote their dual spaces by $X^* := \mathcal{L}(X, \mathbb{R})$ and $Y^* := \mathcal{L}(Y, \mathbb{R})$, and let $A : X \to Y$ be a bounded linear operator. The **dual operator** of A is the linear operator

$$A^* : Y^* \to X^*$$

defined by

$$A^*y^* := y^* \circ A : X \to \mathbb{R} \qquad \text{for } y^* \in Y^*. \tag{4.1.1}$$

Thus, for every bounded linear functional $y^* : Y \to \mathbb{R}$, the bounded linear functional $A^*y^* : X \to \mathbb{R}$ is the composition of the bounded linear operator $A : X \to Y$ with y^*, i.e.

$$\langle A^*y^*, x\rangle = \langle y^*, Ax\rangle \tag{4.1.2}$$

for all $x \in X$.

Lemma 4.1.2. *Let X and Y be real normed vector spaces and let*

$$A : X \to Y$$

be a bounded linear operator. Then the dual operator

$$A^* : Y^* \to X^*$$

is bounded and

$$\|A^*\| = \|A\|.$$

Proof. The operator norm of A^* is given by

$$\begin{aligned}
\|A^*\| &= \sup_{y^*\in Y^*\setminus\{0\}} \frac{\|A^*y^*\|}{\|y^*\|} \\
&= \sup_{y^*\in Y^*\setminus\{0\}} \sup_{x\in X\setminus\{0\}} \frac{|\langle A^*y^*, x\rangle|}{\|y^*\|\,\|x\|} \\
&= \sup_{x\in X\setminus\{0\}} \sup_{y^*\in Y^*\setminus\{0\}} \frac{|\langle y^*, Ax\rangle|}{\|y^*\|\,\|x\|} \\
&= \sup_{x\in X\setminus\{0\}} \frac{\|Ax\|}{\|x\|} \\
&= \|A\|.
\end{aligned}$$

Here the last but one equality follows from the Hahn–Banach Theorem in Corollary 2.3.23. In particular, $\|A^*\| < \infty$ and this proves Lemma 4.1.2. □

Lemma 4.1.3. *Let X, Y, Z be real normed vector spaces and let $A : X \to Y$ and $B : Y \to Z$ be bounded linear operators. Then the following hold.*

(i) $(BA)^* = A^*B^*$ *and* $(\mathbb{1}_X)^* = \mathbb{1}_{X^*}$.

(ii) *The* **bidual operator** $A^{**} : X^{**} \to Y^{**}$ *satisfies* $\iota_Y \circ A = A^{**} \circ \iota_X$, *where* $\iota_X : X \to X^{**}$ *and* $\iota_Y : Y \to Y^{**}$ *are the embeddings of Lemma* 2.4.1.

Proof. This follows directly from the definitions. □

Example 4.1.4. Let (M, d) be a compact metric space and let $\phi : M \to M$ be a homeomorphism. Let $T : C(M) \to C(M)$ be the operator in the proof of Theorem 3.6.9, defined by $Tf := f \circ \phi$ for $f \in C(M)$ (the pullback of f under ϕ). Then, under the identification $C(M)^* \cong \mathcal{M}(M)$ of the dual space of $C(M)$ with the space of signed Borel measures on M, the dual operator of T is the operator $T^* : \mathcal{M}(M) \to \mathcal{M}(M)$, which assigns to every signed Borel measure $\mu : \mathcal{B} \to \mathbb{R}$ its **pushforward** $T^*\mu = \phi_*\mu$ under ϕ. This pushforward is given by $(\phi_*\mu)(B) := \mu(\phi^{-1}(B))$ for every Borel set $B \subset M$.

Example 4.1.5 (Transposed Matrix). A matrix $A \in \mathbb{R}^{m\times n}$ induces a linear map $L_A : \mathbb{R}^n \to \mathbb{R}^m$. Its dual operator corresponds to the transposed matrix under the canonical isomorphisms $\iota_k : R^k \to (\mathbb{R}^k)^*$. This means that $(L_A)^* \circ \iota_m = \iota_n \circ L_{A^T} : \mathbb{R}^m \to (\mathbb{R}^n)^*$.

Example 4.1.6 (Adjoint Operator). Let H be a real Hilbert space, let $A : H \to H$ be a bounded linear operator, and let $A^*_{\text{Banach}} : H^* \to H^*$ be the dual operator of A. In this situation one can identify the Hilbert space H with its own dual space H^* via the isomorphism $I : H \to H^*$ in Theorem 1.4.4. The operator

$$A^*_{\text{Hilbert}} := I^{-1} \circ A^*_{\text{Banach}} \circ I : H \to H$$

is called the **adjoint operator** of A. It is characterized by the formula

$$\langle A^*_{\text{Hilbert}} y, x\rangle = \langle y, Ax\rangle \tag{4.1.3}$$

for all $x, y \in H$, where $\langle\cdot,\cdot\rangle$ denotes the inner product on the Hilbert space H, rather than the pairing between H^* and H as in equation (4.1.2). When working entirely in the Hilbert space setting, it is often convenient to use the notation $A^* := A^*_{\text{Hilbert}}$ for the adjoint operator instead of the dual operator.

Example 4.1.7 (Self-Adjoint Operator). Let $H = \ell^2$ be the Hilbert space in Example 1.4.7 and let $(a_i)_{i\in\mathbb{N}}$ be a bounded sequence of real numbers. Define the bounded linear operator $A : \ell^2 \to \ell^2$ by $Ax := (a_i x_i)_{i\in\mathbb{N}}$ for $x = (x_i)_{i\in\mathbb{N}} \in \ell^2$. This operator is equal to its own adjoint A^*_{Hilbert}. Such an operator is called **self-adjoint** or **symmetric**.

4.1.2. Duality.

Theorem 4.1.8 (Duality). *Let X and Y be real normed vector spaces and let $A : X \to Y$ be a bounded linear operator. Then the following hold.*

(i) $\mathrm{im}(A)^\perp = \ker(A^*)$ *and* ${}^\perp\mathrm{im}(A^*) = \ker(A)$.

(ii) *A has a dense image if and only if A^* is injective.*

(iii) *A is injective if and only if A^* has a weak* dense image.*

Proof. We prove (i). First let $y^* \in Y^*$. Then

$$\begin{aligned} y^* \in \mathrm{im}(A)^\perp \quad &\iff \quad \langle y^*, Ax\rangle = 0 \text{ for all } x \in X \\ &\iff \quad \langle A^*y^*, x\rangle = 0 \text{ for all } x \in X \iff A^*y^* = 0 \end{aligned}$$

and this shows that $\mathrm{im}(A)^\perp = \ker(A^*)$. Now let $x \in X$. Then

$$\begin{aligned} x \in {}^\perp\mathrm{im}(A^*) \quad &\iff \quad \langle A^*y^*, x\rangle = 0 \text{ for all } y^* \in Y^* \\ &\iff \quad \langle y^*, Ax\rangle = 0 \text{ for all } y^* \in Y^* \iff Ax = 0. \end{aligned}$$

The last step uses Corollary 2.3.23. This shows that ${}^\perp\mathrm{im}(A^*) = \ker(A)$.

We prove (ii). The operator A^* is injective if and only if $\ker(A^*) = \{0\}$. This is equivalent to $\mathrm{im}(A)^\perp = \{0\}$ by part (i) and hence to the condition that $\mathrm{im}(A)$ is dense in Y by Corollary 2.3.25.

We prove (iii). The operator A is injective if and only if $\ker(A) = \{0\}$. This is equivalent to ${}^\perp\mathrm{im}(A^*) = \{0\}$ by part (i) and hence to the condition that $\mathrm{im}(A^*)$ is weak* dense in X^* by Corollary 3.1.26. This proves Theorem 4.1.8. □

Example 4.1.9. Define the operator $A : \ell^2 \to \ell^2$ by $Ax := (i^{-1}x_i)_{i\in\mathbb{N}}$ for $x = (x_i)_{i\in\mathbb{N}} \in \ell^2$. This operator is self-adjoint, injective, and has a dense image, but is not surjective. Thus $\mathrm{im}(A) \subsetneq \ell^2 = {}^\perp\ker(A^*)$.

Example 4.1.10. The term "weak* dense" in part (iii) of Theorem 4.1.8 cannot be replaced by "dense". Let $X := \ell^1$ and $Y := c_0$. Then the inclusion $A : \ell^1 \to c_0$ is injective and has a dense image. Moreover, $X^* \cong \ell^\infty$ (Example 1.3.6) and $Y^* \cong \ell^1$ (Example 1.3.7), and $A^* : \ell^1 \to \ell^\infty$ is again the obvious inclusion. Its image is weak* dense (Corollary 3.1.29) but not dense. **Exercise:** The operator $A^{**} : (\ell^\infty)^* \to \ell^\infty$ is not injective.

Example 4.1.11. Let X be a real normed vector space, let $Y \subset X$ be a closed linear subspace, and let $\pi : X \to X/Y$ be the canonical projection. Then the dual operator $\pi^* : (X/Y)^* \to X^*$ is the isometric embedding of Corollary 2.3.26 whose image is the annihilator of Y. The dual operator of the inclusion $\iota : Y \to X$ is a surjective operator $\iota^* : X^* \to Y^*$ with kernel $Y^\perp$. It descends to the isometric isomorphism $X^*/Y^\perp \to Y^*$ in Corollary 2.3.26.

The next two theorems establish a correspondence between an inclusion for the images of two operators with the same target space and an estimate for the dual operators, and vice versa. The main tools for establishing such a correspondence are the Douglas Factorization Theorem (Corollary 2.2.17) and the Hahn–Banach Theorem (Corollary 2.3.4 and Corollary 2.3.26).

Theorem 4.1.12. *Let X, Y, and Z be real normed vector spaces and let $A : X \to Y$ and $B : X \to Z$ be bounded linear operators. Then the following are equivalent.*

(i) $\operatorname{im}(B^*) \subset \operatorname{im}(A^*)$.

(ii) *There exists a constant $c > 0$ such that*

$$\|Bx\|_Z \leq c\,\|Ax\|_Y \qquad \text{for all } x \in X. \tag{4.1.4}$$

Proof. See page 168. □

Theorem 4.1.13. *Let X, Y, Z be real Banach spaces and let $A : X \to Y$ and $B : Z \to Y$ be bounded linear operators. Then the following hold.*

(i) *If* $\operatorname{im}(B) \subset \operatorname{im}(A)$, *then there exists a constant $c > 0$ such that*

$$\|B^*y^*\|_{Z^*} \leq c\,\|A^*y^*\|_{X^*} \qquad \text{for all } y^* \in Y^*. \tag{4.1.5}$$

(ii) *If X is reflexive and* (4.1.5) *holds for some $c > 0$, then* $\operatorname{im}(B) \subset \operatorname{im}(A)$.

Proof. See page 168. □

Example 4.1.14. The hypothesis that X is reflexive cannot be removed in part (ii) of Theorem 4.1.13. However, this hypothesis is not needed when B is bijective (Corollary 4.1.17 below). Let $X := c_0$, $Y := \ell^2$, $Z := \mathbb{R}$, and define $A : c_0 \to \ell^2$, $B : \mathbb{R} \to \ell^2$ by $Ax := (i^{-1}x_i)_{i\in\mathbb{N}}$ for $x = (x_i)_{i\in\mathbb{N}} \in c_0$ and $Bz := (i^{-1}z)_{i\in\mathbb{N}}$ for $z \in \mathbb{R}$. Then (4.1.5) holds and $\operatorname{im}(B) \not\subset \operatorname{im}(A)$.

Lemma 4.1.15. *Let X and Y be real normed vector spaces and let $A : X \to Y$ be a bounded linear operator. Let $x^* \in X^*$. The following are equivalent.*

(i) $x^* \in \operatorname{im}(A^*)$.

(ii) *There is a constant $c \geq 0$ such that $|\langle x^*, x\rangle| \leq c\,\|Ax\|_Y$ for all $x \in X$.*

Proof. If $x^* = A^*y^*$, then $|\langle x^*, x\rangle| = |\langle y^*, Ax\rangle| \leq \|y^*\|_{Y^*}\,\|Ax\|_Y$ for all $x \in X$ and so (ii) holds with $c := \|y^*\|$. Conversely, suppose x^* satisfies (ii) and define the map $\psi : \operatorname{im}(A) \to \mathbb{R}$ as follows. Given $y \in \operatorname{im}(A)$ choose $x \in X$ such that $y = Ax$ and define $\psi(y) := \langle x^*, x\rangle$. By (ii) this number depends only on y, and not on x, and $\psi : \operatorname{im}(A) \to \mathbb{R}$ is a bounded linear functional. By definition, it satisfies $\psi \circ A = x^*$. By Corollary 2.3.4 there exists a $y^* \in Y^*$ such that $y^*|_{\operatorname{im}(A)} = \psi$. It satisfies $x^* = \psi \circ A = y^* \circ A = A^*y^*$ and this proves Lemma 4.1.15. □

Proof of Theorems 4.1.12 and 4.1.13. We prove that (ii) implies (i) in Theorem 4.1.12. Thus assume $A : X \to Y$ and $B : X \to Z$ satisfy (4.1.4) and fix an element $x^* \in \mathrm{im}(B^*)$. By Lemma 4.1.15 there is a constant $b > 0$ such that

$$|\langle x^*, x\rangle| \le b \|Bx\|_Z \le bc \|Ax\|_Y$$

for all $x \in X$. Hence $x^* \in \mathrm{im}(A^*)$ by Lemma 4.1.15. This shows that (ii) implies (i) in Theorem 4.1.12.

We prove part (ii) of Theorem 4.1.13. Thus assume that X is reflexive and the bounded linear operators $A : X \to Y$ and $B : Z \to Y$ satisfy (4.1.5). Since (ii) implies (i) in Theorem 4.1.12, we have $\mathrm{im}(B^{**}) \subset \mathrm{im}(A^{**})$. Now let $z \in Z$ and choose $x^{**} \in X^{**}$ such that $A^{**}x^{**} = B^{**}\iota_Z(z) = \iota_Y(Bz)$ (Lemma 4.1.3). Since X is reflexive there exists an element $x \in X$ such that $\iota_X(x) = x^{**}$. Hence $\iota_Y(Ax) = A^{**}\iota_X(x) = A^{**}x^{**} = \iota_Y(Bz)$ and therefore $Ax = Bz$. This proves part (ii) of Theorem 4.1.13.

We prove part (i) of Theorem 4.1.13. Assume that X, Y, Z are Banach spaces and that the bounded linear operators $A : X \to Y$ and $B : Z \to Y$ satisfy $\mathrm{im}(B) \subset \mathrm{im}(A)$. Define $X_0 := X/\ker(A)$ and let $\pi : X \to X_0$ be the canonical projection. Then $\pi^* : X_0^* \to X^*$ is an isometric embedding with image $\ker(A)^\perp$ (see Corollary 2.3.26). Moreover, the operator $A : X \to Y$ descends to a bounded linear operator $A_0 : X_0 \to Y$ such that $A_0 \circ \pi = A$. It satisfies $A^* = \pi^* \circ A_0^*$ and hence

$$\|A^*y^*\|_{X^*} = \|A_0^*y^*\|_{X_0^*} \qquad \text{for all } y^* \in Y^*. \tag{4.1.6}$$

Since $\mathrm{im}(B) \subset \mathrm{im}(A) = \mathrm{im}(A_0)$ and A_0 is injective, Corollary 2.2.17 asserts that there is a bounded linear operator $T : Z \to X_0$ with $A_0T = B$. Hence

$$\begin{aligned}
\|B^*y^*\|_{Z^*} &= \sup_{z\in Z\setminus\{0\}} \frac{\langle B^*y^*, z\rangle}{\|z\|_Z} \\
&= \sup_{z\in Z\setminus\{0\}} \frac{\langle A_0^*y^*, Tz\rangle}{\|z\|_Z} \\
&\le \sup_{z\in Z\setminus\{0\}} \frac{\|A_0^*y^*\|_{X_0^*}\,\|Tz\|_{X_0}}{\|z\|_Z} \\
&= \|T\|\,\|A^*y^*\|_{X^*}
\end{aligned}$$

for all $y^* \in Y^*$, by (4.1.6). This proves part (i) of Theorem 4.1.13.

We prove that (i) implies (ii) in Theorem 4.1.12. Thus assume that the operators $A : X \to Y$ and $B : X \to Z$ satisfy $\mathrm{im}(B^*) \subset \mathrm{im}(A^*)$. By part (i) of Theorem 4.1.13 there is a $c > 0$ such that $\|B^{**}x^{**}\|_{Z^{**}} \le c\|A^{**}x^{**}\|_{Y^{**}}$ for all $x^{**} \in X^{**}$. Hence, by Lemma 2.4.1 and Lemma 4.1.3, we have

$$\|Bx\|_Z = \|\iota_Z(Bx)\|_{Z^{**}} = \|B^{**}\iota_X(x)\|_{Z^{**}} \le c\|A^{**}\iota_X(x)\|_{Y^{**}} = c\|Ax\|_Y$$

for all $x \in X$. This proves Theorem 4.1.12. □

4.1.3. The Closed Image Theorem. The main theorem of this subsection asserts that a bounded linear operator between two Banach spaces has a closed image if and only if its dual operator has a closed image. A key tool in the proof will be Lemma 2.2.3 which can be viewed as a criterion for surjectivity of a bounded linear operator $A : X \to Y$ between Banach spaces. The criterion is that the closure of the image of the open unit ball in X under A contains a neighborhood of the origin in Y.

Theorem 4.1.16 (Closed Image Theorem). *Let X and Y be Banach spaces, let $A : X \to Y$ be a bounded linear operator, and let $A^* : Y^* \to X^*$ be its dual operator. Then the following are equivalent.*

(i) $\mathrm{im}(A) = {}^\perp\ker(A^*)$.

(ii) *The image of A is a closed subspace of Y.*

(iii) *There exists a constant $c > 0$ such that every $x \in X$ satisfies*

$$\inf_{A\xi=0} \|x + \xi\|_X \le c\,\|Ax\|_Y\,. \tag{4.1.7}$$

Here the infimum runs over all $\xi \in X$ that satisfy $A\xi = 0$.

(iv) $\mathrm{im}(A^*) = \ker(A)^\perp$.

(v) *The image of A^* is a weak* closed subspace of X^*.*

(vi) *The image of A^* is a closed subspace of X^*.*

(vii) *There exists a constant $c > 0$ such that every $y^* \in Y^*$ satisfies*

$$\inf_{A^*\eta^*=0} \|y^* + \eta^*\|_{Y^*} \le c\,\|A^*y^*\|_{X^*}\,. \tag{4.1.8}$$

Here the infimum runs over all $\eta^* \in Y^*$ that satisfy $A^*\eta^* = 0$.

Proof. That (i) implies (ii) follows from the fact that the pre-annihilator of any subset of Y^* is a closed subspace of Y.

We prove that (ii) implies (iii). Define

$$X_0 := X/\ker(A), \qquad Y_0 := \mathrm{im}(A),$$

and let $\pi_0 : X \to X_0$ be the projection which assigns to each element $x \in X$ the equivalence class $\pi_0(x) := [x] := x + \ker(A)$ of x in $X_0 = X/\ker(A)$. Since the kernel of A is closed and X is a Banach space, it follows from Theorem 1.2.14 that the quotient X_0 is a Banach space with

$$\|[x]\|_{X_0} = \inf_{\xi\in\ker(A)} \|x + \xi\|_X \qquad \text{for } x \in X.$$

Since the image of A is closed by (ii), the subspace $Y_0 \subset Y$ is a Banach space. Since the value $Ax \in Y_0 \subset Y$ of an element $x \in X$ under A depends only on the equivalence class of x in the quotient space X_0, there exists a unique linear map $A_0 : X_0 \to Y_0$ such that $A_0[x] = Ax$ for all $x \in X$. The map A_0 is bijective by definition. Moreover,

$$\|Ax\|_Y = \|A(x+\xi)\|_Y \le \|A\| \, \|x+\xi\|_X$$

for all $x \in X$ and all $\xi \in \ker(A)$, and hence

$$\|A_0[x]\|_{Y_0} = \|Ax\|_Y \le \|A\| \inf_{\xi \in \ker(A)} \|x+\xi\|_X = \|A\| \, \|[x]\|_{X_0}$$

for all $x \in X$. This shows that $A_0 : X_0 \to Y_0$ is a bijective bounded linear operator. Hence A_0 is open by the Open Mapping Theorem 2.2.1, so A_0^{-1} is continuous, and therefore A_0^{-1} is bounded by Theorem 1.2.2. Thus there exists a constant $c > 0$ such that $\|A_0^{-1}y\|_{X_0} \le c\|y\|_{Y_0}$ for all $y \in Y_0 \subset Y$. This implies

$$\inf_{\xi \in \ker(A)} \|x+\xi\|_X = \|[x]\|_{X_0} \le c\,\|A_0[x]\|_{Y_0} = c\,\|Ax\|_Y$$

for all $x \in X$. Thus we have proved that (ii) implies (iii).

We prove that (iii) implies (iv). The inclusion $\mathrm{im}(A^*) \subset \ker(A)^\perp$ follows directly from the definitions. To prove the converse inclusion, fix an element $x^* \in \ker(A)^\perp$ so that $\langle x^*, \xi\rangle = 0$ for all $\xi \in \ker(A)$. Then

$$|\langle x^*, x\rangle| = |\langle x^*, x+\xi\rangle| \le \|x^*\|_{X^*} \, \|x+\xi\|_X$$

for all $x \in X$ and all $\xi \in \ker(A)$. Take the infimum over all $\xi \in \ker(A)$ and use the inequality (4.1.7) in (iii) to obtain the estimate

$$|\langle x^*, x\rangle| \le \|x^*\|_{X^*} \inf_{A\xi=0} \|x+\xi\|_X \le c\,\|x^*\|_{X^*}\,\|Ax\|_Y \tag{4.1.9}$$

for all $x \in X$. It follows from (4.1.9) and Lemma 4.1.15 that $x^* \in \mathrm{im}(A^*)$. This shows that (iii) implies (iv).

That (iv) implies (v) follows from the definition of the weak* topology. Namely, the annihilator of any subset of X is a weak* closed subset of X^*. (See the proof of Corollary 3.3.2.)

That (v) implies (vi) follows directly from the fact that every weak* closed subset of X^* is closed with respect to the strong topology induced by the operator norm on the dual space.

That (vi) implies (vii) follows from the fact that (ii) implies (iii) (already proved) with the operator A replaced by its dual operator A^*.

We prove that (vii) implies (i). Assume first that A satisfies (vii) and has a dense image. Then A^* is injective by Theorem 4.1.8 and so the inequality (4.1.8) in part (vii) takes the form

$$\|y^*\|_{Y^*} \le c\,\|A^*y^*\|_{X^*} \qquad \text{for all } y^* \in Y^*. \tag{4.1.10}$$

Define $\delta := c^{-1}$. We prove that

$$\big\{y \in Y \,\big|\, \|y\| \le \delta\big\} \subset \overline{\big\{Ax \,\big|\, x \in X,\ \|x\|_X < 1\big\}}. \tag{4.1.11}$$

To see this, observe that the set

$$K := \overline{\{Ax \mid x \in X,\ \|x\|_X < 1\}}$$

is a closed convex subset of Y. We must show that every $y \in Y \setminus K$ has norm $\|y\|_Y > \delta$. To see this fix an element $y_0 \in Y \setminus K$. By Theorem 2.3.10 there exists a bounded linear functional $y_0^* : Y \to \mathbb{R}$ such that

$$\langle y_0^*, y_0 \rangle > \sup_{y \in K} \langle y_0^*, y \rangle.$$

This implies

$$\begin{aligned} \|A^* y_0^*\|_{X^*} &= \sup_{\substack{x \in X \\ \|x\| < 1}} \langle A^* y_0^*, x \rangle \\ &= \sup_{\substack{x \in X \\ \|x\| < 1}} \langle y_0^*, Ax \rangle \\ &= \sup_{y \in K} \langle y_0^*, y \rangle \\ &< \langle y_0^*, y_0 \rangle \\ &\leq \|y_0\|_Y \|y_0^*\|_{Y^*} \end{aligned}$$

and hence, by (4.1.10),

$$\|y_0\| > \frac{\|A^* y_0^*\|_{X^*}}{\|y_0^*\|_{Y^*}} \geq \frac{1}{c} = \delta.$$

This proves (4.1.11). Hence $\{y \in Y \mid \|y\| < \delta\} \subset \{Ax \mid x \in X,\ \|x\|_X < 1\}$ by Lemma 2.2.3. Thus A is surjective and so $\mathrm{im}(A) = Y = {}^\perp\ker(A^*)$ because A^* is injective. This shows that (vii) implies (i) whenever the operator A has a dense image.

Now suppose A satisfies (vii) and does not have a dense image. Define

$$Y_0 := \overline{\mathrm{im}(A)}, \qquad A_0 := A : X \to Y_0.$$

Thus A_0 is the same operator as A, but viewed as an operator with values in the smaller target space Y_0. The dual operator $A_0^* : Y_0^* \to X^*$ satisfies

$$A_0^*(y^*|_{Y_0}) = A^* y^* \qquad \text{for all } y^* \in Y^* \tag{4.1.12}$$

by definition. Moreover, for all $y^* \in Y^*$, we have

$$\|y^*|_{Y_0}\|_{Y_0^*} = \inf_{\eta^* \in \ker(A^*)} \|y^* + \eta^*\|_{Y^*} \leq c \|A^* y^*\|_{X^*} = c \|A_0^*(y^*|_{Y^0})\|_{X^*}.$$

Here we have used the inequality (4.1.8) in (vii) and equation (4.1.12). Hence it follows from the first part of the proof (the dense image case) that the operator $A_0 : X \to Y_0$ is surjective. Thus

$$\mathrm{im}(A) = \mathrm{im}(A_0) = Y_0 = \overline{\mathrm{im}(A)} = {}^\perp(\mathrm{im}(A)^\perp) = {}^\perp\ker(A^*)$$

by Corollary 3.1.18 and Theorem 4.1.8. This shows that (vii) implies (i) and completes the proof of Theorem 4.1.16. □

Corollary 4.1.17. *Let X and Y be Banach spaces and let $A : X \to Y$ be a bounded linear operator. Then the following hold.*

(i) *The operator A is surjective if and only if A^* is injective and has a closed image. Equivalently, there exists a constant $c > 0$ such that*

$$\|y^*\|_{Y^*} \le c \|A^*y^*\|_{X^*} \qquad \text{for all } y^* \in Y^*. \tag{4.1.13}$$

(ii) *The operator A^* is surjective if and only if A is injective and has a closed image. Equivalently, there exists a constant $c > 0$ such that*

$$\|x\|_X \le c \|Ax\|_Y \qquad \text{for all } x \in X. \tag{4.1.14}$$

Proof. The operator A has a dense image if and only if A^* is injective by Theorem 4.1.8. Hence A is surjective if and only if it has a closed image and A^* is injective. Hence part (i) follows from (4.1.8) in Theorem 4.1.16. Part (ii) is the special case of Theorem 4.1.13 where $Z = X$ and $B = \mathrm{id} : X \to X$. Alternatively, one can argue as in the proof of part (i). The operator A^* has a weak* dense image if and only if A is injective by Theorem 4.1.8. Hence A^* is surjective if and only if it has a weak* closed image and A is injective. Hence part (ii) follows from (4.1.7) in Theorem 4.1.16. This proves Corollary 4.1.17. □

Corollary 4.1.18. *Let X and Y be Banach spaces and let $A : X \to Y$ be a bounded linear operator. Then the following hold.*

(i) *A is bijective if and only if A^* is bijective.*

(ii) *If A is bijective, then $(A^*)^{-1} = (A^{-1})^*$.*

(iii) *A is an isometry if and only if A^* is an isometry.*

Proof. We prove (i). If A is bijective, then A^* is injective by Theorem 4.1.8 and A satisfies the inequality (4.1.14) by Theorem 2.2.1, so A^* is surjective by Corollary 4.1.17. Conversely, if A^* is bijective, then A is injective by Theorem 4.1.8 and A^* satisfies the inequality (4.1.13) by Theorem 2.2.1, so A is surjective by Corollary 4.1.17.

We prove (ii). Assume A is bijective and define $B := A^{-1} : Y \to X$. Then B is a bounded linear operator by Theorem 2.2.1 and

$$AB = \mathrm{id}_Y, \qquad BA = \mathrm{id}_X.$$

Hence $B^*A^* = (AB)^* = (\mathrm{id}_Y)^* = \mathrm{id}_{Y^*}$ and $A^*B^* = (BA)^* = (\mathrm{id}_X)^* = \mathrm{id}_{X^*}$ by Lemma 4.1.3. This shows that $B^* = (A^*)^{-1}$.

We prove (iii). Assume A and A^* are bijective. Then $(A^*)^{-1} = (A^{-1})^*$ by part (ii) and hence $\|A^*\| = \|A\|$ and $\|(A^*)^{-1}\| = \|A^{-1}\|$ by Lemma 4.1.2. With this understood, part (iii) follows from the fact that A is an isometry if and only if $\|A\| = \|A^{-1}\| = 1$. This proves Corollary 4.1.18. □

An example of a Banach space isometry is the pullback under a homeomorphism $\phi : M \to M$ of a compact metric space, acting on the space of continuous functions on M, equipped with the supremum norm. Its dual operator is the pushforward under ϕ, acting on the space of signed Borel measures on M (see Examples 1.3.8 and 4.1.4).

In finite dimensions orthogonal transformations of real vector spaces with inner products and unitary transformations of complex vector spaces with Hermitian inner products are examples of isometries. These examples carry over to infinite-dimensional real and complex Hilbert spaces. In infinite dimensions orthogonal and unitary transformations have many important applications. They arise naturally in the study of certain partial differential equations such as the wave equation and the Schrödinger equation. The functional analytic background for the study of such equations is the theory of strongly continuous semigroups of operators. This is the subject of Chapter 7 below.

4.2. Compact Operators

One of the most important concepts in the study of bounded linear operators is that of a compact operator. The notion of a compact operator can be defined in several equivalent ways. The equivalence of these conditions is the content of the following lemma.

Lemma 4.2.1. *Let X and Y be Banach spaces and let $K : X \to Y$ be a bounded linear operator. Then the following are equivalent.*

(i) *If $(x_n)_{n\in\mathbb{N}}$ is a bounded sequence in X, then the sequence $(Kx_n)_{n\in\mathbb{N}}$ has a Cauchy subsequence.*

(ii) *If $S \subset X$ is a bounded set, then the set $K(S) := \{Kx \,|\, x \in S\}$ has a compact closure.*

(iii) *The set $\overline{\{Kx \,|\, x \in X,\ \|x\|_X \le 1\}}$ is a compact subset of Y.*

Proof. We prove that (i) implies (ii). Thus assume K satisfies (i) and let $S \subset X$ be a bounded set. Then every sequence in $K(S)$ has a Cauchy subsequence by (i). Hence Corollary 1.1.8 asserts that $\overline{K(S)}$ is a compact subset of Y, because Y is complete.

That (ii) implies (iii) is obvious.

We prove that (iii) implies (i). Let $(x_n)_{n\in\mathbb{N}}$ be a bounded sequence and choose $c > 0$ such that $\|x_n\|_X \le c$ for all $n \in \mathbb{N}$. Then $(c^{-1}Kx_n)_{n\in\mathbb{N}}$ has a convergent subsequence $(c^{-1}Kx_{n_i})_{i\in\mathbb{N}}$ by (iii). Hence $(Kx_{n_i})_{i\in\mathbb{N}}$ is the required Cauchy subsequence. This proves Lemma 4.2.1. $\square$

Definition 4.2.2 (Compact Operator). Let X and Y be Banach spaces. A bounded linear operator $K : X \to Y$ is said to be

• **compact** if it satisfies the equivalent conditions of Lemma 4.2.1,

• **of finite rank** if its image is a finite-dimensional subspace of Y,

• **completely continuous** if the image of every weakly convergent sequence in X under K converges in the norm topology on Y.

Lemma 4.2.3. *Let X and Y be Banach spaces. Then the following hold.*

(i) *Every compact operator $K : X \to Y$ is completely continuous.*

(ii) *Assume X is reflexive. Then a bounded linear operator $K : X \to Y$ is compact if and only if it is completely continuous.*

Proof. We prove part (i). Assume K is compact and let $(x_n)_{n\in\mathbb{N}}$ be a sequence in X that converges weakly to $x \in X$. Suppose, by contradiction, that the sequence $(Kx_n)_{n\in\mathbb{N}}$ does not converge to Kx in the norm topology. Then there exist an $\varepsilon > 0$ and a subsequence $(x_{n_i})_{i\in\mathbb{N}}$ such that

$$\|Kx - Kx_{n_i}\|_Y \geq \varepsilon \qquad \text{for all } i \in \mathbb{N}. \tag{4.2.1}$$

Since the sequence $(x_{n_i})_{i\in\mathbb{N}}$ converges weakly, it is bounded by the Uniform Boundedness Theorem 2.1.1. Since K is compact, there exists a further subsequence $(x_{n_{i_k}})_{k\in\mathbb{N}}$ such that the sequence $(Kx_{n_{i_k}})_{k\in\mathbb{N}}$ converges strongly to some element $y \in Y$. This implies

$$\langle y^*, y\rangle = \lim_{k\to\infty} \langle y^*, Kx_{n_{i_k}}\rangle = \lim_{k\to\infty} \langle K^*y^*, x_{n_{i_k}}\rangle = \langle K^*y^*, x\rangle = \langle y^*, Kx\rangle$$

for all $y^* \in Y^*$. Hence $y = Kx$ by Corollary 2.3.23 and so

$$\lim_{k\to\infty} \|Kx_{n_{i_k}} - Kx\|_Y = 0,$$

in contradiction to (4.2.1). This proves (i).

We prove part (ii). Assume X is reflexive and K is completely continuous. Let $(x_n)_{n\in\mathbb{N}}$ be a bounded sequence in X. Since X is reflexive, there exists a weakly convergent subsequence $(x_{n_i})_{i\in\mathbb{N}}$ by Theorem 3.4.1. Let $x \in X$ be the limit of that subsequence. Since K is completely continuous, the sequence $(Kx_{n_i})_{i\in\mathbb{N}}$ converges strongly to Kx. Thus K satisfies condition (i) in Lemma 4.2.1 and hence is compact. This proves Lemma 4.2.3. □

Example 4.2.4. The hypothesis that X is reflexive cannot be removed in part (ii) of Lemma 4.2.3. For example a sequence in ℓ^1 converges weakly if and only if it converges strongly by Exercise 3.1.22. Hence the identity operator $\mathrm{id} : \ell^1 \to \ell^1$ is completely continuous. However, it is not a compact operator by Theorem 1.2.11.

Example 4.2.5. Every finite rank operator is compact.

Example 4.2.6. Let $X := C^1([0,1])$, $Y := C([0,1])$ and let $K : X \to Y$ be the obvious inclusion. Then the image of the closed unit ball is a bounded equi-continuous subset of $C([0,1])$ and hence has a compact closure by the Arzelà–Ascoli Theorem (Corollary 1.1.13). In this example the image of the closed unit ball in X under K is not a closed subset of Y. **Exercise:** If X is reflexive and $K : X \to Y$ is a compact operator, then the image of the closed unit ball $B \subset X$ under K is a closed subset of Y. **Hint:** Every sequence in B has a weakly convergent subsequence by Theorem 3.4.1.

Example 4.2.7. If $K : X \to Y$ is a bounded linear operator between Banach spaces whose image is a closed infinite-dimensional subspace of Y, then K is not compact. Namely, the image of the closed unit ball in X under K contains an open ball in $\mathrm{im}(K)$ by Theorem 4.1.16, and hence does not have a compact closure by Theorem 1.2.11.

Example 4.2.8. Fix a number $1 \le p \le \infty$ and a bounded sequence of real numbers $\lambda = (\lambda_i)_{i\in\mathbb{N}}$. For $i \in \mathbb{N}$ let $e_i := (\delta_{ij})_{j\in\mathbb{N}} \in \ell^p$. Define the bounded linear operator $K_\lambda : \ell^p \to \ell^p$ by

$$K_\lambda x := (\lambda_i x_i)_{i\in\mathbb{N}} \qquad \text{for } x = (x_i)_{i\in\mathbb{N}} \in \ell^p.$$

Then

$$K_\lambda \text{ is compact} \qquad \iff \qquad \lim_{i\to\infty} \lambda_i = 0.$$

The condition $\lim_{i\to\infty} \lambda_i = 0$ is necessary for compactness because, if there exist a constant $\delta > 0$ and a sequence $1 \le n_1 < n_2 < n_3 < \cdots$ such that $|\lambda_{n_k}| \ge \delta$ for all $k \in \mathbb{N}$, then the sequence $Ke_{n_k} = \lambda_{n_k} e_{n_k}$, $k \in \mathbb{N}$, in ℓ^p has no convergent subsequence. The condition $\lim_{i\to\infty} \lambda_i = 0$ implies compactness because then K can be approximated by a sequence of finite rank operators in the norm topology (Example 4.2.5 and Theorem 4.2.10).

Exercise 4.2.9. Find a strongly convergent sequence of compact operators whose limit operator is not compact.

The following theorem shows that the set of compact operators between two Banach spaces is closed with respect to the norm topology.

Theorem 4.2.10 (Compact Operators). *Let X, Y, and Z be Banach spaces. Then the following hold.*

(i) *Let $A : X \to Y$ and $B : Y \to Z$ be bounded linear operators and assume that A is compact or B is compact. Then $BA : X \to Z$ is compact.*

(ii) *Let $K_i : X \to Y$ be a sequence of compact operators that converges to a bounded linear operator $K : X \to Y$ in the norm topology. Then K is compact.*

(iii) *Let $K : X \to Y$ be a bounded linear operator and let $K^* : Y^* \to X^*$ be its dual operator. Then K is compact if and only if K^* is compact.*

Proof. We prove part (i). Let $(x_n)_{n\in\mathbb{N}}$ be a bounded sequence in X. If A is compact, then there exists a subsequence $(x_{n_k})_{k\in\mathbb{N}}$ such that the sequence $(Ax_{n_k})_{k\in\mathbb{N}}$ converges, and so does the subsequence $(BAx_{n_k})_{k\in\mathbb{N}}$. If B is compact, then, since the sequence $(Ax_n)_{n\in\mathbb{N}}$ is bounded, there exists a subsequence $(Ax_{n_k})_{k\in\mathbb{N}}$ such that the sequence $(BAx_{n_k})_{k\in\mathbb{N}}$ converges. This proves (i).

We prove part (ii). Let $(x_n)_{n\in\mathbb{N}}$ be a bounded sequence in X. Then a standard diagonal subsequence argument shows that the sequence $(Kx_n)_{n\in\mathbb{N}}$ has a convergent subsequence. More precisely, since K_1 is compact, there exists a subsequence $(x_{n_{1,k}})_{k\in\mathbb{N}}$ such that the sequence $(K_1x_{n_{1,k}})_{k\in\mathbb{N}}$ converges in Y. Since K_2 is compact there exists a further subsequence $(x_{n_{2,k}})_{k\in\mathbb{N}}$ such that the sequence $(K_2x_{n_{2,k}})_{k\in\mathbb{N}}$ converges in Y. Continue by induction and use the axiom of dependent choice to find a sequence of subsequences $(x_{n_{i,k}})_{k\in\mathbb{N}}$ such that, for each $i\in\mathbb{N}$, the sequence $(x_{n_{i,k+1}})_{k\in\mathbb{N}}$ is a subsequence of $(x_{n_{i,k}})_{k\in\mathbb{N}}$ and the sequence $(K_ix_{n_{i,k}})_{k\in\mathbb{N}}$ converges in Y. Now consider the diagonal subsequence

$$x_{n_k} := x_{n_{k,k}} \qquad \text{for } k\in\mathbb{N}.$$

Then the sequence $(K_ix_{n_k})_{k\in\mathbb{N}}$ converges in Y for every $i\in\mathbb{N}$. We prove that the sequence $(Kx_{n_k})_{k\in\mathbb{N}}$ converges as well. To see this, choose a constant $c>0$ such that

$$\|x_n\|_X \le c \qquad \text{for all } n\in\mathbb{N}.$$

Fix a constant $\varepsilon>0$. Then there exists a positive integer i such that

$$\|K-K_i\| < \frac{\varepsilon}{3c}.$$

Since the sequence $(K_ix_{n_k})_{k\in\mathbb{N}}$ converges, there exists a positive integer k_0 such that all $k,\ell\in\mathbb{N}$ satisfy

$$k,\ell\ge k_0 \qquad \Longrightarrow \qquad \|K_ix_{n_k}-K_ix_{n_\ell}\|_Y < \frac{\varepsilon}{3}.$$

This implies

$$\begin{aligned}
&\|Kx_{n_k}-Kx_{n_\ell}\|_Y \\
&\le \|Kx_{n_k}-K_ix_{n_k}\|_Y + \|K_ix_{n_k}-K_ix_{n_\ell}\|_Y + \|K_ix_{n_\ell}-Kx_{n_\ell}\|_Y \\
&\le \|K-K_i\|\,\|x_{n_k}\|_X + \|K_ix_{n_k}-K_ix_{n_\ell}\|_Y + \|K_i-K\|\,\|x_{n_\ell}\|_X \\
&\le 2c\,\|K-K_i\| + \|K_ix_{n_k}-K_ix_{n_\ell}\|_Y \\
&< \varepsilon
\end{aligned}$$

for all pairs of integers $k,\ell\ge k_0$. Thus $(Kx_{n_k})_{k\in\mathbb{N}}$ is a Cauchy sequence in Y and hence converges, because Y is complete. This shows that K is compact and hence completes the proof of part (ii).

We prove part (iii). Assume first that $K : X \to Y$ is a compact operator. Then the set

$$M := \overline{\{Kx \mid \|x\|_X \leq 1\}} \subset Y$$

is a compact metric space with the distance function determined by the norm on Y. For $y^* \in Y^*$ consider the continuous real valued function

$$f_{y^*} := y^*|_M : M \to \mathbb{R}.$$

Define the set $\mathscr{F} \subset C(M)$ by

$$\mathscr{F} := \left\{ f_{y^*} \,\middle|\, y^* \in Y^*,\ \|y^*\|_{Y^*} \leq 1 \right\}.$$

For each $y^* \in Y^*$ with $\|y^*\|_{Y^*} \leq 1$ the supremum norm of f_{y^*} is given by

$$\begin{aligned} \|f_{y^*}\| &= \sup_{y \in M} |\langle y^*, y\rangle| \\ &= \sup_{x \in X,\, \|x\|_X \leq 1} |\langle y^*, Kx\rangle| \\ &= \sup_{x \in X,\, \|x\|_X \leq 1} |\langle K^*y^*, x\rangle| \\ &= \|K^*y^*\|_{X^*}. \end{aligned} \tag{4.2.2}$$

Thus $\|f\| \leq \|K^*\| = \|K\|$ for all $f \in \mathscr{F}$, so $\mathscr{F}$ is a bounded subset of $C(M)$. Moreover, the set $\mathscr{F}$ is equi-continuous because

$$\begin{aligned} |f_{y^*}(y) - f_{y^*}(y')| &= |\langle y^*, y - y'\rangle| \\ &\leq \|y^*\|_{Y^*} \left\|y - y'\right\|_Y \\ &\leq \left\|y - y'\right\|_Y \end{aligned}$$

for all $y^* \in Y^*$ with $\|y^*\|_{Y^*} \leq 1$ and all $y, y' \in M$. Since M is a compact metric space, it follows from the Arzelà–Ascoli Theorem (Corollary 1.1.13) that $\mathscr{F}$ has a compact closure. This implies that the operator K^* is compact. To see this, let $(y_n^*)_{n\in\mathbb{N}}$ be a sequence in Y^* such that $\|y_n^*\|_{Y^*} \leq 1$ for all $n \in \mathbb{N}$. Then the sequence $(f_{y_n^*})_{n\in\mathbb{N}}$ in $\mathscr{F}$ has a uniformly convergent subsequence $(f_{y_{n_i}^*})_{i\in\mathbb{N}}$. Hence it follows from (4.2.2) that $(K^*y_{n_i}^*)_{i\in\mathbb{N}}$ is a Cauchy sequence in X^* and hence converges. This shows that K^* is a compact operator as claimed.

Conversely, suppose that K^* is compact. Then, by what we have just proved, the bidual operator $K^{**} : X^{**} \to Y^{**}$ is compact. This implies that K is compact. To see this, let $(x_n)_{n\in\mathbb{N}}$ be a bounded sequence in X. Then $(\iota_X(x_n))_{n\in\mathbb{N}}$ is a bounded sequence in X^{**} by Lemma 2.4.1. Since K^{**} is a compact operator, there exists a subsequence $(\iota_X(x_{n_i}))_{i\in\mathbb{N}}$ such that the sequence $K^{**}\iota_X(x_{n_i}) = \iota_Y(Kx_{n_i})$ converges in Y^{**} as i tends to infinity. Hence $(Kx_{n_i})_{i\in\mathbb{N}}$ is a Cauchy sequence in Y by Lemma 2.4.1. Hence K is compact and this proves Theorem 4.2.10. □

It follows from part (ii) of Theorem 4.2.10 that the limit of a sequence of finite rank operators in the norm topology is a compact operator. It is a natural question to ask whether, conversely, every compact operator can be approximated in the norm topology by a sequence of finite rank operators. The answer to this question was an open problem in functional analysis for many years. It was eventually shown that the answer depends on the Banach space in question. Here is a reformulation of the problem due to Grothendieck [**33**].

Exercise 4.2.11. Let Y be a Banach space. Prove that the following are equivalent.

(a) For every Banach space X, every compact operator $K : X \to Y$, and every $\varepsilon > 0$ there is a finite rank operator $T : X \to Y$ such that $\|K - T\| < \varepsilon$.

(b) For every compact subset $C \subset Y$ and every $\varepsilon > 0$ there is a finite rank operator $T : Y \to Y$ such that $\|y - Ty\| < \varepsilon$ for all $y \in C$.

A Banach space Y that satisfies these two equivalent conditions is said to have the **approximation property**.

Exercise 4.2.12. Let Y be a Banach space that has a **Schauder basis** $(e_i)_{i\in\mathbb{N}}$, i.e. for every $y \in Y$ there exists a unique sequence $\lambda = (\lambda_i)_{i\in\mathbb{N}}$ of real numbers such that the sequence $\sum_{i=1}^n \lambda_i e_i$ converges and

$$y = \sum_{i=1}^{\infty} \lambda_i e_i = \lim_{n\to\infty} \sum_{i=1}^{n} \lambda_i e_i.$$

Prove that Y has the approximation property. **Hint:** Let $\Pi_n : Y \to Y$ be the unique projection such that

$$\mathrm{im}(\Pi_n) = \mathrm{span}\{e_1, \ldots, e_n\}, \qquad \Pi_n e_i = 0 \quad \text{for all } i > n.$$

By Exercise 2.5.12, the operators Π_n are uniformly bounded. Prove that

$$\lim_{n\to\infty} \|\Pi_n K - K\| = 0$$

for every compact operator $K : X \to Y$.

The first example of a Banach space without the approximation property was found by Enflo [**27**] in 1973. His example is separable and reflexive. It was later shown by Szankowski in [**82**] that there exist closed linear subspaces of ℓ^p (with $1 \le p < \infty$ and $p \ne 2$) and of c_0 that do not have the approximation property. Another result of Szankovski [**83**] asserts that the Banach space $\mathcal{L}(H)$ of all bounded linear operators from an infinite-dimensional Hilbert space H to itself, equipped with the operator norm, does not have the approximation property.

4.3. Fredholm Operators

Let X and Y be real Banach spaces and let $A : X \to Y$ be a bounded linear operator. Recall that the **kernel**, **image**, and **cokernel** of A are defined by

$$\begin{aligned} \ker(A) &:= \{x \in X \mid Ax = 0\}, \\ \operatorname{im}(A) &:= \{Ax \mid x \in X\}, \\ \operatorname{coker}(A) &:= Y/\operatorname{im}(A). \end{aligned} \tag{4.3.1}$$

If the image of A is a closed subspace of Y, then the cokernel is a Banach space with the norm (1.2.4).

Definition 4.3.1 (Fredholm Operator). Let X and Y be real Banach spaces and let $A : X \to Y$ be a bounded linear operator. A is called a **Fredholm operator** if it has a closed image and its kernel and cokernel are finite-dimensional. If A is a Fredholm operator the difference of the dimensions of its kernel and cokernel is called the **Fredholm index** of A and is denoted by

$$\operatorname{index}(A) := \dim\ker(A) - \dim\operatorname{coker}(A). \tag{4.3.2}$$

The condition that the image of A is closed is actually redundant in Definition 4.3.1. It holds necessarily when the cokernel is finite-dimensional. In other words, while any infinite-dimensional Banach space Y admits linear subspaces $Z \subset Y$ that are not closed and have finite-dimensional quotients Y/Z, such a subspace can never be the image of a bounded linear operator on a Banach space with values in Y.

Lemma 4.3.2. *Let X and Y be Banach spaces and let $A : X \to Y$ be a bounded linear operator with a finite-dimensional cokernel. Then the image of A is a closed subspace of Y.*

Proof. Let $m := \dim\operatorname{coker}(A)$ and choose vectors $y_1, \dots, y_m \in Y$ such that the equivalence classes

$$[y_i] := y_i + \operatorname{im}(A) \in Y/\operatorname{im}(A), \qquad i = 1, \dots, m,$$

form a basis of the cokernel of A. Define

$$\widetilde{X} := X \times \mathbb{R}^m, \qquad \|(x,\lambda)\|_{\widetilde{X}} := \|x\|_X + \|\lambda\|_{\mathbb{R}^m}$$

for $x \in X$ and $\lambda = (\lambda_1, \dots, \lambda_m) \in \mathbb{R}^m$. Then $\widetilde{X}$ is a Banach space. Define the linear operator $\widetilde{A} : \widetilde{X} \to Y$ by

$$\widetilde{A}(x,\lambda) := Ax + \sum_{i=1}^m \lambda_i y_i.$$

Then $\widetilde{A}$ is a surjective bounded linear operator and

$$\ker(\widetilde{A}) = \{(x,\lambda) \in X \times \mathbb{R}^m \mid Ax = 0,\ \lambda = 0\} = \ker(A) \times \{0\}.$$

Since $\widetilde{A}$ is surjective, it follows from Theorem 4.1.16 that there exists a constant $c > 0$ such that

$$\inf_{\xi \in \ker(A)} \|x + \xi\|_X + \|\lambda\|_{\mathbb{R}^m} \le c \left\| Ax + \sum_{i=1}^m \lambda_i y_i \right\|_Y$$

for all $x \in X$ and all $\lambda \in \mathbb{R}^m$. Take $\lambda = 0$ to obtain the inequality

$$\inf_{\xi \in \ker(A)} \|x + \xi\|_X \le c \|Ax\|_Y \qquad \text{for all } x \in X.$$

Thus A has a closed image by Theorem 4.1.16. This proves Lemma 4.3.2. □

Theorem 4.3.3 (Duality for Fredholm Operators). *Let X and Y be Banach spaces and let $A \in \mathcal{L}(X, Y)$. Then the following hold.*

(i) *If A and A^* have closed images, then*

$$\dim \ker(A^*) = \dim \operatorname{coker}(A), \qquad \dim \operatorname{coker}(A^*) = \dim \ker(A).$$

(ii) *A is a Fredholm operator if and only if A^* is a Fredholm operator.*

(iii) *If A is a Fredholm operator, then* $\operatorname{index}(A^*) = -\operatorname{index}(A)$.

Proof. Assume A and A^* have closed images. Then

$$\operatorname{im}(A^*) = \ker(A)^\perp, \qquad \ker(A^*) = \operatorname{im}(A)^\perp$$

by Theorem 4.1.8 and Theorem 4.1.16. Hence it follows from Corollary 2.3.26 that the dual spaces of the linear subspace $\ker(A) \subset X$ and of the quotient space $\operatorname{coker}(A) = Y/\operatorname{im}(A)$ are isomorphic to

$$(\ker(A))^* \cong X^*/\ker(A)^\perp = X^*/\operatorname{im}(A^*) = \operatorname{coker}(A^*),$$
$$(\operatorname{coker}(A))^* = (Y/\operatorname{im}(A))^* \cong \operatorname{im}(A)^\perp = \ker(A^*).$$

This proves part (i). Parts (ii) and (iii) follow directly from (i) and Theorem 4.1.16. This proves Theorem 4.3.3. □

Example 4.3.4. If X and Y are finite-dimensional, then every linear operator $A : X \to Y$ is Fredholm and $\operatorname{index}(A) = \dim X - \dim Y$.

Example 4.3.5. Every bijective bounded linear operator between Banach spaces is a Fredholm operator of index zero.

Example 4.3.6. Consider the Banach space $X = \ell^p$ with $1 \le p \le \infty$ and let $k \in \mathbb{N}$. Define the linear operators $A_k, A_{-k} : \ell^p \to \ell^p$ by

$$\begin{aligned} A_k x &:= (x_{k+1}, x_{k+2}, x_{k+3}, \dots), \\ A_{-k} x &:= (0, \dots, 0, x_1, x_2, x_3, \dots) \end{aligned} \qquad \text{for } x = (x_i)_{i \in \mathbb{N}} \in \ell^p,$$

where x_1 is preceded by k zeros in the formula for A_{-k}. These are Fredholm operators of indices $\operatorname{index}(A_k) = k$ and $\operatorname{index}(A_{-k}) = -k$.

Example 4.3.7. Let X, Y, and Z be Banach spaces and let $A : X \to Y$ and $\Phi : Z \to Y$ be bounded linear operators. Define the bounded linear operator $A \oplus \Phi : X \oplus Z \to Y$ by

$$(A \oplus \Phi)(x, z) := Ax + \Phi z.$$

If A is a Fredholm operator and $\dim Z < \infty$, then $A \oplus \Phi$ is a Fredholm operator and $\mathrm{index}(A \oplus \Phi) = \mathrm{index}(A) + \dim Z$. (Prove this!)

The next theorem characterizes the Fredholm operators as those operators that are invertible modulo the compact operators.

Theorem 4.3.8 (Fredholm and Compact Operators). *Let X and Y be Banach spaces and let $A : X \to Y$ be a bounded linear operator. Then the following are equivalent.*

(i) *A is a Fredholm operator.*

(ii) *There exists a bounded linear operator $F : X \to Y$ such that the operators $\mathbb{1}_X - FA : X \to X$ and $\mathbb{1}_Y - AF : Y \to Y$ are compact.*

Proof. See page 183. □

The proof of Theorem 4.3.8 relies on the following lemma. This lemma also gives a partial answer to the important question of how one can recognize whether a given operator is Fredholm. It characterizes bounded linear operators with a closed image and a finite-dimensional kernel and is a key tool for establishing the Fredholm property for many differential operators.

Lemma 4.3.9 (Main Fredholm Lemma). *Let X and Y be Banach spaces and let $D : X \to Y$ be a bounded linear operator. Then the following are equivalent.*

(i) *D has a finite-dimensional kernel and a closed image.*

(ii) *There exist a Banach space Z, a compact operator $K : X \to Z$, and a constant $c > 0$ such that*

$$\|x\|_X \le c\big(\|Dx\|_Y + \|Kx\|_Z\big) \tag{4.3.3}$$

for all $x \in X$.

Proof. We prove that (i) implies (ii). Thus assume D has a finite-dimensional kernel and a closed image. Define $m := \dim \ker(D)$ and choose a basis $x_1, \dots, x_m$ of $\ker(D)$. By the Hahn–Banach Theorem (Corollary 2.3.4) there exist bounded linear functionals $x_1^*, \dots, x_n^* \in X^*$ such that

$$\langle x_i^*, x_j \rangle = \delta_{ij} = \begin{cases} 1, & \text{if } i = j, \\ 0, & \text{if } i \neq j, \end{cases} \qquad \text{for } i, j = 1, \dots, m.$$

Define the bounded linear operator $K : X \to Z := \ker(D)$ by

$$Kx := \sum_{i=1}^{m} \langle x_i^*, x \rangle x_i.$$

Then K is a compact operator (Example 4.2.5). Moreover, the restriction $K|_{\ker(D)} : \ker(D) \to Z$ is the identity and so is bijective. Hence the operator $X \to Y \times Z : x \mapsto (Dx, Kx)$ is injective and its image $\mathrm{im}(D) \times Z$ is a closed subspace of $Y \times Z$. Hence it follows from Corollary 4.1.17 that there exists a constant $c > 0$ such that (4.3.3) holds.

We prove that (ii) implies (i). Assume D satisfies (ii) and let $K : X \to Z$ and $c > 0$ be as in part (ii). We prove in three steps that D satisfies (i).

Step 1. *Every bounded sequence in* $\ker(D)$ *has a convergent subsequence.*

Let $(x_n)_{n\in\mathbb{N}}$ be a bounded sequence in $\ker(D)$. Since K is a compact operator, there exists a subsequence $(x_{n_i})_{i\in\mathbb{N}}$ such that $(Kx_{n_i})_{i\in\mathbb{N}}$ is a Cauchy sequence in Z. Since $Dx_{n_i} = 0$ for all $i \in \mathbb{N}$, it follows from (4.3.3) that $\|x_{n_i} - x_{n_j}\|_X \le c\|Kx_{n_i} - Kx_{n_j}\|_Z$ for all $i, j \in \mathbb{N}$. Hence $(x_{n_i})_{i\in\mathbb{N}}$ is a Cauchy sequence and therefore converges because X is complete. The limit $x := \lim_{i\to\infty} x_{n_i}$ belongs to the kernel of D and this proves Step 1.

Step 2. *There exists a constant* $C > 0$ *such that*

$$\inf_{\xi\in\ker(D)} \|x + \xi\|_X \le C\,\|Dx\|_Y \qquad \textit{for all } x \in X. \tag{4.3.4}$$

Assume, by contradiction, that there does not exist a constant $C > 0$ such that (4.3.4) holds. Then it follows from the axiom of countable choice that there exists a sequence $(x_n)_{n\in\mathbb{N}}$ in X such that

$$\inf_{\xi\in\ker(D)} \|x_n + \xi\|_X > n\,\|Dx_n\|_Y \qquad \text{for all } n \in \mathbb{N}. \tag{4.3.5}$$

Multiplying each element x_n by a suitable constant and adding to it an element of the kernel of D, if necessary, we may assume that

$$\inf_{\xi\in\ker(D)} \|x_n + \xi\|_X = 1, \qquad 1 \le \|x_n\| \le 2 \qquad \text{for all } n \in \mathbb{N}. \tag{4.3.6}$$

Then $\|Dx_n\|_Y < 1/n$ by (4.3.5) and (4.3.6) and so $\lim_{n\to\infty} Dx_n = 0$. Moreover, since K is compact, there is a subsequence $(x_{n_i})_{i\in\mathbb{N}}$ such that $(Kx_{n_i})_{i\in\mathbb{N}}$ is a Cauchy sequence in Z. Since $(Dx_{n_i})_{i\in\mathbb{N}}$ and $(Kx_{n_i})_{i\in\mathbb{N}}$ are Cauchy sequences, it follows from (4.3.3) that $(x_{n_i})_{i\in\mathbb{N}}$ is a Cauchy sequence in X. This sequence converges because X is complete. Define $x := \lim_{i\to\infty} x_{n_i}$. Then $Dx = \lim_{i\to\infty} x_{n_i} = 0$ and hence, by (4.3.6),

$$1 = \inf_{\xi\in\ker(D)} \|x_{n_i} + \xi\|_X \le \|x_{n_i} - x\|_X \qquad \text{for all } i \in \mathbb{N}.$$

Since $\lim_{i\to\infty}\|x_{n_i} - x\|_X = 0$, this is a contradiction. This proves Step 2.

Step 3. *D satisfies* (i).

It follows from Step 1 and Theorem 1.2.11 that $\dim \ker(D) < \infty$. It follows from Step 2 and Theorem 4.1.16 that the operator $D : X \to Y$ has a closed image. This proves Step 3 and Lemma 4.3.9. □

Proof of Theorem 4.3.8. We prove that (i) implies (ii). Thus assume that $A : X \to Y$ is a Fredholm operator and define

$$X_0 := \ker(A), \qquad Y_1 := \operatorname{im}(A).$$

Then, by Lemma 2.3.30, there exist closed linear subspaces

$$X_1 \subset X, \qquad Y_0 \subset Y$$

such that

$$X = X_0 \oplus X_1, \qquad Y = Y_0 \oplus Y_1.$$

This implies that the bounded linear operator

$$A_1 := A|_{X_1} : X_1 \to Y_1$$

is bijective. Hence $A_1^{-1} : Y_1 \to X_1$ is bounded by the Inverse Operator Theorem 2.2.5. Define the bounded linear operator $F : Y \to X$ by

$$F(y_0 + y_1) := A_1^{-1} y_1 \qquad \text{for } y_0 \in Y_0 \text{ and } y_1 \in Y_1.$$

Then $AF(y_0 + y_1) = y_1$ and $FA(x_0 + x_1) = x_1$ and hence

$$(\mathbb{1}_Y - AF)(y_0 + y_1) = y_0, \qquad (\mathbb{1}_X - FA)(x_0 + x_1) = x_0$$

for all $x_0 \in X_0$, $x_1 \in X_1$, $y_0 \in Y_0$, and $y_1 \in Y_1$. Since X_0 and Y_0 are finite-dimensional, the operators $\mathbb{1}_Y - AF$ and $\mathbb{1}_X - FA$ have finite rank and are therefore compact (see Example 4.2.5).

We prove that (ii) implies (i). Thus assume that there exists a bounded linear operator $F : Y \to X$ such that the operators $K := \mathbb{1}_X - FA : X \to X$ and $L := \mathbb{1}_Y - AF : Y \to Y$ are compact. Then

$$\|x\|_X = \|FAx + Kx\|_X \le c\big(\|Ax\|_Y + \|Kx\|_X\big)$$

for all $x \in X$, where $c := \max\{1, \|F\|\}$. Hence A has a finite-dimensional kernel and a closed image by Lemma 4.3.9. Moreover, $L^* : Y^* \to Y^*$ is a compact operator by Theorem 4.2.10 and

$$\|y^*\|_{Y^*} = \|F^*A^*y^* + L^*y^*\|_{Y^*} \le c\big(\|A^*y^*\|_{Y^*} + \|L^*y^*\|_{Y^*}\big)$$

for all $y^* \in Y^*$. Hence A^* has a finite-dimensional kernel by Lemma 4.3.9 and so A has a finite-dimensional cokernel by Theorem 4.3.3. This proves Theorem 4.3.8. □

4.4. Composition and Stability

Theorem 4.4.1 (Composition of Fredholm Operators). *Let X, Y, Z be Banach spaces and let $A : X \to Y$ and $B : Y \to Z$ be Fredholm operators. Then $BA : X \to Z$ is a Fredholm operator and*

$$\operatorname{index}(BA) = \operatorname{index}(A) + \operatorname{index}(B).$$

Proof. By Theorem 4.3.8 there exist bounded linear operators $F : Y \to X$ and $G : Z \to Y$ such that the operators $\mathbb{1}_X - FA$, $\mathbb{1}_Y - AF$, $\mathbb{1}_Y - GB$, and $\mathbb{1}_Z - BG$ are compact. Define $H := FG : Z \to X$. Then the operators

$$\begin{aligned} \mathbb{1}_X - HBA &= F(\mathbb{1}_Y - GB)A + \mathbb{1}_X - FA, \\ \mathbb{1}_Z - BAH &= B(\mathbb{1}_Y - AF)G + \mathbb{1}_Z - BG \end{aligned}$$

are compact by part (i) of Theorem 4.2.10. Hence BA is a Fredholm operator by Theorem 4.3.8.

To prove the index formula, consider the operators

$$A_0 : \frac{\ker(BA)}{\ker(A)} \to \ker(B), \qquad A_0[x] := Ax,$$

$$B_0 : \frac{Y}{\operatorname{im}(A)} \to \frac{\operatorname{im}(B)}{\operatorname{im}(BA)}, \qquad B_0[y] := [By].$$

These are well defined linear operators between finite-dimensional real vector spaces. The operator A_0 is injective and B_0 is surjective by definition. Second, $\operatorname{im}(A_0) = \operatorname{im}(A) \cap \ker(B)$ and hence

$$\operatorname{coker}(A_0) = \frac{\ker(B)}{\operatorname{im}(A) \cap \ker(B)}.$$

Third,

$$\begin{aligned} \ker(B_0) &= \big\{[y] \in Y/\operatorname{im}(A) \,\big|\, By \in \operatorname{im}(BA)\big\} \\ &= \big\{[y] \in Y/\operatorname{im}(A) \,\big|\, \exists\, x \in X \text{ such that } B(y - Ax) = 0\big\} \\ &= \big\{[y] \in Y/\operatorname{im}(A) \,\big|\, y \in \operatorname{im}(A) + \ker(B)\big\} \\ &= \frac{\operatorname{im}(A) + \ker(B)}{\operatorname{im}(A)} \\ &\cong \frac{\ker(B)}{\operatorname{im}(A) \cap \ker(B)} \\ &= \operatorname{coker}(A_0). \end{aligned}$$

Hence, by Example 4.3.4, we have

$$\begin{aligned}
0 &= \operatorname{index}(A_0) + \operatorname{index}(B_0) \\
&= \dim\left(\frac{\ker(BA)}{\ker(A)}\right) - \dim\ker(B) + \dim\operatorname{coker}(A) - \dim\left(\frac{\operatorname{im}(B)}{\operatorname{im}(BA)}\right) \\
&= \dim\ker(BA) - \dim\ker(A) - \dim\ker(B) \\
&\quad + \dim\operatorname{coker}(A) + \dim\operatorname{coker}(B) - \dim\operatorname{coker}(BA) \\
&= \operatorname{index}(BA) - \operatorname{index}(A) - \operatorname{index}(B).
\end{aligned}$$

This proves Theorem 4.4.1. □

Theorem 4.4.2 (Stability of the Fredholm Index). *Let X and Y be Banach spaces and let $D : X \to Y$ be a Fredholm operator.*

(i) *If $K : X \to Y$ is a compact operator, then $D + K$ is a Fredholm operator and* $\operatorname{index}(D + K) = \operatorname{index}(D)$.

(ii) *There is a constant $\varepsilon > 0$ such that the following holds. If $P : X \to Y$ is a bounded linear operator such that $\|P\| < \varepsilon$, then $D + P$ is a Fredholm operator and* $\operatorname{index}(D + P) = \operatorname{index}(D)$.

Proof. We prove the Fredholm property in part (i). Thus let $D : X \to Y$ be a Fredholm operator and let $K : X \to Y$ be a compact operator. By Theorem 4.3.8 there exists a bounded linear operator $T : Y \to X$ such that the operators $1\!\!1_X - TD$ and $1\!\!1_Y - DT$ are compact. Hence so are the operators $1\!\!1_X - T(D + K)$ and $1\!\!1_Y - (D + K)T$ by Theorem 4.2.10, so $D + K$ is a Fredholm operator by Theorem 4.3.8.

We prove the Fredholm property in part (ii). Let $D : X \to Y$ be a Fredholm operator. By Lemma 4.3.9 there exist a compact operator $K : X \to Z$ and a constant $c > 0$ such that $\|x\|_X \le c(\|Dx\|_Y + \|Kx\|_Z)$ for all $x \in X$. Now let $P : X \to Y$ be a bounded linear operator with the operator norm

$$\|P\| < \frac{1}{c}.$$

Then, for all $x \in X$, we have

$$\begin{aligned}
\|x\|_X &\le c\big(\|Dx\|_Y + \|Kx\|_Z\big) \\
&\le c\big(\|Dx + Px\|_Y + \|Px\|_Y + \|Kx\|_Z\big) \\
&\le c\big(\|(D + P)x\|_Y + \|Kx\|_Z\big) + c\,\|P\|\,\|x\|_X
\end{aligned}$$

and hence

$$(1 - c\|P\|)\|x\|_X \le c(\|(D + P)x\|_Y + \|Kx\|_Z).$$

So $D+P$ has a closed image and a finite-dimensional kernel by Lemma 4.3.9. The same argument for the dual operators shows that $D^* + P^*$ has a finite-dimensional kernel whenever $\|P^*\| = \|P\|$ is sufficiently small, and so $D + P$ has a finite-dimensional cokernel by Theorem 4.3.3.

We prove the index formula in part (ii). As in the proof of Theorem 4.3.8, define $X_0 := \ker(A)$ and $Y_1 := \mathrm{im}(A)$ and use Lemma 2.3.30 to find closed linear subspaces $X_1 \subset X$ and $Y_0 \subset Y$ such that

$$X = X_0 \oplus X_1, \qquad Y = Y_0 \oplus Y_1.$$

For $i,j \in \{0,1\}$ define $P_{ji} : X_i \to Y_j$ as the composition of $P|_{X_i} : X_i \to Y$ with the projection $Y = Y_0 \oplus Y_1 \to Y_j : y_0 + y_1 \mapsto y_j$. Let $D_{11} : X_1 \to Y_1$ be the restriction of D to X_1 with values in $Y_1 = \mathrm{im}(D)$. Then D_{11} is bijective. We prove the following.

Claim. *Assume the operator* $D_{11} + P_{11} : X_1 \to Y_1$ *is bijective and define*

$$A_0 := P_{00} - P_{01}\big(D_{11} + P_{11}\big)^{-1} P_{10} : X_0 \to Y_0.$$

Then $\mathrm{index}(D + P) = \mathrm{index}(A_0)$.

The claim shows that

$$\mathrm{index}(D + P) = \mathrm{index}(A_0) = \dim X_0 - \dim Y_0 = \mathrm{index}(D)$$

whenever the operator $D_{11} + P_{11}$ is bijective. By Corollary 1.5.7, this holds whenever $\|P_{11}\|\|(D_{11})^{-1}\| < 1$, and hence when $\|P\|$ is sufficiently small.

To prove the claim, observe that the equation

$$(D + P)(x_0 + x_1) = y_0 + y_1 \tag{4.4.1}$$

can be written as

$$\begin{aligned} y_0 &= P_{00}x_0 + P_{01}x_1, \\ y_1 &= P_{10}x_0 + \big(D_{11} + P_{11}\big)x_1 \end{aligned} \tag{4.4.2}$$

for $x_0 \in X_0$, $x_1 \in X_1$ and $y_0 \in Y_0$, $y_1 \in Y_1$. Since $D_{11} + P_{11}$ is bijective, the equations in (4.4.2) can be written in the form

$$\begin{aligned} A_0 x_0 &= y_0 - P_{01}\big(D_{11} + P_{11}\big)^{-1} y_1, \\ x_1 &= \big(D_{11} + P_{11}\big)^{-1}(y_1 - P_{10}x_0). \end{aligned} \tag{4.4.3}$$

This shows that

$$x_0 + x_1 \in \ker(D + P) \qquad \iff \qquad \left\{ \begin{array}{l} x_0 \in \ker(A_0), \\ x_1 = -\big(D_{11} + P_{11}\big)^{-1} P_{10} x_0 \end{array} \right.$$

for $x_i \in X_i$ and so $\ker(D + P) \cong \ker(A_0)$. Equation (4.4.3) also shows that

$$y_0 + y_1 \in \mathrm{im}(D + P) \qquad \iff \qquad y_0 - P_{01}\big(D_{11} + P_{11}\big)^{-1} y_1 \in \mathrm{im}(A_0)$$

for $y_i \in Y_i$. Thus the map $Y \to Y_0 : y_0 + y_1 \mapsto y_0 - P_{01}\big(D_{11} + P_{11}\big)^{-1} y_1$ descends to an isomorphism from $Y/\mathrm{im}(D + P)$ to $Y_0/\mathrm{im}(A_0)$. Hence

$$\mathrm{coker}(D + P) \cong \mathrm{coker}(A_0).$$

This proves the claim and the index formula in part (ii).

It remains to prove the index formula in part (i). Let $K : X \to Y$ be a compact operator and define $I := \{t \in \mathbb{R} \mid \text{index}(D + tK) = \text{index}(D)\}$. By part (ii) the set $\mathcal{F}_k(X,Y) \subset \mathcal{L}(X,Y)$ of Fredholm operators of index k is open for every $k \in \mathbb{Z}$, and so is their union $\mathcal{F}(X,Y) := \bigcup_{k\in\mathbb{Z}} \mathcal{F}_k(X,Y)$. Moreover, the map $\mathbb{R} \to \mathcal{F}(X,Y) : t \mapsto D + tK$ is continuous and hence the pre-image of $\mathcal{F}_k(X,Y)$ under this map is open for every $k \in \mathbb{Z}$. Thus the set $I_k := \{t \in \mathbb{R} \mid \text{index}(D + tK) = k\}$ is open for all $k \in \mathbb{Z}$ and $\mathbb{R} = \bigcup_{k\in\mathbb{Z}} I_k$. Since $I_k = I$ for $k = \text{index}(D)$ it follows that I and $\mathbb{R} \setminus I = \bigcup_{\ell \neq k} I_\ell$ are open. Since $0 \in I$ and $\mathbb{R}$ is connected, it follows that $I = \mathbb{R}$, thus $1 \in I$, and so $\text{index}(D + K) = \text{index}(D)$. This proves Theorem 4.4.2. □

Remark 4.4.3 (Fredholm Alternative). It is interesting to consider the special case where $X = Y$ is a Banach space and $K : X \to X$ is a compact operator. Then Theorem 4.4.2 asserts that $\mathbb{1} - K$ is a Fredholm operator of index zero. This gives rise to the so-called **Fredholm alternative**. It asserts that either the inhomogeneous linear equation

$$x - Kx = y$$

has a solution $x \in X$ for every $y \in X$, or the corresponding homogeneous equation $x - Kx = 0$ has a nontrivial solution. This is simply a consequence of the fact that the kernel and cokernel of the operator $\mathbb{1} - K$ have the same dimension, and hence are either both trivial or both nontrivial.

Remark 4.4.4 (Calkin Algebra). Let X be a Banach space, denote by $\mathcal{L}(X)$ the Banach space of bounded linear operators from X to itself, denote by $\mathcal{F}(X) \subset \mathcal{L}(X)$ the subset of all Fredholm operators, and denote by $\mathcal{K}(X) \subset \mathcal{L}(X)$ the subset of all compact operators. By part (ii) of Theorem 4.2.10 the linear subspace $\mathcal{K}(X) \subset \mathcal{L}(X)$ is closed and, by part (i) of Theorem 4.2.10, the quotient space

$$\mathcal{L}(X)/\mathcal{K}(X)$$

is a Banach algebra, called the **Calkin algebra**. By part (ii) of Theorem 4.4.2, the set $\mathcal{F}(X)$ of Fredholm operators is an open subset of $\mathcal{L}(X)$ and, by part (i) of Theorem 4.4.2, this open set is invariant under the equivalence relation. By Theorem 4.3.8 the corresponding open subset

$$\mathcal{F}(X)/\mathcal{K}(X) \subset \mathcal{L}(X)/\mathcal{K}(X)$$

of the quotient space is the group of invertible elements in the Calkin algebra. By part (i) of Theorem 4.4.2 the Fredholm index gives rise to a well-defined map

$$\mathcal{F}(X)/\mathcal{K}(X) \to \mathbb{Z} : [D] \mapsto \text{index}(D). \tag{4.4.4}$$

By Theorem 4.4.1 this map is a group homomorphism.

Remark 4.4.5 (Fredholm Operators and K-theory). Let H be an infinite-dimensional separable Hilbert space. A theorem of Kuiper [**53**] asserts that the group

$$\mathrm{Aut}(H) := \{A : H \to H \,|\, A \text{ is a bijective bounded linear operator}\}$$

is contractible. This can be used to prove that the space $\mathcal{F}(H)$ of Fredholm operators from H to itself is a **classifying space for K-theory**. The starting point is the observation that, if M is a compact Hausdorff space and $A : M \to \mathcal{F}(H)$ is a continuous map such that the operator $A(p)$ is surjective for all $p \in M$, then the kernels of these operators determine a vector bundle E over M, defined by

$$E := \{(p, x) \in M \times H \mid A(p)x = 0\}. \tag{4.4.5}$$

More generally, any continuous map $A : M \to \mathcal{F}(H)$, defined on a compact Hausdorff space M, determines a so-called **K-theory class on M** (an equivalence class of pairs of vector bundles under the equivalence relation $(E, F) \sim (E', F')$ iff $E \oplus F' \cong E' \oplus F$), the K-theory classes associated to two such maps agree if and only if the maps are homotopic, and every K-theory class on a compact Hausdorff space can be obtained this way. This is the **Atiyah–Jänich Theorem** [**5**–**7**, **42**]. In particular, when M is a single point, the theorem asserts that the space $\mathcal{F}_k(H)$ of Fredholm operators of index k is nonempty and connected for all $k \in \mathbb{Z}$.

Remark 4.4.6 (Banach Hyperplane Problem). In 1932 Banach [**8**] asked the question of whether every infinite-dimensional real Banach space X is isomorphic to $X \times \mathbb{R}$ or, equivalently, whether every closed codimension one subspace of X is isomorphic to X (see Exercise 4.5.9). This question was answered by Gowers [**31**] in 1994. He constructed an infinite-dimensional real Banach space X that is not isomorphic to any of its proper subspaces and so every Fredholm operator on X has Fredholm index zero. This example was later refined by Argyros and Haydon [**4**]. The **Argyros–Haydon space** is an infinite-dimensional real Banach space X such that every bounded linear operator $A : X \to X$ has the form $A = \lambda \mathbb{1} + K$, where λ is a real number and $K : X \to X$ is a compact operator. Thus every bounded linear operator on X is either a compact operator or a Fredholm operator of index zero, the open set $\mathcal{F}(X) = \mathcal{F}_0(X) = \mathcal{L}(X) \setminus \mathcal{K}(X)$ of Fredholm operators on X has two connected components, and the Calkin algebra is isomorphic to the real numbers, i.e.

$$\mathcal{L}(X)/\mathcal{K}(X) \cong \mathbb{R}.$$

This shows that the Hilbert space H in the Atiyah–Jänich Theorem cannot be replaced by an arbitrary Banach space (see Remark 4.4.5). The details of the constructions of Gowers and Argyros–Haydon go far beyond the scope of the present book.

4.5. Problems

Exercise 4.5.1 (Injections and Surjections). Let X and Y be Banach spaces. Prove the following.

(a) The set of all surjective bounded linear operators $A : X \to Y$ is an open subset of $\mathcal{L}(X, Y)$ with respect to the norm topology.

(b) The set of all injective bounded linear operators $A : X \to Y$ is not necessarily an open subset of $\mathcal{L}(X, Y)$ with respect to the norm topology.

(c) The set of all injective bounded linear operators $A : X \to Y$ with closed image is an open subset of $\mathcal{L}(X, Y)$ with respect to the norm topology.

Exercise 4.5.2 (The Image of a Compact Operator). Let X and Y be Banach spaces and let $K : X \to Y$ be a compact operator. Prove the following.

(a) If K has a closed image, then $\dim \operatorname{im}(K) < \infty$.

(b) The image of K is a separable subspace of Y.

(c) If Y is separable, then there exist a Banach space X and a compact operator $K : X \to Y$ with a dense image.

Exercise 4.5.3 (Compact Subsets of Banach Spaces). *Let X be a Banach space and let $C \subset X$ be a closed subset. Then the following are equivalent.*

(i) *C is compact.*

(ii) *There exists a sequence $x_n \in C$ such that*

$$\lim_{n\to\infty} \|x_n\| = 0, \qquad C \subset \overline{\operatorname{conv}}(\{x_n \,|\, n \in \mathbb{N}\}). \tag{4.5.1}$$

Hint 1: To prove that (ii) implies (i) observe that

$$\overline{\operatorname{conv}}(\{x_n \,|\, n \in \mathbb{N}\}) = \left\{ \sum_{n=1}^{\infty} \lambda_n x_n \,\middle|\, \lambda_n \geq 0, \sum_{n=1}^{\infty} \lambda_n = 1 \right\} \tag{4.5.2}$$

whenever $\lim_{n\to\infty} \|x_n\| = 0$.

Hint 2: To prove that (i) implies (ii), choose a sequence of compact sets $C_k \subset X$ and a sequence of finite subsets $A_k \subset C_k$ such that $C_1 = C$ and

$$2C_k \subset \bigcup_{x\in A_k} \overline{B}_{4^{-k}}(x), \qquad C_{k+1} := \bigcup_{x\in A_k} \Big(\big(2C \cap \overline{B}_{4^{-k}}(x)\big) - x \Big)$$

for $k \in \mathbb{N}$. Prove that, for every $c \in C$, there is a sequence $x_k \in A_k$ such that $x = \sum_{k=1}^{\infty} 2^{-k} x_k$. Note that $\|x\| \leq 4^{-k}$ for all $x \in A_{k+1}$ and all $k \in \mathbb{N}$.

Exercise 4.5.4 (Continuity). Let X and Y be normed vector spaces.

(a) A linear operator $A : X \to Y$ is bounded if and only if it is continuous with respect to the weak topologies on X and Y.

(b) A linear operator $B : Y^* \to X^*$ is continuous with respect to the weak* topologies on Y^* and X^* if and only if there exists a bounded linear operator $A : X \to Y$ such that $B = A^*$.

(c) A linear operator $A : X \to Y$ is continuous with respect to the weak topology on X and the norm topology on Y if and only if it is bounded and has finite rank.

(d) Suppose X and Y are Banach spaces and denote by $B^* \subset Y^*$ the closed unit ball. Then a bounded linear operator $A : X \to Y$ is compact if and only if $A^*|_{B^*} : B^* \to X^*$ is continuous with respect to the weak* topology on B^* and the norm topology on X^*.

(e) Suppose X and Y are reflexive Banach spaces and denote by $B \subset X$ the closed unit ball. Then a bounded linear operator $A : X \to Y$ is compact if and only if $A|_B : B \to X$ is continuous with respect to the weak topology on B and the norm topology on Y.

Exercise 4.5.5 (Gantmacher's Theorem). Let X and Y be Banach spaces and let $A : X \to Y$ be a bounded linear operator. Then the following are equivalent.

(i) A is **weakly compact**, i.e. if $B \subset X$ is a bounded set, then the weak closure of $A(B)$ is a weakly compact subset of Y.

(ii) If $(x_n)_{n\in\mathbb{N}}$ is a bounded sequence in X, then the sequence $(Ax_n)_{n\in\mathbb{N}}$ in Y has a weakly convergent subsequence.

(iii) $A^{**}(X^{**}) \subset \iota_Y(Y)$.

(iv) *$A^* : Y^* \to X^*$ is continuous with respect to the weak* topology on Y^* and the weak topology on X^*.*

(v) *The dual operator $A^* : Y^* \to X^*$ is weakly compact.*

Hint: To prove that (i) implies (iii) denote by

$$B \subset X, \qquad B^{**} \subset X^{**}$$

the closed unit balls and denote by $C \subset Y$ the weak closure of $A(B)$. If (i) holds, then $\iota_Y(C)$ is a weak* compact subset of Y^{**}. Use Goldstine's Theorem (Corollary 3.1.29) to prove that

$$A^{**}(B^{**}) \subset \iota_Y(C).$$

(See Exercise 3.7.8.)

Exercise 4.5.6 (Pitt's Theorem). Let $1 \le p < q < \infty$. Then every bounded linear operator $A : \ell^q \to \ell^p$ is compact.

(a) Fix a bounded linear operator $A : \ell^q \to \ell^p$ such that $\|A\| = 1$ and a sequence $(x_n)_{n\in\mathbb{N}}$ in ℓ^q that converges weakly to zero. It suffices to prove

$$\lim_{n\to\infty} \|Ax_n\|_p = 0.$$

Hint: Use Theorem 3.4.1 and part (e) of Exercise 4.5.4.

(b) If $(y_n)_{n\in\mathbb{N}}$ is a sequence in ℓ^p that converges weakly to zero, then

$$\limsup_{n\to\infty} \|y + y_n\|_p^p = \|y\|_p^p + \limsup_{n\to\infty} \|y_n\|_p^p \tag{4.5.3}$$

for every $y \in \ell^p$. **Hint:** Assume first that y has finite support.

(c) Let x_n be as in (a), fix a constant $\varepsilon > 0$, and choose $x \in \ell^q$ such that

$$\|x\|_q = 1, \qquad 1 - \varepsilon < \|Ax\|_p < 1. \tag{4.5.4}$$

Then

$$\left(\|Ax\|_p^p + \lambda^p \limsup_{n\to\infty} \|Ax_n\|_p^p\right)^{1/p} \le \left(\|x\|_q^q + \lambda^q \limsup_{n\to\infty} \|x_n\|_q^q\right)^{1/q} \tag{4.5.5}$$

for all $\lambda > 0$. **Hint:** Use the equation (4.5.3) in part (b) with

$$y_n := \lambda A x_n$$

and the inequality $\|Ax + \lambda A x_n\|_p \le \|x + \lambda x_n\|_q$.

(d) There exists a constant $C > 0$ such that

$$\limsup_{n\to\infty} \|Ax_n\|_p^p \le \frac{(1 + \lambda^q C^q)^{p/q} - (1-\varepsilon)^p}{\lambda^p} \tag{4.5.6}$$

for all $\lambda > 0$ and all $\varepsilon > 0$. **Hint:** Take $C \ge \sup_{n\in\mathbb{N}} \|x_n\|_q$ and use the inequalities (4.5.4) and (4.5.5) in part (c).

(e) Choose $\lambda := C^{-1}\varepsilon^{1/q}$ in (4.5.6) to obtain

$$\limsup_{n\to\infty} \|Ax_n\|_p^p \le C^p \varepsilon^{1-p/q} \left(\frac{(1+\varepsilon)^{p/q} - 1}{\varepsilon} + \frac{1 - (1-\varepsilon)^p}{\varepsilon}\right) \tag{4.5.7}$$

for all $\varepsilon > 0$. Take the limit $\varepsilon \to 0$ in (4.5.7) to obtain $\lim_{n\to\infty} \|Ax_n\|_p = 0$.

Exercise 4.5.7 (Existence of Fredholm Operators). Let X and Y be Banach spaces and suppose that there exists a Fredholm operator from X to Y. Prove the following.

(a) X is reflexive if and only if Y is reflexive.

(b) X is separable if and only if Y is separable.

Exercise 4.5.8 (Codimension One Subspaces). Let X be a real Banach space. Prove that any two closed codimension one subspaces of X are isomorphic to one another. **Hint:** If Y and Z are distinct closed codimension one subspaces of X, then each of them is isomorphic to $(Y \cap Z) \times \mathbb{R}$.

Exercise 4.5.9 (Existence of Index One Fredholm Operators). Let X be an infinite-dimensional real Banach space. Prove that the following are equivalent.

(i) X is isomorphic to $X \times \mathbb{R}$.

(ii) There exists a codimension one subspace of X that is isomorphic to X.

(iii) Every closed codimension one subspace of X is isomorphic to X.

(iv) There exists a Fredholm operator $A : X \to X$ of index one.

(v) The homomorphism (4.4.4) is surjective.

Exercise 4.5.10 (Existence of Index Zero Fredholm Operators).

(a) Let X and Y be Banach spaces and suppose that there exists an index zero Fredholm operator from X to Y. Prove that X and Y are isomorphic.

(b) Let X be a Banach space and let $Y \subset X$ be a closed codimension one subspace. Prove that there is an index one Fredholm operator $A : X \to Y$. If X is not isomorphic to any proper closed subspace of X, prove that every Fredholm operator from X to Y has index one.

Exercise 4.5.11 (Fredholm Operators Between ℓ^p Spaces).

(a) Let $1 \le p \le \infty$. For every integer $n \in \mathbb{Z}$ construct a Fredholm operator $A : \ell^p \to \ell^p$ of index n.

(b) Construct a family of examples in (a) that are neither injective nor surjective.

(c) Let $1 \le p, q \le \infty$ and $p \ne q$. Does there exist a Fredholm operator from ℓ^p to ℓ^q?

Exercise 4.5.12 (Fredholm Operators and Vector Bundles). Let H be a separable infinite-dimensional Hilbert space and, for $k \in \mathbb{Z}$, denote by $\mathcal{F}_k(H)$ the space of Fredholm operators $A : H \to H$ of index k. Find a continuous map

$$A : S^1 \to \mathcal{F}_1(H)$$

such that the Fredholm operator $A(z) : H \to H$ is surjective for all $z \in S^1$, and the vector bundle

$$E := \{(z, \xi) \in S^1 \times H \mid A(z)\xi = 0\}$$

over S^1 is a Möbius band.

Exercise 4.5.13 (Fredholm Alternative). Fix an interval $I := [a,b]$ with $a < b$, let $f_i, g_i \in \mathcal{L}^2(I)$ for $i = 1, \dots, n$, and define

$$K(x,y) := \sum_{i=1}^{n} f_i(x) g_i(y) \qquad \text{for } a \le x, y \le b.$$

For $h \in L^2(I)$ consider the equation

$$u(x) + \int_a^b K(x,y) u(y)\, dy = h(x) \qquad \text{for } a \le x \le b. \tag{4.5.8}$$

Prove that equation (4.5.8) either has a unique solution $u \in L^2(I)$ for every h, or the homogeneous equation with $h = 0$ has a nonzero solution u.

Exercise 4.5.14 (Hilbert Spheres).

(a) The unit sphere

$$S := \left\{ x \in \ell^2 \mid \|x\|_2 = 1 \right\}$$

is contractible, i.e. there exists a continuous map $f : [0,1] \times S \to S$ and an element $e \in S$ such that

$$f(0,x) = e, \qquad f(1,x) = x$$

for all $x \in S$.

Hint: Let $e_1, e_2, e_3, \dots$ be the standard orthonormal basis of ℓ^2 and define the shift operator $T : \ell^2 \to \ell^2$ by

$$T(x_1, x_2, x_3, \dots) := (0, x_1, x_2, x_3, \dots) \qquad \text{for } x = (x_i)_{i \in \mathbb{N}} \in \ell^2.$$

Then $Te_n = e_{n+1}$ for all $n \in \mathbb{N}$. Consider the maps $g : [0,1] \times \ell^2 \to \ell^2$ and $h : [0,1] \times \ell^2 \to \ell^2$ defined by

$$g(t,x) := (1-t)e_1 + tTx, \qquad h(t,x) := (1-t)Tx + tx$$

for $0 \le t \le 1$ and $x \in \ell^2$. Use these maps to show that $\ell^2 \setminus \{0\}$ is contractible and then normalize to deduce that S is contractible.

(b) Refine the construction in (a) to obtain a map $f : [0,1] \times S \to S$ that satisfies

$$f(0,x) = e, \qquad f(1,x) = x, \qquad f(t,e) = e$$

for all $x \in S$ and all $t \in [0,1]$. This means that the singleton $\{e\}$ is a **deformation retract of** S.

(c) Prove that the unit sphere in any infinite-dimensional Hilbert space is contractible.

Exercise 4.5.15 (Fredholm Intersection Theory). Let X be a Banach space and let $X_1, X_2 \subset X$ be closed subspaces. The triple (X, X_1, X_2) is called a **Fredholm triple** if the subspace $X_1 + X_2$ is closed, and the spaces $X_1 \cap X_2$ and $X/(X_1 + X_2)$ are finite-dimensional. The **Fredholm index** of a Fredhom triple (X, X_1, X_2) is defined by

$$\operatorname{index}(X, X_1, X_2) := \dim(X_1 \cap X_2) - \dim(X/(X_1 + X_2)). \tag{4.5.9}$$

(a) Prove that (X, X_1, X_2) is a Fredholm triple if and only if the operator

$$X_1 \times X_2 \to X : (x_1, x_2) \mapsto x_1 + x_2$$

is Fredholm. Show that the Fredholm indices agree. **Hint:** Corollary 2.2.9.

(b) Assume $X_1 + X_2$ has finite codimension in X. Prove that $X_1 + X_2$ is a closed subspace of X. **Hint:** Lemma 4.3.2.

(c) Assume (X, X_1, X_2) is a Fredholm triple. Prove that the subspaces X_1 and X_2 are complemented.

(d) Define the notion of a *small deformation* of a complemented subspace.

(e) Prove that the Fredholm property and the Fredholm index of a Fredholm triple (X, X_1, X_2) are stable under small deformations of the subspaces X_1 and X_2. **Hint:** Theorem 4.4.2.

Exercise 4.5.16 (Rellich's Theorem). Let $I := [0, 1] \subset \mathbb{R}$ be the unit interval and fix a real number $p \geq 1$. Denote by

$$W^{1,p}(I) := \left\{ f : I \to \mathbb{R} \,\middle|\, \begin{array}{l} f \text{ is absolutely continuous} \\ \text{and } \int_0^1 |f'(t)|^p \, dt < \infty \end{array} \right\} \tag{4.5.10}$$

the Sobolev space of $W^{1,p}$-functions on I with the norm

$$\|f\|_{W^{1,p}} := \left(\int_0^1 \left(|f(t)|^p + |f'(t)|^p \right) dt \right)^{1/p} \tag{4.5.11}$$

for $f \in W^{1,p}(I, \mathbb{R})$. In particular, $W^{1,1}(I)$ is the Banach space of absolutely continuous functions.

(a) Prove that $W^{1,p}(I)$ is a Banach space with the norm (4.5.11). **Hint:** Use [**75**, Thm. 6.19] or Theorem 7.5.18 with $X = \mathbb{R}$.

(b) Prove that the inclusion of $W^{1,p}(I)$ into the Banach space $C(I)$ of continuous functions $f : I \to \mathbb{R}$, equipped with the supremum norm, is a bounded linear operator.

(c) Prove that the inclusion $W^{1,p}(I) \hookrightarrow C(I)$ is a compact operator for $p > 1$ but not for $p = 1$. **Hint:** Show that the unit ball in $W^{1,p}(I)$ is equicontinuous for $p > 1$ and use the Arzelà–Ascoli Theorem (Corollary 1.1.13). For $p = 1$ consider the functions $f_n(t) := t^n$.

Exercise 4.5.17 (Fredholm Theory and Homological Algebra).

(a) Exact Sequences. A finite sequence

$$0 \longrightarrow V_0 \xrightarrow{d_0} V_1 \xrightarrow{d_1} V_2 \xrightarrow{d_2} \cdots \longrightarrow V_{n-1} \xrightarrow{d_{n-1}} V_n \longrightarrow 0$$

of vector spaces and linear maps is called **exact** if d_0 is injective, d_{n-1} is surjective, and $\ker(d_k) = \mathrm{im}(d_{k-1})$ for $k = 1, \dots, n-1$. If the sequence is exact and the vector spaces V_k are all finite-dimensional, then its **Euler characteristic** vanishes, i.e. $\sum_{k=0}^{n}(-1)^k \dim V_k = 0$.

(b) Two linear operators $A : X \to Y$ and $B : Y \to Z$ between vector spaces determine a natural **long exact sequence**

$$0 \longrightarrow \ker(A) \longrightarrow \ker(BA) \longrightarrow \ker(B) \xrightarrow{\delta}$$
$$\xrightarrow{\delta} \mathrm{coker}(A) \longrightarrow \mathrm{coker}(BA) \longrightarrow \mathrm{coker}(B) \longrightarrow 0,$$

where the map $\delta : \ker(B) \to \mathrm{coker}(A)$ assigns to an element $y \in \ker(B)$ the equivalence class of y in the quotient space $Y/\mathrm{im}(A) = \mathrm{coker}(A)$.

(c) If the vector spaces X, Y, Z in (b) are Banach spaces and two out of the three operators A, B, BA are Fredholm operators, then so is the third and $\mathrm{index}(BA) = \mathrm{index}(A) + \mathrm{index}(B)$. (See also Theorem 4.4.1.)

(d) The Snake Lemma. Consider a commutative diagram

$$\begin{array}{ccccccccc} 0 & \longrightarrow & U & \longrightarrow & V & \longrightarrow & W & \longrightarrow & 0 \\ & & \downarrow{\scriptstyle A} & & \downarrow{\scriptstyle B} & & \downarrow{\scriptstyle C} & & \\ 0 & \longrightarrow & X & \longrightarrow & Y & \longrightarrow & Z & \longrightarrow & 0 \end{array}$$

of vector spaces and linear operators such that the horizontal rows are short exact sequences. Then there is a natural long exact sequence

$$0 \longrightarrow \ker(A) \longrightarrow \ker(B) \longrightarrow \ker(C) \xrightarrow{\delta}$$
$$\xrightarrow{\delta} \mathrm{coker}(A) \longrightarrow \mathrm{coker}(B) \longrightarrow \mathrm{coker}(C) \longrightarrow 0,$$

where the boundary map

$$\delta : \ker(C) \to \mathrm{coker}(A)$$

is defined as follows. Let $w \in \ker(C)$ and choose an element $v \in V$ that maps to w under the surjection $V \to W$. Then $Bv \in Y$ belongs to the kernel of the map $Y \to Z$; so there is a unique element $x \in X$ that maps to Bv under the injection $X \to Y$ and $\delta w := [x] \in X/\mathrm{im}(A) = \mathrm{coker}(A)$ is independent of the choice of v.

(e) Deduce from the Snake Lemma that, if U, V, W, X, Y, Z are Banach spaces and two out of the three operators A, B, C are Fredholm operators, then so is the third and $\mathrm{index}(B) = \mathrm{index}(A) + \mathrm{index}(C)$.

Chapter 5

Spectral Theory

The purpose of the present chapter is to study the spectrum of a bounded linear operator on a real or complex Banach space. In linear algebra a real matrix may have complex eigenvalues and the situation is analogous in infinite dimensions. To define the eigenvalues and, more generally, the spectral values of a bounded real linear operator on a real Banach space it will be necessary to complexify real Banach spaces. Complex Banach spaces and the complexifications of real Banach spaces are discussed in a first preparatory Section 5.1. Other topics in the first section are the integral of a continuous Banach space valued function on a compact interval and holomorphic operator valued functions. These are elementary but important tools in spectral theory. Section 5.2 introduces the spectrum of a bounded linear operator, examines its elementary properties, shows that the spectral radius is the supremum of the moduli of the spectral values, examines the spectrum of a compact operator, and establishes the holomorphic functional calculus. The remainder of this chapter deals exclusively with operators on Hilbert spaces. Section 5.3 discusses complex Hilbert spaces and examines the elementary properties of the spectra of normal and self-adjoint operators. Section 5.4 introduces C* algebras and establishes the continuous functional calculus for self-adjoint operators. It takes the form of an isomorphism from the C* algebra of complex valued continuous functions on the spectrum to the smallest C* algebra containing the given operator. Section 5.5 introduces the Gelfand spectrum of a commutative unital Banach algebra and uses it to extend the continuous functional calculus to normal operators. Section 5.6 shows that every normal operator can be represented by a projection valued measure on the spectrum. Section 5.7 shows that every self-adjoint operator is isomorphic to a direct sum of multiplication operators on L^2 spaces.

5.1. Complex Banach Spaces

5.1.1. Definition and Examples.

Definition 5.1.1. **(i)** A **complex normed vector space** is a complex vector space X, equipped with a norm function $X \to \mathbb{R} : x \mapsto \|x\|$ as in Definition 1.1.2 that satisfies in addition

$$\|\lambda x\| = |\lambda| \, \|x\| \qquad \text{for all } x \in X \text{ and all } \lambda \in \mathbb{C}.$$

A complex normed vector space $(X, \|\cdot\|)$ is called a **complex Banach space** if it is complete with respect to the metric (1.1.1).

(ii) Let X and Y be complex Banach spaces and denote by

$$\mathcal{L}^c(X, Y) := \{A : X \to Y \mid A \text{ is complex linear and bounded}\}$$

the space of bounded complex linear operators from X to Y (see Definition 1.2.1). Then $\mathcal{L}^c(X, Y)$ is a complex Banach space with the operator norm (1.2.2). In the case $X = Y$ we abbreviate $\mathcal{L}^c(X) := \mathcal{L}^c(X, X)$.

(iii) The **(complex) dual space** of a complex Banach space X is the space

$$X^* := \mathcal{L}^c(X, \mathbb{C})$$

of bounded complex linear functionals $\Lambda : X \to \mathbb{C}$. If X and Y are complex Banach spaces and $A : X \to Y$ is a bounded complex linear operator, then the **(complex) dual operator of** A is the bounded complex linear operator $A^* : Y^* \to X^*$ defined by $A^*y^* := y^* \circ A : X \to \mathbb{C}$ for $y^* \in Y^*$.

Remark 5.1.2. A complex normed vector space X can be viewed as a real normed vector space, equipped with a linear map $J : X \to X$ such that

$$J^2 = -\mathbb{1} \tag{5.1.1}$$

and

$$\|\cos(\theta)x + \sin(\theta)Jx\| = \|x\| \qquad \text{for all } \theta \in \mathbb{R} \text{ and all } x \in X. \tag{5.1.2}$$

If $J : X \to X$ is a linear map that satisfies (5.1.1) and (5.1.2), then X has a unique structure of a complex normed vector space such that multiplication by the complex number $\mathbf{i}$ is given by the linear operator J. Scalar multiplication is then given by the formula

$$(s + \mathbf{i}t)x := sx + tJx \qquad \text{for } s, t \in \mathbb{R} \text{ and } x \in X. \tag{5.1.3}$$

In this notation a complex linear operator from X to itself is a real linear operator that commutes with J.

Remark 5.1.3. Let X be a complex Banach space and let $\Lambda : X \to \mathbb{C}$ be a bounded complex linear functional. Then $\mathrm{Re}\Lambda : X \to \mathbb{R}$ is a bounded real linear functional of the same norm as Λ, i.e.

$$\|\Lambda\| = \sup_{x \in X\setminus\{0\}} \frac{|\Lambda(x)|}{\|x\|} = \sup_{x \in X\setminus\{0\}} \frac{|\mathrm{Re}\Lambda(x)|}{\|x\|} = \|\mathrm{Re}\Lambda\|.$$

To see this, let $x \in X$ and choose $\theta \in \mathbb{R}$ such that $e^{\mathbf{i}\theta}\Lambda(x) \in \mathbb{R}$. Then

$$|\Lambda(x)| = |e^{\mathbf{i}\theta}\Lambda(x)| = |\mathrm{Re}\Lambda(e^{\mathbf{i}\theta}x)| \leq \|\mathrm{Re}\Lambda\|\|e^{\mathbf{i}\theta}x\| = \|\mathrm{Re}\Lambda\|\|x\|.$$

Hence $\|\Lambda\| \leq \|\mathrm{Re}\Lambda\|$ and the converse inequality is obvious. Thus the map

$$\mathcal{L}^c(X, \mathbb{C}) \to \mathcal{L}(X, \mathbb{R}) : \Lambda \mapsto \mathrm{Re}\Lambda$$

is a Banach space isometry. Its inverse sends a bounded real linear functional $\Lambda_0 : X \to \mathbb{R}$ to the bounded complex linear functional $\Lambda : X \to \mathbb{C}$ defined by $\Lambda(x) := \Lambda_0(x) - \mathbf{i}\Lambda_0(\mathbf{i}x)$ for $x \in X$. This shows that all the results about dual spaces and dual operators proved in Chapters 2, 3, and 4 carry over verbatim to the complex setting. In particular, the complex dual operator A^* has the same operator norm as A by Lemma 4.1.2.

The reader is cautioned that for the complex dual space X^* and the complex dual operator A^* the same notation is used as in the setting of real Banach spaces although the meanings are different. It should always be clear from the context which dual space or dual operator is used in the text. We emphasize that the examples in Subsection 1.1.1 all have natural complex analogues. Here is a list.

Example 5.1.4. (i) The vector space $\mathbb{C}^n$ of all n-tuples $x = (x_1, \dots, x_n)$ of complex numbers is a complex Banach space with each of the norms

$$\|x\|_p := \left(\sum_{i=1}^n |x_i|^p\right)^{1/p}, \qquad \|x\|_\infty := \max_{i=1,\dots,n} |x_i|$$

for $1 \leq p < \infty$ and $x = (x_1, \dots, x_n) \in \mathbb{C}^n$.

(ii) For $1 \leq p < \infty$ the set $\ell^p(\mathbb{N}, \mathbb{C})$ of p-summable sequences $x = (x_i)_{i\in\mathbb{N}}$ of complex numbers is a complex Banach space with the norm

$$\|x\|_p := \left(\sum_{i=1}^\infty |x_i|^p\right)^{1/p}$$

for $x = (x_i)_{i\in\mathbb{N}} \in \ell^p(\mathbb{N}, \mathbb{C})$. Likewise, the space $\ell^\infty(\mathbb{N}, \mathbb{C})$ of bounded sequences of complex numbers is a complex Banach space with the norm

$$\|x\|_\infty := \sup_{i\in\mathbb{N}} |x_i|$$

for $x = (x_i)_{i\in\mathbb{N}} \in \ell^\infty(\mathbb{N}, \mathbb{C})$.

(iii) Let $(M, \mathcal{A}, \mu)$ be a measure space, fix a constant $1 \le p < \infty$, and denote the space of p-integrable complex valued functions on M by $\mathcal{L}^p(\mu, \mathbb{C})$. The function

$$\mathcal{L}^p(\mu, \mathbb{C}) \to \mathbb{R} : f \mapsto \|f\|_p := \left(\int_M |f|^p \, d\mu \right)^{1/p}$$

descends to the quotient space

$$L^p(\mu, \mathbb{C}) := \mathcal{L}^p(\mu, \mathbb{C})/\sim,$$

where $f \sim g$ iff the function $f - g$ vanishes almost everywhere. This quotient is a complex Banach space.

(iv) Let $(M, \mathcal{A}, \mu)$ be a measure space and denote by $\mathcal{L}^\infty(\mu, \mathbb{C})$ the space of complex valued bounded measurable functions $f : M \to \mathbb{C}$. As in part (iii) denote by $\sim$ the equivalence relation on $\mathcal{L}^\infty(\mu, \mathbb{C})$ given by equality almost everywhere. Then the quotient space

$$L^\infty(\mu, \mathbb{C}) := \mathcal{L}^\infty(\mu, \mathbb{C})/\sim$$

is a complex Banach space with the norm defined by (1.1.3).

(v) Let M be a compact topological space. Then the space $C(M, \mathbb{C})$ of bounded continuous functions $f : M \to \mathbb{C}$ is a complex Banach space with the supremum norm

$$\|f\|_\infty := \sup_{p \in M} |f(p)|$$

for $f \in C(M, \mathbb{C})$.

(vi) Let $(M, \mathcal{A})$ be a measurable space, i.e. M is a set and $\mathcal{A} \subset 2^M$ is a σ-algebra. A **complex measure** on $(M, \mathcal{A})$ is a function

$$\mu : \mathcal{A} \to \mathbb{C}$$

that satisfies $\mu(\emptyset) = 0$ and is σ-additive, i.e.

$$\mu \left(\bigcup_{i=1}^\infty A_i \right) = \sum_{i=1}^\infty \mu(A_i) = \lim_{n \to \infty} \sum_{i=1}^n \mu(A_i)$$

for every sequence of pairwise disjoint measurable sets $A_i \in \mathcal{A}$. The space

$$\mathcal{M}(M, \mathcal{A}, \mathbb{C}) := \left\{ \mu : \mathcal{A} \to \mathbb{C} \,\middle|\, \mu \text{ is a complex measure} \right\}$$

of complex measures on $(M, \mathcal{A})$ is a Banach space with the norm given by

$$\|\mu\| := \sup \left\{ \sum_{i=1}^n |\mu(A_i)| \;\middle|\; \begin{array}{l} n \in \mathbb{N},\ A_1, \dots, A_n \in \mathcal{A}, \\ A_i \cap A_j = \emptyset \text{ for } i \ne j, \\ \bigcup_{i=1}^n A_i = M \end{array} \right\} \tag{5.1.4}$$

for $\mu \in \mathcal{M}(M, \mathcal{A}, \mathbb{C})$.

The next goal is to show that every real Banach space can be complexified. Recall first that the **complexification** of a real vector space is the complex vector space

$$X^c := X \times X \cong X \otimes_{\mathbb{R}} \mathbb{C},$$

equipped with the scalar multiplication $(s + \mathbf{i}t) \cdot (x, y) := (sx - ty, tx + sy)$ for $\lambda = s + \mathbf{i}t \in \mathbb{C}$ and $z = (x, y) \in X^c$. With slight abuse of notation, we write $x + \mathbf{i}y := (x, y)$, $x := x + \mathbf{i}0 = (x, 0)$, $\mathbf{i}y := 0 + \mathbf{i}y = (0, y)$ for $x, y \in X$. Thus we do not distinguish in notation between an element $x \in X$ and the corresponding element $(x, 0) \in X^c$. In other words, the vector spaces X and $\mathbf{i}X$ are viewed as real linear subspaces of the complex vector space X^c via the embeddings $X \to X^c : x \mapsto (x, 0)$ and $\mathbf{i}X \to X^c : \mathbf{i}y \mapsto (0, y)$. Then

$$X^c = X \oplus \mathbf{i}X$$

and scalar multiplication is given by the familiar formula

$$(s + \mathbf{i}t)(x + \mathbf{i}y) := (sx - ty) + \mathbf{i}(tx + sy)$$

for $s + \mathbf{i}t \in \mathbb{C}$ and $x + \mathbf{i}y \in X^c$. If $z = x + \mathbf{i}y \in X^c$ with $x, y \in X$, then the vector $x =: \mathrm{Re}(z) \in X$ is called the **real part of** z and $y =: \mathrm{Im}(z) \in X$ is called the **imaginary part of** z.

Exercise 5.1.5. Let X be a real normed vector space and define

$$\|z\|_{X^c} := \sup_{\theta \in \mathbb{R}} \sqrt{\|\mathrm{Re}(e^{\mathbf{i}\theta} z)\|_X^2 + \|\mathrm{Im}(e^{\mathbf{i}\theta} z)\|_X^2} \qquad \text{for } z \in X^c. \tag{5.1.5}$$

Prove the following.

(i) $(X^c, \|\cdot\|_{X^c})$ is a complex normed vector space.

(ii) The natural inclusions $X \to X^c$ and $\mathbf{i}X \to X^c$ are isometric embeddings.

(iii) If X is a Banach space, then so is X^c. **Hint:** For all $z \in X^c$

$$\sqrt{\|\mathrm{Re}(z)\|_X^2 + \|\mathrm{Im}(z)\|_X^2} \le \|z\|_{X^c} \le \sqrt{2\,\|\mathrm{Re}(z)\|_X^2 + 2\,\|\mathrm{Im}(z)\|_X^2}.$$

(iv) If Y is another real normed vector space, $A : X \to Y$ is a bounded real linear operator, and the **complexified operator** $A^c : X^c \to Y^c$ is defined by $A^c(x_1 + \mathbf{i}x_2) := Ax_1 + \mathbf{i}Ax_2$ for $x_1 + \mathbf{i}x_2 \in X^c$, then A^c is a bounded complex linear operator and $\|A^c\| = \|A\|$.

(v) If $A : X \to X$ is a bounded linear operator, then A and A^c have the same spectral radius (see Definition 1.5.6).

The norm (5.1.5) on the complexified Banach space X^c is a very general construction that applies to any real Banach space, but it is not necessarily the most useful norm in each explicit example, as the next exercise shows.

Exercise 5.1.6. Let (M,d) be a nonempty compact metric space. The complexification of the space $C(M)$ of continuous real valued functions on M is the space $C(M,\mathbb{C})$ of continuous complex valued functions on M. Show that the supremum norm on $C(M,\mathbb{C})$ does not agree with the norm in (5.1.5) unless M is a singleton. Show that both norms are equivalent.

Exercise 5.1.7. Let X be a real Banach space. Prove that the *complexification of the dual space*, $\mathcal{L}(X,\mathbb{R})^c$, is isomorphic to the *dual space of the complexification*, $\mathcal{L}^c(X^c,\mathbb{C})$. **Hint:** The isomorphism assigns to each element $\Lambda_1 + \mathbf{i}\Lambda_2 \in \mathcal{L}(X,\mathbb{R})^c$ a complex linear functional $\Lambda^c : X^c \to \mathbb{C}$ via

$$\Lambda^c(x+\mathbf{i}y) := \Lambda_1(x) - \Lambda_2(y) + \mathbf{i}(\Lambda_2(x) + \Lambda_1(y)) \quad \text{for } x, y \in X.$$

Prove that the isomorphism $\mathcal{L}(X,\mathbb{R})^c \to \mathcal{L}^c(X^c,\mathbb{C})$ is an isometry whenever X is a Hilbert space, but not in general.

5.1.2. Integration. It is often useful to integrate continuous functions on a compact interval with values in a Banach space. Assuming the Riemann integral for real or complex valued functions, the integral is defined as follows.

Lemma 5.1.8 (Integral of a Continuous Function). *Let X be a real or complex Banach space, fix two real numbers $a < b$, and let $x : [a,b] \to X$ be a continuous function. Then there exists a unique vector $\xi \in X$ such that*

$$\langle x^*, \xi \rangle = \int_a^b \langle x^*, x(t) \rangle \, dt \qquad \text{for all } x^* \in X^*. \tag{5.1.6}$$

Proof. For $n \in \mathbb{N}$ define $\xi_n \in X$ and $\delta_n \geq 0$ by

$$\xi_n := \sum_{k=0}^{2^n-1} \frac{b-a}{2^n} x\left(a + k\frac{b-a}{2^n}\right), \qquad \delta_n := \sup_{|s-t| \leq 2^{-n}(b-a)} \|x(s) - x(t)\|.$$

Here the supremum runs over all $s, t \in [a,b]$ such that $|s-t| \leq 2^{-n}(b-a)$. Then $\lim_{n\to\infty} \delta_n = 0$ because x is uniformly continuous. Moreover,

$$\|\xi_{n+m} - \xi_n\| \leq (b-a)\delta_n \qquad \text{for all } m, n \in \mathbb{N}.$$

Hence $(\xi_n)_{n\in\mathbb{N}}$ is a Cauchy sequence in X. Since X is complete, this sequence converges. Denote its limit by $\xi := \lim_{n\to\infty} \xi_n$. Then

$$\langle x^*, \xi \rangle = \lim_{n\to\infty} \sum_{k=0}^{2^n-1} \frac{b-a}{2^n} \left\langle x^*, x\left(a + k\frac{b-a}{2^n}\right)\right\rangle = \int_a^b \langle x^*, x(t) \rangle \, dt$$

for all $x^* \in X^*$, by the convergence theorem for Riemann sums. This proves existence. Uniqueness follows from the Hahn–Banach Theorem (Corollary 2.3.4 and Corollary 2.3.5). This proves Lemma 5.1.8. □

Definition 5.1.9 (Integral). Let X be a real or complex Banach space and suppose that $x : [a,b] \to X$ is a continuous function on a compact interval $[a,b] \subset \mathbb{R}$. The vector $\xi \in X$ in Lemma 5.1.8 is called the **integral of x over** $[a,b]$ and will be denoted by $\int_a^b x(t)\,dt := \xi$. Thus the integral of x over $[a,b]$ is the unique element $\int_a^b x(t)\,dt \in X$ that satisfies the equation

$$\left\langle x^*, \int_a^b x(t)\,dt \right\rangle := \int_a^b \langle x^*, x(t)\rangle\,dt \qquad \text{for all } x^* \in X^*. \tag{5.1.7}$$

With this definition in place all the main results about the one-dimensional Riemann integral in first year analysis carry over to vector valued integrals.

Lemma 5.1.10 (Properties of the Integral). *Let X be a real or complex Banach space, fix two real numbers $a < b$, and let $x, y : [a,b] \to X$ be continuous functions. Then the following hold.*

(i) *The integral is a linear operator $C([a,b],X) \to X$. In particular,*

$$\int_a^b \big(x(t) + y(t)\big)\,dt = \int_a^b x(t)\,dt + \int_a^b y(t)\,dt.$$

(ii) *If $a < c < b$, then*

$$\int_a^b x(t)\,dt = \int_a^c x(t)\,dt + \int_c^b x(t)\,dt.$$

(iii) *If Y is another (real or complex) Banach space and $A : X \to Y$ is a bounded (real or complex) linear operator, then*

$$\int_a^b Ax(t)\,dt = A\int_a^b x(t)\,dt.$$

(iv) *Assume $x : [a,b] \to X$ is continuously differentiable, i.e. the limit*

$$\dot{x}(t) := \lim_{h\to 0} \frac{x(t+h) - x(t)}{h}$$

exists for all $t \in [a,b]$ and the **derivative** *$\dot{x} : [a,b] \to X$ is continuous. Then*

$$\int_a^b \dot{x}(t)\,dt = x(b) - x(a).$$

(v) *If $\alpha < \beta$ and $\phi : [\alpha,\beta] \to [a,b]$ is a diffeomorphism, then*

$$\int_a^b x(t)\,dt = \int_\alpha^\beta x(\phi(s))\dot{\phi}(s)\,ds.$$

(vi) *The integral satisfies the* **mean value inequality**

$$\left\| \int_a^b x(t)\, dt \right\| \leq \int_a^b \|x(t)\|\, dt.$$

(vii) *Let $x_0 \in X$ and assume*

$$x(t) = x_0 + \int_a^t y(s)\, ds \qquad \text{for } a \leq t \leq b.$$

Then x is continuously differentiable and $\dot{x}(t) = y(t)$ for all $t \in [a, b]$.

Proof. Parts (i), (ii), (iii) follow directly from the definitions, the additivity of the Riemann integral, and the Hahn–Banach Theorem. Part (iv) follows from the Fundamental Theorem of Calculus and the Hahn–Banach Theorem, and part (v) follows from change of variables for the Riemann integral and the Hahn–Banach Theorem. To prove part (vi), observe that

$$\begin{aligned}\left| \left\langle x^*, \int_a^b x(t)dt \right\rangle \right| &= \left| \int_a^b \langle x^*, x(t)\rangle dt \right| \\ &\leq \int_a^b |\langle x^*, x(t)\rangle|\, dt \\ &\leq \|x^*\| \int_a^b \|x(t)\|\, dt\end{aligned}$$

for all $x^* \in X^*$ and hence, by Lemma 2.4.1,

$$\left\| \int_a^b x(t)\, dt \right\| = \sup_{x^* \in X^* \setminus \{0\}} \frac{\left| \langle x^*, \int_a^b x(t)\, dt\rangle \right|}{\|x^*\|} \leq \int_a^b \|x(t)\|\, dt.$$

This proves (vi). Now let x, y be as in (vii) and let $a \leq t < t + h \leq b$. Then

$$\left\| y(t) - \tfrac{1}{h} \int_t^{t+h} y(s)ds \right\| \leq \tfrac{1}{h} \int_t^{t+h} \|y(t) - y(s)\|\, ds \leq \sup_{t \leq s \leq t+h} \|y(t) - y(s)\|$$

by (vi). Since $y : [a, b] \to X$ is continuous, this implies

$$y(t) = \lim_{h \to 0, h > 0} \frac{1}{h} \int_t^{t+h} y(s)ds = \lim_{h \to 0, h > 0} \frac{x(t+h) - x(t)}{h}$$

for $a \leq t < b$. Here the second equation follows from (ii). Likewise,

$$y(t) = \lim_{h \to 0, h > 0} \frac{x(t) - x(t-h)}{h}$$

for $a < t \leq b$. This proves part (vii) and Lemma 5.1.10. □

5.1.3. Holomorphic Functions. This is another preparatory subsection. It discusses holomorphic functions on an open subset of the complex plane with values in a complex Banach space. The most important examples in spectral theory are operator valued holomorphic functions.

Definition 5.1.11 (Holomorphic Function). Let $\Omega \subset \mathbb{C}$ be an open set, let X be a complex Banach space, and let $f : \Omega \to X$ be a continuous function.

(i) The function f is called **holomorphic** if the limit

$$f'(z) := \lim_{h \to 0} \frac{f(z+h) - f(z)}{h}$$

exists for all $z \in \Omega$ and the function $f' : \Omega \to X$ is continuous.

(ii) Let $\gamma : [a, b] \to \Omega$ be a continuously differentiable function on a compact interval $[a, b] \subset \mathbb{R}$. The vector

$$\int_\gamma f\,dz := \int_a^b f(\gamma(t))\dot{\gamma}(t)\,dt \tag{5.1.8}$$

in X is called the **integral of f over γ**.

The next lemma characterizes operator valued holomorphic functions. It shows, in particular, that every weakly holomorphic operator valued function is continuous in the norm topology.

Lemma 5.1.12 (Characterization of Holomorphic Functions). *Let X and Y be complex Banach spaces and let $A : \Omega \to \mathcal{L}^c(X, Y)$ be a weakly continuous function, defined on an open set $\Omega \subset \mathbb{C}$. Then the following are equivalent.*

(i) *The function A is holomorphic.*

(ii) *The function*

$$\Omega \to \mathbb{C} : z \mapsto \langle y^*, A(z)x \rangle$$

is holomorphic for every $x \in X$ and every $y^* \in Y^*$.

(iii) *Let $z_0 \in \Omega$ and $r > 0$ such that*

$$\overline{B_r(z_0)} = \{z \in \mathbb{C} \mid |z - z_0| \le r\} \subset \Omega.$$

Define the loop $\gamma : [0, 1] \to \Omega$ by

$$\gamma(t) := z_0 + re^{2\pi \mathbf{i} t} \qquad \text{for } 0 \le t \le 1.$$

Then, for all $x \in X$, all $y^* \in Y^*$, and all $w \in \mathbb{C}$, we have

$$|w - z_0| < r \quad \Longrightarrow \quad \langle y^*, A(w)x \rangle = \frac{1}{2\pi \mathbf{i}} \int_\gamma \frac{\langle y^*, A(z)x \rangle}{z - w}\,dz. \tag{5.1.9}$$

Proof. That (i) implies (ii) follows directly from the definitions and that (ii) implies (iii) is Cauchy's integral formula for complex valued holomorphic functions (see [**1**, p. 119]).

We prove that (iii) implies (i) by extending the standard argument for holomorphic functions to operator valued functions. For each $w \in \mathbb{C}$ with $|w - z_0| < r$, define $B(w) \in \mathcal{L}^c(X, Y)$ and $c \geq 0$ by

$$B(w)x := \frac{1}{2\pi\mathbf{i}} \int_\gamma \frac{A(z)x}{(z-w)^2}\, dz, \qquad c := \sup_{|z-z_0|=r} \|A(z)\|\,. \tag{5.1.10}$$

Then c is finite by Theorem 2.1.1. For $h \in \mathbb{C}$ with $0 < |h| < r - |w|$ we prove

$$\left\| \frac{A(w+h) - A(w)}{h} - B(w) \right\| \leq \frac{cr|h|}{(r-|w|)^2(r-|w|-|h|)}. \tag{5.1.11}$$

To see this, let $x \in X$ and $y^* \in Y^*$. Then, by (5.1.9) and (5.1.10),

$$\begin{aligned}
&\left\langle y^*, \frac{A(w+h)x - A(w)x}{h} - B(w)x \right\rangle \\
&\quad = \frac{1}{2\pi\mathbf{i}} \int_\gamma \left(\frac{1}{h}\left(\frac{1}{z-w-h} - \frac{1}{z-w} \right) - \frac{1}{(z-w)^2} \right) \langle y^*, A(z)x\rangle\, dz \\
&\quad = \frac{1}{2\pi\mathbf{i}} \int_\gamma \frac{h\langle y^*, A(z)x\rangle}{(z-w)^2(z-w-h)}\, dz.
\end{aligned}$$

The absolute value of the integral of a function over a curve is bounded above by the supremum norm of the function times the length of the curve. In the case at hand the length is $2\pi r$. Hence

$$\begin{aligned}
&\left| \left\langle y^*, \frac{A(w+h)x - A(w)x}{h} - B(w)x \right\rangle \right| \\
&\quad = \frac{1}{2\pi} \left| \int_\gamma \frac{h\langle y^*, A(z)x\rangle}{(z-w)^2(z-w-h)}\, dz \right| \\
&\quad \leq \sup_{|z-z_0|=r} \frac{r|h|\,|\langle y^*, A(z)x\rangle|}{|z-w|^2|z-w-h|} \\
&\quad \leq \frac{cr|h|\|y^*\|\|x\|}{(r-|w|)^2(r-|w|-|h|)}
\end{aligned}$$

for all $x \in X$ and all $y^* \in Y^*$. Thus the estimate (5.1.11) follows from the Hahn–Banach Theorem (Corollary 2.3.5).

By (5.1.11) the function $A : \Omega \to \mathcal{L}^c(X, Y)$ is differentiable at each point $w \in B_r(z_0)$ and its derivative at w is equal to $B(w)$. Thus A is continuous in the norm topology and so is the function $B : B_r(z_0) \to \mathcal{L}^c(X, Y)$ by (5.1.10). Hence A is holomorphic and this proves Lemma 5.1.12. □

The next three exercises show that many of the familiar results in complex analysis carry over to the present setting.

Exercise 5.1.13 (Holomorphic Functions are Smooth). Let X be a complex Banach space, let $\Omega \subset \mathbb{C}$ be an open subset, and let $f : \Omega \to X$ be a holomorphic function.

(i) Prove that its derivative $f' : \Omega \to X$ is again holomorphic. **Hint:** Use the equivalence of (i) and (ii) in Lemma 5.1.12 and use [**1**, Lemma 3, p. 121].

(ii) Prove that f is smooth. **Hint:** Induction.

(iii) Let $z_0 \in \Omega$ and $r > 0$ such that $\overline{B_r(z_0)} \subset \Omega$ and define $\gamma(t) := z_0 + re^{2\pi \mathbf{i} t}$ for $0 \le t \le 1$. Prove that the nth complex derivative of f at $w \in B_r(z_0)$ is given by the **Cauchy integral formula**

$$f^{(n)}(w) = \frac{n!}{2\pi \mathbf{i}} \int_\gamma \frac{f(z)}{(z-w)^{n+1}}\, dz. \tag{5.1.12}$$

Hint: Use the Hahn–Banach Theorem (Corollary 2.3.5) and the Cauchy integral formula for derivatives (see [**1**, p. 120] or [**74**, p. 60]).

Exercise 5.1.14 (Power Series). Let X be a complex Banach space and let $(a_n)_{n\in\mathbb{N}_0}$ be a sequence in X such that

$$\rho := \frac{1}{\limsup_{n\to\infty} \|a_n\|^{1/n}} > 0.$$

Prove that the power series

$$f(z) := \sum_{n=0}^\infty a_n z^n$$

converges for every complex number $z \in \mathbb{C}$ with $|z| < \rho$ and defines a holomorphic function $f : B_\rho(0) \to X$. Choose a number $0 < r < \rho$ and define the loop $\gamma : \mathbb{R}/\mathbb{Z} \to \mathbb{C}$ by $\gamma(t) := re^{2\pi \mathbf{i} t}$ for $t \in \mathbb{R}$. For $n \in \mathbb{N}_0$ prove that the coefficient $a_n \in X$ is given by

$$a_n = \frac{f^{(n)}(0)}{n!} = \frac{1}{2\pi \mathbf{i}} \int_\gamma \frac{f(z)}{z^{n+1}}\, dz. \tag{5.1.13}$$

Hint: Use the Hahn–Banach Theorem (Corollary 2.3.5) and the familiar results about power series in complex analysis (see [**1**, p. 38]).

Exercise 5.1.15 (Unique Continuation). Let $\Omega \subset \mathbb{C}$ be a connected open set, fix an element $z_0 \in \Omega$, and let $f, g : \Omega \to X$ be holomorphic functions with values in a complex Banach space X. Prove that $f \equiv g$ if and only if $f^{(n)}(z_0) = g^{(n)}(z_0)$ for all $n \in \mathbb{N}_0$.

The archetypal example of an operator valued holomorphic function is given by $z \mapsto (z\mathbb{1} - A)^{-1}$, where $A : X \to X$ is a bounded complex linear operator on a complex Banach space X. It takes values in the space $\mathcal{L}^c(X)$ of bounded complex linear endomorphisms of X and is defined on the open set of all complex numbers $z \in \mathbb{C}$ such that the operator $z\mathbb{1} - A$ is invertible.

5.2. Spectrum

5.2.1. The Spectrum of a Bounded Linear Operator.

Definition 5.2.1 (Spectrum). Let X be a complex Banach space and let $A \in \mathcal{L}^c(X)$. The **spectrum of** A is the set

$$\begin{aligned}\sigma(A) &:= \left\{\lambda \in \mathbb{C} \,\middle|\, \text{the operator } \lambda 1\!\!1 - A \text{ is not bijective}\right\} \\ &= \mathrm{P}\sigma(A) \cup \mathrm{R}\sigma(A) \cup \mathrm{C}\sigma(A).\end{aligned} \tag{5.2.1}$$

Here $\mathrm{P}\sigma(A)$ is the **point spectrum**, $\mathrm{R}\sigma(A)$ is the **residual spectrum**, and $\mathrm{C}\sigma(A)$ is the **continuous spectrum**. These are defined by

$$\begin{aligned}\mathrm{P}\sigma(A) &:= \left\{\lambda \in \mathbb{C} \,\middle|\, \text{the operator } \lambda 1\!\!1 - A \text{ is not injective}\right\}, \\ \mathrm{R}\sigma(A) &:= \left\{\lambda \in \mathbb{C} \,\middle|\, \begin{array}{l}\text{the operator } \lambda 1\!\!1 - A \text{ is injective} \\ \text{and its image is not dense}\end{array}\right\}, \\ \mathrm{C}\sigma(A) &:= \left\{\lambda \in \mathbb{C} \,\middle|\, \begin{array}{l}\text{the operator } \lambda 1\!\!1 - A \text{ is injective} \\ \text{and its image is dense,} \\ \text{but it is not surjective}\end{array}\right\}.\end{aligned} \tag{5.2.2}$$

The **resolvent set of** A is the complement of the spectrum. It is denoted by

$$\rho(A) := \mathbb{C} \setminus \sigma(A) = \left\{\lambda \in \mathbb{C} \,\middle|\, \text{the operator } \lambda 1\!\!1 - A \text{ is bijective}\right\}. \tag{5.2.3}$$

A complex number λ belongs to the point spectrum $\mathrm{P}\sigma(A)$ if and only if there exists a nonzero vector $x \in X$ such that

$$Ax = \lambda x.$$

The elements $\lambda \in \mathrm{P}\sigma(A)$ are called **eigenvalues** of A and the nonzero vectors $x \in \ker(\lambda 1\!\!1 - A)$ are called **eigenvectors**. When X is a real Banach space and $A \in \mathcal{L}(X)$ we denote by $\sigma(A) := \sigma(A^c)$ the spectrum of the complexified operator A^c and similarly for the point, continuous, and residual spectra.

Example 5.2.2. If $\dim X = n < \infty$, then $\sigma(A) = \mathrm{P}\sigma(A)$ is the set of eigenvalues and $\#\sigma(A) \leq n$. If $X = \{0\}$, then $\sigma(A) = \emptyset$.

Example 5.2.3. Let $X = \ell^2$ and define the operators $A, B : \ell^2 \to \ell^2$ by

$$Ax := (x_2, x_3, x_4, \dots), \qquad Bx := (0, x_1, x_2, x_3, \dots)$$

for $x = (x_i)_{i \in \mathbb{N}} \in \ell^2$. Then

$$\sigma(A) = \sigma(B) = \mathbb{D}$$

is the closed unit disc in $\mathbb{C}$ and

$$\begin{aligned}&\mathrm{P}\sigma(A) = \mathrm{int}(\mathbb{D}), \qquad &&\mathrm{R}\sigma(A) = \emptyset, \qquad &&\mathrm{C}\sigma(A) = S^1, \\ &\mathrm{P}\sigma(B) = \emptyset, &&\mathrm{R}\sigma(B) = \mathrm{int}(\mathbb{D}), &&\mathrm{C}\sigma(B) = S^1.\end{aligned}$$

Example 5.2.4. Let $X = \ell^2$ and let $(\lambda_i)_{i\in\mathbb{N}}$ be a bounded sequence of complex numbers. Define the bounded linear operator $A : \ell^2 \to \ell^2$ by

$$Ax := (\lambda_i x_i)_{i\in\mathbb{N}} \qquad \text{for } x = (x_i)_{i\in\mathbb{N}} \in \ell^2.$$

Then

$$\sigma(A) = \overline{\{\lambda_i \,|\, i \in \mathbb{N}\}}, \qquad \mathrm{P}\sigma(A) = \{\lambda_i \,|\, i \in \mathbb{N}\}, \qquad \mathrm{R}\sigma(A) = \emptyset.$$

Thus every nonempty compact subset of $\mathbb{C}$ is the spectrum of a bounded linear operator on an infinite-dimensional Hilbert space.

Lemma 5.2.5 (Spectrum). *Let $A : X \to X$ be a bounded complex linear operator on a complex Banach space X and denote by $A^* : X^* \to X^*$ the complex dual operator. Then the following hold.*

(i) *The spectrum $\sigma(A)$ is a compact subset of $\mathbb{C}$.*

(ii) $\sigma(A^*) = \sigma(A)$.

(iii) *The point, residual, and continuous spectra of A and A^* are related by*

$$\begin{array}{ll} \mathrm{P}\sigma(A^*) \subset \mathrm{P}\sigma(A) \cup \mathrm{R}\sigma(A), & \mathrm{P}\sigma(A) \subset \mathrm{P}\sigma(A^*) \cup \mathrm{R}\sigma(A^*), \\ \mathrm{R}\sigma(A^*) \subset \mathrm{P}\sigma(A) \cup \mathrm{C}\sigma(A), & \mathrm{R}\sigma(A) \subset \mathrm{P}\sigma(A^*), \\ \mathrm{C}\sigma(A^*) \subset \mathrm{C}\sigma(A), & \mathrm{C}\sigma(A) \subset \mathrm{R}\sigma(A^*) \cup \mathrm{C}\sigma(A^*). \end{array}$$

(iv) *If X is reflexive, then $\mathrm{C}\sigma(A^*) = \mathrm{C}\sigma(A)$ and*

$$\begin{array}{ll} \mathrm{P}\sigma(A^*) \subset \mathrm{P}\sigma(A) \cup \mathrm{R}\sigma(A), & \mathrm{P}\sigma(A) \subset \mathrm{P}\sigma(A^*) \cup \mathrm{R}\sigma(A^*), \\ \mathrm{R}\sigma(A^*) \subset \mathrm{P}\sigma(A), & \mathrm{R}\sigma(A) \subset \mathrm{P}\sigma(A^*). \end{array}$$

Proof. The spectrum is a bounded subset of $\mathbb{C}$ and its complement is an open subset of $\mathbb{C}$ by Theorem 1.5.5. This proves (i). Part (ii) follows from Corollary 4.1.18 and the identity $(\lambda 1\!\!1_X - A)^* = \lambda 1\!\!1_{X^*} - A^*$.

To prove part (iii), assume first that $\lambda \in \mathrm{P}\sigma(A^*)$. Then $\lambda 1\!\!1 - A^*$ is not injective, hence $\lambda 1\!\!1 - A$ does not have a dense image by Theorem 4.1.8, and hence $\lambda \in \mathrm{P}\sigma(A) \cup \mathrm{R}\sigma(A)$. Next assume $\lambda \in \mathrm{R}\sigma(A^*)$. Then $\lambda 1\!\!1 - A^*$ is injective, hence $\lambda 1\!\!1 - A$ has a dense image, and hence $\lambda \in \mathrm{P}\sigma(A) \cup \mathrm{C}\sigma(A)$. Third, assume $\lambda \in \mathrm{C}\sigma(A^*)$. Then $\lambda 1\!\!1 - A^*$ is injective and has a dense image and therefore also has a weak* dense image. Thus it follows from Theorem 4.1.8 that $\lambda 1\!\!1 - A$ is injective and has a dense image, so $\lambda \in \mathrm{C}\sigma(A)$. It follows from these three inclusions that $\mathrm{P}\sigma(A)$ is disjoint from $\mathrm{C}\sigma(A^*)$, that $\mathrm{C}\sigma(A)$ is disjoint from $\mathrm{P}\sigma(A^*)$, and that $\mathrm{R}\sigma(A)$ is disjoint from $\mathrm{R}\sigma(A^*) \cup \mathrm{C}\sigma(A^*)$. This proves part (iii).

To prove part (iv) observe that in the reflexive case a linear subspace of X^* is weak* dense if and only if it is dense. Hence it follows from Theorem 4.1.8 that $\mathrm{C}\sigma(A) = \mathrm{C}\sigma(A^*)$ whenever X is reflexive. With this understood, the remaining assertions of part (iv) follow directly from part (iii). This proves Lemma 5.2.5. □

Lemma 5.2.6 (Resolvent Identity). *Let X be a complex Banach space and let $A \in \mathcal{L}^c(X)$. Then the resolvent set $\rho(A) \subset \mathbb{C}$ is open. For $\lambda \in \rho(A)$ define the* **resolvent operator** $R_\lambda(A) \in \mathcal{L}^c(X)$ *by*

$$R_\lambda(A) := (\lambda \mathbb{1} - A)^{-1}. \tag{5.2.4}$$

Then the map $\rho(A) \to \mathcal{L}^c(X) : \lambda \mapsto R_\lambda(A)$ is holomorphic and satisfies

$$R_\lambda(A) - R_\mu(A) = (\mu - \lambda) R_\lambda(A) R_\mu(A) \tag{5.2.5}$$

for all $\lambda, \mu \in \rho(A)$. Equation (5.2.5) is called the **resolvent identity**.

Proof. We prove the resolvent identity. Let $\lambda, \mu \in \rho(A)$. Then

$$(\lambda \mathbb{1} - A)\big(R_\lambda(A) - R_\mu(A)\big)(\mu \mathbb{1} - A) = (\mu \mathbb{1} - A) - (\lambda \mathbb{1} - A) = (\mu - \lambda)\mathbb{1}.$$

Multiply by $R_\lambda(A)$ on the left and by $R_\mu(A)$ on the right to obtain the resolvent identity (5.2.5).

We prove that $\rho(A)$ is open and the map $\rho(A) \to \mathcal{L}^c(X) : \lambda \mapsto R_\lambda(A)$ is continuous. Fix an element $\lambda \in \rho(A)$ and choose $\mu \in \mathbb{C}$ such that

$$|\mu - \lambda| \, \|R_\lambda(A)\| < 1.$$

Then Corollary 1.5.7 asserts that the operator

$$(\mu \mathbb{1} - A) R_\lambda(A) = \mathbb{1} - (\lambda - \mu) R_\lambda(A)$$

is bijective and

$$\big((\mu \mathbb{1} - A) R_\lambda(A)\big)^{-1} = \sum_{k=0}^\infty (\lambda - \mu)^k R_\lambda(A)^k.$$

Hence $\mu \in \rho(A)$ and

$$R_\mu(A) = \sum_{k=0}^\infty (\lambda - \mu)^k R_\lambda(A)^{k+1} = R_\lambda(A) + \sum_{k=1}^\infty (\lambda - \mu)^k R_\lambda(A)^{k+1},$$

and hence

$$\begin{aligned} \|R_\mu(A) - R_\lambda(A)\| &\le \sum_{k=1}^\infty |\lambda - \mu|^k \, \|R_\lambda(A)\|^{k+1} \\ &= \frac{|\mu - \lambda| \, \|R_\lambda(A)\|^2}{1 - |\mu - \lambda| \, \|R_\lambda(A)\|}. \end{aligned}$$

This proves that $\rho(A)$ is open and the map $\rho(A) \to \mathcal{L}^c(X) : \lambda \mapsto R_\lambda(A)$ is continuous. That it is holomorphic follows from the equation

$$\lim_{\mu \to \lambda} \frac{R_\mu(A) - R_\lambda(A)}{\mu - \lambda} = - \lim_{\mu \to \lambda} R_\lambda(A) R_\mu(A) = -R_\lambda(A)^2$$

for $\lambda \in \rho(A)$ and the fact that the map $\lambda \mapsto R_\lambda(A)^2$ is continuous. This proves Lemma 5.2.6. □

5.2.2. The Spectral Radius. Recall from Definition 1.5.6 that the spectral radius of a bounded linear operator $A : X \to X$ on a real or complex Banach space is the real number

$$r_A := \inf_{n\in\mathbb{N}} \|A^n\|^{1/n} = \lim_{n\to\infty} \|A^n\|^{1/n} \le \|A\| .$$

If A is a bounded linear operator on a real Banach space, then its complexification A^c has the same spectral radius as A by Exercise 5.1.5. The reason for the terminology *spectral radius* is the next theorem.

Theorem 5.2.7 (Spectral Radius). *Let X be a nonzero complex Banach space and let $A \in \mathcal{L}^c(X)$. Then $\sigma(A) \neq \emptyset$ and*

$$r_A := \lim_{n\to\infty} \|A^n\|^{1/n} = \sup_{\lambda\in\sigma(A)} |\lambda|. \tag{5.2.6}$$

Proof. Let $\lambda \in \mathbb{C}$ such that $|\lambda| > r_A$. Then $r_{\lambda^{-1}A} = |\lambda|^{-1} r_A < 1$ and hence the operator $\mathbb{1} - \lambda^{-1}A$ is invertible by Corollary 1.5.7. Thus the operator $\lambda\mathbb{1} - A = \lambda(\mathbb{1} - \lambda^{-1}A)$ is bijective and so $\lambda \notin \sigma(A)$. Hence

$$\sup_{\lambda\in\sigma(A)} |\lambda| \le r_A. \tag{5.2.7}$$

To prove the converse inequality, define the set $\Omega \subset \mathbb{C}$ by

$$\Omega := \left\{ z \in \mathbb{C} \,|\, z = 0 \text{ or } z^{-1} \in \rho(A) \right\}$$

and define the map $R : \Omega \to \mathcal{L}^c(X)$ by $R(0) := 0$ and by

$$R(z) := (z^{-1}\mathbb{1} - A)^{-1} \qquad \text{for } z \in \Omega \setminus \{0\}.$$

Then Ω is an open subset of $\mathbb{C}$ and the restriction of R to $\Omega \setminus \{0\}$ is holomorphic by Lemma 5.2.6. Moreover, Ω contains the open disc of radius r_A^{-1} centered at the origin and it follows from Corollary 1.5.7 that

$$R(z) = z(\mathbb{1} - zA)^{-1} = \sum_{k=0}^{\infty} z^{k+1} A^k \tag{5.2.8}$$

for all $z \in \mathbb{C}$ such that $r_A|z| < 1$. Hence R is holomorphic by Lemma 5.1.12. By Exercise 5.1.13 the nth derivative $R^{(n)} : \Omega \to \mathcal{L}^c(X)$ of R is holomorphic for every $n \in \mathbb{N}$.

Now let $r > \sup_{\lambda\in\sigma(A)}|\lambda|$, so the closed disc of radius r^{-1} centered at the origin is contained in Ω. Let $x \in X$ and $x^* \in X^*$ and apply the Cauchy integral formula in (5.1.12) or [**1**, p. 120] to the power series

$$\langle x^*, R(z)x\rangle = \sum_{k=1}^{\infty} \langle x^*, A^{k-1}x\rangle z^k$$

and the loop

$$\gamma(t) := \frac{e^{2\pi\mathbf{i}t}}{r}, \qquad 0 \le t \le 1.$$

Then, for each $n \in \mathbb{N}$, we have

$$\langle x^*, A^{n-1}x\rangle = \frac{1}{n!}\frac{d^n}{dz^n}\bigg|_{z=0} \langle x^*, R(z)x\rangle = \frac{1}{2\pi\mathbf{i}}\int_\gamma \frac{\langle x^*, R(z)x\rangle}{z^{n+1}} dz.$$

By the Hahn–Banach Theorem (Corollary 2.3.5), this implies

$$A^n = \frac{1}{2\pi\mathbf{i}}\int_\gamma \frac{R(z)}{z^{n+2}}\,dz = \frac{1}{2\pi\mathbf{i}}\int_0^1 \frac{\dot\gamma(t)R(\gamma(t))}{\gamma(t)^{n+2}}\,dt = \int_0^1 \frac{R(\gamma(t))}{\gamma(t)^{n+1}}\,dt.$$

Hence, by part (vi) of Lemma 5.1.10, we have

$$\begin{aligned} \|A^n\| &\le \int_0^1 \frac{\|R(\gamma(t))\|}{|\gamma(t)|^{n+1}}\,dt \\ &= r^{n+1}\int_0^1 \|R(\gamma(t))\|\,dt \\ &\le r^{n+1}\sup_{0\le t\le 1}\|R(\gamma(t))\| \\ &= r^{n+1}\sup_{|\lambda|=r}\left\|(\lambda\mathbb{1}-A)^{-1}\right\| \end{aligned}$$

for all $n \in \mathbb{N}$. Abbreviate

$$c := \sup_{|\lambda|=r}\|(\lambda\mathbb{1}-A)^{-1}\|.$$

Then $\|A^n\|^{1/n} \le r(rc)^{1/n}$ for all $n \in \mathbb{N}$ and hence

$$r_A = \lim_{n\to\infty}\|A^n\|^{1/n} \le r\lim_{n\to\infty}(rc)^{1/n} = r.$$

This holds for all $r > \sup_{\lambda\in\sigma(A)}|\lambda|$, so

$$r_A \le \sup_{\lambda\in\sigma(A)}|\lambda|$$

as claimed. By (5.2.7) this proves equation (5.2.6).

We prove that $\sigma(A) \neq \emptyset$. Suppose, by contradiction, that $\sigma(A) = \emptyset$ and so, in particular, A is invertible. Choose any nonzero element $x \in X$. Then $A^{-1}x \neq 0$ and so, by Corollary 2.3.5, there exists an element $x^* \in X^*$ such that $\langle x^*, A^{-1}x\rangle = -1$. Define the function $f : \mathbb{C} \to \mathbb{C}$ by

$$f(\lambda) := \langle x^*, (\lambda\mathbb{1}-A)^{-1}x\rangle \qquad \text{for } \lambda \in \mathbb{C} = \rho(A).$$

Then f is holomorphic by Lemma 5.2.6, $f(0) = 1$ by definition, and

$$|f(\lambda)| \le \|x^*\|\|x\|\|(\lambda\mathbb{1}-A)^{-1}\| \le \frac{\|x^*\|\,\|x\|}{|\lambda|-\|A\|}$$

for all $\lambda \in \mathbb{C}$ such that $|\lambda| > \|A\|$. Thus f is a nonconstant bounded holomorphic function on $\mathbb{C}$, in contradiction to Liouville's Theorem. Hence the spectrum of A is nonempty and this proves Theorem 5.2.7. □

5.2.3. The Spectrum of a Compact Operator. The spectral theory of compact operators is considerably simpler than that of general bounded linear operators. In particular, every nonzero spectral value is an eigenvalue, the generalized eigenspaces are all finite-dimensional, and zero is the only possible accumulation point of the spectrum (i.e. each nonzero spectral value is an isolated point of the spectrum). All these observations are fairly direct consequences of the results in Chapter 4.

Let X be a complex Banach space and let $A \in \mathcal{L}^c(X)$ be a bounded complex linear operator. Then $\ker(\lambda\mathbb{1} - A)^k \subset \ker(\lambda\mathbb{1} - A)^{k+1}$ for all $\lambda \in \mathbb{C}$ and all $k \in \mathbb{N}$. Moreover, if $\ker(\lambda\mathbb{1} - A)^m = \ker(\lambda\mathbb{1} - A)^{m+1}$ for some integer $m \geq 1$, then $\ker(\lambda\mathbb{1} - A)^m = \ker(\lambda\mathbb{1} - A)^{m+k}$ for all $k \in \mathbb{N}$. The union of these subspaces is called the **generalized eigenspace** of A associated to the eigenvalue $\lambda \in \mathrm{P}\sigma(A)$ and will be denoted by

$$E_\lambda := E_\lambda(A) := \bigcup_{m=1}^{\infty} \ker(\lambda\mathbb{1} - A)^m. \tag{5.2.9}$$

Theorem 5.2.8 (Spectrum of a Compact Operator). *Let X be a nonzero complex Banach space and let $A \in \mathcal{L}^c(X)$ be a compact operator. Then the following hold.*

(i) *If $\lambda \in \sigma(A)$ and $\lambda \neq 0$, then λ is an eigenvalue of A, $\dim E_\lambda(A) < \infty$, and there exists an integer $m \in \mathbb{N}$ such that*

$$E_\lambda(A) = \ker(\lambda\mathbb{1} - A)^m, \qquad X = \ker(\lambda\mathbb{1} - A)^m \oplus \mathrm{im}(\lambda\mathbb{1} - A)^m.$$

(ii) *Nonzero eigenvalues of A are isolated, i.e. for every $\lambda \in \sigma(A)\setminus\{0\}$ there exists a constant $\varepsilon > 0$ such that every $\mu \in \mathbb{C}$ satisfies*

$$0 < |\lambda - \mu| < \varepsilon \qquad \Longrightarrow \qquad \mu \in \rho(A).$$

Proof. We prove part (i). Fix a nonzero complex number λ. Then $\lambda\mathbb{1} - A$ is a Fredholm operator of index zero by part (i) of Theorem 4.4.2. Hence

$$\dim\ker(\lambda\mathbb{1} - A) = \dim\mathrm{coker}(\lambda\mathbb{1} - A)$$

and so $\lambda\mathbb{1} - A$ is either bijective, in which case $\lambda \notin \sigma(A)$, or not injective, in which case $\lambda \in \mathrm{P}\sigma(A)$.

Now fix an element

$$\lambda \in \mathrm{P}\sigma(A) \setminus \{0\}$$

and define

$$K := \lambda^{-1}A, \qquad E_n := \ker(\mathbb{1} - K)^n = \ker(\lambda\mathbb{1} - A)^n \qquad \text{for } n \in \mathbb{N}.$$

Since K is a compact operator, it follows from Theorem 4.4.1 and Theorem 4.4.2 that $(\mathbb{1} - K)^n$ is a Fredholm operator and hence has a finite-dimensional kernel for all $n \in \mathbb{N}$. Thus $\dim(E_n) < \infty$ for all $n \in \mathbb{N}$.

Next we prove that there exists an integer $m \in \mathbb{N}$ such that $E_m = E_{m+1}$. Suppose, by contradiction, that this is not the case. Then $E_{n-1} \subsetneq E_n$ for all $n \in \mathbb{N}$. Hence it follows from Lemma 1.2.12 and the axiom of countable choice that there exists a sequence $(x_n)_{n\in\mathbb{N}}$ in X such that, for all $n \in \mathbb{N}$,

$$x_n \in E_n, \qquad \|x_n\| = 1, \qquad \inf_{x\in E_{n-1}} \|x_n - x\| \geq \frac{1}{2}. \tag{5.2.10}$$

Fix two integers $n > m > 0$. Then $Kx_m \in E_{n-1}$ and $x_n - Kx_n \in E_{n-1}$, so

$$\|Kx_n - Kx_m\| = \|x_n - (Kx_m + x_n - Kx_n)\| \geq \frac{1}{2}.$$

Hence the sequence $(Kx_n)_{n\in\mathbb{N}}$ does not have a convergent subsequence, in contradiction to the fact that the operator K is compact.

Thus we have proved that there exists an integer $m \in \mathbb{N}$ such that

$$\ker(\lambda\mathbb{1} - A)^m = \ker(\lambda\mathbb{1} - A)^{m+k} \qquad \text{for all } k \in \mathbb{N}.$$

Define

$$X_0 := \ker(\lambda\mathbb{1} - A)^m, \qquad X_1 := \mathrm{im}(\lambda\mathbb{1} - A)^m.$$

Since $(\lambda\mathbb{1} - A)^m$ is a Fredholm operator these subspaces are both closed and X_0 is finite-dimensional. Moreover, these subspaces are both invariant under A. We prove that

$$X = X_0 \oplus X_1. \tag{5.2.11}$$

If $x \in X_0 \cap X_1$, then $(\lambda\mathbb{1} - A)^m x = 0$ and there exists an element $\xi \in X$ such that $x = (\lambda\mathbb{1} - A)^m\xi$. Hence $\xi \in \ker(\lambda\mathbb{1} - A)^{2m} = \ker(\lambda\mathbb{1} - A)^m$ and so $x = (\lambda\mathbb{1} - A)^m\xi = 0$. The annihilator of $X_0 \oplus X_1$ in $X^* = \mathcal{L}^c(X, \mathbb{C})$ is

$$\begin{aligned}(X_0 \oplus X_1)^\perp &= (\ker(\lambda\mathbb{1} - A)^m)^\perp \cap (\mathrm{im}(\lambda\mathbb{1} - A)^m)^\perp \\ &= \mathrm{im}(\lambda\mathbb{1} - A^*)^m \cap \ker(\lambda\mathbb{1} - A^*)^m \\ &= \{0\}.\end{aligned}$$

Here the second equation follows from Theorem 4.1.8 and Theorem 4.1.16. The last equation follows from the fact that the kernels of the linear operators $(\lambda\mathbb{1} - A)^k$ and $(\lambda\mathbb{1} - A^*)^k$ have the same dimension for all $k \in \mathbb{N}$ and so $\ker(\lambda\mathbb{1} - A^*)^{2m} = \ker(\lambda\mathbb{1} - A^*)^m$. Now it follows from Corollary 2.3.5 that $X_0 \oplus X_1$ is dense in X and therefore is equal to X. This proves (5.2.11) and part (i).

Now the operator $\lambda\mathbb{1} - A : X_1 \to X_1$ is bijective. Hence Theorem 2.2.5 asserts that there exists a constant $\varepsilon > 0$ such that $\varepsilon\|x_1\| \leq \|\lambda x_1 - Ax_1\|$ for all $x_1 \in X_1$. Hence, by Corollary 1.5.7 the operator $\mu\mathbb{1} - A : X_1 \to X_1$ is invertible for all $\mu \in \mathbb{C}$ with $|\mu - \lambda| < \varepsilon$. Moreover, if $\mu \neq \lambda$, then the operator $\mu\mathbb{1} - A : X_0 \to X_0$ is bijective, because λ is the only eigenvalue of $A|_{X_0}$. Hence $\mu\mathbb{1} - A$ is bijective for all $\mu \in \mathbb{C}$ such that $0 < |\mu - \lambda| < \varepsilon$. This proves part (ii) and Theorem 5.2.8. □

Example 5.2.9. Let X be the complexification of the Argyros–Haydon space (Remark 4.4.6). Then every bounded linear operator $A : X \to X$ has the form

$$A = \lambda\mathbb{1} + K,$$

where $\lambda \in \mathbb{C}$ and $K : X \to X$ is a compact operator. By Theorem 5.2.8, the spectrum of K is either a finite set or a sequence that converges to zero (exercise). Hence the spectrum of every bounded linear operator on X is either a finite set or a convergent sequence. This is in sharp contrast to infinite-dimensional Hilbert spaces where every nonempty compact subset of the complex plane is the spectrum of some bounded linear operator (see Example 5.2.4).

Remark 5.2.10 (Spectral Projection). Let X be a complex Banach space, let $A \in \mathcal{L}^c(X)$ be a compact operator, let $\lambda \in \sigma(A)$ be a nonzero eigenvalue of A, and choose $m \in \mathbb{N}$ such that

$$E_\lambda := \ker(\lambda\mathbb{1} - A)^m = \ker(\lambda\mathbb{1} - A)^{m+1}.$$

By Theorem 5.2.8 such an integer m exists, E_λ is a finite-dimensional linear subspace of X, the operator $(\lambda\mathbb{1} - A)^m$ has a closed image, and

$$X = \ker(\lambda\mathbb{1} - A)^m \oplus \operatorname{im}(\lambda\mathbb{1} - A)^m.$$

Hence the formula

$$P_\lambda(x_0 + x_1) := x_0$$

for $x_0 \in \ker(\lambda\mathbb{1} - A)^m$ and $x_1 \in \operatorname{im}(\lambda\mathbb{1} - A)^m$ defines a bounded linear operator $P_\lambda : X \to X$ which is an A-invariant projection onto E_λ, i.e.

$$P_\lambda^2 = P_\lambda, \qquad P_\lambda A = AP_\lambda, \qquad \operatorname{im}(P_\lambda) = E_\lambda. \tag{5.2.12}$$

The operator P_λ is uniquely determined by (5.2.12) and is called the **spectral projection** associated to λ. It can also be written in the form

$$P_\lambda = \frac{1}{2\pi\mathbf{i}} \int_\gamma (z\mathbb{1} - A)^{-1} dz. \tag{5.2.13}$$

Here $r > 0$ is chosen such that

$$\overline{B_r(\lambda)} \cap \sigma(A) = \{\lambda\}$$

(see part (ii) of Theorem 5.2.8) and the loop $\gamma : [0,1] \to \rho(A)$ is defined by

$$\gamma(t) := \lambda + re^{2\pi\mathbf{i}t} \qquad \text{for } 0 \le t \le 1.$$

Equation (5.2.13) is a special case of part (vi) of Theorem 5.2.12 below.

5.2.4. Holomorphic Functional Calculus. Let X be a nonzero complex Banach space and let $A \in \mathcal{L}^c(X)$ be a bounded complex linear operator. Then the spectrum of A is a nonempty compact subset of the complex plane by Lemma 5.2.6 and Theorem 5.2.7. The holomorphic functional calculus assigns a bounded linear operator $f(A) \in \mathcal{L}^c(X)$ to every holomorphic function $f : U \to \mathbb{C}$ on an open set $U \subset \mathbb{C}$ containing $\sigma(A)$. The operator $f(A)$ is defined as the *Dunford integral* of the resolvent operators along a cycle in $U \setminus \sigma(A)$ encircling the spectrum.

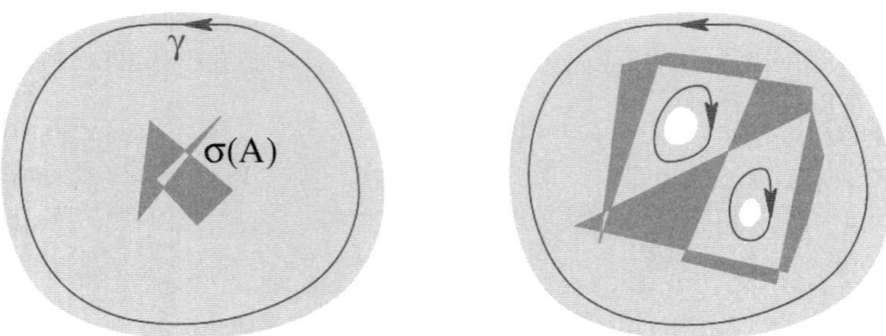

Figure 5.2.1. A cycle encircling the spectrum.

Definition 5.2.11 (Dunford Integral). Let X be a nonzero complex Banach space and let $A \in \mathcal{L}^c(X)$. Let $U \subset \mathbb{C}$ be an open set such that

$$\sigma(A) \subset U$$

and let $\gamma = (\gamma_1, \dots, \gamma_m)$ be a collection of smooth loops $\gamma_i : \mathbb{R}/\mathbb{Z} \to U \setminus \sigma(A)$ with **winding numbers**

$$\mathrm{w}(\gamma, \lambda) := \frac{1}{2\pi\mathbf{i}} \sum_{i=1}^{m} \int_{\gamma_i} \frac{dz}{z - \lambda} = \begin{cases} 1, & \text{for } \lambda \in \sigma(A), \\ 0, & \text{for } \lambda \in \mathbb{C} \setminus U. \end{cases} \tag{5.2.14}$$

(See Figure 5.2.1.) The collection γ is called a **cycle in** $U \setminus \sigma(A)$ and the **image of the cycle** γ is the set $\mathrm{im}(\gamma) := \bigcup_{i=1}^{n} \gamma_i(\mathbb{R}/\mathbb{Z})$. For the existence of γ see [**1**, p. 139] or [**74**, p. 90]. The operator $f(A) \in \mathcal{L}^c(X)$ is defined by

$$\begin{aligned} f(A) &:= \frac{1}{2\pi\mathbf{i}} \int_{\gamma} f(z)(z\mathbb{1} - A)^{-1}\, dz \\ &= \frac{1}{2\pi\mathbf{i}} \sum_{i=1}^{m} \int_{\gamma_i} f(z)(z\mathbb{1} - A)^{-1}\, dz. \end{aligned} \tag{5.2.15}$$

The integral in (5.2.15) is called the **Dunford integral**.

The next theorem establishes the basic properties of the operators $f(A)$ and examines their spectra.

Theorem 5.2.12 (Holomorphic Functional Calculus). *Let X be a nonzero complex Banach space and let $A \in \mathcal{L}^c(X)$. Then the following hold.*

(i) *The operator $f(A)$ is independent of the choice of the cycle γ in $U \setminus \sigma(A)$ satisfying (5.2.14) that is used to define it.*

(ii) *Let $U \subset \mathbb{C}$ be an open set such that $\sigma(A) \subset U$ and let $f, g : U \to \mathbb{C}$ be holomorphic. Then*

$$(f+g)(A) = f(A) + g(A), \qquad (fg)(A) = f(A)g(A). \tag{5.2.16}$$

(iii) *If $p(z) = \sum_{k=0}^n a_k z^k$ is a polynomial, then $p(A) = \sum_{k=0}^n a_k A^k$.*

(iv) *Let $U \subset \mathbb{C}$ be an open set such that $\sigma(A) \subset U$ and let $f : U \to \mathbb{C}$ be holomorphic. Then*

$$\sigma(f(A)) = f(\sigma(A)). \tag{5.2.17}$$

This assertion is the **Spectral Mapping Theorem**.

(v) *Let $U, V \subset \mathbb{C}$ be open sets such that $\sigma(A) \subset U$ and let $f : U \to V$ and $g : V \to \mathbb{C}$ be holomorphic functions. Then*

$$g(f(A)) = (g \circ f)(A). \tag{5.2.18}$$

(vi) *Let $\Sigma_0, \Sigma_1 \subset \sigma(A)$ be disjoint compact sets such that $\Sigma_0 \cup \Sigma_1 = \sigma(A)$ and let $U_0, U_1 \subset \mathbb{C}$ be disjoint open sets such that $\Sigma_i \subset U_i$ for $i = 0, 1$. Define the function $f : U := U_0 \cup U_1 \to \mathbb{C}$ by $f|_{U_0} := 0$ and $f|_{U_1} := 1$, and define $P := f(A) \in \mathcal{L}^c(X)$. Then P is a projection and commutes with A, i.e. $P^2 = P$ and $PA = AP$. Thus $X_0 := \ker(P)$ and $X_1 := \mathrm{im}(P)$ are closed A-invariant subspaces of X such that $X = X_0 \oplus X_1$. The spectrum of the operator $A_i := A|_{X_i} : X_i \to X_i$ is given by $\sigma(A_i) = \Sigma_i$ for $i = 0, 1$.*

Proof. We prove part (i). Let β and γ be two collections of loops in $U \setminus \sigma(A)$ that satisfy (5.2.14). Then their difference $\gamma - \beta$, understood as a cycle in $U \setminus \sigma(A)$, is **homologous to zero**, in that its winding number about every point in the complement of $U \setminus \sigma(A)$ is zero. Hence the Cauchy integral formula [**1**, Thm. 14, p. 141] asserts that the integral of every holomorphic function on $U \setminus \sigma(A)$ over $\gamma - \beta$ must vanish. This implies

$$\int_\beta f(z)\langle x^*, (z\mathbb{1} - A)^{-1}x\rangle \, dz = \int_\gamma f(z)\langle x^*, (z\mathbb{1} - A)^{-1}x\rangle \, dz$$

for every holomorphic function $f : U \to \mathbb{C}$ and all $x \in X$ and all $x^* \in X^*$. Hence it follows from the Hahn–Banach Theorem (Corollary 2.3.5) that the integrals of the operator valued function $U \setminus \sigma(A) \to \mathcal{L}^c(X) : z \mapsto f(z)(z\mathbb{1} - A)^{-1}$ over β and γ agree for every holomorphic function $f : U \to \mathbb{C}$. This proves part (i).

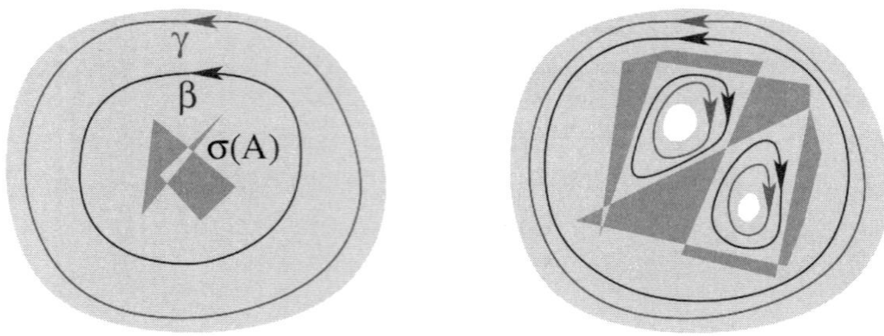

Figure 5.2.2. Two cycles encircling the spectrum.

We prove part (ii). The assertion about the sum follows directly from the definition. To prove the assertion about the product, choose two cycles β and γ in $U \setminus \sigma(A)$ that both satisfy (5.2.14), have disjoint images so that $\mathrm{im}(\beta) \cap \mathrm{im}(\gamma) = \emptyset$, and such that the image of β is encircled by γ, i.e.

$$\begin{aligned} \mathrm{w}(\gamma, w) &= 1 \qquad \text{for all } w \in \mathrm{im}(\beta), \\ \mathrm{w}(\beta, z) &= 0 \qquad \text{for all } z \in \mathrm{im}(\gamma). \end{aligned} \tag{5.2.19}$$

(See Figure 5.2.2.) Then, by the resolvent identity in Lemma 5.2.6, we have

$$\begin{aligned} f(A)g(A) &= \frac{1}{2\pi\mathbf{i}}\int_\beta f(w)R_w(A)\,dw \frac{1}{2\pi\mathbf{i}}\int_\gamma g(z)R_z(A)\,dz \\ &= \frac{1}{2\pi\mathbf{i}}\frac{1}{2\pi\mathbf{i}}\int_\beta\int_\gamma f(w)g(z)\frac{R_w(A)-R_z(A)}{z-w}\,dz\,dw \\ &= \frac{1}{2\pi\mathbf{i}}\int_\beta f(w)\left(\frac{1}{2\pi\mathbf{i}}\int_\gamma \frac{g(z)\,dz}{z-w}\right)R_w(A)\,dw \\ &\quad + \frac{1}{2\pi\mathbf{i}}\int_\gamma g(z)\left(\frac{1}{2\pi\mathbf{i}}\int_\beta \frac{f(w)\,dw}{w-z}\right)R_z(A)\,dz \\ &= \frac{1}{2\pi\mathbf{i}}\int_\beta f(w)g(w)R_w(A)\,dw \\ &= (fg)(A). \end{aligned}$$

Here the penultimate step uses (5.2.19). This proves part (ii).

We prove part (iii). In view of part (ii) it suffices to prove the equations

$$1(A) = 1\!\!1_X, \qquad \mathrm{id}(A) = A, \tag{5.2.20}$$

associated to the holomorphic functions $f(z) = 1$ and $f(z) = z$. In these cases we can choose $U = \mathbb{C}$ and $\gamma_r(t) := re^{2\pi\mathbf{i}t}$ with $r > \|A\|$. Then

$$f(A) = \frac{1}{2\pi\mathbf{i}}\int_{\gamma_r} f(z)(z1\!\!1 - A)^{-1}\,dz = \int_0^1 f(re^{2\pi\mathbf{i}t})(1\!\!1 - r^{-1}e^{-2\pi\mathbf{i}t}A)^{-1}\,dt.$$

For $f \equiv 1$ it follows from Corollary 1.5.7 that the integrand converges uniformly to $\mathbb{1}$ as r tends to ∞ and so $1(A) = \mathbb{1}$. In the case $f(z) = z$ we obtain

$$\begin{aligned} \mathrm{id}(A) &= \frac{1}{2\pi\mathbf{i}}\int_{\gamma_r} z(z\mathbb{1} - A)^{-1}\,dz \\ &= \frac{1}{2\pi\mathbf{i}}\int_{\gamma_r} A(z\mathbb{1} - A)^{-1}\,dz \\ &= A \circ 1(A) \\ &= A. \end{aligned}$$

Here the difference of the second and third terms vanishes because it is the integral of the constant operator valued function $z \mapsto \mathbb{1}$ over a cycle in U that is homologous to zero by (5.2.14). This proves part (iii).

We prove part (iv). Fix a spectral value $\lambda \in \sigma(A)$. Then there exists a holomorphic function $g : U \to \mathbb{C}$ such that

$$f(z) - f(\lambda) = (z - \lambda)g(z) \qquad \text{for all } z \in U.$$

By part (ii) this implies

$$f(\lambda)\mathbb{1} - f(A) = (\lambda\mathbb{1} - A)g(A) = g(A)(\lambda\mathbb{1} - A).$$

Hence $f(\lambda)\mathbb{1} - f(A)$ cannot be bijective and so

$$f(\lambda) \in \sigma(f(A)).$$

This shows that

$$f(\sigma(A)) \subset \sigma(f(A)).$$

To prove the converse inclusion, fix an element $\lambda \in \mathbb{C} \setminus f(\sigma(A))$. Then

$$V := U \setminus f^{-1}(\lambda)$$

is an open neighborhood of $\sigma(A)$. Define $g_\lambda : V \to \mathbb{C}$ by

$$g_\lambda(z) := \frac{1}{\lambda - f(z)} \qquad \text{for } z \in V = U \setminus f^{-1}(\lambda).$$

Then g_λ is holomorphic, and it follows from parts (ii) and (iii) that

$$g_\lambda(A)(\lambda\mathbb{1} - f(A)) = (\lambda\mathbb{1} - f(A))g_\lambda(A) = 1(A) = \mathbb{1}.$$

Hence $\lambda\mathbb{1} - f(A)$ is invertible and so

$$\lambda \in \mathbb{C} \setminus \sigma(f(A)).$$

This shows that

$$\sigma(f(A)) \subset f(\sigma(A))$$

and proves part (iv).

To prove part (v), note first that the operator $g(f(A))$ is well defined, because $\sigma(f(A)) = f(\sigma(A)) \subset V$ by part (iv). Choose a cycle β in $U \setminus \sigma(A)$ such that $\mathrm{w}(\beta, \lambda) = 1$ for $\lambda \in \sigma(A)$ and $\mathrm{w}(\beta, \lambda) = 0$ for $\lambda \in \mathbb{C} \setminus U$. Then

$$K := \mathrm{im}(\beta) \cup \{w \in U \setminus \mathrm{im}(\beta) \,|\, \mathrm{w}(\beta, w) \neq 0\}$$

is a compact neighborhood of $\sigma(A)$. Then, for every $z \in \mathbb{C} \setminus f(K)$, the functions $w \mapsto (z - f(w))^{-1}$ and $w \mapsto z - f(w)$ are holomorphic in an open neighborhood of K and their product is the constant function 1. Hence it follows from parts (ii), (iii), and (iv) that

$$(5.2.21) \qquad (z\mathbb{1} - f(A))^{-1} = \frac{1}{2\pi\mathbf{i}} \int_\beta \frac{(w\mathbb{1} - A)^{-1}}{z - f(w)}\, dw \quad \text{for } z \in \mathbb{C} \setminus f(K).$$

Choose a cycle γ in $V \setminus f(K)$ such that

$$(5.2.22) \qquad \mathrm{w}(\gamma, \mu) = \begin{cases} 1, & \text{for } \mu \in f(K), \\ 0, & \text{for } \mu \in \mathbb{C} \setminus V. \end{cases}$$

Then

$$\begin{aligned} g(f(A)) &= \frac{1}{2\pi\mathbf{i}} \int_\gamma g(z)(z\mathbb{1} - f(A))^{-1}\, dz \\ &= \frac{1}{2\pi\mathbf{i}} \int_\gamma g(z) \left(\frac{1}{2\pi\mathbf{i}} \int_\beta \frac{(w\mathbb{1} - A)^{-1}}{z - f(w)}\, dw \right) dz \\ &= \frac{1}{2\pi\mathbf{i}} \int_\beta \left(\frac{1}{2\pi\mathbf{i}} \int_\gamma \frac{g(z)}{z - f(w)}\, dz \right) (w\mathbb{1} - A)^{-1}\, dw \\ &= \frac{1}{2\pi\mathbf{i}} \int_\beta g(f(w))(w\mathbb{1} - A)^{-1}\, dw \\ &= (g \circ f)(A). \end{aligned}$$

Here the second step uses (5.2.21) and the fourth step uses (5.2.22) and the Cauchy integral formula. This proves part (v).

We prove part (vi). Since $f^2 = f$ it follows from (ii) that $P^2 = P$. Moreover P commutes with A by definition. Define $g : U \to \mathbb{C}$ by $g(z) = z$ for $z \in U$ and let $c \in \mathbb{C}$. Then, by parts (ii) and (iii), we have

$$c\mathbb{1}_{X_0} \oplus A_1 = c(\mathbb{1}_X - P) + AP = (c(1 - f) + gf)(A)$$

and hence $\sigma(c\mathbb{1}_{X_0} \oplus A_1) = \{c\} \cup \Sigma_1$ by part (iv). If $\lambda \in \mathbb{C} \setminus \Sigma_1$, it follows that the operator $(\lambda - c)\mathbb{1}_{X_0} \oplus (\lambda\mathbb{1}_{X_1} - A_1)$ is bijective for $c \neq \lambda$ and so the operator $\lambda\mathbb{1}_{X_1} - A_1$ is bijective. Conversely, suppose that $\lambda \in \Sigma_1$. Then the operator $(\lambda - c)\mathbb{1}_{X_0} \oplus (\lambda\mathbb{1}_{X_1} - A_1)$ is not bijective and, for $c \neq \lambda$, this implies that the operator $\lambda\mathbb{1}_{X_1} - A_1$ is not bijective. Thus $\sigma(A_1) = \Sigma_1$. The equation $\sigma(A_0) = \Sigma_0$ follows by interchanging Σ_0 and Σ_1. This proves Theorem 5.2.12. □

Exercise 5.2.13 (Exponential Map). Let X be a nonzero complex Banach space and let $A \in \mathcal{L}^c(X)$ be a bounded complex linear operator. Choose a real number $r > \|A\|$ and define $\gamma_r(\theta) := re^{2\pi \mathbf{i}\theta}$ for $0 \leq \theta \leq 1$. Prove that

$$e^A := \sum_{k=0}^{\infty} \frac{A^k}{k!} = \frac{1}{2\pi\mathbf{i}} \int_{\gamma_r} e^z (z\mathbb{1} - A)^{-1}\, dz.$$

Prove that

$$\sigma(e^A) = \left\{ e^\lambda \,|\, \lambda \in \sigma(A) \right\}$$

and, for all $s, t \in \mathbb{R}$,

$$e^{(s+t)A} = e^{sA}e^{tA}, \qquad e^{0A} = \mathbb{1}$$

and

$$\frac{d}{dt} e^{tA} = Ae^{tA} = e^{tA}A.$$

Exercise 5.2.14 (Logarithm). Let X be a nonzero complex Banach space and let $T \in \mathcal{L}^c(X)$ be a bounded complex linear operator such that

$$\mathrm{Re}(\lambda) > 0 \qquad \text{for all } \lambda \in \sigma(T).$$

Choose a smooth curve $\gamma : \mathbb{R}/\mathbb{Z} \to \mathbb{C} \setminus \sigma(T)$ such that $\mathrm{Re}(\gamma(t)) > 0$ for all t and $\mathrm{w}(\gamma, \lambda) = 1$ for all $\lambda \in \sigma(T)$. Denote by $\log : \{z \in \mathbb{C} \,|\, \mathrm{Re}(z) > 0\} \to \mathbb{C}$ the branch of the logarithm with $\log(1) = 0$. Define

$$\log(T) := \frac{1}{2\pi\mathbf{i}} \int_{\gamma} \log(z)(z\mathbb{1} - T)^{-1}\, dz.$$

Prove that

$$e^{\log(T)} = T, \qquad \log(e^A) = A$$

for all $A \in \mathcal{L}^c(X)$. Let $n \in \mathbb{N}$ and deduce that the operator $S := e^{\log(T)/n}$ satisfies $S^n = T$.

Exercise 5.2.15 (Inverse). Let X be a nonzero complex Banach space and let $A \in \mathcal{L}^c(X)$ be a bijective bounded complex linear operator. Choose real numbers ε and r such that

$$0 < \varepsilon < \|A^{-1}\|^{-1} \leq \|A\| < r.$$

Show that $\varepsilon < |\lambda| < r$ for all $\lambda \in \sigma(A)$. With $\gamma_r, \gamma_\varepsilon$ as in Exercise 5.2.13, show that

$$A^{-1} = \frac{1}{2\pi\mathbf{i}} \int_{\gamma_r} \frac{(z\mathbb{1} - A)^{-1}}{z}\, dz - \frac{1}{2\pi\mathbf{i}} \int_{\gamma_\varepsilon} \frac{(z\mathbb{1} - A)^{-1}}{z}\, dz.$$

Exercise 5.2.16 (Spectral Projection). Verify the formula (5.2.13).

5.3. Operators on Hilbert Spaces

The remainder of this chapter discusses the spectral theory of operators on Hilbert spaces. The present section begins with an introduction to complex Hilbert spaces (Subsection 5.3.1) and the adjoint operator (Subsection 5.3.2). It then moves on to examine the properties of the spectra of normal operators (Subsection 5.3.3) and self-adjoint operators (Subsection 5.3.4). The next two sections establish the continuous functional calculus for self-adjoint operators (Section 5.4) and normal operators (Section 5.5). Section 5.6 introduces the spectral measure of a normal operator and Section 5.7 examines cyclic vectors of self-adjoint operators.

5.3.1. Complex Hilbert Spaces.

Definition 5.3.1 (Hermitian Inner Product). Let H be a complex vector space. A **Hermitian inner product** on H is a real bilinear map

$$H \times H \to \mathbb{C} : (x, y) \mapsto \langle x, y \rangle \tag{5.3.1}$$

that satisfies the following three axioms.

(a) The map (5.3.1) is complex anti-linear in the first variable and is complex linear in the second variable, i.e.

$$\langle \lambda x, y \rangle = \overline{\lambda} \langle x, y \rangle, \qquad \langle x, \lambda y \rangle = \lambda \langle x, y \rangle$$

for all $x, y \in H$ and all $\lambda \in \mathbb{C}$.

(b) $\langle x, y \rangle = \overline{\langle y, x \rangle}$ for all $x, y \in H$.

(c) The map (5.3.1) is positive definite, i.e. $\langle x, x \rangle > 0$ for all $x \in H \setminus \{0\}$.

It is sometimes convenient to denote the Hermitian inner product by $\langle \cdot, \cdot \rangle_{\mathbb{C}}$, to distinguish it from the real inner product in Definition 1.4.1.

Assume H is a complex vector space equipped with a Hermitian inner product (5.3.1). Then the real part of the Hermitian inner product is a real inner product as in Definition 1.4.3 and so the formula

$$H \to \mathbb{R} : x \mapsto \|x\| := \sqrt{\langle x, x \rangle} \tag{5.3.2}$$

defines a norm on H. The next lemma shows that Hermitian inner products satisfy a stronger form of the Cauchy–Schwarz inequality. It is proved by the same argument as in Lemma 1.4.2.

Lemma 5.3.2 (Complex Cauchy–Schwarz Inequality). *Let H be a complex vector space equipped with Hermitian inner product* (5.3.1) *and the associated norm* (5.3.2). *Then the Hermitian inner product and norm satisfy the* **complex Cauchy–Schwarz inequality**

$$|\langle x, y \rangle| \le \|x\| \, \|y\| \qquad \text{for all } x, y \in H. \tag{5.3.3}$$

Proof. The Cauchy–Schwarz inequality is obvious when $x = 0$ or $y = 0$. Hence assume $x \neq 0$ and $y \neq 0$ and define

$$\xi := \|x\|^{-1} x, \qquad \eta := \|y\|^{-1} y.$$

Then $\|\xi\| = \|\eta\| = 1$ and $\langle \eta, \xi - \langle \eta, \xi \rangle \eta \rangle = \langle \eta, \xi \rangle - \langle \eta, \xi \rangle \|\eta\|^2 = 0$, and hence

$$\begin{aligned} 0 &\leq \|\xi - \langle \eta, \xi \rangle \eta\|^2 \\ &= \langle \xi, \xi - \langle \eta, \xi \rangle \eta \rangle \\ &= \langle \xi, \xi \rangle - \langle \eta, \xi \rangle \langle \xi, \eta \rangle \\ &= 1 - |\langle \xi, \eta \rangle|^2 . \end{aligned}$$

Thus $|\langle \xi, \eta \rangle| \leq 1$ and so $|\langle x, y \rangle| \leq \|x\| \, \|y\|$. This proves Lemma 5.3.2. □

Definition 5.3.3 (Complex Hilbert Space). A **complex Hilbert space** is a complex vector space H equipped with a Hermitian inner product (5.3.1) such that the norm (5.3.2) is complete.

Remark 5.3.4. (i) Let $(H, \langle \cdot, \cdot \rangle_{\mathbb{C}})$ be a complex Hilbert space. Then H is also a real Hilbert space with the inner product

$$\langle x, y \rangle_{\mathbb{R}} := \mathrm{Re}\langle x, y \rangle_{\mathbb{C}}. \tag{5.3.4}$$

Hence all results about real Hilbert spaces, such as Theorem 1.4.4 and Theorem 1.4.5, continue to hold for complex Hilbert spaces.

(ii) If H is a complex Hilbert space, then the Hermitian inner product and the real inner product (5.3.4) are related by the formula

$$\langle x, y \rangle_{\mathbb{C}} = \langle x, y \rangle_{\mathbb{R}} + \mathbf{i}\langle \mathbf{i}x, y \rangle_{\mathbb{R}} \qquad \text{for all } x, y \in H. \tag{5.3.5}$$

(iii) Conversely, suppose that $(H, \langle \cdot, \cdot \rangle_{\mathbb{R}})$ is a real Hilbert space and that $J : H \to H$ is a linear map such that

$$J^2 = -\mathbb{1}, \qquad \|Jx\| = \|x\| \qquad \text{for all } x \in H.$$

Then H carries a unique structure of a complex Hilbert space such that multiplication by $\mathbf{i}$ is the operator J, and $\langle \cdot, \cdot \rangle_{\mathbb{R}}$ is the real part of the Hermitian inner product. The scalar multiplication is defined by $(s + \mathbf{i}t)x := sx + tJx$ for $s+\mathbf{i}t \in \mathbb{C}$ and $x \in H$, and the Hermitian inner product is given by (5.3.5).

(iv) Let $(H, \langle \cdot, \cdot \rangle)$ be a real Hilbert space. Then its **complexification**

$$H^c := H \oplus \mathbf{i}H$$

is a complex Hilbert space with the Hermitian inner product

$$\langle x + \mathbf{i}y, \xi + \mathbf{i}\eta \rangle^c := \langle x, \xi \rangle + \langle y, \eta \rangle + \mathbf{i}\big(\langle x, \eta \rangle - \langle y, \xi \rangle\big) \tag{5.3.6}$$

for $x, y, \xi, \eta \in H$.

Exercise 5.3.5. **(i)** Verify parts (iii) and (iv) of Remark 5.3.4.

(ii) Let $(M, \mathcal{A}, \mu)$ be a measure space. Prove that $L^2(\mu, \mathbb{C})$ is a complex Hilbert space with the Hermitian inner product

$$\langle f, g\rangle := \int_M \overline{f} g \, d\mu \qquad \text{for } f, g \in \mathcal{L}^2(\mu, \mathbb{C}). \tag{5.3.7}$$

(iii) Prove that $\ell^2(\mathbb{N}, \mathbb{C})$ is a complex Hilbert space with

$$\langle x, y\rangle := \sum_{i=1}^{\infty} \overline{x}_i y_i \qquad \text{for } x = (x_i)_{i\in\mathbb{N}}, y = (y_i)_{i\in\mathbb{N}} \in \ell^2(\mathbb{N}, \mathbb{C}). \tag{5.3.8}$$

(iv) Prove that $L^2(\mu, \mathbb{C})$ is the complexification of $L^2(\mu, \mathbb{R})$ and $\ell^2(\mathbb{N}, \mathbb{C})$ is the complexification of $\ell^2(\mathbb{N}, \mathbb{R})$.

The next theorem shows that a complex Hilbert space is isomorphic to its complex dual space. An important caveat is that the isomorphism is necessarily complex anti-linear. The result is a direct consequence of the Riesz Representation Theorem 1.4.4.

Theorem 5.3.6 (Riesz). *Let H be a complex Hilbert space and denote by $H^* := \mathcal{L}^c(H, \mathbb{C})$ its complex dual space. Define the map $\iota : H \to H^*$ by*

$$\langle \iota(x), y\rangle_{H^*,H} := \langle x, y\rangle \qquad \text{for } x, y \in H. \tag{5.3.9}$$

Then ι is a complex anti-linear isometric isomorphism.

Proof. It follows directly from the definitions that the map $\iota : H \to H^*$ is complex anti-linear, i.e. $\iota(\lambda x) = \overline{\lambda}\iota(x)$ for all $x \in H$ and all $\lambda \in \mathbb{C}$. That it is an isometry follows from the complex Cauchy–Schwarz inequality in Lemma 5.3.2, namely

$$\|x\| = \frac{|\langle x, x\rangle|}{\|x\|} \le \|\iota(x)\| = \sup_{y\in H\setminus\{0\}} \frac{|\langle x, y\rangle|}{\|y\|} \le \|x\|$$

for all $x \in H \setminus \{0\}$ and so $\|\iota(x)\| = \|x\|$ for all $x \in H$. In particular, ι is injective. To prove that it is surjective, fix a bounded complex linear functional $\Lambda : H \to \mathbb{C}$. Then $\mathrm{Re}\Lambda : H \to \mathbb{R}$ is a bounded real linear functional. Hence Theorem 1.4.4 asserts that there exists a unique element $x \in H$ such that $\mathrm{Re}\Lambda(y) = \mathrm{Re}\langle x, y\rangle$ for all $y \in H$. This implies

$$\begin{aligned}
\Lambda(y) &= \mathrm{Re}\Lambda(y) + \mathbf{i}\mathrm{Im}\Lambda(y) = \mathrm{Re}\Lambda(y) - \mathbf{i}\mathrm{Re}\Lambda(\mathbf{i}y) \\
&= \mathrm{Re}\langle x, y\rangle - \mathbf{i}\mathrm{Re}\langle x, \mathbf{i}y\rangle = \mathrm{Re}\langle x, y\rangle + \mathbf{i}\mathrm{Im}\langle x, y\rangle \\
&= \langle x, y\rangle
\end{aligned}$$

for all $y \in H$. Here the last equation follows from (5.3.5). Thus ι is surjective and this proves Theorem 5.3.6. □

5.3.2. The Adjoint Operator. Let $A : X \to Y$ be a bounded complex linear operator between complex Hilbert spaces. Then the dual operator of A is the bounded linear operator $A^*_{\mathrm{Banach}} : Y^* \to X^*$ between the complex dual spaces, introduced in part (iii) of Definition 5.1.1. In the Hilbert space setting one can use the isomorphisms of Theorem 5.3.6 to replace the dual operator A^*_{Banach} by the operator

$$A^*_{\mathrm{Hilbert}} := \iota_X^{-1} \circ A^*_{\mathrm{Banach}} \circ \iota_Y : Y \to X$$

between the original Hilbert spaces which is called the adjoint operator of A. Thus the dual operator and the adjoint operator are related by the commutative diagram

$$\begin{array}{ccc} Y & \xrightarrow{A^*_{\mathrm{Hilbert}}} & X \\ {\scriptstyle \iota_Y}\downarrow & & \downarrow{\scriptstyle \iota_X} \\ Y^* & \xrightarrow{A^*_{\mathrm{Banach}}} & X^* \end{array}$$

From now on we drop the subscripts *"Banach"* and *"Hilbert"* and work exclusively with the adjoint operator. Thus, throughout the remainder of this chapter, the notation A^* acquires a new meaning and will denote the adjoint operator of a bounded complex linear operator between complex Hilbert spaces. The dual operator of the Banach space setting will no longer be used.

Definition 5.3.7 (Adjoint Operator). Let X and Y be complex Hilbert spaces and let $A \in \mathcal{L}^c(X, Y)$ be a bounded complex linear operator. The **adjoint operator of** A is the unique operator $A^* : Y \to X$ that satisfies the equation

$$\langle A^*y, x\rangle_X = \langle y, Ax\rangle_Y$$

for all $x \in X$ and all $y \in Y$. It is well defined by Theorem 5.3.6 and it agrees with the adjoint operator in Example 4.1.6 associated to the real parts of the Hermitian inner products on X and Y.

If H is a complex Hilbert space, then the **complex orthogonal complement** of a subset $S \subset H$ is denoted by

$$S^\perp := \{x \in H \mid \langle x, y\rangle = 0 \text{ for all } y \in S\}.$$

The complex orthogonal complement of any subset $S \subset H$ is a closed complex linear subspace. It is isomorphic to the complex annihilator of S under the isomorphism $\iota : H \to H^*$ in Theorem 5.3.6 and, in general, it differs from the orthogonal complement of S with respect to the real inner product. The real and complex orthogonal complements agree whenever the subset S is invariant under multiplication by $\mathbf{i}$. The next two lemmas summarize the properties of the orthogonal complement and the adjoint operator.

Lemma 5.3.8. *Let H be a complex Hilbert space and let $E \subset H$ be a complex linear subspace. Then $\overline{E} = E^{\perp\perp}$ and so E is closed if and only if $E = E^{\perp\perp}$.*

Proof. By definition the orthogonal complement of the orthogonal complement of E agrees with the pre-annihilator of the annihilator of E. Hence the assertion follows from the complex analogue of Corollary 2.3.24. (See also Corollary 3.1.18.) □

Lemma 5.3.9. *Let X, Y, Z be complex Hilbert spaces and let $A \in \mathcal{L}^c(X, Y)$ and $B \in \mathcal{L}^c(Y, Z)$. Then the following hold.*

(i) *A^* is a bounded complex linear operator and $\|A^*\| = \|A\|$.*

(ii) *$(AB)^* = B^*A^*$ and $(\lambda\mathbb{1})^* = \overline{\lambda}\mathbb{1}$ for all $\lambda \in \mathbb{C}$.*

(iii) *$A^{**} = A$.*

(iv) *$\ker(A^*) = \operatorname{im}(A)^\perp$ and $\overline{\operatorname{im}(A^*)} = \ker(A)^\perp$.*

(v) *If A has a closed image, then A^* has a closed image.*

(vi) *If A is bijective, then so is A^* and $(A^*)^{-1} = (A^{-1})^*$.*

(vii) *If A is an isometry, then so is A^*.*

(viii) *If A is compact, then so is A^*.*

(ix) *If A is Fredholm, then so is A^* and $\operatorname{index}(A^*) = -\operatorname{index}(A)$.*

(x) *Assume $X = Y = H$. Then*

$$\sigma(A^*) = \left\{\overline{\lambda} \,|\, \lambda \in \sigma(A)\right\}$$

and

$$\begin{aligned} \mathrm{P}\sigma(A^*) &\subset \left\{\overline{\lambda} \,|\, \lambda \in \mathrm{P}\sigma(A) \cup \mathrm{R}\sigma(A)\right\}, \\ \mathrm{R}\sigma(A^*) &\subset \left\{\overline{\lambda} \,|\, \lambda \in \mathrm{P}\sigma(A)\right\}, \\ \mathrm{C}\sigma(A^*) &= \left\{\overline{\lambda} \,|\, \lambda \in \mathrm{C}\sigma(A)\right\}. \end{aligned}$$

Proof. Part (i) follows from the same argument as in Lemma 4.1.2 and parts (ii) and (iii) follow directly from the definitions (see also Lemma 4.1.3). Part (iv) follows from Theorem 4.1.8 and Lemma 5.3.8. Part (v) follows from Theorem 4.1.16, parts (vi) and (vii) follow from Corollary 4.1.18, part (viii) follows from Theorem 4.2.10, and part (ix) follows from Theorem 4.3.3. Part (x) follows from parts (iv) and (vi) and the fact that

$$(\lambda\mathbb{1} - A)^* = \overline{\lambda}\mathbb{1} - A^*$$

by part (ii) (see also Lemma 5.2.5). This proves Lemma 5.3.9. □

5.3.3. The Spectrum of a Normal Operator.

Definition 5.3.10 (Normal Operator). Let H be a complex Hilbert space. A bounded complex linear operator $A : H \to H$ is called

- **normal** if $A^*A = AA^*$,
- **unitary** if $A^*A = AA^* = \mathbb{1}$,
- **self-adjoint** if $A^* = A$.

Thus every self-adjoint operator and every unitary operator is normal.

Exercise 5.3.11. Let H be a complex Hilbert space and let

$$A = A^* : H \to H$$

be a self-adjoint operator. Prove that

$$A = 0 \qquad \iff \qquad \langle x, Ax\rangle = 0 \quad \text{for all } x \in H.$$

Example 5.3.12. Consider the complex Hilbert space $H := \ell^2(\mathbb{N}, \mathbb{C})$ and choose a bounded sequence $(\lambda_i)_{i\in\mathbb{N}}$ of complex numbers. Then the operator

$$A_\lambda : \ell^2(\mathbb{N}, \mathbb{C}) \to \ell^2(\mathbb{N}, \mathbb{C}),$$

defined by

$$A_\lambda x := (\lambda_i x_i)_{i\in\mathbb{N}} \qquad \text{for } x = (x_i)_{i\in\mathbb{N}} \in \ell^2(\mathbb{N}, \mathbb{C}),$$

is normal and its adjoint operator is given by

$$A_\lambda^* x := (\overline{\lambda}_i x_i)_{i\in\mathbb{N}} \qquad \text{for } x = (x_i)_{i\in\mathbb{N}} \in \ell^2(\mathbb{N}, \mathbb{C}).$$

Thus A_λ is self-adjoint if and only if $\lambda_i \in \mathbb{R}$ for all i, and A_λ is unitary if and only if $|\lambda_i| = 1$ for all i.

Example 5.3.13. Define the bounded complex linear operator

$$A : \ell^2(\mathbb{N}, \mathbb{C}) \to \ell^2(\mathbb{N}, \mathbb{C})$$

by

$$Ax := (0, x_1, x_2, x_3, \dots) \qquad \text{for } x = (x_i)_{i\in\mathbb{N}} \in \ell^2(\mathbb{N}, \mathbb{C}).$$

Then

$$A^*x := (x_2, x_3, x_4, \dots) \qquad \text{for } x = (x_i)_{i\in\mathbb{N}} \in \ell^2(\mathbb{N}, \mathbb{C})$$

and hence

$$A^*A = \mathbb{1} \neq AA^*.$$

Thus A is not normal. It is an isometric embedding but is not unitary.

Lemma 5.3.14 (Characterization of Normal Operators). *Let H be a complex Hilbert space and let $A : H \to H$ be a bounded complex linear operator. Then the following hold.*

(i) *A is normal if and only if $\|A^*x\| = \|Ax\|$ for all $x \in H$.*

(ii) *A is unitary if and only if $\|A^*x\| = \|Ax\| = \|x\|$ for all $x \in H$.*

(iii) *A is self-adjoint if and only if $\langle x, Ax\rangle \in \mathbb{R}$ for all $x \in H$.*

Proof. We prove part (i). If A is normal, then

$$\|Ax\|^2 = \langle Ax, Ax\rangle = \langle x, A^*Ax\rangle = \langle x, AA^*x\rangle = \|A^*x\|^2$$

for all $x \in X$. Conversely, assume $\|A^*x\| = \|Ax\|$ for all $x \in X$. Then, for all $x, y \in H$, we have

$$\begin{aligned} \mathrm{Re}\langle Ax, Ay\rangle &= \tfrac{1}{4}\big(\|Ax + Ay\|^2 - \|Ax - Ay\|^2\big) \\ &= \tfrac{1}{4}\big(\|A^*x + A^*y\|^2 - \|A^*x - A^*y\|^2\big) = \mathrm{Re}\langle A^*x, A^*y\rangle \end{aligned}$$

and so $\mathrm{Im}\langle Ax, Ay\rangle = \mathrm{Re}\langle A\mathbf{i}x, Ay\rangle = \mathrm{Re}\langle A^*\mathbf{i}x, A^*y\rangle = \mathrm{Im}\langle A^*x, A^*y\rangle$. Thus

$$\langle A^*Ax, y\rangle = \langle Ax, Ay\rangle = \langle A^*x, A^*y\rangle = \langle AA^*x, y\rangle$$

for all $x, y \in H$ and hence $A^*A = AA^*$. This proves (i).

We prove part (ii). If A is unitary, then

$$\|Ax\|^2 = \langle Ax, Ax\rangle = \langle x, A^*Ax\rangle = \langle x, x\rangle = \|x\|^2$$

and, by an analogous argument, $\|A^*x\| = \|x\|$ for all $x \in X$. Conversely, assume $\|Ax\| = \|A^*x\| = \|x\|$ for all $x \in X$. Then, for all $x, y \in H$, we have

$$\begin{aligned} \mathrm{Re}\langle Ax, Ay\rangle &= \tfrac{1}{4}\big(\|Ax + Ay\|^2 - \|Ax - Ay\|^2\big) \\ &= \tfrac{1}{4}\big(\|x + y\|^2 - \|x - y\|^2\big) = \mathrm{Re}\langle x, y\rangle \end{aligned}$$

and so $\mathrm{Im}\langle Ax, Ay\rangle = \mathrm{Re}\langle A\mathbf{i}x, Ay\rangle = \mathrm{Re}\langle \mathbf{i}x, y\rangle = \mathrm{Im}\langle x, y\rangle$. Thus

$$\langle A^*Ax, y\rangle = \langle Ax, Ay\rangle = \langle x, y\rangle$$

for all $x, y \in H$ and hence $A^*A = \mathbb{1}$. The same argument with A and A^* interchanged shows that $AA^* = \mathbb{1}$. Thus A is unitary and this proves (ii).

We prove (iii). If A is self-adjoint, then $\overline{\langle x, Ax\rangle} = \langle Ax, x\rangle = \langle x, Ax\rangle$ and so $\langle x, Ax\rangle \in \mathbb{R}$ for all $x \in X$. Conversely, assume $\langle x, Ax\rangle \in \mathbb{R}$ for all $x \in X$. Then, for all $x, y \in H$, we have

$$\begin{aligned} \mathrm{Im}\langle x, Ay\rangle - \mathrm{Im}\langle Ax, y\rangle &= \mathrm{Im}\big(\langle x, Ay\rangle + \langle y, Ax\rangle\big) \\ &= \tfrac{1}{2}\mathrm{Im}\big(\langle x + y, Ax + Ay\rangle - \langle x - y, Ax - Ay\rangle\big) = 0 \end{aligned}$$

and so $\mathrm{Re}\langle x, Ay\rangle - \mathrm{Re}\langle Ax, y\rangle = \mathrm{Im}\langle x, A\mathbf{i}y\rangle - \mathrm{Im}\langle Ax, \mathbf{i}y\rangle = 0$. Thus

$$\langle A^*x, y\rangle = \langle x, Ay\rangle = \langle Ax, y\rangle$$

for all $x, y \in H$ and hence $A^* = A$. This proves Lemma 5.3.14. $\square$

Theorem 5.3.15 (Spectrum of a Normal Operator). *Let H be a nonzero complex Hilbert space and let $A \in \mathcal{L}^c(H)$ be a normal operator. Then the following hold.*

(i) $\|A^n\| = \|A\|^n$ *for all* $n \in \mathbb{N}$.

(ii) $\|A\| = \sup_{\lambda\in\sigma(A)}|\lambda|$.

(iii) $\mathrm{R}\sigma(A^*) = \mathrm{R}\sigma(A) = \emptyset$ *and* $\mathrm{P}\sigma(A^*) = \{\overline{\lambda}\,|\,\lambda \in \mathrm{P}\sigma(A)\}$.

(iv) *If A is unitary, then* $\sigma(A) \subset S^1$.

(v) *Assume A is compact. Then H admits an orthonormal basis of eigenvectors of A. More precisely, there exist a set $I \subset \mathbb{N}$, either equal to $\mathbb{N}$ or finite, an orthonormal sequence $(e_i)_{i\in I}$ in H, and a map $I \to \mathbb{C}\setminus\{0\} : i \mapsto \lambda_i$ such that $\lim_{i\to\infty} \lambda_i = 0$ when $I = \mathbb{N}$ and*

$$Ax = \sum_{i\in I} \lambda_i \langle e_i, x\rangle e_i \qquad \text{for all } x \in H.$$

Proof. If $x \in H$ is a unit vector, then, by Lemma 5.3.14,

$$\|Ax\|^2 = \langle Ax, Ax\rangle = \langle x, A^*Ax\rangle \le \|A^*Ax\| = \|A^2x\|.$$

Hence

$$\|A^2\| \le \|A\|^2 = \sup_{\|x\|=1}\|Ax\|^2 \le \sup_{\|x\|=1}\|A^2x\| = \|A^2\|$$

and so $\|A^2\| = \|A\|^2$. Hence it follows by induction that $\|A^{2^m}\| = \|A\|^{2^m}$ for all $m \in \mathbb{N}$. Given any integer $n \ge 1$, choose $m \in \mathbb{N}$ such that $n < 2^m$, and deduce that

$$\|A\|^{2^m-n}\,\|A\|^n = \|A^{2^m}\| \le \|A^n\|\,\|A\|^{2^m-n}.$$

Hence $\|A\|^n \le \|A^n\| \le \|A\|^n$ and so $\|A^n\| = \|A\|^n$. This proves part (i).

Part (ii) follows from part (i) and Theorem 5.2.7.

To prove part (iii), fix an element $\lambda \in \mathbb{C}$. Then $(\lambda 1\!\!1 - A)^* = \overline{\lambda}1\!\!1 - A^*$ by part (ii) of Lemma 5.3.9. Hence $\lambda 1\!\!1 - A$ is normal and it follows from part (i) of Lemma 5.3.14 that $\ker(\overline{\lambda}1\!\!1 - A^*) = \ker(\lambda 1\!\!1 - A)$. Hence, by part (iv) of Lemma 5.3.9, we have

$$\overline{\mathrm{im}(\lambda 1\!\!1 - A)} = \ker(\overline{\lambda}1\!\!1 - A^*)^\perp = \ker(\lambda 1\!\!1 - A)^\perp.$$

By Lemma 5.3.8 this shows that the operator $\lambda 1\!\!1 - A$ is injective if and only if it has a dense image. Thus $\mathrm{R}\sigma(A) = \emptyset$ and so $\mathrm{P}\sigma(A^*) = \{\overline{\lambda}\,|\,\lambda \in \mathrm{P}\sigma(A)\}$ by part (x) of Lemma 5.3.9. This proves part (iii).

To prove part (iv), assume A is unitary and let $\lambda \in \sigma(A)$. Then $|\lambda| \le 1$ by Theorem 5.2.7. Moreover, $\lambda \ne 0$ because A is invertible, and the operator $\lambda^{-1}1\!\!1 - A^{-1} = (\lambda A)^{-1}(A - \lambda 1\!\!1)$ is not invertible. Hence $\lambda^{-1} \in \sigma(A^{-1})$ and so $|\lambda|^{-1} \le \|A^{-1}\| = \|A^*\| = \|A\| = 1$. This proves part (iv).

We prove part (v) in three steps. The first step shows that the eigenspaces are pairwise orthogonal, the second step shows that each generalized eigenvector is an eigenvector, and the third step shows that the orthogonal complement of the direct sum of all the eigenspaces associated to the nonzero eigenvalues is the kernel of A.

Step 1. *If $\lambda, \mu \in \sigma(A)$ such that $\lambda \neq \mu$ and $x, y \in H$ such that $Ax = \lambda x$ and $Ay = \mu y$, then $\langle x, y \rangle = 0$.*

By Lemma 5.3.14, $\ker(\lambda \mathbb{1} - A) = \ker(\lambda \mathbb{1} - A)^* = \ker(\overline{\lambda} - A^*)$. Hence

$$(\lambda - \mu)\langle x, y \rangle = \langle \overline{\lambda} x, y \rangle - \langle x, \mu y \rangle = \langle A^* x, y \rangle - \langle x, Ay \rangle = 0$$

and this proves Step 1.

Step 2. *Let $\lambda \in \sigma(A)$ and $n \in \mathbb{N}$. Then $\ker(\lambda \mathbb{1} - A)^n = \ker(\lambda \mathbb{1} - A)$.*

Let $x \in \ker(\lambda \mathbb{1} - A)^2$. Then $(\overline{\lambda} \mathbb{1} - A^*)(\lambda x - Ax) = 0$ by Lemma 5.3.14, hence

$$\|\lambda x - Ax\|^2 = \langle \lambda x - Ax, \lambda x - Ax \rangle = \langle x, (\overline{\lambda} \mathbb{1} - A^*)(\lambda x - Ax) \rangle = 0,$$

and hence $x \in \ker(\lambda \mathbb{1} - A)$. Thus

$$\ker(\lambda \mathbb{1} - A)^2 = \ker(\lambda \mathbb{1} - A)$$

and this implies $\ker(\lambda \mathbb{1} - A)^n = \ker(\lambda \mathbb{1} - A)$ for all $n \in \mathbb{N}$.

Step 3. *Define $E_\lambda := \ker(\lambda \mathbb{1} - A)$ for $\lambda \in \sigma(A) \setminus \{0\}$. Then*

$$x \perp E_\lambda \quad \textit{for all } \lambda \in \sigma(A) \setminus \{0\} \qquad \Longleftrightarrow \qquad Ax = 0$$

for all $x \in H$.

If $x \in \ker(A)$, then $x \perp E_\lambda$ for all $\lambda \in \sigma(A) \setminus \{0\}$ by Step 1. To prove the converse, define

$$H_0 := \{x \in H \,|\, x \perp E_\lambda \text{ for all } \lambda \in \sigma(A) \setminus \{0\}\}.$$

Then H_0 is a closed A-invariant subspace of H and

$$A_0 := A|_{H_0} : H_0 \to H_0$$

is a compact normal operator. Suppose, by contradiction, that $A_0 \neq 0$. Then it follows from Theorem 5.2.8 and part (ii) that A_0 has a nonzero eigenvalue. This contradicts the definition of H_0 and proves Step 3.

By Theorem 5.2.8 the set $\sigma(A) \setminus \{0\}$ is either finite or is a sequence converging to zero and $\dim E_\lambda < \infty$ for all $\lambda \in \sigma(A) \setminus \{0\}$. Hence part (v) follows from Step 1, Step 2, and Step 3 by choosing orthonormal bases of the eigenspaces E_λ for all $\lambda \in \sigma(A) \setminus \{0\}$. This proves Theorem 5.3.15. □

5.3.4. The Spectrum of a Self-Adjoint Operator. Let X and Y be real Hilbert spaces and let $T : X \to Y$ be a bounded linear operator. Then

$$\|T\|^2 = \sup_{\|x\|=1} \|Tx\|_Y^2 = \sup_{\|x\|=1} \langle x, T^*Tx\rangle_X \le \|T^*T\| \le \|T^*\|\,\|T\| = \|T\|^2$$

and hence

$$\|T\|^2 = \sup_{\|x\|=1} \langle x, T^*Tx\rangle_X = \|T^*T\|\,. \tag{5.3.10}$$

This formula is the special case $A = T^*T$ of Theorem 5.3.16 below. It can sometimes be used to compute the norm of an operator (Exercise 5.8.9).

Theorem 5.3.16 (Spectrum of a Self-Adjoint Operator). *Let H be a nonzero complex Hilbert space and let $A \in \mathcal{L}^c(H)$ be a self-adjoint operator. Then the following hold.*

(i) $\sigma(A) \subset \mathbb{R}$.

(ii) $\sup \sigma(A) = \sup_{\|x\|=1} \langle x, Ax\rangle$.

(iii) $\inf \sigma(A) = \inf_{\|x\|=1} \langle x, Ax\rangle$.

(iv) $\|A\| = \sup_{\|x\|=1} |\langle x, Ax\rangle|$.

(v) *Assume A is compact. Then H admits an orthonormal basis of eigenvectors of A. More precisely, there exist a set $I \subset \mathbb{N}$, either equal to $\mathbb{N}$ or finite, an orthonormal sequence $(e_i)_{i\in I}$ in H, and a map $I \to \mathbb{R}\setminus\{0\} : i \mapsto \lambda_i$ such that $\lim_{i\to\infty} \lambda_i = 0$ when $I = \mathbb{N}$ and*

$$Ax = \sum_{i\in I} \lambda_i \langle e_i, x\rangle e_i$$

for all $x \in H$.

Proof. We prove part (i). Let $\lambda \in \mathbb{C} \setminus \mathbb{R}$. Then, for all $x \in H$,

$$\begin{aligned}
\|\lambda x - Ax\|^2 &= \langle \lambda x - Ax, \lambda x - Ax\rangle \\
&= |\lambda|^2 \|x\|^2 - \lambda\langle Ax, x\rangle - \overline{\lambda}\langle x, Ax\rangle + \|Ax\|^2 \\
&= |\mathrm{Im}\lambda|^2 \|x\|^2 + |\mathrm{Re}\lambda|^2 \|x\|^2 - 2(\mathrm{Re}\lambda)\langle Ax, x\rangle + \|Ax\|^2 \\
&= |\mathrm{Im}\lambda|^2 \|x\|^2 + \|(\mathrm{Re}\lambda)x - Ax\|^2 \\
&\ge |\mathrm{Im}\lambda|^2 \|x\|^2 .
\end{aligned}$$

This shows that $\lambda\mathbb{1} - A$ is injective and has a closed image (Theorem 4.1.16). Replace λ by $\overline{\lambda}$ to deduce that the adjoint operator

$$(\lambda\mathbb{1} - A)^* = \overline{\lambda}\mathbb{1} - A^* = \overline{\lambda}\mathbb{1} - A$$

is also injective and so $\lambda\mathbb{1} - A$ has a dense image by part (iv) of Lemma 5.3.9. Hence $\lambda\mathbb{1} - A$ is bijective and this proves (i).

We prove part (ii). It suffices to assume

$$\langle x, Ax\rangle \geq 0 \qquad \text{for all } x \in H. \tag{5.3.11}$$

(Otherwise replace A by $A + a1\!\!1$ for a suitable constant $a > 0$.) Under this assumption we prove that

$$\sigma(A) \subset [0,\infty), \qquad \|A\| = \sup_{\|x\|=1} \langle x, Ax\rangle. \tag{5.3.12}$$

To see this, let $\varepsilon > 0$. Then

$$\varepsilon \|x\|^2 = \langle x, \varepsilon x\rangle \leq \langle x, \varepsilon x + Ax\rangle \leq \|x\| \|\varepsilon x + Ax\|$$

and so $\varepsilon \|x\| \leq \|\varepsilon x + Ax\|$ for all $x \in X$. Hence $\varepsilon 1\!\!1 + A$ is injective and has a closed image by Theorem 4.1.16. Thus $\mathrm{im}(\varepsilon 1\!\!1 + A) = (\ker(\varepsilon 1\!\!1 + A))^\perp = H$ by part (iv) of Lemma 5.3.9, so $\varepsilon 1\!\!1 + A$ is bijective. Hence $-\varepsilon \notin \sigma(A)$. Since the spectrum of A is real by part (i), this proves the first assertion in (5.3.12). Next define

$$a := \sup_{\|x\|=1} \langle x, Ax\rangle.$$

If $x \in H$ satisfies $\|x\| = 1$, then

$$\langle x, Ax\rangle \leq \|x\| \|Ax\| \leq \|A\| \|x\|^2 = \|A\|.$$

Thus $a \leq \|A\|$. To prove the converse inequality observe that, for all $x, y \in H$, we have $\mathrm{Re}\langle x, Ay\rangle = \frac{1}{4}\langle x+y, A(x+y)\rangle - \frac{1}{4}\langle x-y, A(x-y)\rangle$ and hence

$$-\frac{1}{4}\langle x-y, A(x-y)\rangle \leq \mathrm{Re}\langle x, Ay\rangle \leq \frac{1}{4}\langle x+y, A(x+y)\rangle.$$

If $\|x\| = \|y\| = 1$, it follows that

$$\begin{aligned} -a \leq -\frac{a}{4}\|x-y\|^2 &\leq -\frac{1}{4}\langle x-y, A(x-y)\rangle \\ &\leq \mathrm{Re}\langle x, Ay\rangle \leq \frac{1}{4}\langle x+y, A(x+y)\rangle \leq \frac{a}{4}\|x+y\|^2 \leq a. \end{aligned}$$

Thus $|\mathrm{Re}\langle x, Ay\rangle| \leq a$ for all $x, y \in H$ with $\|x\| = \|y\| = 1$ and hence

$$\|A\| = \sup_{\|x\|=\|y\|=1} |\mathrm{Re}\langle x, Ay\rangle| \leq a.$$

This proves (5.3.12). It follows from (5.3.12) and part (ii) of Theorem 5.3.15 that

$$\sup \sigma(A) = \sup_{\lambda\in\sigma(A)} |\lambda| = \|A\| = \sup_{\|x\|=1} \langle x, Ax\rangle$$

for every self-adjoint operator $A = A^* \in \mathcal{L}^c(H)$ that satisfies (5.3.11) and this proves (ii).

Part (iii) follows from (ii) by replacing A with $-A$, part (iv) follows from (ii), (iii), and Theorem 5.3.15, and part (v) follows from (i) and Theorem 5.3.15. This proves Theorem 5.3.16. □

Definition 5.3.17 (Singular value). Let X and Y be complex Hilbert spaces and let $T \in \mathcal{L}^c(X, Y)$. A real number $\lambda \geq 0$ is called a **singular value** of T if $\lambda^2 \in \sigma(T^*T)$.

Thus the singular values of T are the square roots of the (nonnegative) spectral values of the self-adjoint operator $T^*T : X \to X$. Equation (5.3.10) shows that the supremum of the singular values is the norm of T.

Corollary 5.3.18 (Compact Operators). *Let X and Y be complex Hilbert spaces and let $0 \neq K \in \mathcal{L}^c(X, Y)$. Then the following are equivalent.*

(i) *K is compact.*

(ii) *There exist a set $I \subset \mathbb{N}$, either equal to $\mathbb{N}$ or equal to $\{1, \ldots, n\}$ for some $n \in \mathbb{N}$, and orthonormal sequences $(x_i)_{i \in I}$ in X and $(y_i)_{i \in I}$ in Y, and a sequence $(\lambda_i)_{i \in I}$ of positive real numbers, such that $\lim_{i \to \infty} \lambda_i = 0$ in the case $I = \mathbb{N}$ and*

$$Kx = \sum_{i \in I} \lambda_i \langle x_i, x \rangle y_i \qquad \text{for all } x \in X. \tag{5.3.13}$$

Proof. That (ii) implies (i) follows from Theorem 4.2.10. To prove the converse, consider the operator

$$A := K^*K : X \to X.$$

This operator is self-adjoint by Lemma 5.3.9 and is compact by Theorem 4.2.10. Hence $\sigma(K^*K) \setminus \{0\}$ is a discrete subset of the positive real axis $(0, \infty)$ by Theorems 5.2.8 and 5.3.16. Write $\sigma(K^*K) \setminus \{0\} = \{\lambda_i^2 \,|\, i \in I\}$, where $I = \mathbb{N}$ when the spectrum is infinite and $I = \{1, \ldots, n\}$ otherwise, the λ_i are chosen positive, and $\#\{i \in I \,|\, \lambda_i = \lambda\} = \dim \ker(\lambda^2 \mathbb{1} - K^*K)$ for all $\lambda > 0$. Choose an orthonormal sequence $(x_i)_{i \in I}$ in X such that

$$K^*Kx_i = \lambda_i^2 x_i \qquad \text{for all } i \in I$$

and define $y_i := \lambda_i^{-1} K x_i$. Then $\langle y_i, y_j \rangle_Y = (\lambda_i \lambda_j)^{-1} \langle x_i, K^*Kx_j \rangle = \delta_{ij}$ for all $i, j \in I$. Moreover, $K^*Kx = \sum_{i \in I} \lambda_i^2 \langle x_i, x \rangle x_i$ and hence

$$\|Kx\|^2 = \langle x, K^*Kx \rangle = \sum_{i \in I} \lambda_i^2 \left| \langle x_i, x \rangle \right|^2$$

for all $x \in X$. Since $K^*y_i = \lambda_i x_i$ for all $i \in I$, this implies

$$\begin{aligned} \Big\| Kx - \sum_{i \in I} \lambda_i \langle x_i, x \rangle y_i \Big\|^2 &= \|Kx\|^2 + \sum_{i \in I} \lambda_i^2 \left| \langle x_i, x \rangle \right|^2 \\ &\quad - 2 \sum_{i \in I} \lambda_i \mathrm{Re}\big(\langle x_i, x \rangle \langle Kx, y_i \rangle \big) = 0. \end{aligned}$$

Since K is compact, the sequence $(\lambda_i)_{i \in \mathbb{N}}$ converges to zero whenever $I = \mathbb{N}$. This proves Corollary 5.3.18. □

5.4. Functional Calculus for Self-Adjoint Operators

In Subsection 5.2.4 we have introduced the holomorphic functional calculus for general bounded linear operators on complex Banach spaces. In the special case of normal operators on Hilbert spaces this functional calculus extends to arbitrary complex valued continuous functions on the spectrum. The complex valued continuous functions on any compact Hausdorff space Σ form a C* algebra $C(\Sigma)$ as do the bounded complex linear operators on a complex Hilbert space. The continuous functional calculus assigns to every normal operator $A \in \mathcal{L}^c(H)$ on a complex Hilbert space H the unique C* algebra homomorphism $\Phi_A : C(\sigma(A)) \to \mathcal{L}^c(H)$ that satisfies $\Phi_A(\mathrm{id}) = A$. The Spectral Mapping Theorem asserts that the spectrum of the image of a function $f \in C(\sigma(A))$ under this homomorphism is the image of the spectrum under f. We prove this in Subsection 5.4.3 for self-adjoint operators and in Subsection 5.5.3 for normal operators.

5.4.1. C* Algebras. Recall the definition of a complex Banach algebra in Definition 1.5.2.

Definition 5.4.1. (i) A **(unital) C* algebra** is a complex unital Banach algebra $\mathcal{A}$, equipped with an anti-linear involution $\mathcal{A} \to \mathcal{A} : a \mapsto a^*$ that reverses the product and satisfies the **C* identy**, i.e.

$$(ab)^* = b^*a^*, \qquad \mathbb{1}^* = \mathbb{1}, \qquad a^{**} = a, \qquad (\lambda a)^* = \overline{\lambda}a^*, \qquad \|a^*a\| = \|a\|^2$$

for all $a, b \in \mathcal{A}$ and all $\lambda \in \mathbb{C}$, where $a^{**} := (a^*)^*$.

(ii) A C* algebra $\mathcal{A}$ is called **commutative** if $ab = ba$ for all $a, b \in \mathcal{A}$.

(iii) Let $\mathcal{A}$ and $\mathcal{B}$ be unital C* algebras. A **C* algebra homomorphism** is a bounded complex linear operator $\Phi : \mathcal{A} \to \mathcal{B}$ such that

$$\Phi(\mathbb{1}_{\mathcal{A}}) = \mathbb{1}_{\mathcal{B}}, \qquad \Phi(aa') = \Phi(a)\Phi(a'), \qquad \Phi(a^*) = \Phi(a)^*$$

for all $a, a' \in \mathcal{A}$.

Example 5.4.2. Let M be a nonempty compact Hausdorff space. Then the space $C(M) := C(M, \mathbb{C})$ of complex valued continuous functions on M with the supremum norm is a commutative C* algebra. The complex anti-linear involution $C(M) \to C(M) : f \mapsto \overline{f}$ is given by complex conjugation.

Example 5.4.3. Let H be a nonzero complex Hilbert space. Then the space $\mathcal{L}^c(H)$ of bounded complex linear operators $A : H \to H$ with the operator norm is a C* algebra. The complex anti-linear involution is the map $\mathcal{L}^c(H) \to \mathcal{L}^c(H) : A \mapsto A^*$ which assigns to each operator $A \in \mathcal{L}^c(H)$ its adjoint operator A^* (see Definition 5.3.7 and (5.3.10)). The C* algebra $\mathcal{L}^c(H)$ is commutative if and only if H has complex dimension one.

The goal of the present section is to show that, for every self-adjoint operator $A \in \mathcal{L}^c(H)$ on a nonzero complex Hilbert space H, there exists a unique C* algebra homomorphism $\Phi_A : C(\sigma(A)) \to \mathcal{L}^c(H)$ such that $\Phi_A(\mathrm{id}) = A$. This homomorphism is an isometric embedding and its image is the smallest C* algebra $\mathcal{A} \subset \mathcal{L}^c(H)$ that contains A. The first step is the next lemma.

Lemma 5.4.4. *Let H be a nonzero complex Hilbert space and let $A \in \mathcal{L}^c(H)$ be a bounded complex linear operator. For a polynomial $p(z) = \sum_{k=0}^n a_k z^k$ with complex coefficients $a_0, a_1, \dots, a_n \in \mathbb{C}$ define*

$$p(A) := \sum_{k=0}^n a_k A^k \in \mathcal{L}^c(H).$$

Then the following hold for any two polynomials $p, q : \mathbb{C} \to \mathbb{C}$.

(i) $(p+q)(A) = p(A) + q(A)$ *and* $(pq)(A) = p(A)q(A)$.

(ii) $\sigma(p(A)) = p(\sigma(A))$.

(iii) *If A is normal, then so is $p(A)$ and*

$$\|p(A)\| = \sup_{\lambda \in \sigma(A)} |p(\lambda)|. \tag{5.4.1}$$

Proof. Assertion (i) follows directly from the definitions and assertion (ii) follows from parts (iii) and (iv) of Theorem 5.2.12 (see also Exercise 5.8.3). To prove (iii), consider the polynomial

$$q(z) := \sum_{k=0}^n \overline{a}_k z^k$$

and recall that $(A^k)^* = (A^*)^k$ and $(\lambda A)^* = \overline{\lambda} A^*$ for all $k \in \mathbb{N}$ and all $\lambda \in \mathbb{C}$ by Lemma 5.3.9. Hence

$$p(A)^* = \left(\sum_{k=0}^n a_k A^k \right)^* = \sum_{k=0}^n \overline{a}_k (A^*)^k = q(A^*).$$

Now assume A is normal. Then

$$p(A)q(A^*) = q(A^*)p(A)$$

and therefore

$$p(A)^* p(A) = q(A^*)p(A) = p(A)q(A^*) = p(A)p(A)^*.$$

Thus $p(A)$ is normal and so (5.4.1) follows from (ii) and Theorem 5.3.15. This proves Lemma 5.4.4. □

5.4.2. The Stone–Weierstraß Theorem. The second ingredient in the construction of the C* algebra homomorphism from $C(\sigma(A))$ to $\mathcal{L}^c(H)$ is the Stone–Weierstraß Theorem.

Theorem 5.4.5 (Stone–Weierstraß). *Let M be a nonempty compact Hausdorff space and let $\mathcal{A} \subset C(M)$ be a subalgebra of the algebra of complex valued continuous functions on M that satisfies the following axioms.*

(SW1) *Each constant function is an element of $\mathcal{A}$.*

(SW2) *$\mathcal{A}$* **separates points**, *i.e. for all $x, y \in M$ such that $x \neq y$ there exists a function $f \in \mathcal{A}$ such that $f(x) \neq f(y)$.*

(SW3) *If $f \in \mathcal{A}$, then $\overline{f} \in \mathcal{A}$.*

Then $\mathcal{A}$ is dense in $C(M)$.

Proof. The proof is taken from [**18**]. The real subalgebra

$$\mathcal{A}_{\mathbb{R}} := \mathcal{A} \cap C(M, \mathbb{R})$$

contains the real and imaginary parts of every function $f \in \mathcal{A}$ by (SW3). Hence it contains the constant functions by (SW1) and separates points by (SW2). We prove in six steps that $\mathcal{A}_{\mathbb{R}}$ is dense in $C(M, \mathbb{R})$. Then $\mathcal{A}$ is dense in $C(M) = C(M, \mathbb{C})$ by (SW1). Denote the closure of $\mathcal{A}_{\mathbb{R}}$ with respect to the supremum norm by $\overline{\mathcal{A}}_{\mathbb{R}} \subset C(M, \mathbb{R})$.

Step 1. *$\overline{\mathcal{A}}_{\mathbb{R}}$ is a subalgebra of $C(M, \mathbb{R})$ that contains the constant functions and separates points.*

This follows directly from the assumptions.

Step 2. *There exists a sequence of polynomials*

$$P_n : [-1, 1] \to [0, 1]$$

such that

$$\lim_{n \to \infty} P_n(s) = |s| \qquad \text{for all } s \in [-1, 1] \tag{5.4.2}$$

and the convergence is uniform on the interval $[-1, 1]$.

The existence of such a sequence follows from the Weierstraß Approximation Theorem. More explicitly, one can use the ancient **Babylonian method for constructing square roots**. Define a sequence of polynomials $p_n : [0, 1] \to [0, 1]$ with real coefficients by the recursion formula

$$p_0(t) := 0, \qquad p_n(t) := \frac{t + p_{n-1}(t)^2}{2} \qquad \text{for } n \in \mathbb{N}. \tag{5.4.3}$$

Then each p_n is monotonically increasing on the interval $[0,1]$ and

$$
\begin{aligned}
p_{n+1}(t) - p_n(t) &= \frac{p_n(t)^2 - p_{n-1}(t)^2}{2} \\
&= \frac{(p_n(t) - p_{n-1}(t))(p_n(t) + p_{n-1}(t))}{2}
\end{aligned}
\tag{5.4.4}
$$

for each integer $n \geq 2$ and each $t \in [0,1]$. This implies, by induction, that

$$p_{n+1}(t) \geq p_n(t)$$

for all $n \in \mathbb{N}$ and all $t \in [0,1]$. Hence the sequence $(p_n(t))_{n\in\mathbb{N}}$ converges for all $t \in [0,1]$ and it follows from the recursion formula (5.4.3) that the limit $r(t) := \lim_{n\to\infty} p_n(t) \in [0,1]$ satisfies the equation $2r(t) = t + r(t)^2$ and therefore $(1 - r(t))^2 = 1 - t$. Thus

$$\lim_{n\to\infty} \bigl(1 - p_n(t)\bigr) = \sqrt{1-t} \qquad \text{for all } t \in [0,1]. \tag{5.4.5}$$

The formula (5.4.4) also shows that the polynomial $p_{n+1} - p_n : [0,1] \to [0,1]$ is nondecreasing for all $n \in \mathbb{N}$. Hence $p_{n+1}(t) - p_n(t) \leq p_{n+1}(1) - p_n(1)$ and thus $p_m(t) - p_n(t) \leq p_m(1) - p_n(1)$ for all $m > n$ and all $t \in [0,1]$. Take the limit $m \to \infty$ to obtain

$$0 \leq 1 - p_n(t) - \sqrt{1-t} \leq 1 - p_n(1) \qquad \text{for all } n \in \mathbb{N} \text{ and all } t \in [0,1].$$

This shows that the convergence in (5.4.5) is uniform on $[0,1]$. Hence

$$\lim_{n\to\infty} \bigl(1 - p_n(1-s^2)\bigr) = \sqrt{s^2} = |s| \qquad \text{for all } s \in [-1,1] \tag{5.4.6}$$

and the convergence is uniform on the interval $[-1,1]$. This proves Step 2.

Step 3. *If $f \in \overline{\mathcal{A}}_{\mathbb{R}}$, then $|f| \in \overline{\mathcal{A}}_{\mathbb{R}}$.*

Let $f \in \overline{\mathcal{A}}_{\mathbb{R}} \setminus \{0\}$ and $\varepsilon > 0$. Then the function

$$h := \frac{f}{\|f\|} \in \overline{\mathcal{A}}_{\mathbb{R}}$$

takes values in the interval $[-1,1]$. Moreover, by Step 2, there exists a polynomial $P : [-1,1] \to [0,1]$ with real coefficients such that

$$\sup_{|s|\leq 1} \Bigl| |s| - P(s) \Bigr| < \frac{\varepsilon}{\|f\|}.$$

This implies

$$\Bigl\| |f| - \|f\|\, P \circ h \Bigr\| = \|f\| \sup_{x\in M} \Bigl| |h(x)| - P(h(x)) \Bigr| < \varepsilon.$$

Since $\|f\|\, P \circ h \in \overline{\mathcal{A}}_{\mathbb{R}}$, this proves Step 3.

Step 4. *If $f, g \in \overline{\mathcal{A}}_{\mathbb{R}}$, then*

$$\max\{f,g\} \in \overline{\mathcal{A}}_{\mathbb{R}}, \qquad \min\{f,g\} \in \overline{\mathcal{A}}_{\mathbb{R}}.$$

This follows from Step 3 and the equations

$$\max\{f,g\} = \frac{1}{2}\Big(f + g + |f-g|\Big),$$
$$\min\{f,g\} = \frac{1}{2}\Big(f + g - |f-g|\Big).$$

Step 5. *If $f \in C(M,\mathbb{R})$ and $x, y \in M$, then there exists an element $g \in \overline{\mathcal{A}}_{\mathbb{R}}$ such that*

$$g(x) = f(x), \qquad g(y) = f(y).$$

This follows from the fact that $\overline{\mathcal{A}}_{\mathbb{R}}$ contains the constant functions and separates points. Namely, choose $h \in \overline{\mathcal{A}}_{\mathbb{R}}$ with

$$h(x) \neq h(y)$$

and define $g \in \overline{\mathcal{A}}_{\mathbb{R}}$ by

$$g(z) := \frac{h(z)-h(y)}{h(x)-h(y)} f(x) + \frac{h(z)-h(x)}{h(y)-h(x)} f(y)$$

for $z \in M$.

Step 6. $\overline{\mathcal{A}}_{\mathbb{R}} = C(M,\mathbb{R})$.

Let $f \in C(M,\mathbb{R})$. By Step 5 and the axiom of choice, there exists a collection of functions $g_{x,y} \in \overline{\mathcal{A}}_{\mathbb{R}}$, one for each pair $x, y \in M$, such that

$$g_{x,y}(x) = f(x), \qquad g_{x,y}(y) = f(y)$$

for all $x, y \in M$. Let $\varepsilon > 0$ and, for $x, y \in M$, define

$$\begin{aligned} U_{x,y} &:= \{z \in M \mid g_{x,y}(z) > f(z) - \varepsilon\}, \\ V_{x,y} &:= \{z \in M \mid g_{x,y}(z) < f(z) + \varepsilon\}. \end{aligned} \tag{5.4.7}$$

These sets are open and

$$\{x,y\} \subset U_{x,y} \cap V_{x,y}$$

for all $x, y \in M$. Fix an element $y \in M$. Then $\{U_{x,y}\}_{x\in M}$ is an open cover of M. Since M is compact, there are finitely many elements $x_1, \dots, x_m \in M$ such that

$$M = \bigcup_{i=1}^{m} U_{x_i,y}.$$

For $y, z \in M$ define

$$g_y(z) := \max_{i=1,\dots,m} g_{x_i,y}(z), \qquad V_y := V_{x_1,y} \cap V_{x_2,y} \cap \cdots \cap V_{x_m,y}.$$

Then $g_y \in \overline{\mathcal{A}}_{\mathbb{R}}$ by Step 4 and V_y is an open neighborhood of y by definition. Moreover, for every $z \in M$, there exists an index $i \in \{1, \dots, m\}$ such that $z \in U_{x_i,y}$ and so

$$g_y(z) \geq g_{x_i,y}(z) > f(z) - \varepsilon$$

by (5.4.7). Also, if $z \in V_y$, then $z \in V_{x_i,y}$ and hence $g_{x_i,y}(z) < f(z) + \varepsilon$ for all $i \in \{1, \dots, m\}$ by (5.4.7), and therefore

$$g_y(z) < f(z) + \varepsilon.$$

To sum up, we have proved that

$$\begin{aligned} g_y(z) &> f(z) - \varepsilon \qquad \text{for all } z \in M, \\ g_y(z) &< f(z) + \varepsilon \qquad \text{for all } z \in V_y. \end{aligned} \tag{5.4.8}$$

Since $\{V_y\}_{y \in M}$ is an open cover of M, there exist elements $y_1, \dots, y_n \in M$ such that $M = \bigcup_{j=1}^n V_{y_j}$. Define the function $g : M \to \mathbb{R}$ by

$$g(z) := \min_{j=1,\dots,n} g_{y_j}(z)$$

for $z \in M$. Then $g \in \overline{\mathcal{A}}_{\mathbb{R}}$ by Step 4 and it follows from (5.4.8) that

$$f(z) - \varepsilon < g(z) < f(z) + \varepsilon$$

for all $z \in M$. This shows that for every $\varepsilon > 0$ there exists a $g \in \overline{\mathcal{A}}_{\mathbb{R}}$ such that $\|f - g\| < \varepsilon$. Thus $f \in \overline{\mathcal{A}}_{\mathbb{R}}$ for all $f \in C(M, \mathbb{R})$. This proves Step 6 and Theorem 5.4.5. □

Example 5.4.6 (Hardy Space). The hypothesis (SW3) cannot be removed in Theorem 5.4.5. Let $M = S^1 \subset \mathbb{C}$ be the unit circle and define

$$\mathcal{H} := \left\{ f : S^1 \to \mathbb{C} \,\middle|\, \begin{array}{l} f \text{ is continuous and} \\ \int_0^1 e^{2\pi \mathbf{i} kt} f(e^{2\pi \mathbf{i} t})\, dt = 0 \text{ for all } k \in \mathbb{N} \end{array} \right\}.$$

This is the **Hardy space**. A continuous function $f : S^1 \to \mathbb{C}$ belongs to $\mathcal{H}$ if and only if its Fourier expansion has the form

$$f(e^{2\pi \mathbf{i} t}) = \sum_{k=0}^{\infty} a_k e^{2\pi \mathbf{i} kt} \qquad \text{for } t \in \mathbb{R},$$

where

$$a_k := \int_0^1 e^{-2\pi \mathbf{i} kt} f(e^{2\pi \mathbf{i} t})\, dt$$

for $k \in \mathbb{N}_0$. This means that f extends to a continuous function $u : \mathbb{D} \to \mathbb{C}$ on the closed unit disc $\mathbb{D} \subset \mathbb{C}$ that is holomorphic in the interior of $\mathbb{D}$. The Hardy space $\mathcal{H}$ contains the constant functions and separates points because it contains the identity map on S^1. However, it is not invariant under complex conjugation and the only real valued functions in $\mathcal{H}$ are the constant ones. Thus $\mathcal{H}$ is not dense in $C(S^1)$.

5.4.3. Continuous Functional Calculus.

Theorem 5.4.7 (Continuous Functional Calculus). *Let H be a nonzero complex Hilbert space and let $A : H \to H$ be a bounded self-adjoint complex linear operator. Denote its spectrum by $\Sigma := \sigma(A) \subset \mathbb{R}$. Then there exists a bounded complex linear operator*

$$C(\Sigma) \to \mathcal{L}^c(H) : f \mapsto f(A) \tag{5.4.9}$$

that satisfies the following axioms.

(Product) $1(A) = \mathbb{1}$ *and* $(fg)(A) = f(A)g(A)$ *for all* $f, g \in C(\Sigma)$.

(Conjugation) $\overline{f}(A) = f(A)^*$ *for all* $f \in C(\Sigma)$.

(Normalization) *If* $f(\lambda) = \lambda$ *for all* $\lambda \in \Sigma$, *then* $f(A) = A$.

(Isometry) $\|f(A)\| = \sup_{\lambda\in\Sigma}|f(\lambda)| =: \|f\|$ *for all* $f \in C(\Sigma)$.

(Commutative) *If* $B \in \mathcal{L}^c(H)$ *satisfies* $AB = BA$, *then* $f(A)B = Bf(A)$ *for all* $f \in C(\Sigma)$.

(Image) *The image* $\mathcal{A} := \{f(A) \,|\, f \in C(\Sigma)\}$ *of the linear operator* (5.4.9) *is the smallest* C^* *subalgebra of* $\mathcal{L}^c(H)$ *that contains the operator* A.

(Eigenvector) *If* $\lambda \in \Sigma$ *and* $x \in H$ *satisfy* $Ax = \lambda x$, *then* $f(A)x = f(\lambda)x$ *for all* $f \in C(\Sigma)$.

(Spectrum) $f(A)$ *is normal and* $\sigma(f(A)) = f(\sigma(A))$ *for all* $f \in C(\Sigma)$.

(Composition) *If* $f \in C(\Sigma, \mathbb{R})$ *and* $g \in C(f(\Sigma))$, *then* $(g \circ f)(A) = g(f(A))$.

The bounded complex linear operator (5.4.9) *is uniquely determined by the (Product) and (Normalization) axioms. The (Product) and (Conjugation) axioms assert that* (5.4.9) *is a* C^* *algebra homomorphism.*

Proof. See page 241. □

The (Eigenvector) and (Spectrum) axioms in Theorem 5.4.7 are called the **Spectral Mapping Theorem**. Theorem 5.4.7 carries over verbatim to normal operators, with the caveat that $\Sigma = \sigma(A)$ is then an arbitrary nonempty compact subset of the complex plane (see Theorem 5.5.14 below). One approach is to replace polynomials in one real variable by polynomials p in z and $\bar{z}$ and show that $\sigma(p(A)) = p(\sigma(A))$ for every such polynomial. In the simple case $p(z) = z + \bar{z}$ this is the identity $\sigma(A + A^*) = \{\lambda + \overline{\lambda} \,|\, \lambda \in \sigma(A)\}$ and to verify this already requires some effort (see Exercise 5.8.2). Once the formula $\sigma(p(A)) = p(\sigma(A))$ has been established for all polynomials in z and $\bar{z}$ the proof proceeds essentially as in the self-adjoint case. Another approach via Gelfand representations is explained in Section 5.5.

Proof of Theorem 5.4.7. Denote the space of polynomials in one real variable with complex coefficients by

$$\mathbb{C}[t] := \left\{ p : \mathbb{R} \to \mathbb{C} \;\middle|\; \begin{array}{l} \text{there exist an } n \in \mathbb{N} \text{ and complex} \\ \text{numbers } a_0, a_1, \dots, a_n \text{ such that} \\ p(t) = \sum_{k=0}^{n} a_k t^k \text{ for all } t \in \mathbb{R} \end{array} \right\}.$$

Thus a polynomial $p \in \mathbb{C}[t]$ is thought of as a continuous function from $\mathbb{R}$ to $\mathbb{C}$ for the purpose of this proof. Since A is self-adjoint, its spectrum

$$\Sigma = \sigma(A)$$

is a nonempty compact subset of the real axis by Theorem 5.3.16. Define the subalgebra $\mathcal{P}(\Sigma) \subset C(\Sigma)$ by

$$\mathcal{P}(\Sigma) := \{p|_\Sigma \,|\, p \in \mathbb{C}[t]\} \subset C(\Sigma).$$

This subalgebra contains the constant functions, is invariant under conjugation, and separates points because it contains the identity map on Σ. Hence $\mathcal{P}(\Sigma)$ is dense in $C(\Sigma)$ by the Stone–Weierstraß Theorem 5.4.5. With this understood, the proof has five steps.

Step 1. *There exists a unique bounded complex linear operator*

$$\Phi_A : C(\Sigma) \to \mathcal{L}^c(H)$$

such that

$$\Phi_A(p|_\Sigma) = p(A)$$

for all $p \in \mathbb{C}[t]$.

The map $\mathbb{C}[t] \to \mathcal{P}(\Sigma) : p \mapsto p|_\Sigma$ need not be injective. Its kernel

$$\mathcal{I}(\Sigma) := \{p \in \mathbb{C}[t] \,|\, p|_\Sigma = 0\}$$

is an ideal in $\mathbb{C}[t]$, which is nontrivial if and only if Σ is a finite set. The algebra homomorphism $\mathbb{C}[t] \to \mathcal{P}(\Sigma) : p \mapsto p|_\Sigma$ descends to an algebra isomorphism $\mathbb{C}[t]/\mathcal{I}(\Sigma) \to \mathcal{P}(\Sigma)$. Given a polynomial

$$p = \sum_{k=0}^{n} a_k t^k$$

with complex coefficients consider the bounded complex linear operator

$$p(A) := \sum_{k=0}^{n} a_k A^k \in \mathcal{L}^c(H).$$

This operator is normal and $\sigma(p(A)) = p(\sigma(A))$ by Lemma 5.4.4. Hence

$$\|p(A)\| = \sup_{\mu \in \sigma(p(A))} |\mu| = \sup_{\lambda \in \sigma(A)} |p(\lambda)| = \|p|_\Sigma\| \tag{5.4.10}$$

by Theorem 5.3.15.

Equation (5.4.10) shows that the kernel of the complex linear operator

$$\mathbb{C}[t] \to \mathcal{L}^c(H) : p \mapsto p(A)$$

agrees with the kernel $\mathcal{I}(\Sigma)$ of the surjective complex linear operator

$$\mathbb{C}[t] \to \mathcal{P}(\Sigma) : p \mapsto p|_\Sigma.$$

Hence there is a unique map

$$\Phi_A : \mathcal{P}(\Sigma) \to \mathcal{L}^c(H)$$

such that

$$\Phi_A(p|_\Sigma) = p(A) \qquad \text{for all } p \in \mathbb{C}[t]. \tag{5.4.11}$$

In other words, if $p, q \in \mathbb{C}[t]$ are two polynomials such that $p(\lambda) = q(\lambda)$ for all $\lambda \in \Sigma$, then $\|p(A) - q(A)\| = \|p|_\Sigma - q|_\Sigma\| = 0$ by (5.4.10) and so $p(A) = q(A)$. Thus the operator $p(A) \in \mathcal{L}^c(H)$ depends only on the restriction of p to Σ, and this shows that there is a unique map $\Phi_A : \mathcal{P}(\Sigma) \to \mathcal{L}^c(H)$ that satisfies (5.4.11). Equation (5.4.11) asserts that the following diagram commutes:

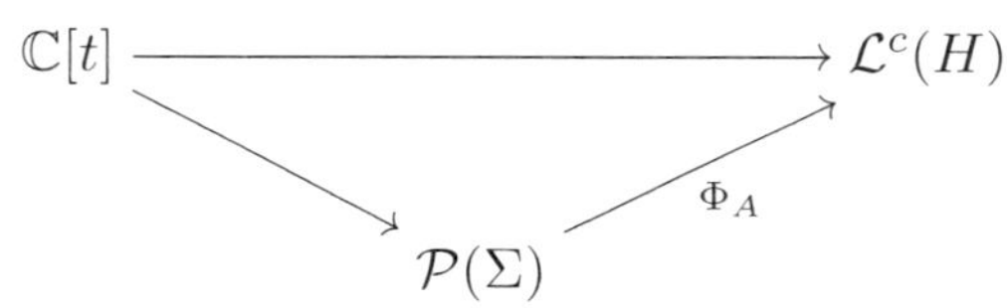

The operator $\Phi_A : \mathcal{P}(\Sigma) \to \mathcal{L}^c(H)$ is complex linear by definition and is an isometric embedding by (5.4.10). Since $\mathcal{P}(\Sigma)$ is a dense subspace of $C(\Sigma)$, it extends uniquely to an isometric embedding of $C(\Sigma)$ into $\mathcal{L}^c(H)$, still denoted by Φ_A. More precisely, fix a continuous function $f : \Sigma \to \mathbb{C}$. By the Stone–Weierstraß Theorem 5.4.5 there exists a sequence of polynomials $p_n \in \mathbb{C}[t]$ such that the sequence $p_n|_\Sigma$ converges uniformly to f. Then $p_n(A) \in \mathcal{L}^c(H)$ is a Cauchy sequence by (5.4.10). Since $\mathcal{L}^c(H)$ is complete by Theorem 1.3.1 the sequence $p_n(A)$ converges. Denote the limit by

$$\Phi_A(f) := \lim_{n\to\infty} p_n(A).$$

It is independent of the choice of the sequence of polynomials $p_n \in \mathbb{C}[t]$ used to define it. Namely, let $q_n \in \mathbb{C}[t]$ be another sequence of polynomials such that $q_n|_\Sigma$ converges uniformly to f. Then $p_n|_\Sigma - q_n|_\Sigma$ converges uniformly to zero, hence

$$\lim_{n\to\infty} \|p_n(A) - q_n(A)\| = \lim_{n\to\infty} \|p_n|_\Sigma - q_n|_\Sigma\| = 0$$

by (5.4.10), and so

$$\lim_{n\to\infty} p_n(A) = \lim_{n\to\infty} q_n(A).$$

This proves Step 1.

Step 2. *The map $\Phi_A : C(\Sigma) \to \mathcal{L}^c(H)$ in Step 1 satisfies the (Product), (Conjugation), (Normalization), (Isometry), (Commutative), (Image), and (Eigenvector) axioms.*

The map satisfies the (Normalization) and (Isometry) axioms by its definition in Step 1. To prove the (Product) axiom, let $f, g \in C(\Sigma)$ and choose two sequences of polynomials $p_n, q_n \in \mathbb{C}[t]$ such that $p_n|_\Sigma$ converges uniformly to f and $q_n|_\Sigma$ converges uniformly to g as n tends to infinity. Then $p_n q_n|_\Sigma$ converges uniformly to fg as n tends to infinity and hence

$$\Phi_A(fg) = \lim_{n\to\infty} \Phi_A(p_n q_n) = \lim_{n\to\infty} \Phi_A(p_n)\Phi_A(q_n) = \Phi_A(f)\Phi_A(g).$$

Likewise $\overline{p}_n$ converges uniformly to $\overline{f}$ and hence

$$\Phi_A(\overline{f}) = \lim_{n\to\infty} \Phi_A(\overline{p}_n) = \lim_{n\to\infty} \Phi_A(p_n)^* = \Phi_A(f)^*.$$

This proves the (Conjugation) axiom. The (Commutative) and (Eigenvector) axioms hold for all functions in $\mathcal{P}(\Sigma)$ by definition and hence the same approximation argument as above shows that they hold for all $f \in C(\Sigma)$.

To prove the (Image) axiom, denote by $\mathcal{A} \subset \mathcal{L}^c(H)$ the smallest C* subalgebra containing A. Then $\Phi_A(\mathcal{P}(\Sigma)) \subset \mathcal{A}$ because $\mathcal{A}$ is a C* subalgebra containing A. Moreover, $C(\Sigma)$ is the closure of $\mathcal{P}(\Sigma)$ and so $\Phi_A(C(\Sigma)) \subset \mathcal{A}$ because $\mathcal{A}$ is closed. Conversely, $\mathcal{A} \subset \Phi_A(C(\Sigma))$ because $\Phi_A(C(\Sigma))$ is a C* subalgebra of $\mathcal{L}^c(H)$ that contains A. This proves Step 2.

Step 3. *The map Φ_A in Step 1 satisfies the (Spectrum) axiom.*

Fix a continuous function $f : \Sigma \to \mathbb{C}$. Then

$$f(A)^* f(A) = \overline{f}(A) f(A) = |f|^2(A) = f(A)\overline{f}(A) = f(A) f(A)^*$$

by the (Product) and (Conjugation) axioms and hence $f(A)$ is normal. To prove the assertion about the spectrum we first show that

$$\sigma(f(A)) \subset f(\Sigma).$$

To see this, let $\mu \in \mathbb{C} \setminus f(\Sigma)$ and define the function $g : \Sigma \to \mathbb{C}$ by

$$g(\lambda) := \frac{1}{\mu - f(\lambda)}$$

for $\lambda \in \Sigma$. This function is continuous and satisfies

$$g(\mu - f) = (\mu - f)g = 1.$$

Hence

$$g(A)(\mu\mathbb{1} - f(A)) = (\mu\mathbb{1} - f(A))g(A) = \mathbb{1}$$

by the (Product) axiom. Thus the operator $\mu\mathbb{1} - f(A)$ is bijective and this shows that $\mu \notin \sigma(f(A))$.

To prove the converse inclusion $f(\Sigma) \subset \sigma(f(A))$, fix a spectral value $\lambda \in \Sigma = \sigma(A)$ and define $\mu := f(\lambda)$. We must prove that $\mu \in \sigma(f(A))$. Suppose, by contradiction, that $\mu \notin \sigma(f(A))$. Then the operator $\mu\mathbb{1} - f(A)$ is bijective. Choose a sequence $p_n \in \mathbb{C}[t]$ such that the sequence $p_n|_\Sigma$ converges uniformly to f. Then the sequence of operators $p_n(\lambda)\mathbb{1} - p_n(A)$ converges to $\mu\mathbb{1} - f(A)$ in the norm topology. Hence the operator $p_n(\lambda)\mathbb{1} - p_n(A)$ is bijective for n sufficiently large by the Open Mapping Theorem 2.2.1 and Corollary 1.5.7. Hence $p_n(\lambda) \notin \sigma(p_n(A))$ for large n, contradicting part (ii) of Lemma 5.4.4. This proves Step 3.

Step 4. *The map Φ_A in Step 1 satisfies the (Composition) axiom.*

Let $f \in C(\Sigma, \mathbb{R})$ and let $g \in C(f(\Sigma))$. Assume first that

$$g = q|_{f(\Sigma)}$$

for a polynomial $q : \mathbb{R} \to \mathbb{C}$. Choose a sequence of polynomials $p_n : \mathbb{R} \to \mathbb{R}$ with real coefficients such that $p_n|_\Sigma$ converges uniformly to f. Then $q \circ p_n|_\Sigma$ converges uniformly to $q \circ f$ and

$$(q \circ p_n|_\Sigma)(A) = q(p_n(A))$$

for all $n \in \mathbb{N}$. Hence

$$(q \circ f)(A) = \lim_{n\to\infty} (q \circ p_n)(A) = \lim_{n\to\infty} q(p_n(A)) = q(f(A)).$$

Here the last step follows from the definition of $q(B)$ for $B \in \mathcal{L}^c(H)$ and the fact that $p_n(A)$ converges to $f(A)$ in the norm topology as n tends to infinity.

Now let $g : f(\Sigma) \to \mathbb{C}$ be any continuous function and choose a sequence of polynomials $q_n : \mathbb{R} \to \mathbb{C}$ such that the sequence $q_n|_{f(\Sigma)}$ converges uniformly to g as n tends to infinity. Then $q_n \circ f$ converges uniformly to $g \circ f$ as n tends to infinity and

$$(q_n \circ f)(A) = q_n(f(A))$$

for all $n \in \mathbb{N}$ by what we have proved above. Hence

$$(g \circ f)(A) = \lim_{n\to\infty} (q_n \circ f)(A) = \lim_{n\to\infty} q_n(f(A)) = g(f(A)).$$

This proves Step 4.

Step 5. *The map Φ_A in Step 1 is uniquely determined by the (Product) and (Normalization) axioms.*

Let $\Psi : C(\Sigma) \to \mathcal{L}^c(H)$ be any bounded complex linear operator that satisfies the (Product) and (Normalization) axioms. Then $\Psi(f) = \Phi_A(f)$ for all $f \in \mathcal{P}(\Sigma)$. Since $\mathcal{P}(\Sigma)$ is dense in $C(\Sigma)$ it follows from the continuity of Ψ and Φ_A that $\Psi(f) = \Phi_A(f)$ for all $f \in C(\Sigma)$. This proves Step 5 and Theorem 5.4.7. □

Definition 5.4.8 (Positive Semidefinite Operator). Let H be a complex Hilbert space. A self-adjoint operator $A = A^* \in \mathcal{L}^c(H)$ is called **positive semidefinite** if $\langle x, Ax \rangle \geq 0$ for all $x \in H$. The notation $A \geq 0$ or $A = A^* \geq 0$ signifies that A is a positive semidefinite self-adjoint operator.

Corollary 5.4.9 (Square Root). *Let H be a complex Hilbert space, let $A = A^* \in \mathcal{L}^c(H)$ be a self-adjoint operator, and let $f \in C(\sigma(A))$. Then the following hold.*

(i) *$f(A) = f(A)^*$ if and only if $f(\sigma(A)) \subset \mathbb{R}$.*

(ii) *Assume $f(\sigma(A)) \subset \mathbb{R}$. Then $f(A) \geq 0$ if and only if $f \geq 0$.*

(iii) *Assume $A \geq 0$. Then there exists a unique positive semidefinite self-adjoint operator $B = B^* \in \mathcal{L}^c(H)$ such that $B^2 = A$.*

Proof. Assume without loss of generality that $H \neq \{0\}$.

We prove part (i). Since $f(A) - f(A)^* = (f - \overline{f})(A) = 2\mathbf{i}(\mathrm{Im} f)(A)$ by the (Conjugation) axiom, we have $\|f(A) - f(A)^*\| = 2\sup_{\lambda \in \sigma(A)} |\mathrm{Im} f(\lambda)|$ by the (Isometry) axiom. This proves (i).

We prove part (ii). Thus assume $f(\sigma(A)) \subset \mathbb{R}$. Then it follows from Theorem 5.3.16 and Theorem 5.4.7 that $\inf_{\|x\|=1} \langle x, f(A)x \rangle = \inf_{\lambda \in \sigma(A)} f(\lambda)$. This proves (ii).

We prove existence in (iii). Since A is positive semidefinite we have $\sigma(A) \subset [0, \infty)$ by Theorem 5.3.16. Define $f : \sigma(A) \to [0, \infty)$ by $f(\lambda) := \sqrt{\lambda}$ for $\lambda \in \sigma(A)$. Then the operator $B := f(A) \in \mathcal{L}^c(H)$ is self-adjoint by part (i), is positive semidefinite by part (ii), and $B^2 = f(A)^2 = f^2(A) = \mathrm{id}(A) = A$ by the (Product) and (Normalization) axioms. This proves existence.

We prove uniqueness in (iii). Assume that $C \in \mathcal{L}^c(H)$ is any positive semidefinite self-adjoint operator such that $C^2 = A$. Then $CA = C^3 = AC$ and hence it follows from the (Commutative) axiom that $CB = BC$. This implies $(B + C)(B - C) = B^2 - C^2 = 0$ and hence

$$\begin{aligned} 0 &= \langle Bx - Cx, (B + C)(Bx - Cx) \rangle \\ &= \langle Bx - Cx, B(Bx - Cx) \rangle + \langle Bx - Cx, C(Bx - Cx) \rangle \end{aligned}$$

for all $x \in H$. Since both summands on the right are nonnegative, we have

$$\langle Bx - Cx, B(Bx - Cx) \rangle = \langle Bx - Cx, C(Bx - Cx) \rangle = 0$$

for all $x \in H$. Hence $\langle x, (B - C)^3 x \rangle = 0$ for all $x \in H$. Since $(B - C)^3$ is self-adjoint, it follows from Theorem 5.3.16 that

$$0 = \|(B - C)^3\| = \|B - C\|^3.$$

Here the last equation follows from part (i) of Theorem 5.3.15. Thus $C = B$ and this proves Corollary 5.4.9. □

5.5. Gelfand Spectrum and Normal Operators

This section extends the continuous functional calculus for self-adjoint operators, developed in Section 5.4, to normal operators, following the elegant approach of Schwartz [**78**, pp. 155–161] and Yosida [**88**, pp. 294–309].

5.5.1. The Gelfand Representation. Recall the definition of a complex commutative unital Banach algebra as a complex Banach space $\mathcal{A}$, equipped with an associative and commutative bilinear map $\mathcal{A} \times \mathcal{A} \to \mathcal{A} : (a, b) \mapsto ab$ that satisfies the inequality

$$\|ab\| \leq \|a\|\|b\|$$

for all $a, b \in \mathcal{A}$ and a nonzero element $\mathbb{1} \in \mathcal{A}$ that satisfies

$$\mathbb{1}a = a\mathbb{1} = a$$

for all $a \in \mathcal{A}$ (Definition 1.5.2).

Definition 5.5.1 (Ideal). Let $\mathcal{A}$ be a complex commutative unital Banach algebra such that

$$\|\mathbb{1}\| = 1.$$

An **ideal in** $\mathcal{A}$ is a complex linear subspace $\mathcal{J} \subset \mathcal{A}$ such that

$$a \in \mathcal{A}, \quad b \in \mathcal{J} \qquad \Longrightarrow \qquad ab \in \mathcal{J}.$$

An ideal $\mathcal{J} \subset \mathcal{A}$ is called **nontrivial** if $\mathcal{J} \neq \mathcal{A}$. It is called **maximal** if it is nontrivial and if it is not contained in any other nontrivial ideal. The set

$$\mathrm{Spec}(\mathcal{A}) := \{\mathcal{J} \subset \mathcal{A} \,|\, \mathcal{J} \text{ is a maximal ideal}\}$$

is called the **Gelfand spectrum of** $\mathcal{A}$. The **Jacobson radical of** $\mathcal{A}$ is the ideal

$$\mathcal{R}(\mathcal{A}) := \bigcap_{\mathcal{J} \in \mathrm{Spec}(\mathcal{A})} \mathcal{J}.$$

The Banach algebra $\mathcal{A}$ is called **semisimple** if $\mathcal{R} = \{0\}$. The **spectrum of an element** $a \in \mathcal{A}$ is the set

$$\sigma(a) := \{\lambda \in \mathbb{C} \,|\, \lambda\mathbb{1} - a \text{ is not invertible}\}.$$

If M is a nonempty compact Hausdorff space, then the space $\mathcal{A} := C(M)$ of continuous complex valued functions on M is a complex commutative unital Banach algebra, the spectrum of an element $f \in C(M)$ is its image $\sigma(f) = f(M)$, every maximal ideal has the form $\mathcal{J} = \{f \in C(M) \,|\, f(p) = 0\}$ for some element $p \in M$, and so the set $\mathrm{Spec}(\mathcal{A})$ can be naturally identified with M. The only maximal ideal in $\mathcal{A} := \mathbb{C}$ is $\mathcal{J} = \{0\}$. In these examples the quotient algebra $\mathcal{A}/\mathcal{J}$ is isomorphic to $\mathbb{C}$ for every maximal ideal $\mathcal{J} \subset \mathcal{A}$. The next theorem shows that this continues to hold in general.

Theorem 5.5.2 (Maximal Ideals). *Let $\mathcal{A}$ be a complex commutative unital Banach algebra such that $\|\mathbb{1}\| = 1$. Then the following hold.*

(i) *Every nontrivial ideal in $\mathcal{A}$ is contained in a maximal ideal.*

(ii) *An element $a \in \mathcal{A}$ is invertible if and only if it is not contained in any maximal ideal.*

(iii) *Every maximal ideal is a closed linear subspace of $\mathcal{A}$.*

(iv) *$\sigma(a) \neq \emptyset$ for all $a \in \mathcal{A}$.*

(v) *If $\mathcal{J} \subset \mathcal{A}$ is a maximal ideal, then $\mathcal{A}/\mathcal{J}$ is isomorphic to $\mathbb{C}$ and*

$$\inf_{a \in \mathcal{J}} \|\lambda\mathbb{1} - a\| = |\lambda| \tag{5.5.1}$$

for all $\lambda \in \mathbb{C}$.

Proof. We prove (i). The set

$$\mathscr{J} := \{\mathcal{J} \subset \mathcal{A} \,|\, \mathcal{J} \text{ is an ideal and } \mathcal{J} \subsetneq \mathcal{A}\}$$

of nontrivial ideals is nonempty because $\{0\} \in \mathscr{J}$ and is partially ordered by inclusion. If

$$\mathscr{C} \subset \mathscr{J}$$

is a nonempty chain, then the set

$$\mathcal{J} := \bigcup_{\mathcal{I} \in \mathscr{C}} \mathcal{I} \subset \mathcal{A}$$

is an ideal, and $\mathcal{J} \neq \mathcal{A}$ because otherwise there would exist an element $\mathcal{I} \in \mathscr{C}$ containing $\mathbb{1}$, in contradiction to the fact that $\mathcal{I} \subsetneq \mathcal{A}$. Thus $\mathcal{J} \in \mathscr{J}$ and so every nonempty chain in $\mathscr{J}$ has a supremum. Hence part (i) follows from the Lemma of Zorn.

We prove (ii). Let $a_0 \in \mathcal{A}$ and define

$$\mathcal{J}_0 := \{aa_0 \,|\, a \in \mathcal{A}\}.$$

Then $\mathcal{J}_0$ is an ideal and every ideal $\mathcal{J} \subset \mathcal{A}$ that contains a_0 also contains $\mathcal{J}_0$. If a_0 is invertible, then $\mathcal{J}_0 = \mathcal{A}$ and so a_0 is not contained in any maximal ideal. If a_0 is not invertible, then $\mathcal{J}_0$ is a nontrivial ideal and hence there exists a maximal ideal $\mathcal{J}$ containing $\mathcal{J}_0$ by part (i). This proves part (ii).

We prove (iii). The group $\mathcal{G} \subset \mathcal{A}$ of invertible elements is an open subset of $\mathcal{A}$ by Theorem 1.5.5. Let $\mathcal{J} \subset \mathcal{A}$ be a maximal ideal and denote by $\overline{\mathcal{J}}$ the closure of $\mathcal{J}$. Then $\mathcal{J} \cap \mathcal{G} = \emptyset$ by part (ii) and hence

$$\overline{\mathcal{J}} \cap \mathcal{G} = \emptyset$$

because $\mathcal{G}$ is open. Hence $\overline{\mathcal{J}}$ is a nontrivial ideal and so $\mathcal{J} = \overline{\mathcal{J}}$ because $\mathcal{J}$ is maximal. This proves part (iii).

We prove (iv). Fix an element $a \in \mathcal{A}$ and assume, by contradiction, that $\sigma(a) = \emptyset$. In particular, a is invertible and, by Corollary 2.3.5, there exists a bounded complex linear functional $\Lambda : \mathcal{A} \to \mathbb{C}$ such that $\Lambda(a^{-1}) = 1$. Since $\lambda\mathbb{1} - a$ is invertible for all $\lambda \in \mathbb{C}$, the same argument as in the proof of Lemma 5.2.6 shows that the map $\mathbb{C} \to \mathcal{A} : \lambda \mapsto (\lambda\mathbb{1} - a)^{-1}$ is holomorphic. Moreover, by part (iii) of Theorem 1.5.5,

$$\|(\lambda\mathbb{1} - a)^{-1}\| \le \frac{1}{|\lambda| - \|a\|}$$

for all $\lambda \in \mathbb{C}$ with $|\lambda| > \|a\|$. Hence the function

$$\mathbb{C} \to \mathbb{C} : \lambda \mapsto f(\lambda) := \Lambda((\lambda\mathbb{1} - a)^{-1})$$

is holomorphic and bounded. Thus it is constant by Liouville's Theorem, and this is impossible because $\lim_{|\lambda|\to\infty}|f(\lambda)| = 0$ and $f(0) = 1$. This contradiction proves part (iv).

We prove (v). Let $\mathcal{J} \subset \mathcal{A}$ be a maximal ideal and consider the quotient space $\mathcal{B} := \mathcal{A}/\mathcal{J}$ with the norm

$$\|[a]_{\mathcal{J}}\| := \inf_{b\in\mathcal{J}} \|a + b\| \qquad \text{for } [a]_{\mathcal{J}} := a + \mathcal{J} \in \mathcal{A}/\mathcal{J}.$$

By part (iii) and Theorem 1.2.14 this is a Banach space and, since J is an ideal, the product in $\mathcal{A}$ descends to the quotient. It satisfies the inequalities $\|[ab]_{\mathcal{J}}\| \le \|[a]_{\mathcal{J}}\|\|[b_{\mathcal{J}}]\|$ for all $a, b \in \mathcal{A}$ and $\|[\mathbb{1}]_{\mathcal{J}}\| \le \|\mathbb{1}\| = 1$ by definition. Moreover $\|[\mathbb{1}]_{\mathcal{J}}\| = 1$, because otherwise there would exist an element $a \in \mathcal{J}$ such that $\|\mathbb{1} - a\| < 1$, so a would be invertible by Theorem 1.5.5, in contradiction to part (ii). This shows that $\mathcal{B}$ is a complex commutative unital Banach algebra whose unit $[\mathbb{1}]_{\mathcal{J}}$ has norm one. Thus

$$|\lambda| = \|[\lambda\mathbb{1}]_{\mathcal{J}}\| = \inf_{a\in\mathcal{J}} \|\lambda\mathbb{1} - a\|$$

for all $\lambda \in \mathbb{C}$ and this proves (5.5.1).

Next we observe that every nonzero element $[a]_{\mathcal{J}} \in \mathcal{B} = \mathcal{A}/\mathcal{J}$ is invertible in $\mathcal{B}$. To see this, let $a \in \mathcal{A} \setminus \mathcal{J}$. Then the set

$$\mathcal{J}_a := \{ab + c \,|\, b \in \mathcal{A},\, c \in \mathcal{J}\}$$

is an ideal such that $\mathcal{J} \subsetneq \mathcal{J}_a$ and so $\mathcal{J}_a = \mathcal{A}$. Thus there exists a $b \in \mathcal{A}$ such that $ab - \mathbb{1} \in \mathcal{J}$. Hence $[a]_{\mathcal{J}}$ is invertible in $\mathcal{B}$ and $[a]_{\mathcal{J}}^{-1} = [b]_{\mathcal{J}}$.

Now the **Gelfand–Mazur Theorem** asserts that every complex commutative unital Banach algebra $\mathcal{B}$ in which every nonzero element is invertible and whose unit has norm one is isometrically isomorphic to $\mathbb{C}$. To prove it, fix an element $b \in \mathcal{B}$. Then $\sigma(b) \ne \emptyset$ by part (iv). Choose an element $\lambda \in \sigma(b)$. Then $\lambda\mathbb{1} - b$ is not invertible and so $b = \lambda\mathbb{1}$. Hence the map $\mathbb{C} \to \mathcal{B} : \lambda \mapsto \lambda\mathbb{1}$ is an isometric isomorphism of Banach algebras. This proves the Gelfand–Mazur Theorem, part (v), and Theorem 5.5.2. □

Definition 5.5.3 (Gelfand Representation). Let $\mathcal{A}$ be a complex commutative unital Banach algebra such that $\|\mathbb{1}\| = 1$. By Theorem 5.5.2 there exists a unique function

$$\mathcal{A} \times \mathrm{Spec}(\mathcal{A}) \to \mathbb{C} : (a, \mathcal{J}) \mapsto f_a(\mathcal{J}) \tag{5.5.2}$$

such that

$$f_a(\mathcal{J})\mathbb{1} - a \in \mathcal{J} \qquad \text{for all } a \in \mathcal{A} \text{ and all } \mathcal{J} \in \mathrm{Spec}(\mathcal{A}). \tag{5.5.3}$$

The map $a \mapsto f_a$ is called the **Gelfand representation** or the **Gelfand transform**. It assigns to each element $a \in \mathcal{A}$ a function $f_a : \mathrm{Spec}(\mathcal{A}) \to \mathbb{C}$. The **Gelfand topology** on $\mathrm{Spec}(\mathcal{A})$ is the weakest topology such that f_a is continuous for every element $a \in \mathcal{A}$.

To understand the Gelfand topology on $\mathrm{Spec}(\mathcal{A})$ it will be convenient to change the point of view by fixing a maximal ideal $\mathcal{J} \in \mathrm{Spec}(\mathcal{A})$ and considering the function $\mathcal{A} \to \mathbb{C} : a \mapsto f_a(\mathcal{J})$. Lemma 5.5.6 below shows that this construction gives rise to a one-to-one correspondence between maximal ideals and unital algebra homomorphisms $\Lambda : \mathcal{A} \to \mathbb{C}$.

Definition 5.5.4. Let $\mathcal{A}$ be a complex commutative unital Banach algebra such that $\|\mathbb{1}\| = 1$. A map $\Lambda : \mathcal{A} \to \mathbb{C}$ is called a **unital algebra homomorphism** if it satisfies the conditions

$$\Lambda(a+b) = \Lambda(a) + \Lambda(b), \qquad \Lambda(ab) = \Lambda(a)\Lambda(b), \qquad \Lambda(z\mathbb{1}) = z$$

for all $a, b \in \mathcal{A}$ and all $z \in \mathbb{C}$. Define

$$\widehat{\mathcal{A}} := \left\{ \Lambda : \mathcal{A} \to \mathbb{C} \;\middle|\; \begin{array}{l} \Lambda \text{ is a unital} \\ \text{algebra homomorphism} \end{array} \right\}.$$

The next two lemmas show that every unital algebra homomorphism is a bounded linear functional of norm one. Hence $\widehat{\mathcal{A}}$ is a subset of the unit sphere in the complex dual space $\mathcal{A}^* = \mathcal{L}^c(\mathcal{A}, \mathbb{C})$.

Lemma 5.5.5. *Let $\mathcal{A}$ be a complex commutative unital Banach algebra with $\|\mathbb{1}\| = 1$, let $\mathcal{J} \in \mathrm{Spec}(\mathcal{A})$, and define the map $\Lambda_{\mathcal{J}} : \mathcal{A} \to \mathbb{C}$ by*

$$\Lambda_{\mathcal{J}}(a) := f_a(\mathcal{J}) \qquad \text{for } a \in \mathcal{A}. \tag{5.5.4}$$

Then the following hold.

(i) *$\Lambda_{\mathcal{J}}$ is a unital algebra homomorphism.*

(ii) $\ker(\Lambda_{\mathcal{J}}) = \mathcal{J}$.

(iii) *$\Lambda_{\mathcal{J}}$ is a bounded linear functional of norm one, i.e.*

$$|\Lambda_{\mathcal{J}}(a)| = |f_a(\mathcal{J})| \le \|a\| \tag{5.5.5}$$

for all $a \in \mathcal{A}$ and equality holds for $a = \mathbb{1}$.

Proof. We prove (i). Let $a, b \in \mathcal{A}$ and define $\lambda := f_a(\mathcal{J})$ and $\mu := f_b(\mathcal{J})$. Then $\lambda\mathbb{1} - a \in \mathcal{J}$ and $\mu\mathbb{1} - b \in \mathcal{J}$ and hence

$$(\lambda + \mu)\mathbb{1} - (a + b) = (\lambda\mathbb{1} - a) + (\mu\mathbb{1} - b) \in \mathcal{J}$$

and

$$\lambda\mu\mathbb{1} - ab = (\lambda\mathbb{1} - a)b + \lambda(\mu\mathbb{1} - b) \in \mathcal{J}.$$

Thus $f_{a+b}(\mathcal{J}) = \lambda + \mu$ and $f_{ab}(\mathcal{J}) = \lambda\mu$. Since $f_{\mathbb{1}}(\mathcal{J}) = 1$, this proves (i).

We prove (ii). Let $a \in \mathcal{A}$. Then we have $\Lambda_{\mathcal{J}}(a) = f_a(\mathcal{J}) = 0$ if and only if $a \in \mathcal{J}$, by definition of the map f_a in (5.5.4). Hence $\ker \Lambda_{\mathcal{J}} = \mathcal{J}$ and this proves (ii).

We prove (iii). Observe that

$$|\Lambda_{\mathcal{J}}(a)| = |f_a(\mathcal{J})| = \inf_{b \in \mathcal{J}} \|f_a(\mathcal{J})\mathbb{1} - b\| = \inf_{b \in \mathcal{J}} \|a - b\| \leq \|a\|$$

for all $a \in \mathcal{A}$ and all $\mathcal{J} \in \mathrm{Spec}(\mathcal{A})$. Here the second equality follows from (5.5.1) and the third from the fact that $f_a(\mathcal{J})\mathbb{1} - a \in \mathcal{J}$. This proves equation (5.5.5) and thus $\|\Lambda_{\mathcal{J}}\| \leq 1$. Since $\Lambda_{\mathcal{J}}(\mathbb{1}) = 1$, we have $\|\Lambda_{\mathcal{J}}\| = 1$. This proves (iii) and Lemma 5.5.5. □

Lemma 5.5.6. *Let $\mathcal{A}$ be a complex commutative unital Banach algebra with $\|\mathbb{1}\| = 1$. Then* $\mathrm{Spec}(\mathcal{A})$ *is a compact Hausdorff space with the Gelfand topology, $\widehat{\mathcal{A}}$ is a weak* compact subset of the dual space $\mathcal{A}^*$, and the map*

$$\mathrm{Spec}(\mathcal{A}) \to \widehat{\mathcal{A}} : \mathcal{J} \mapsto \Lambda_{\mathcal{J}} \tag{5.5.6}$$

defined by (5.5.4) *is a homeomorphism.*

Proof. We prove that the map (5.5.6) is bijective. Let $\Lambda \in \widehat{\mathcal{A}}$ and define $\mathcal{J} := \ker(\Lambda)$. Then $\mathcal{J}$ is a linear subspace of $\mathcal{A}$. Moreover, if $a \in \mathcal{A}$ and $b \in \mathcal{J}$, then we have $\Lambda(ab) = \Lambda(a)\Lambda(b) = 0$ and so $ab \in \mathcal{J}$. Thus $\mathcal{J}$ is an ideal of codimension one and hence is a maximal ideal. Now let $a \in \mathcal{A}$ and define $\lambda := f_a(\mathcal{J})$. Then $\lambda\mathbb{1} - a \in \mathcal{J} = \ker(\Lambda)$ and hence

$$\Lambda(a) = \Lambda(\lambda\mathbb{1}) = \lambda \cdot \Lambda(\mathbb{1}) = \lambda = f_a(\mathcal{J}).$$

Thus $\Lambda = \Lambda_{\mathcal{J}}$ and so the map (5.5.6) is surjective. To prove that it is injective, fix two distinct maximal ideals $\mathcal{I}, \mathcal{J} \in \mathrm{Spec}(\mathcal{A})$ and choose an element $a \in \mathcal{I} \setminus \mathcal{J}$. Then $\Lambda_{\mathcal{I}}(a) = 0$ and $\Lambda_{\mathcal{J}}(a) \neq 0$, and so $\Lambda_{\mathcal{I}} \neq \Lambda_{\mathcal{J}}$.

Since the map (5.5.6) is bijective, it follows from part (iii) of Lemma 5.5.5 that $\widehat{\mathcal{A}}$ is contained in the unit sphere of the complex dual space $\mathcal{A}^*$. Moreover, $\widehat{\mathcal{A}}$ is a weak* closed subset of $\mathcal{A}^*$ by definition of a unital algebra homomorphism. Hence $\widehat{\mathcal{A}}$ is a weak* compact subset of $\mathcal{A}^*$ by Theorem 3.2.5. Now the Gelfand topology on $\mathrm{Spec}(\mathcal{A})$ is, by definition, induced by the weak* topology on $\mathcal{A}^*$ under the inclusion $\mathrm{Spec}(\mathcal{A}) \cong \widehat{\mathcal{A}} \subset \mathcal{A}^*$. Thus the map (5.5.6) is a homeomorphism and this proves Lemma 5.5.6. □

Denote by $C(\mathrm{Spec}(\mathcal{A}))$ the space of complex valued continuous functions on the compact Hausdorff space $\mathrm{Spec}(\mathcal{A})$ equipped with the Gelfand topology of Definition 5.5.3. Then $C(\mathrm{Spec}(\mathcal{A}))$ is a unital Banach algebra with the supremum norm and the unit (the constant function one) has norm one. The next theorem shows that the Gelfand representation

$$\mathcal{A} \to C(\mathrm{Spec}(\mathcal{A})) : a \mapsto f_a \tag{5.5.7}$$

defined by (5.5.3) is a unital algebra homomorphism.

Theorem 5.5.7 (Gelfand). *Let $\mathcal{A}$ be a complex commutative unital Banach algebra such that $\|\mathbb{1}\| = 1$. Then the following hold.*

(i) *The Gelfand representation* (5.5.7) *is a unital algebra homomorphism and a bounded complex linear operator of norm one.*

(ii) *Every $a \in \mathcal{A}$ satisfies*

$$\sigma(a) = f_a(\mathrm{Spec}(\mathcal{A})) \tag{5.5.8}$$

and

$$\lim_{n\to\infty} \|a^n\|^{1/n} = \inf_{n\in\mathbb{N}} \|a^n\|^{1/n} = \|f_a\|. \tag{5.5.9}$$

(iii) *The kernel of the Gelfand representation* (5.5.7) *is the Jacobson radical*

$$\mathcal{R}(\mathcal{A}) = \bigcap_{\mathcal{J}\in\mathrm{Spec}(\mathcal{A})} \mathcal{J} = \{a \in \mathcal{A} \,|\, f_a = 0\}. \tag{5.5.10}$$

(iv) *The image of the Gelfand representation* (5.5.7) *is a subalgebra of the space $C(\mathrm{Spec}(\mathcal{A}))$ that separates points and contains the constant functions.*

(v) *The Gelfand representation* (5.5.7) *is an isometric embedding if and only if*

$$\|a^2\| = \|a\|^2$$

for all $a \in \mathcal{A}$.

Proof. We prove part (i). That (5.5.7) is a unital algebra homomorphism follows from part (i) of Lemma 5.5.5 and that it is a bounded linear operator of norm one follows from part (iii) of Lemma 5.5.5. This proves (i).

We prove part (ii). Fix an element $a \in \mathcal{A}$ and a complex number λ. If $\lambda \in \sigma(a)$, then $\lambda\mathbb{1} - a$ is not invertible, hence part (ii) of Theorem 5.5.2 asserts that there exists a maximal ideal $\mathcal{J}$ such that $\lambda\mathbb{1} - a \in \mathcal{J}$, and hence $f_a(\mathcal{J}) = \lambda$. Conversely, suppose that $\lambda = f_a(\mathcal{J})$ for some maximal ideal $\mathcal{J}$. Then $\lambda\mathbb{1} - a \in \mathcal{J}$ by definition of f_a, hence $\lambda\mathbb{1} - a$ is not invertible by part (ii) of Theorem 5.5.2, and hence $\lambda \in \sigma(a)$. This proves (5.5.8).

To prove (5.5.9), recall that

$$r := \lim_{n\to\infty} \|a^n\|^{1/n} = \inf_{n\in\mathbb{N}} \|a^n\|^{1/n}$$

by Theorem 1.5.5. Now the proof of Theorem 5.2.7 carries over verbatim to complex unital Banach algebras with $\|\mathbb{1}\| = 1$. Hence, by (5.5.8),

$$r = \sup_{\lambda\in\sigma(a)} |\lambda| = \sup_{\mathcal{J}\in\mathrm{Spec}(\mathcal{A})} |f_a(\mathcal{J})| = \|f_a\|$$

and this proves (5.5.9) and part (ii).

Part (iii) follows from the fact that an element $a \in \mathcal{A}$ satisfies $f_a = 0$ if and only if $a \in \mathcal{J}$ for all $\mathcal{J} \in \mathrm{Spec}(\mathcal{A})$.

Part (iv) follows from the fact that $f_{\mathbb{1}}(\mathcal{J}) = 1$ for all $\mathcal{J} \in \mathrm{Spec}(\mathcal{A})$ and that the map $\mathrm{Spec}(\mathcal{A}) \to \widehat{\mathcal{A}} : \mathcal{J} \mapsto \Lambda_{\mathcal{J}}$ in Lemma 5.5.6 is injective. Namely, if $\mathcal{I}, \mathcal{J} \in \mathrm{Spec}(\mathcal{A})$ are two distinct maximal ideals, then $\Lambda_{\mathcal{I}} \neq \Lambda_{\mathcal{J}}$, and hence there exists an element $a \in \mathcal{A}$ such that

$$f_a(\mathcal{I}) = \Lambda_{\mathcal{I}}(a) \neq \Lambda_{\mathcal{J}}(a) = f_a(\mathcal{J}).$$

This proves (iv).

We prove part (v). If the Gelfand representation

$$\mathcal{A} \to C(\mathrm{Spec}(\mathcal{A})) : a \mapsto f_a$$

is an isometric embedding, then

$$\|a\| = \|f_a\| = \inf_{n\in\mathbb{N}} \|a^n\|^{1/n} \qquad \text{for all } a \in \mathcal{A}$$

by (5.5.9) and hence

$$\|a^n\| = \|a\|^n \qquad \text{for all } a \in \mathcal{A} \text{ and all } n \in \mathbb{N}.$$

Conversely, suppose that

$$\|a^2\| = \|a\|^2 \qquad \text{for all } a \in \mathcal{A}.$$

Then one shows as in the proof of Theorem 5.3.15 that $\|a^n\| = \|a\|^n$ for all $a \in \mathcal{A}$ and all $n \in \mathbb{N}$. Hence $\|f_a\| = \|a\|$ for all $a \in \mathcal{A}$ and so the Gelfand representation is an isometric embedding. This proves part (v) and Theorem 5.5.7. □

In view of Theorem 5.5.7 it is a natural question to ask under which conditions the Gelfand representation (5.5.7) is an isometric isomorphism of commutative unital Banach algebras. For C* algebras (Definition 5.4.1) the next theorem gives an affirmative answer to this question.

Theorem 5.5.8 (Gelfand). *Let $\mathcal{A}$ be a commutative C^* algebra, so that*

$$\|a^*a\| = \|a\|^2 \qquad \text{for all } a \in \mathcal{A}. \tag{5.5.11}$$

Then $\|1\!\!1\| = 1$ and the Gelfand representation $\mathcal{A} \to C(\mathrm{Spec}(\mathcal{A})) : a \mapsto f_a$ in (5.5.7) is an isometric C^* algebra isomorphism. In particular,

$$f_{a^*} = \overline{f}_a \qquad \text{for all } a \in \mathcal{A}.$$

Proof. See page 254. □

Lemma 5.5.9. *Let $\mathcal{A}$ be a commutative C^* algebra $\|1\!\!1\| = 1$. Then the following are equivalent.*

(i) *Every maximal ideal is invariant under the involution $\mathcal{A} \to \mathcal{A} : a \mapsto a^*$.*

(ii) *If $a \in \mathcal{A}$ satisfies $a = a^*$, then $f_a(\mathcal{J}) \in \mathbb{R}$ for all $\mathcal{J} \in \mathrm{Spec}(\mathcal{A})$.*

(iii) *$f_{a^*} = \overline{f}_a$ for all $a \in \mathcal{A}$.*

Proof. We prove that (i) implies (ii). Fix an element $a = a^* \in \mathcal{A}$ and a maximal ideal $\mathcal{J} \in \mathrm{Spec}(\mathcal{A})$ and define

$$\lambda := f_a(\mathcal{J}).$$

Then $\lambda 1\!\!1 - a \in \mathcal{J}$ and hence

$$\overline{\lambda}1\!\!1 - a = \overline{\lambda}1\!\!1 - a^* = (\lambda 1\!\!1 - a)^* \in \mathcal{J}$$

by part (i). This implies $\lambda = \overline{\lambda} \in \mathbb{R}$. Thus (ii) holds.

We prove that (ii) implies (iii). Let $a \in \mathcal{A}$ and define $b, c \in \mathcal{A}$ by

$$b := \frac{1}{2}(a + a^*), \qquad c := \frac{1}{2\mathbf{i}}(a - a^*).$$

Then $b = b^*$ and $c = c^*$ and $a = b + \mathbf{i}c$ and $a^* = b - \mathbf{i}c$. Hence f_b and f_c are real valued functions on $\mathrm{Spec}(\mathcal{A})$ by part (ii) and therefore

$$f_{a^*} = f_b - \mathbf{i}f_c = \overline{f_b + \mathbf{i}f_c} = \overline{f}_a.$$

Thus (iii) holds.

We prove that (iii) implies (i). Fix a maximal ideal $\mathcal{J} \subset \mathcal{A}$ and define the function $\Lambda : \mathcal{A} \to \mathbb{C}$ by

$$\Lambda(a) := f_a(\mathcal{J})$$

for $a \in \mathcal{A}$. By part (iii) it satisfies $\Lambda(a^*) = \overline{\Lambda(a)}$ for all $a \in \mathcal{A}$. Since

$$\ker(\Lambda) = \mathcal{J}$$

this shows that $\mathcal{J}$ is invariant under the involution $a \mapsto a^*$. Thus (i) holds. This proves Lemma 5.5.9. □

Proof of Theorem 5.5.8. Let $\mathcal{A}$ be a commutative C* algebra. Following Schwartz [**78**, pp. 159–161], we prove in four steps that the Gelfand representation is an isometric C* algebra isomorphism.

Step 1. *$\|f_a\| = \|a\|$ for all $a \in \mathcal{A}$. In particular, $\|\mathbb{1}\| = 1$.*

By (5.5.11), every $a \in \mathcal{A}$ satisfies

$$\left\|a^2\right\|^2 = \left\|(a^2)^* a^2\right\| = \|(a^*a)^*(a^*a)\| = \|a^*a\|^2 = \|a\|^4$$

and so $\|a^2\| = \|a\|^2$. Hence Step 1 follows from part (v) of Theorem 5.5.7.

Step 2. *$f_{e^{\mathbf{i}a}}(\mathcal{J}) = e^{\mathbf{i}f_a(\mathcal{J})}$ for all $a \in \mathcal{A}$ and all $\mathcal{J} \in \mathrm{Spec}(\mathcal{A})$.*

This follows directly from the fact that the Gelfand representation is a continuous homomorphism of complex Banach algebras.

Step 3. *If $a \in \mathcal{A}$ satisfies $a = a^*$, then $f_a(\mathcal{J}) \in \mathbb{R}$ for all $\mathcal{J} \in \mathrm{Spec}(\mathcal{A})$.*

Let $a \in \mathcal{A}$ such that $a = a^*$. Then

$$(e^{\mathbf{i}a})^* e^{\mathbf{i}a} = e^{-\mathbf{i}a^*} e^{\mathbf{i}a} = e^{\mathbf{i}(a-a^*)} = \mathbb{1}$$

and hence

$$\|e^{\mathbf{i}a}\|^2 = \|(e^{\mathbf{i}a})^* e^{\mathbf{i}a}\| = 1$$

by (5.5.11) and Step 1. Thus

$$|f_{e^{\mathbf{i}a}}(\mathcal{J})| \le \|e^{\mathbf{i}a}\| = 1$$

and, likewise, $|f_{e^{-\mathbf{i}a}}(\mathcal{J})| \le 1$ for all $\mathcal{J} \in \mathrm{Spec}(\mathcal{A})$. Hence

$$1 = |f_{\mathbb{1}}(\mathcal{J})| = |f_{e^{\mathbf{i}a}}(\mathcal{J}) f_{e^{-\mathbf{i}a}}(\mathcal{J})| = |f_{e^{\mathbf{i}a}}(\mathcal{J})|\,|f_{e^{-\mathbf{i}a}}(\mathcal{J})| \le 1$$

and therefore, by Step 2,

$$|e^{\mathbf{i}f_a(\mathcal{J})}| = |f_{e^{\mathbf{i}a}}(\mathcal{J})| = 1 \qquad \text{for all } \mathcal{J} \in \mathrm{Spec}(\mathcal{A}).$$

Hence $f_a(\mathcal{J}) \in \mathbb{R}$ for all $\mathcal{J} \in \mathrm{Spec}(\mathcal{A})$. This proves Step 3.

Step 4. *The Gelfand representation is an isometric C* algebra isomorphism.*

By Step 1 and part (v) of Theorem 5.5.7, the Gelfand representation is an isometric embedding and, by Step 3 and Lemma 5.5.9, it is a C* algebra homomorphism. We must prove that it is surjective. To see this, define $\mathscr{F}_{\mathcal{A}} := \{f_a \,|\, a \in \mathcal{A}\}$. This is a closed subspace of $C(\mathrm{Spec}(\mathcal{A}))$ by Step 1, it is a subalgebra of $C(\mathrm{Spec}(\mathcal{A}))$ that contains the constant functions and separates points by part (iv) of Theorem 5.5.7, and it is invariant under complex conjugation by Step 3 and Lemma 5.5.9. Hence the set $\mathscr{F}_{\mathcal{A}}$ satisfies the requirements of the Stone–Weierstraß Theorem 5.4.5 and therefore is dense in $C(\mathrm{Spec}(\mathcal{A}))$. Thus $\mathscr{F}_{\mathcal{A}} = C(\mathrm{Spec}(\mathcal{A}))$. This proves Step 4 and Theorem 5.5.8. □

5.5.2. C* Algebras of Normal Operators. The construction of the continuous functional calculus for normal operators is based on several lemmas. Assume throughout that H is a nonzero complex Hilbert space and that $A_0 \in \mathcal{L}^c(H)$ is a normal operator. Let $\mathcal{A}_0 \subset \mathcal{L}^c(H)$ be the smallest (unital) C* subalgebra that contains A_0.

Lemma 5.5.10. *$\mathcal{A}_0$ is commutative and every operator $A \in \mathcal{A}_0$ is normal. Moreover, if $B \in \mathcal{L}^c(H)$ satisfies $BA_0 = A_0B$ and $BA_0^* = A_0^*B$, then B commutes with every element of $\mathcal{A}_0$.*

Proof. Define

$$\mathcal{B} := \{B \in \mathcal{L}^c(H) \,|\, A_0B = BA_0 \text{ and } BA_0^* = A_0^*B\}.$$

Then $\mathcal{B}$ is a closed subspace of $\mathcal{L}^c(H)$ that contains the identity and is invariant under composition. Moreover, $A_0 \in \mathcal{B}$ because A_0 and A_0^* commute, and $B \in \mathcal{B}$ implies $B^* \in \mathcal{B}$. Hence $\mathcal{B}$ is a C* subalgebra of $\mathcal{L}^c(H)$ that contains A_0. Hence the set

$$\mathcal{C} := \{C \in \mathcal{L}^c(H) \,|\, BC = CB \text{ for all } B \in \mathcal{B}\}$$

is also a C* subalgebra of $\mathcal{L}^c(H)$ that contains A_0. Moreover, since A_0 and A_0^* are elements of $\mathcal{B}$, we have $\mathcal{C} \subset \mathcal{B}$. Hence $\mathcal{C}$ is commutative, and therefore every element $C \in \mathcal{C}$ is normal. Since $\mathcal{C}$ is a C* subalgebra of $\mathcal{L}^c(H)$ and $A_0 \in \mathcal{C}$, we have $\mathcal{A}_0 \subset \mathcal{C}$ and this proves Lemma 5.5.10. □

Lemma 5.5.11. *Let* $\mathrm{Spec}(\mathcal{A}_0)$ *be the set of maximal ideals in $\mathcal{A}_0$. Then, for each $A \in \mathcal{A}_0$, there is a unique function $f_A : \mathrm{Spec}(\mathcal{A}_0) \to \mathbb{C}$ such that*

$$f_A(\mathcal{J})\mathbb{1} - A \in \mathcal{J} \tag{5.5.12}$$

for all $\mathcal{J} \in \mathrm{Spec}(\mathcal{A}_0)$. Equip $\mathrm{Spec}(\mathcal{A}_0)$ *with the weakest topology such that f_A is continuous for every $A \in \mathcal{A}_0$. Then* $\mathrm{Spec}(\mathcal{A}_0)$ *is a compact Hausdorff space, the Gelfand representation*

$$\mathcal{A}_0 \to C(\mathrm{Spec}(\mathcal{A}_0)) : A \mapsto f_A \tag{5.5.13}$$

is an isometric C algebra isomorphism, and*

$$f_A(\mathrm{Spec}(\mathcal{A}_0)) = \sigma(A) \qquad \text{for all } A \in \mathcal{A}_0. \tag{5.5.14}$$

Proof. Existence and uniqueness of the f_A follow from Theorem 5.5.2, the topology on $\mathrm{Spec}(\mathcal{A}_0)$ is compact and Hausdorff by Lemma 5.5.6, the map (5.5.13) is a unital algebra homomorphism by part (i) of Theorem 5.5.7, and (5.5.14) holds by part (ii) of Theorem 5.5.7. By Theorem 5.3.16, we have

$$\|A^*A\| = \sup_{\|x\|=1} \langle x, A^*Ax\rangle = \sup_{\|x\|=1} \|Ax\|^2 = \|A\|^2$$

for all $A \in \mathcal{A}_0$. Hence the Gelfand representation (5.5.13) is an isometric C* algebra isomorphism by Theorem 5.5.8. This proves Lemma 5.5.11. □

Lemma 5.5.12. *Let $A \in \mathcal{A}_0$. Then*

$$A = A^* \quad \Longleftrightarrow \quad f_A(\mathcal{J}) \in \mathbb{R} \text{ for all } \mathcal{J} \in \operatorname{Spec}(\mathcal{A}_0) \tag{5.5.15}$$

and

$$A = A^* \geq 0 \quad \Longleftrightarrow \quad f_A(\mathcal{J}) \geq 0 \text{ for all } \mathcal{J} \in \operatorname{Spec}(\mathcal{A}_0). \tag{5.5.16}$$

Proof. If $A \in \mathcal{A}_0$ is self-adjoint, then

$$\overline{f_A(\mathcal{J})} = f_{A^*}(J) = f_A(\mathcal{J})$$

by Lemma 5.5.11 and so $f_A(\mathcal{J}) \in \mathbb{R}$ for all $\mathcal{J} \in \operatorname{Spec}(\mathcal{A}_0)$. Alternatively, it follows from (5.5.14) and Theorem 5.3.16 that $f_A(\operatorname{Spec}(\mathcal{A}_0)) = \sigma(A) \subset \mathbb{R}$.

Conversely, let $A \in \mathcal{A}_0$ such that $f_A(\operatorname{Spec}(\mathcal{A}_0)) \subset \mathbb{R}$. Then

$$f_{A-A^*}(\mathcal{J}) = f_A(\mathcal{J}) - f_{A^*}(\mathcal{J}) = f_A(\mathcal{J}) - \overline{f_A(\mathcal{J})} = 0$$

for all $\mathcal{J} \in \operatorname{Spec}(\mathcal{A}_0)$ and so

$$\|A - A^*\| = \|f_{A-A^*}\| = 0$$

by Lemma 5.5.11. Hence $A = A^*$ and this proves (5.5.15).

To prove (5.5.16), fix an element $A \in \mathcal{A}_0$. If A is self-adjoint and positive semidefinite, then

$$f_A(\operatorname{Spec}(\mathcal{A}_0)) = \sigma(A) \subset [0, \infty)$$

by (5.5.14) and Theorem 5.3.16. Conversely, assume $f_A(\operatorname{Spec}(\mathcal{A}_0)) \subset [0, \infty)$. Then A is self-adjoint by (5.5.15) and $\sigma(A) \subset [0, \infty)$ by (5.5.14). Hence A is positive semidefinite by Theorem 5.3.16. This proves Lemma 5.5.12. □

Lemma 5.5.13. *The function $f_{A_0} : \operatorname{Spec}(\mathcal{A}_0) \to \sigma(A_0)$ is a homeomorphism.*

Proof. By (5.5.14) we have $f_{A_0}(\operatorname{Spec}(\mathcal{A}_0)) = \sigma(A_0)$. We prove that f_{A_0} is injective. Assume, by contradiction, that there exist two distinct maximal ideals $\mathcal{I}, \mathcal{J} \in \operatorname{Spec}(\mathcal{A}_0)$ such that $f_{A_0}(\mathcal{I}) = f_{A_0}(\mathcal{J}) =: \lambda$. Then $\lambda \in \sigma(A_0)$ and $\lambda\mathbb{1} - A_0 \in \mathcal{I} \cap \mathcal{J}$. Define $\mathcal{A}_1 := \{z\mathbb{1} + A \,|\, z \in \mathbb{C},\, A \in \mathcal{I} \cap \mathcal{J}\}$. This set is a proper C* subalgebra of $\mathcal{A}_0$ that contains A_0, in contradiction to the definition of $\mathcal{A}_0$. This contradiction shows that the map

$$f_{A_0} : \operatorname{Spec}(\mathcal{A}_0) \to \sigma(A_0)$$

is bijective. Since f_{A_0} is continuous, its domain is compact, and its target space is Hausdorff, it is a homeomorphism. This proves Lemma 5.5.13. □

5.5.3. Functional Calculus for Normal Operators. With these preparations in place we are ready to establish the continuous functional calculus for normal operators on Hilbert spaces.

Theorem 5.5.14 **(Continuous Functional Calculus).** *Let H be a nonzero complex Hilbert space, let*

$$A \in \mathcal{L}^c(H)$$

be a bounded normal operator, and let

$$\Sigma := \sigma(A) \subset \mathbb{C}$$

be the spectrum of A. Then there exists a bounded complex linear operator

$$C(\Sigma) \to \mathcal{L}^c(H) : f \mapsto f(A) \tag{5.5.17}$$

that satisfies the following axioms.

(Product) $1(A) = \mathbb{1}$ *and* $(fg)(A) = f(A)g(A)$ *for all* $f, g \in C(\Sigma)$.

(Conjugation) $\overline{f}(A) = f(A)^*$ *for all* $f \in C(\Sigma)$.

(Positive) *Let* $f \in C(\Sigma, \mathbb{R})$. *Then* $f \geq 0$ *if and only if* $f(A) = f(A)^* \geq 0$.

(Normalization) *If* $f(\lambda) = \lambda$ *for all* $\lambda \in \Sigma$, *then* $f(A) = A$.

(Isometry) $\|f(A)\| = \sup_{\lambda \in \Sigma} |f(\lambda)| =: \|f\|$ *for all* $f \in C(\Sigma)$.

(Commutative) *If* $B \in \mathcal{L}^c(H)$ *satisfies* $AB = BA$ *and* $A^*B = BA^*$, *then*

$$f(A)B = Bf(A) \qquad \text{for all } f \in C(\Sigma).$$

(Image) *The image*

$$\mathcal{A} := \{f(A) \mid f \in C(\Sigma)\}$$

of the linear operator (5.5.17) *is the smallest* C^* *subalgebra of* $\mathcal{L}^c(H)$ *that contains the operator* A.

(Eigenvector) *If* $\lambda \in \Sigma$ *and* $x \in H$ *satisfy* $Ax = \lambda x$, *then*

$$f(A)x = f(\lambda)x \qquad \text{for all } f \in C(\Sigma).$$

(Spectrum) *For every* $f \in C(\Sigma)$ *the operator* $f(A)$ *is normal and*

$$\sigma(f(A)) = f(\sigma(A)).$$

(Composition) *If* $f \in C(\Sigma)$ *and* $g \in C(f(\Sigma))$, *then* $(g \circ f)(A) = g(f(A))$.

The bounded complex linear operator (5.5.17) *is uniquely determined by the (Product), (Conjugation), and (Normalization) axioms. The (Product) and (Conjugation) axioms assert that* (5.5.17) *is a* C^* *algebra homomorphism.*

Proof. Fix a normal operator $A_0 \in \mathcal{L}^c(H)$ and denote by

$$\mathcal{A}_0 \subset \mathcal{L}^c(H)$$

the smallest C* subalgebra that contains A_0, as in Subsection 5.5.2. Denote the spectrum of the operator A_0 by

$$\Sigma_0 := \sigma(A_0) \subset \mathbb{C}.$$

Then the Gelfand representation

$$\mathcal{A}_0 \to C(\mathrm{Spec}(\mathcal{A}_0)) : A \mapsto f_A, \tag{5.5.18}$$

introduced in Definition 5.5.3, is an isometric C* algebra isomorphism by Lemma 5.5.11. Moreover, the map

$$f_{A_0} : \mathrm{Spec}(\mathcal{A}_0) \to \Sigma_0$$

is a homeomorphism by Lemma 5.5.13. These two observations give rise to an isometric C* algebra isomorphism

$$C(\Sigma_0) \to \mathcal{A}_0 : f \mapsto f(A_0), \tag{5.5.19}$$

defined as the composition of the isometric C* algebra isomorphism

$$C(\Sigma_0) \to C(\mathrm{Spec}(\mathcal{A}_0)) : f \mapsto f \circ f_{A_0}$$

with the inverse of the isomorphism (5.5.18). Thus

$$A = f(A_0) \qquad \iff \qquad f_A = f \circ f_{A_0} \tag{5.5.20}$$

for all $A \in \mathcal{A}_0$ and all $f \in C(\Sigma_0)$. The resulting C* algebra isomorphism (5.5.19) satisfies the (Positive) axiom by Lemma 5.5.12, the (Normalization) and (Image) axioms by definition, the (Isometry) axiom because the Gelfand representation (5.5.18) is an isometry, the (Commutative) axiom by Lemma 5.5.10, and the (Spectrum) axiom by equation (5.5.14) in Lemma 5.5.11.

We prove that the C* algebra homomorphism (5.5.19) is uniquely determined by continuity and the (Normalization) axiom $\mathrm{id}(A_0) = A_0$. To see this, let $\Psi : C(\Sigma_0) \to \mathcal{L}^c(H)$ be any continuous C* algebra homomorphism such that $\Psi(\mathrm{id}) = A_0$ and let

$$\mathcal{P}(\Sigma_0) \subset C(\Sigma_0)$$

be the space of all functions $p : \Sigma_0 \to \mathbb{C}$ that can be expressed as polynomials in z and $\overline{z}$. Then $\mathcal{P}(\Sigma_0)$ is a subalgebra of $C(\Sigma_0)$ that contains the constant functions, separates points because it contains the identity map, and is invariant under complex conjugation. Hence $\mathcal{P}(\Sigma_0)$ is a dense subspace of $C(\Sigma_0)$ by Theorem 5.4.5. Moreover, $\Psi(p) = p(A)$ for all $p \in \mathcal{P}(\Sigma_0)$ by linearity and the (Product), (Conjugation), and (Normalization) axioms. Since the map $C(\Sigma_0) \to \mathcal{L}^c(H) : f \mapsto \Psi(f) - f(A)$ is continuous and $\mathcal{P}(\Sigma_0)$

is dense in $C(\Sigma_0)$, it follows that $\Psi(f) = f(A)$ for all $f \in C(\Sigma_0)$. This proves uniqueness of the continuous functional calculus for normal operators.

We prove the (Eigenvector) axiom. Fix an eigenvalue

$$\lambda \in \mathrm{P}\sigma(A_0)$$

and choose a nonzero vector $x \in H$ such that

$$A_0 x = \lambda x.$$

Then $\|\overline{\lambda}x - A_0^* x\| = \|\lambda x - A_0 x\| = 0$ by Lemma 5.3.14 and hence

$$A_0^* x = \overline{\lambda} x.$$

This implies

$$p(A_0)x = p(\lambda)x$$

for every polynomial $p \in \mathcal{P}(\Sigma_0)$ in z and $\overline{z}$. Now let $f \in C(\Sigma_0)$ and choose a sequence $p_n \in \mathcal{P}(\Sigma_0)$ that converges uniformly to f. Then

$$\lim_{n\to\infty} \|f(A_0) - p_n(A_0)\| = \lim_{n\to\infty} \|f - p_n\| = 0$$

by the (Isometry) axiom, and hence

$$f(A_0)x = \lim_{n\to\infty} p_n(A_0)x = \lim_{n\to\infty} p_n(\lambda)x = f(\lambda)x.$$

This proves the (Eigenvector) axiom.

We prove the (Composition) axiom. Fix a continuous function

$$f : \Sigma_0 \to \mathbb{C}.$$

Then $f(A_0) \in \mathcal{L}^c(H)$ is a normal operator whose spectrum is

$$\sigma(f(A_0)) = f(\Sigma_0)$$

by the (Spectrum) axiom. Now consider the map

$$C(f(\Sigma_0)) \to \mathcal{L}^c(H) : g \mapsto (g \circ f)(A_0).$$

This map is a continuous C* algebra homomorphism and it sends the identity map $g = \mathrm{id} : f(\Sigma_0) \to \mathbb{C}$ to the operator $f(A_0)$. Hence it follows from the uniqueness statement, with A_0 replaced by $f(A_0)$, that

$$(g \circ f)(A_0) = g(f(A_0)) \qquad \text{for all } g \in C(f(\Sigma_0)).$$

This proves Theorem 5.5.14. □

In Theorem 5.4.7 the continuous functional calculus was established for self-adjoint operators. Theorem 5.5.14 extends this result to normal operators and at the same time provides an alternative proof. The next goal is to extend the continuous functional calculus further to the C* algebra of complex valued bounded measurable functions on the spectrum. Taking the characteristic functions of Borel sets one then obtains the spectral measure associated to a normal operator. This is the content of Section 5.6 below.

Remark 5.5.15. Let H be an infinite-dimensional complex Hilbert space. It is useful to examine the special case of Theorem 5.4.7 where the normal operator $A \in \mathcal{L}^c(H)$ is compact, which we now assume.

(i) By part (v) of Theorem 5.3.16 the Hilbert space H admits an orthonormal basis $\{e_i\}_{i\in I}$ of eigenvectors of A. Here I is an infinite index set, uncountable whenever H is not separable, and

$$\langle e_i, e_j \rangle = \delta_{ij} \qquad \text{for all } i, j \in I.$$

Here $\langle e_i, e_j\rangle$ denotes the Hermitian inner product and the e_i are linearly independent over the complex numbers. There exists a map $I \to \mathbb{C} : i \mapsto \lambda_i$ such that $Ae_i = \lambda_i e_i$ for all $i \in I$ and hence

$$Ax = \sum_{i\in I} \lambda_i \langle e_i, x\rangle e_i \qquad \text{for all } x \in H. \tag{5.5.21}$$

The numbers λ_i are the eigenvalues of A and $\sigma(A) = \{\lambda_i \,|\, i \in I\} \cup \{0\}$. Thus we have $\sup_{i\in I}|\lambda_i| < \infty$. Moreover, the set $\{i \in I \,|\, |\lambda_i| > \varepsilon\}$ is finite for every $\varepsilon > 0$, because A is compact. The eigenvalues λ_i appear with the multiplicities

$$\#\{i \in I \,|\, \lambda_i = \lambda\} = \dim\ker(\lambda\mathbb{1} - A) \qquad \text{for all } \lambda \in \mathbb{R}.$$

If $f : \sigma(A) \to \mathbb{C}$ is any continuous function, then the operator $f(A) \in \mathcal{L}^c(H)$ is given by

$$f(A)x = \sum_{i\in I} f(\lambda_i)\langle e_i, x\rangle e_i \qquad \text{for all } x \in H. \tag{5.5.22}$$

Note that $f(A)$ is compact if and only if $f(0) = 0$.

(ii) It is also useful to rewrite the formula (5.5.22) in terms of the spectral projections. Let $\sigma(A) = \{\lambda_0, \lambda_1, \lambda_2, \dots\}$ where $\lambda_i \neq \lambda_j$ for $i \neq j$ and $\lambda_0 = 0$. For each i let $P_i \in \mathcal{L}^c(H)$ be the orthogonal projection onto the eigenspace of λ_i, i.e.

$$P_i^2 = P_i = P_i^*, \quad \operatorname{im}(P_i) = E_i := \ker(\lambda_i\mathbb{1} - A), \quad \ker(P_i) = E_i^\perp. \tag{5.5.23}$$

Then $P_iP_j = 0$ for $i \neq j$ and

$$x = \sum_i P_i x, \qquad Ax = \sum_i \lambda_i P_i x, \qquad f(A)x = \sum_i f(\lambda_i) P_i x \tag{5.5.24}$$

for all $x \in H$. Here the sums may be either finite or infinite, depending on whether or not $\sigma(A)$ is a finite set. If $\sigma(A)$ is an infinite set, we emphasize that the sequence of projections $\sum_{i=0}^n P_i$ converges to the identity in the strong operator topology, but not in the norm topology, because $\|\mathbb{1} - \sum_{i=0}^n P_i\| = 1$ for all $n \in \mathbb{N}$. However, the sequence $\sum_{i=0}^n \lambda_i P_i$ converges to A in the norm topology because $\lim_{i\to\infty} \lambda_i = 0$.

5.6. Spectral Measures

Assume that H is a nonzero complex Hilbert space and $A \in \mathcal{L}^c(H)$ is a normal operator. Then the spectrum

$$\Sigma := \sigma(A) \subset \mathbb{C}$$

is a nonempty compact subset of the complex plane by Theorem 5.3.15. Let

$$C(\Sigma) \to \mathcal{L}^c(H) : f \mapsto f(A)$$

be the C* algebra homomorphism introduced in Theorem 5.5.14. The purpose of the present section is to assign to A a Borel measure on Σ with values in the space of orthogonal projections on H, called the **spectral measure of** A. When A is a compact operator this measure assigns to each Borel set $\Omega \subset \Sigma$ the spectral projection

$$P_\Omega := \sum_{\lambda \in \sigma(A) \cap \Omega} P_\lambda$$

associated to all the eigenvalues of A in Ω (see Remark 5.5.15). The general construction of the spectral measure is considerably more subtle and is closely related to an extension of the homomorphism in Theorem 5.5.14 to the C* algebra $B(\Sigma)$ of all bounded Borel measurable functions on Σ. The starting point for the construction of this extension and the spectral measure is the observation that every element $x \in H$ determines a conjugation equivariant bounded linear functional $\Lambda_x : C(\Sigma) \to \mathbb{C}$ via the formula

$$\Lambda_x(f) := \langle x, f(A)x \rangle \qquad \text{for } f \in C(\Sigma). \tag{5.6.1}$$

Since $\Lambda_x(\overline{f}) = \overline{\Lambda_x(f)}$ for all $f \in C(\Sigma)$, the functional Λ_x is uniquely determined by its restriction to the subspace $C(\Sigma, \mathbb{R})$ of real valued continuous functions. This restriction takes values in $\mathbb{R}$ and the restricted functional $\Lambda_x : C(\Sigma, \mathbb{R}) \to \mathbb{R}$ is positive by Theorem 5.5.14, i.e. for all $f \in C(\Sigma, \mathbb{R})$,

$$f \geq 0 \qquad \Longrightarrow \qquad \Lambda_x(f) \geq 0.$$

Hence the Riesz Representation Theorem asserts that Λ_x can be represented by a Borel measure. Namely, let $\mathcal{B} \subset 2^\Sigma$ be the Borel σ-algebra. Then, for every $x \in \Sigma$, there exists a unique Borel measure $\mu_x : \mathcal{B} \to [0, \infty)$ such that

$$\int_\Sigma f \, d\mu_x = \langle x, f(A)x \rangle \qquad \text{for all } f \in C(\Sigma, \mathbb{R}). \tag{5.6.2}$$

(See [**75**, Cor. 3.19].) These Borel measures can be used to define the desired extension of the C* algebra homomorphism $C(\Sigma) \to \mathcal{L}^c(H)$ to $B(\Sigma)$ as well as the spectral measure of A.

5.6.1. Projection Valued Measures.

Definition 5.6.1 (Projection Valued Measure). Let H be a complex Hilbert space, let

$$\Sigma \subset \mathbb{C}$$

be a nonempty closed subset, and denote by $\mathcal{B} \subset 2^{\Sigma}$ the Borel σ-algebra. A **projection valued Borel measure** on Σ is a map

$$\mathcal{B} \to \mathcal{L}^c(H) : \Omega \to P_\Omega \tag{5.6.3}$$

which assigns to every Borel set $\Omega \subset \Sigma$ a bounded complex linear operator $P_\Omega : H \to H$ and satisfies the following axioms.

(Projection) For every Borel set $\Omega \subset \Sigma$ the operator P_Ω is an orthogonal projection, i.e.

$$P_\Omega^2 = P_\Omega = P_\Omega^*. \tag{5.6.4}$$

(Normalization) The projections associated to $\Omega = \emptyset$ and $\Omega = \Sigma$ are

$$P_\emptyset = 0, \qquad P_\Sigma = 1\!\!1. \tag{5.6.5}$$

(Intersection) If $\Omega_1, \Omega_2 \subset \Sigma$ are two Borel sets, then

$$P_{\Omega_1 \cap \Omega_2} = P_{\Omega_1} P_{\Omega_2} = P_{\Omega_2} P_{\Omega_1}. \tag{5.6.6}$$

(σ-Additive) If $(\Omega_i)_{i\in\mathbb{N}}$ is a sequence of pairwise disjoint Borel sets in Σ so that $\Omega_i \cap \Omega_j = \emptyset$ for $i \neq j$ and $\Omega := \bigcup_{i=1}^\infty \Omega_i$, then

$$P_\Omega x = \lim_{n\to\infty} \sum_{i=1}^n P_{\Omega_i} x \tag{5.6.7}$$

for all $x \in H$.

For every nonempty compact Hausdorff space Σ define

$$B(\Sigma) := \{ f : \Sigma \to \mathbb{C} \,|\, f \text{ is bounded and Borel measurable} \}.$$

This space is a C* algebra with the supremum norm

$$\|f\| := \sup_{\lambda\in\Sigma} |f(\lambda)|$$

for $f \in B(\Sigma)$, and with the complex anti-linear isometric involution given by complex conjugation. The next theorem shows that, if Σ is a closed subset of $\mathbb{C}$ and $\mathcal{B} \subset 2^\Sigma$ is the Borel σ-algebra, then every projection valued measure $\mathcal{B} \to \mathcal{L}^c(H) : \Omega \mapsto P_\Omega$ gives rise to a C* algebra homomorphism from $B(\Sigma)$ to $\mathcal{L}^c(H)$.

Theorem 5.6.2. *Let $H, \Sigma, \mathcal{B}$ be as in Definition 5.6.1 and fix a projection valued measure (5.6.3). Denote by $B(\Sigma)$ the C^* algebra of complex valued bounded Borel measurable functions on Σ, equipped with the supremum norm. For $x, y \in H$ define the signed Borel measure $\mu_{x,y} : \mathcal{B} \to \mathbb{R}$ by*

$$\mu_{x,y}(\Omega) := \mathrm{Re}\langle x, P_\Omega y\rangle \qquad \text{for } \Omega \in \mathcal{B}. \tag{5.6.8}$$

Then, for each $f \in B(\Sigma)$, there exists a unique operator $\Psi(f) \in \mathcal{L}^c(H)$ such that

$$\mathrm{Re}\langle x, \Psi(f)y\rangle = \int_\Sigma \mathrm{Re} f \, d\mu_{x,y} + \int_\Sigma \mathrm{Im} f \, d\mu_{x,\mathbf{i}y} \quad \text{for all } x, y \in H. \tag{5.6.9}$$

The resulting map $\Psi : B(\Sigma) \to \mathcal{L}^c(H)$ is a C^* algebra homomorphism and a bounded linear operator and it satisfies $\sigma(\Psi(f)) \subset \overline{f(\Sigma)}$ for all $f \in B(\Sigma)$.

Proof. See page 264. □

Assume the situation of Theorem 5.6.2 and suppose, in addition, that Σ is compact. Since the map $\Psi : B(\Sigma) \to \mathcal{L}^c(H)$ is a C* algebra homomorphism, the operator $\Psi(f)$ is normal for every $f \in B(\Sigma)$. Thus every projection valued measure on Σ determines a normal operator $A := \Psi(\mathrm{id})$ associated to the identity map and the spectrum of A is contained in Σ. Conversely, every normal operator $A \in \mathcal{L}^c(H)$ gives rise to a unique projection valued measure in H with support on its spectrum $\Sigma := \sigma(A)$. Thus there is a one-to-one correspondence between compactly supported projection valued measures on $\mathbb{C}$ and bounded normal operators on H.

Theorem 5.6.3 (Spectral Measure). *Let H be a nonzero complex Hilbert space, let $A \in \mathcal{L}^c(H)$ be a normal operator, let $\Sigma := \sigma(A) \subset \mathbb{C}$ be its spectrum, and denote by $\mathcal{B} \subset 2^\Sigma$ the Borel σ-algebra. Then there exists a unique projection valued Borel measure*

$$\mathcal{B} \to \mathcal{L}^c(H) : \Omega \mapsto P_\Omega \tag{5.6.10}$$

such that

$$\int_\Sigma \mathrm{Re}\lambda \, d\mu_{x,y}(\lambda) + \int_\Sigma \mathrm{Im}\lambda \, d\mu_{x,\mathbf{i}y}(\lambda) = \mathrm{Re}\langle x, Ay\rangle \tag{5.6.11}$$

for all $x, y \in H$, where the signed Borel measures $\mu_{x,y} : \mathcal{B} \to \mathbb{R}$ are given by

$$\mu_{x,y}(\Omega) := \mathrm{Re}\langle x, P_\Omega y\rangle \tag{5.6.12}$$

for $x, y \in H$ and $\Omega \in \mathcal{B}$. The projection valued measure (5.6.10) is called the **spectral measure of** *A.*

Proof. See page 273. □

Example 5.6.4. Let $\Sigma \subset \mathbb{C}$ be a nonempty compact set, equipped with the Lebesgue measure, and let $H := L^2(\Sigma)$ be the Hilbert space of complex valued L^2 functions on Σ. For a Borel set $\Omega \subset \Sigma$ define $P_\Omega : H \to H$ by $P_\Omega \psi := \chi_\Omega \psi$ for $\psi \in H$, where $\chi_\Omega : \Sigma \to \mathbb{R}$ denotes the characteristic function of Ω. These operators define a projection valued measure on H and the operator $\Psi : B(\Sigma) \to \mathcal{L}^c(L^2(\Sigma))$ in Theorem 5.6.3 is given by

$$\Psi(f)\psi = f\psi$$

for all $f \in B(\Sigma)$ and all $\psi \in L^2(\Sigma)$. In the case $\Sigma = [0,1] \subset \mathbb{R}$ and

$$f(\lambda) := \begin{cases} \lambda, & \text{for } 0 \le \lambda < 1, \\ 2, & \text{for } \lambda = 1, \end{cases}$$

we obtain $f(\Sigma) = [0,1) \cup \{2\}$ and $\sigma(\Psi(f)) = [0,1]$. Thus $f(\Sigma)$ is not closed and $f(\Sigma) \not\subset \sigma(\Psi(f)) \subsetneq \overline{f(\Sigma)}$.

The proof of Theorem 5.6.2 is carried out in the present subsection, while the proof of Theorem 5.6.3 is postponed to Subsection 5.6.2. As in part (vi) of Example 1.1.3, denote by $\mathcal{M}(\Sigma)$ the Banach space of signed Borel measures $\mu : \mathcal{B} \to \mathbb{R}$ with the norm

$$\|\mu\| = \sup_{\Omega \in \mathcal{B}} \big(\mu(\Omega) - \mu(\Sigma \setminus \Omega)\big)$$

for $\mu \in \mathcal{M}(\Sigma)$.

Proof of Theorem 5.6.2. The proof has five steps.

Step 1. *The map $H \times H \to \mathcal{M}(\Sigma) : (x,y) \mapsto \mu_{x,y}$ is real bilinear and symmetric and satisfies the inequality*

$$\|\mu_{x,y}\| \le \|x\| \, \|y\|$$

for all $x, y \in H$.

That the map is real bilinear and symmetric follows directly from the definition of $\mu_{x,y}$. Moreover, for all $x, y \in H$ and all $\Omega \in \mathcal{B}$ we have $P_\Omega P_{\Sigma\setminus\Omega} = 0$, hence the vectors $P_\Omega y$ and $P_{\Sigma\setminus\Omega} y$ are orthogonal to each other, hence

$$\begin{aligned} \left\|P_\Omega y - P_{\Sigma\setminus\Omega} y\right\|^2 &= \|P_\Omega y\|^2 + \|P_{\Sigma\setminus\Omega} y\|^2 \\ &= \left\|P_\Omega y + P_{\Sigma\setminus\Omega} y\right\|^2 \\ &= \|P_\Sigma y\|^2 \\ &= \|y\|^2, \end{aligned}$$

and hence

$$\mu_{x,y}(\Omega) - \mu_{x,y}(\Sigma \setminus \Omega) = \operatorname{Re}\langle x, (P_\Omega - P_{\Sigma\setminus\Omega}) y \rangle \le \|x\| \|y\|$$

by (5.6.8). This proves Step 1.

Step 2. *Let $B \in \mathcal{L}^c(H)$ such that $P_\Omega B = BP_\Omega$ for all $\Omega \in \mathcal{B}$. Then*

$$\mu_{x,By} = \mu_{B^*x,y}$$

for all $x, y \in H$.

Let $\Omega \in \mathcal{B}$ and $x, y \in H$. Then, by (5.6.8) we have

$$\mu_{x,By}(\Omega) = \mathrm{Re}\langle x, P_\Omega By\rangle = \mathrm{Re}\langle x, BP_\Omega y\rangle = \mathrm{Re}\langle B^*x, P_\Omega y\rangle = \mu_{B^*x,y}(\Omega).$$

This proves Step 2.

Step 3. *For every $f \in B(\Sigma)$ there exists a unique bounded complex linear operator $\Psi(f) : H \to H$ that satisfies (5.6.9). Moreover, $\Psi(\overline{f}) = \Psi(f)^*$ for all $f \in B(\Sigma)$ and the map $\Psi : B(\Sigma) \to \mathcal{L}^c(H)$ is a bounded complex linear operator.*

Let $f \in B(\Sigma, \mathbb{R})$ and define the real bilinear form $B_f : H \times H \to \mathbb{R}$ by

$$B_f(x, y) := \int_\Sigma f \, d\mu_{x,y}.$$

Then, for all $x, y \in H$, we have

$$|B_f(x, y)| \le \|f\|\|\mu_{x,y}\| \le \|f\|\|x\|\|y\|$$

by Step 1 and [**75**, Exercise 5.35 (i)]. Hence there exists a unique bounded real linear operator $\Psi(f) \in \mathcal{L}(H)$ such that

$$\mathrm{Re}\langle x, \Psi(f)y\rangle = B_f(x, y) \qquad \text{for all } x, y \in H.$$

This operator is self-adjoint because B_f is symmetric by Step 1, and

$$\|\Psi(f)\| \le \|f\|.$$

Hence the resulting map $\Psi : B(\Sigma, \mathbb{R}) \to \mathcal{L}(H)$ is a bounded linear operator. Moreover, $B_f(x, \mathbf{i}y) = -B_f(\mathbf{i}x, y)$ by Step 2 with $B = \mathbf{i}\mathbb{1}$, and hence

$$\begin{aligned}
\mathrm{Re}\langle x, \Psi(f)\mathbf{i}y\rangle &= B_f(x, \mathbf{i}y) \\
&= -B_f(\mathbf{i}x, y) \\
&= -\mathrm{Re}\langle \mathbf{i}x, \Psi(f)y\rangle \\
&= \mathrm{Re}\langle x, \mathbf{i}\Psi(f)y\rangle
\end{aligned}$$

for all $x, y \in H$. Thus the operator $\Psi(f) : H \to H$ is complex linear. For $f \in B(\Sigma)$ define

$$\Psi(f) := \Psi(\mathrm{Re} f) + \mathbf{i}\Psi(\mathrm{Im} f) \in \mathcal{L}^c(H).$$

Then $\Psi(f)$ satisfies condition (5.6.9) and is uniquely determined by this equation. Moreover, the map $B(\Sigma) \to \mathcal{L}^c(H) : f \mapsto \Psi(f)$ is complex linear and the formula $\Psi(\overline{f}) = \Psi(f)^*$ follows from the fact that the operators $\Psi(\mathrm{Re} f)$ and $\Psi(\mathrm{Im} f)$ are self-adjoint. This proves Step 3.

Step 4. *Let $\Psi : B(\Sigma) \to \mathcal{L}^c(H)$ be as in Step 3. Then*

$$\Psi(fg) = \Psi(f)\Psi(g)$$

for all $f, g \in B(\Sigma)$.

Since the operator $\Psi : B(\Sigma) \to \mathcal{L}^c(H)$ is complex linear it suffices to verify the equation $\Psi(fg) = \Psi(f)\Psi(g)$ for real valued functions $f, g \in B(\Sigma, \mathbb{R})$. Assume first that $g = \chi_\Omega$ for some Borel set $\Omega \subset \Sigma$. Then

$$\begin{aligned}\mu_{P_\Omega x,y}(\Omega') &= \mathrm{Re}\langle P_\Omega x, P_{\Omega'} y\rangle \\ &= \mathrm{Re}\langle x, P_\Omega P_{\Omega'} y\rangle \\ &= \mathrm{Re}\langle x, P_{\Omega\cap\Omega'} y\rangle \\ &= \mu_{x,y}(\Omega \cap \Omega') \\ &= \int_{\Omega'} \chi_\Omega \, d\mu_{x,y}\end{aligned}$$

for all $\Omega' \in \mathcal{B}$. By [**75**, Thm. 1.40] this implies

$$\begin{aligned}\int_\Omega g\, d\mu_{x,y} &= \int_\Sigma g\chi_\Omega \, d\mu_{x,y} \\ &= \int_\Sigma g\, d\mu_{P_\Omega x,y} \\ &= \mathrm{Re}\langle P_\Omega x, \Psi(g)y\rangle \\ &= \mathrm{Re}\langle x, P_\Omega \Psi(g)y\rangle \\ &= \mu_{x,\Psi(g)y}(\Omega)\end{aligned}$$

for all $g \in B(\Sigma, \mathbb{R})$. Apply [**75**, Thm. 1.40] again to obtain

$$\mathrm{Re}\langle x, \Psi(fg)y\rangle = \int_\Sigma fg\, d\mu_{x,y} = \int_\Sigma f\, d\mu_{x,\Psi(g)y} = \mathrm{Re}\langle x, \Psi(f)\Psi(g)y\rangle$$

for all $f, g \in B(\Sigma, \mathbb{R})$ and all $x, y \in H$. This proves Step 4.

Step 5. $\sigma(\Psi(f)) \subset \overline{f(\Sigma)}$ *for all $f \in B(\Sigma)$.*

Let $f \in B(\Sigma)$ and $\lambda \in \mathbb{C} \setminus \overline{f(\Sigma)}$ and define the function $g : \Sigma \to \mathbb{C}$ by

$$g(\mu) := (\lambda - f(\mu))^{-1} \qquad \text{for } \mu \in \Sigma.$$

Then $g \in B(\Sigma)$ and $g(\lambda - f) = (\lambda - f)g = 1$. Hence

$$\Psi(g)(\lambda\mathbb{1} - \Psi(f)) = (\lambda\mathbb{1} - \Psi(f))\Psi(g) = \Psi(1) = \mathbb{1}.$$

Thus $\lambda\mathbb{1} - \Psi(f)$ is invertible and so $\lambda \in \rho(\Psi(f))$. This proves Step 5 and Theorem 5.6.2. □

5.6.2. Measurable Functional Calculus. The next theorem extends the continuous functional calculus for normal operators, established in Theorem 5.5.14, to bounded measurable functions. The new ingredients are the (Convergence) axiom, the (Contraction) axiom in place of the (Isometry) axiom, and the modified (Image) and (Spectrum) axioms.

Theorem 5.6.5 (Measurable Functional Calculus). *Let H be a nonzero complex Hilbert space, let $A \in \mathcal{L}^c(H)$ be a normal operator, and let $\Sigma := \sigma(A)$. Then there exists a complex linear operator*

$$B(\Sigma) \to \mathcal{L}^c(H) : f \mapsto f(A) \tag{5.6.13}$$

that satisfies the following axioms.

(Product) $1(A) = 1\!\!1$ *and* $(fg)(A) = f(A)g(A)$ *for all* $f, g \in B(\Sigma)$.

(Conjugation) $\overline{f}(A) = f(A)^*$ *for all* $f \in B(\Sigma)$.

(Positive) *If* $f \in B(\Sigma, \mathbb{R})$ *and* $f \geq 0$, *then* $f(A) = f(A)^* \geq 0$.

(Normalization) *If* $f(\lambda) = \lambda$ *for all* $\lambda \in \Sigma$, *then* $f(A) = A$.

(Contraction) $\|f(A)\| \leq \sup_{\lambda \in \Sigma} |f(\lambda)| = \|f\|$ *for all* $f \in B(\Sigma)$.

(Convergence) *Let* $f_i \in B(\Sigma)$ *be a sequence such that* $\sup_{i \in \mathbb{N}} \|f_i\| < \infty$ *and let* $f \in B(\Sigma)$ *such that* $\lim_{i \to \infty} f_i(\lambda) = f(\lambda)$ *for all* $\lambda \in \Sigma$. *Then*

$$\lim_{i \to \infty} f_i(A)x = f(A)x \qquad \text{for all } x \in H.$$

(Commutative) *If* $B \in \mathcal{L}^c(H)$ *satisfies* $AB = BA$ *and* $A^*B = BA^*$, *then* $f(A)B = Bf(A)$ *for all* $f \in B(\Sigma)$.

(Image) *The image of the operator* (5.6.13) *is the smallest* C^* *subalgebra of* $\mathcal{L}^c(H)$ *that contains* A *and is closed under strong convergence.*

(Eigenvector) *If* $\lambda \in \Sigma$ *and* $x \in H$ *satisfy* $Ax = \lambda x$, *then*

$$f(A)x = f(\lambda)x$$

for all $f \in B(\Sigma)$.

(Spectrum) *If* $f \in B(\Sigma)$, *then* $f(A)$ *is normal and* $\sigma(f(A)) \subset \overline{f(\Sigma)}$. *Moreover,* $\sigma(f(A)) = f(\Sigma)$ *for all* $f \in C(\Sigma)$.

(Composition) *If* $f \in C(\Sigma)$ *and* $g \in B(f(\Sigma))$, *then* $(g \circ f)(A) = g(f(A))$.

The complex linear operator (5.6.13) *is uniquely determined by the (Product), (Conjugation), (Normalization), and (Convergence) axioms. The (Product) and (Conjugation) axioms assert that it is a* C^* *algebra homomorphism.*

Proof. See page 276. □

The proofs of both Theorems 5.6.3 and 5.6.5 will be based on a series of lemmas. Assume throughout that H is a nonzero complex Hilbert space, that $A \in \mathcal{L}^c(H)$ is a normal operator with spectrum $\Sigma := \sigma(A) \subset \mathbb{C}$, and that $\mathcal{B} \subset 2^\Sigma$ is the Borel σ-algebra. The starting point is the Riesz Representation Theorem which asserts that, for every positive linear functional $\Lambda : C(\Sigma, \mathbb{R}) \to \mathbb{R}$, there exists a unique Borel measure $\mu : \mathcal{B} \to [0, \infty)$ such that $\Lambda(f) = \int_\Sigma f \, d\mu$ for all $f \in C(\Sigma, \mathbb{R})$ (see [**75**, Cor. 3.19]). By Theorem 5.5.14, this implies that, for each $x \in H$, there exists a unique Borel measure $\mu_x : \mathcal{B} \to [0, \infty)$ that satisfies (5.6.2), i.e.

$$\int_\Sigma f \, d\mu_x = \langle x, f(A)x \rangle \qquad \text{for all } f \in C(\Sigma, \mathbb{R}).$$

For $x, y \in H$ define the signed measure $\mu_{x,y} : \mathcal{B} \to \mathbb{R}$ by

$$\mu_{x,y} := \frac{1}{4}\big(\mu_{x+y} - \mu_{x-y}\big). \tag{5.6.14}$$

The next lemma summarizes some basic properties of these signed measures.

Lemma 5.6.6. **(i)** *The map*

$$H \times H \to \mathcal{M}(\Sigma) : (x, y) \mapsto \mu_{x,y} \tag{5.6.15}$$

defined by (5.6.14) *is real bilinear and symmetric.*

(ii) *The signed measures* $\mu_{x,y}$ *satisfy*

$$\int_\Sigma f \, d\mu_{x,y} = \mathrm{Re}\langle x, f(A)y \rangle \tag{5.6.16}$$

for all $x, y \in H$ *and all* $f \in C(\Sigma, \mathbb{R})$.

(iii) *Let* $B \in \mathcal{L}^c(H)$ *such that* $AB = BA$ *and* $A^*B = BA^*$. *Then*

$$\mu_{x,By} = \mu_{B^*x,y} \tag{5.6.17}$$

and, in particular,

$$\mu_{x,\mathbf{i}y} = -\mu_{\mathbf{i}x,y}$$

for all $x, y \in H$.

(iv) *The signed measures* $\mu_{x,y}$ *satisfy*

$$\|\mu_{x,y}\| \le \|x\| \, \|y\| \tag{5.6.18}$$

for all $x, y \in H$.

Proof. Equation (5.6.16) follows from (5.6.2) and the definition of $\mu_{x,y}$ in (5.6.14). It implies that the map (5.6.15) is real bilinear and symmetric. This proves parts (i) and (ii).

To prove part (iii), assume $B \in \mathcal{L}^c(H)$ commutes with A and A^* and let $x, y \in H$. If $f \in C(\Sigma, \mathbb{R})$, then $f(A)B = Bf(A)$ by the (Commutative) axiom in Theorem 5.5.14. Hence it follows from (ii) that

$$\int_\Sigma f\mu_{x,By} = \mathrm{Re}\langle x, f(A)By\rangle = \mathrm{Re}\langle B^*x, f(A)y\rangle = \int_\Sigma f\mu_{B^*x,y}$$

for all $f \in C(\Sigma, \mathbb{R})$. This implies $\mu_{x,By} = \mu_{B^*x,y}$ by uniqueness in the Riesz Representation Theorem. This proves part (iii).

We prove part (iv). The Hahn Decomposition Theorem asserts that, for every $\mu \in \mathcal{M}(\Sigma)$, there exists a Borel set $P \subset \Sigma$ such that $\mu(\Omega \cap P) \geq 0$ and $\mu(\Omega \setminus P) \leq 0$ for every Borel set $\Omega \subset \Sigma$ (see [**75**, Thm. 5.19]). The norm of μ is then given by

$$\begin{aligned} \|\mu\| &= \mu(P) - \mu(\Sigma \setminus P) \\ &= \sup_{f \in C(\Sigma,\mathbb{R})} \frac{\int_\Sigma f\,d\mu}{\|f\|} \\ &= \sup_{f \in B(\Sigma,\mathbb{R})} \frac{\int_\Sigma f\,d\mu}{\|f\|}. \end{aligned} \tag{5.6.19}$$

(See [**75**, Exercise 5.35 (i)].) Hence

$$\begin{aligned} \|\mu_{x,y}\| &= \sup_{f \in C(\Sigma,\mathbb{R})} \frac{\int_M f\,d\mu_{x,y}}{\|f\|} \\ &= \sup_{f \in C(\Sigma,\mathbb{R})} \frac{\mathrm{Re}\langle x, f(A)y\rangle}{\|f\|} \\ &\leq \sup_{f \in C(\Sigma,\mathbb{R})} \frac{\|x\|\,\|f(A)\|\,\|y\|}{\|f\|} \\ &= \|x\|\,\|y\| \end{aligned}$$

for all $x, y \in H$. Here the first step follows from (5.6.19) and the last step follows from the identity $\|f(A)\| = \|f\|$ for $f \in C(\Sigma, \mathbb{R})$ (see Theorem 5.5.14). This proves Lemma 5.6.6. □

Lemma 5.6.6 allows us to define the map $B(\Sigma) \to \mathcal{L}^c(H) : f \mapsto f(A)$ in Theorem 5.6.5 and the map $\mathcal{B} \to \mathcal{L}^c(H) : \Omega \to P_\Omega$ in Theorem 5.6.3. This is the content of Lemma 5.6.7 below. The task at hand will then be to verify that these maps satisfy all the axioms in Theorems 5.6.3 and 5.6.5 and, finally, to prove the uniqueness statements. A key step for verifying the properties of these maps will be the proof of the (Product) axiom in Theorem 5.6.5. This is the content of Lemma 5.6.8 below. The (Convergence) axiom will be verified in Lemma 5.6.9.

Lemma 5.6.7 (The Operator Ψ_A). *There exists a unique bounded complex linear operator $\Psi_A : B(\Sigma) \to \mathcal{L}^c(H)$ such that*

$$\mathrm{Re}\langle x, \Psi_A(f)y\rangle = \int_\Sigma f\, d\mu_{x,y} \tag{5.6.20}$$

for all $x, y \in H$ and all $f \in B(\Sigma, \mathbb{R})$, where $\mu_{x,y} \in \mathcal{M}(\Sigma)$ is defined by (5.6.14). *The operator Ψ_A satisfies the (Conjugation), (Positive), (Normalization), (Contraction), and (Commutative) axioms in Theorem* 5.6.5. *Its restriction to $C(\Sigma)$ is the operator* (5.5.17) *in Theorem* 5.5.14.

Proof. Fix a bounded real valued Borel measurable function $f : \Sigma \to \mathbb{R}$ and define the map $B_f : H \times H \to \mathbb{R}$ by

$$B_f(x, y) := \int_\Sigma f\, d\mu_{x,y} \qquad \text{for } x, y \in H.$$

This map is real bilinear and symmetric by part (i) of Lemma 5.6.6 and

$$|B_f(x, y)| \le \|f\|\, \|\mu_{x,y}\| \le \|f\|\, \|x\|\, \|y\| \qquad \text{for all } x, y \in H \tag{5.6.21}$$

by (5.6.19) and part (iv) of Lemma 5.6.6. Hence, by Theorem 1.4.4, there exists a unique bounded real linear operator $\Psi_A(f) : H \to H$ such that

$$\mathrm{Re}\langle x, \Psi_A(f)y\rangle = B_f(x, y) = \int_\Sigma f\, d\mu_{x,y}$$

for all $x, y \in H$. Since B_f is symmetric the operator $\Psi_A(f)$ is self-adjoint. Moreover, $\|\Psi_A(f)\| \le \|f\|$ by (5.6.21). Since

$$\begin{aligned}\mathrm{Re}\langle x, \Psi_A(f)\mathbf{i}y\rangle &= \int_\Sigma f\, d\mu_{x,\mathbf{i}y} = -\int_\Sigma f\, d\mu_{\mathbf{i}x,y} \\ &= -\mathrm{Re}\langle \mathbf{i}x, \Psi_A(f)y\rangle = \mathrm{Re}\langle x, \mathbf{i}\Psi_A(f)y\rangle\end{aligned}$$

for all $x, y \in H$ by part (iii) of Lemma 5.6.6, the operator $\Psi_A(f)$ is complex linear. The resulting map $\Psi_A : B(\Sigma, \mathbb{R}) \to \mathcal{L}^c(H)$ extends uniquely to a bounded complex linear operator $\Psi_A : B(\Sigma) \to \mathcal{L}^c(H)$ via

$$\Psi_A(f) := \Psi_A(\mathrm{Re} f) + \mathbf{i}\Psi_A(\mathrm{Im} f) \qquad \text{for } f \in B(\Sigma).$$

By definition, this operator satisfies (5.6.20) as well as the (Conjugation), (Positive), (Normalization), and (Contraction) axioms. If $B \in \mathcal{L}^c(H)$ commutes with A and A^*, then

$$\mathrm{Re}\langle x, \Psi_A(f)By\rangle = \int_\Sigma f\, d\mu_{x,By} = \int_\Sigma f\, d\mu_{B^*x,y} = \mathrm{Re}\langle B^*x, \Psi_A(f)y\rangle$$

for all $x, y \in H$ by part (iii) of Lemma 5.6.6 and so $\Psi_A(f)B = B\Psi_A(f)$. Thus Ψ_A satisfies the (Commutative) axiom. That Ψ_A is uniquely determined by (5.6.20) is obvious and that $\Psi_A(f) = f(A)$ is the operator in Theorem 5.5.14 for every $f \in C(\Sigma)$ follows from part (ii) of Lemma 5.6.6. This proves Lemma 5.6.7. □

Lemma 5.6.8 (Product Axiom). *The map $\Psi_A : B(\Sigma) \to \mathcal{L}^c(H)$ in Lemma* 5.6.7 *satisfies the (Product) axiom in Theorem* 5.6.5.

Proof. Assume first that $f : \Sigma \to [0, \infty)$ is continuous and let $x \in H$. Then it follows from the (Product) axiom in Theorem 5.5.14 that

$$\begin{aligned}\int_\Sigma g\, d\mu_{x,f(A)x} &= \mathrm{Re}\langle x, g(A)f(A)x\rangle \\ &= \mathrm{Re}\langle x, (gf)(A)x\rangle \\ &= \int_\Sigma gf\, d\mu_x\end{aligned}$$

for all $g \in C(\Sigma, \mathbb{R})$. The last term on the right is the integral of g with respect to the Borel measure

$$\mathcal{B} \to [0, \infty) : \Omega \mapsto \int_\Omega f\, d\mu_x$$

by [**75**, Thm. 1.40]. Hence it follows from uniqueness in the Riesz Representation Theorem that

$$\mu_{x,f(A)x}(\Omega) = \int_\Omega f\, d\mu_x \tag{5.6.22}$$

for every Borel set $\Omega \subset \Sigma$. Now let $g \in B(\Sigma, \mathbb{R})$. Then, for all $x \in H$,

$$\begin{aligned}\mathrm{Re}\langle x, \Psi_A(g)\Psi_A(f)x\rangle &= \int_\Sigma g\, d\mu_{x,f(A)x} \\ &= \int_\Sigma gf\, d\mu_x \\ &= \mathrm{Re}\langle x, \Psi_A(gf)x\rangle.\end{aligned} \tag{5.6.23}$$

Here the second equality follows from (5.6.22) and [**75**, Thm. 1.40]. Moreover, the operator $\Psi_A(f) = f(A)$ commutes with A and A^* by Theorem 5.5.14, and hence $\Psi_A(f)$ commutes with $\Psi_A(g)$ by Lemma 5.6.7. Since both operators are self-adjoint, so is their composition as is $\Psi_A(gf)$. Hence it follows from (5.6.23) that

$$\Psi_A(f)\Psi_A(g) = \Psi_A(g)\Psi_A(f) = \Psi_A(gf)$$

whenever $f : \Sigma \to [0, \infty)$ is continuous and $g : \Sigma \to \mathbb{R}$ is bounded and Borel measurable. Now take differences and multiply by $\mathbf{i}$, to obtain the (Product) axiom for all $f \in C(\Sigma)$ and all $g \in B(\Sigma)$.

Now fix any bounded measurable function $f : \Sigma \to [0, \infty)$ and repeat the above argument to obtain that (5.6.22) holds for all $\Omega \in \mathcal{B}$ and hence (5.6.23) holds for all $g \in B(\Sigma, \mathbb{R})$. Then the (Product) axiom holds for all $f, g \in B(\Sigma)$ by taking differences and multiplying by $\mathbf{i}$. This proves Lemma 5.6.8. □

Lemma 5.6.9 (Convergence Axiom). *The map $\Psi_A : B(\Sigma) \to \mathcal{L}^c(H)$ in Lemma 5.6.7 satisfies the (Convergence) axiom in Theorem 5.6.5.*

Proof. It suffices to establish the convergence axiom for real valued functions. Thus assume that

$$f_i : \Sigma \to \mathbb{R}, \qquad i \in \mathbb{N},$$

is a sequence of bounded Borel measurable functions that satisfies

$$\sup_{i\in\mathbb{N}} \|f_i\| < \infty$$

and converges pointwise to a Borel measurable function

$$f : \Sigma \to \mathbb{R},$$

i.e.

$$\lim_{i\to\infty} f_i(\lambda) = f(\lambda) \qquad \text{for all } \lambda \in \Sigma.$$

Fix an element $x \in H$. Then it follows from equation (5.6.20) in Lemma 5.6.7 and the Lebesgue Dominated Convergence Theorem [**75**, Thm. 1.45] that

$$\begin{aligned}
\mathrm{Re}\langle y, \Psi_A(f)x\rangle &= \int_\Sigma f\,d\mu_{y,x} \\
&= \lim_{i\to\infty} \int_\Sigma f_i\,d\mu_{y,x} \\
&= \lim_{i\to\infty} \mathrm{Re}\langle y, \Psi_A(f_i)x\rangle
\end{aligned}$$

for all $y \in H$. Replace f_i by f_i^2 and use Lemma 5.6.8 to obtain

$$\begin{aligned}
\|\Psi_A(f)x\|^2 &= \langle \Psi_A(f)x, \Psi_A(f)x\rangle \\
&= \langle x, \Psi_A(f^2)x\rangle \\
&= \lim_{i\to\infty} \langle x, \Psi_A(f_i^2)x\rangle \\
&= \lim_{i\to\infty} \langle \Psi_A(f_i)x, \Psi_A(f_i)x\rangle \\
&= \lim_{i\to\infty} \|\Psi_A(f_i)x\|^2 .
\end{aligned}$$

Thus we have proved that the sequence

$$(\Psi_A(f_i)x)_{i\in\mathbb{N}}$$

in H converges weakly to $\Psi_A(f)x$ and the norm of the limit is the limit of the norms. Hence it follows from Exercise 3.7.1 that

$$\lim_{i\to\infty} \|\Psi_A(f_i)x - \Psi_A(f)x\| = 0.$$

This proves Lemma 5.6.9. □

Proof of Theorem 5.6.3. Denote the characteristic function of $\Omega \subset \Sigma$ by

$$\chi_\Omega : \Sigma \to \mathbb{R}, \qquad \chi_\Omega(\lambda) := \begin{cases} 1, & \text{for } \lambda \in \Omega, \\ 0, & \text{for } \lambda \in \Sigma \setminus \Omega. \end{cases}$$

Let $\Psi_A : B(\Sigma) \to \mathcal{L}^c(H)$ be the bounded complex linear operator introduced in Lemma 5.6.7 and define the map $\mathcal{B} \to \mathcal{L}^c(H) : \Omega \mapsto P_\Omega$ by

$$P_\Omega := \Psi_A(\chi_\Omega) \qquad \text{for } \Omega \in \mathcal{B}. \tag{5.6.24}$$

Since χ_Ω is real valued the operator P_Ω is self-adjoint and, since

$$\chi_\Omega \chi_{\Omega'} = \chi_{\Omega \cap \Omega'},$$

it follows from Lemma 5.6.8 that

$$P_\Omega P_{\Omega'} = \Psi_A(\chi_\Omega)\Psi_A(\chi_{\Omega'}) = \Psi_A(\chi_\Omega \chi_{\Omega'}) = \Psi_A(\chi_{\Omega\cap\Omega'}) = P_{\Omega\cap\Omega'}$$

for all $\Omega, \Omega' \in \mathcal{B}$. Thus P_Ω is an orthogonal projection for every $\Omega \in \mathcal{B}$. Moreover,

$$P_\emptyset = \Psi_A(\chi_\emptyset) = \Psi_A(0) = 0, \qquad P_\Sigma = \Psi_A(\chi_\Sigma) = \Psi_A(1) = 1\!\!1.$$

Now let $(\Omega_i)_{i\in\mathbb{N}}$ be a sequence of pairwise disjoint Borel subsets of Σ and define

$$\Omega := \bigcup_{i=1}^\infty \Omega_i.$$

Then

$$f_n := \sum_{i=1}^n \chi_{\Omega_i} : \Sigma \to \mathbb{R}$$

is a sequence of bounded Borel measurable functions that satisfies $\|f_n\| \le 1$ for all n and that converges pointwise to

$$f := \chi_\Omega.$$

Hence, by Lemma 5.6.9, we have

$$P_\Omega x = \Psi(\chi_\Omega)x = \lim_{n\to\infty} \Psi(f_n)x = \lim_{n\to\infty} \sum_{i=1}^n \Psi(\chi_{\Omega_i})x = \lim_{n\to\infty} \sum_{i=1}^n P_{\Omega_i} x$$

for all $x \in H$. This shows that the map (5.6.24) satisfies all the axioms in Definition 5.6.1 and hence is a projection valued Borel measure on Σ. By definition of Ψ_A in Lemma 5.6.7, this projection valued measure satisfies

$$\operatorname{Re}\langle x, P_\Omega y\rangle = \operatorname{Re}\langle x, \Psi_A(\chi_\Omega)y\rangle = \int_\Sigma \chi_\Omega \, d\mu_{x,y} = \mu_{x,y}(\Omega)$$

for all $x, y \in H$ and all $\Omega \in \mathcal{B}$, where the signed measures $\mu_{x,y} \in \mathcal{M}(\Sigma)$ are given by (5.6.14). Thus the signed measures $\mu_{x,y}$ are related to the projection valued Borel measure $\{P_\Omega\}_{\Omega\in\mathcal{B}}$ via equation (5.6.12). Hence (5.6.11) follows the (Normalization) axiom $\Psi_A(\mathrm{id}) = A$ and (5.6.20) with $f(\lambda) = \mathrm{Re}\lambda$ and $f(\lambda) = \mathrm{Im}\lambda$. This proves existence.

To prove uniqueness, fix any projection valued measure

$$\mathcal{B} \to \mathcal{L}^c(H) : \Omega \mapsto \widetilde{P}_\Omega,$$

define the signed Borel measures $\widetilde{\mu}_{x,y} : \mathcal{B} \to \mathbb{R}$ by equation (5.6.12), i.e.

$$\widetilde{\mu}_{x,y}(\Omega) := \mathrm{Re}\langle x, \widetilde{P}_\Omega y\rangle$$

for $x, y \in H$ and $\Omega \in \mathcal{B}$, and suppose that (5.6.11) holds, i.e.

$$\int_\Sigma \mathrm{Re}\lambda\, d\widetilde{\mu}_{x,y}(\lambda) + \int_\Sigma \mathrm{Im}\lambda\, d\widetilde{\mu}_{x,\mathbf{i}y}(\lambda) = \mathrm{Re}\langle x, Ay\rangle \tag{5.6.25}$$

for all $x, y \in H$. Let $\widetilde{\Psi} : B(\Sigma) \to \mathcal{L}^c(H)$ be the continuous C* algebra homomorphism associated to $\{\widetilde{P}_\Omega\}_{\Omega\in\mathcal{B}}$ in Theorem 5.6.2, i.e.

$$\mathrm{Re}\langle x, \widetilde{\Psi}(f)y\rangle = \int_\Sigma \mathrm{Re}f\, d\widetilde{\mu}_{x,y} + \int_\Sigma \mathrm{Im}f\, d\widetilde{\mu}_{x,\mathbf{i}y} \tag{5.6.26}$$

for all $x, y \in H$ and all $f \in B(\Sigma)$. Then the restriction of $\widetilde{\Psi}$ to $C(\Sigma)$ is a continuous C* algebra homomorphism from $C(\Sigma)$ to $\mathcal{L}^c(H)$ by Theorem 5.6.2, and it follows from (5.6.25) and (5.6.26) that

$$\widetilde{\Psi}(\mathrm{id}) = A.$$

Hence it follows from the uniqueness statement in Theorem 5.5.14 that

$$\widetilde{\Psi}(f) = f(A) \qquad \text{for all } f \in C(\Sigma),$$

where $C(\Sigma) \to \mathcal{L}^c(H) : f \mapsto f(A)$ is the C* algebra homomorphism associated to A in Theorem 5.5.14. By (5.6.26) this implies

$$\int_\Sigma f\, d\widetilde{\mu}_{x,x} = \langle x, \widetilde{\Psi}(f)x\rangle = \langle x, f(A)x\rangle = \int_\Sigma f d\mu_x$$

for all $f \in C(\Sigma, \mathbb{R})$ and all $x \in H$. Here the Borel measures

$$\mu_x : \mathcal{B} \to [0, \infty)$$

are defined by (5.6.2). Hence it follows from uniqueness in the Riesz Representation Theorem (see [**75**, Cor. 3.19]) that

$$\widetilde{\mu}_{x,x} = \mu_x$$

for all $x \in H$. This implies

$$\langle x, \widetilde{P}_\Omega x\rangle = \widetilde{\mu}_{x,x}(\Omega) = \mu_x(\Omega) = \langle x, \Psi_A(\chi_\Omega)x\rangle$$

for every $x \in H$ and every Borel set $\Omega \subset \Sigma$. Here $\Psi_A : B(\Sigma) \to \mathcal{L}^c(H)$ is the complex linear operator of Lemma 5.6.7. Since the operators $\widetilde{P}_\Omega$ and $\Psi_A(\chi_\Omega)$ are self-adjoint, we obtain for each Borel set $\Omega \subset \Sigma$ that

$$\widetilde{P}_\Omega = \Psi_A(\chi_\Omega).$$

This proves the uniqueness statement in Theorem 5.6.3. □

The next lemma is useful in preparation for the proof of Theorem 5.6.5.

Lemma 5.6.10. *Let Σ be a nonempty compact Hausdorff space such that every open subset of Σ is σ-compact. Let $B(\Sigma)$ be the Banach space of bounded Borel measurable complex valued functions on Σ equipped with the supremum norm and let $C(\Sigma) \subset B(\Sigma)$ be the subalgebra of continuous functions. Let $\mathcal{F} \subset B(\Sigma)$ be a subset that satisfies the following conditions.*

(a) *$\mathcal{F}$ is a complex unital subalgebra of $B(\Sigma)$.*

(b) *The subalgebra $\mathcal{F} \cap C(\Sigma)$ separates points.*

(c) *If $f \in \mathcal{F}$, then $\overline{f} \in \mathcal{F}$.*

(d) *If $(f_i)_{i\in\mathbb{N}}$ is a sequence in $\mathcal{F}$ and $f \in B(\Sigma)$ such that $\sup_{i\in\mathbb{N}} \|f_i\| < \infty$ and $\lim_{i\to\infty} f_i(\lambda) = f(\lambda)$ for all $\lambda \in \Sigma$, then $f \in \mathcal{F}$.*

Then $\mathcal{F} = B(\Sigma)$.

Proof. By (a), (b), (c), and Theorem 5.4.5 the set $\mathcal{F}\cap C(\Sigma)$ is dense in $C(\Sigma)$ with respect to the supremum norm. By (d) this subset is also closed and hence $C(\Sigma) \subset \mathcal{F}$.

Let $\mathcal{B} \subset 2^\Sigma$ be the Borel σ-algebra and define

$$\mathcal{B}_\mathcal{F} := \{\Omega \in \mathcal{B} \,|\, \chi_\Omega \in \mathcal{F}\}.$$

We prove that $\mathcal{B}_\mathcal{F}$ is a σ-algebra. First, $\emptyset, \Sigma \in \mathcal{B}_\mathcal{F}$ by (a) because the characteristic functions $\chi_\emptyset = 0$ and $\chi_\Sigma = 1$ are constant. Second, if $\Omega_1, \Omega_2 \in \mathcal{B}_\mathcal{F}$, then $\chi_{\Omega_1\setminus\Omega_2} = \chi_{\Omega_1}(1 - \chi_{\Omega_2}) \in \mathcal{F}$ by (a) and so $\Omega_1 \setminus \Omega_2 \in \mathcal{B}_\mathcal{F}$. Third, if Ω_i is a pairwise disjoint sequence of Borel sets in $\mathcal{B}_\mathcal{F}$ and $\Omega := \bigcup_{i=1}^\infty \Omega_i$, then the sequence $\sum_{i=1}^n \chi_{\Omega_i}$ belongs to $\mathcal{F}$ by (a) and converges pointwise to χ_Ω. Hence $\chi_\Omega \in \mathcal{F}$ by (d) and so $\Omega \in \mathcal{B}_\mathcal{F}$. This shows that $\mathcal{B}_\mathcal{F}$ is a σ-algebra.

Let $U \subset \Sigma$ be open. Since U is σ-compact, there is a sequence of compact sets $K_i \subset \Sigma$ such that $K_i \subset K_{i+1}$ for all i and $U = \bigcup_{i=1}^\infty K_i$. By Urysohn's Lemma, there is a sequence of continuous functions $f_i : \Sigma \to [0, 1]$ such that

$$f_i(\lambda) = \begin{cases} 1, & \text{for all } x \in K_i, \\ 0, & \text{for all } x \in \Sigma \setminus U. \end{cases}$$

This sequence converges pointwise to the characteristic function χ_U of U. Since $f_i \in C(\Sigma) \subset \mathcal{F}$ for all i, it follows that $\chi_U \in \mathcal{F}$ by (d) and so $U \in \mathcal{B}_\mathcal{F}$. This shows that $\mathcal{B}_\mathcal{F} \subset \mathcal{B}$ is a σ-algebra that contains all open sets, so $\mathcal{B}_\mathcal{F} = \mathcal{B}$. Thus $\chi_\Omega \in \mathcal{F}$ for all $\Omega \in \mathcal{B}$.

Now let $f : \Sigma \to \mathbb{C}$ be any bounded Borel measurable function. Then there exists a sequence of Borel measurable step functions $f_i : \Sigma \to \mathbb{C}$ (whose images are finite sets) such that f_i converges pointwise to f and $\|f_i\| \le \|f\|$ for all i (see [**75**, Thm. 1.26]). Hence it follows from (d) that $f \in \mathcal{F}$. This proves Lemma 5.6.10. □

Proof of Theorem 5.6.5. Let $\Psi_A : B(\Sigma) \to \mathcal{L}^c(H)$ be the bounded complex linear operator introduced in Lemma 5.6.7. It satisfies the (Conjugation), (Normalization), (Positive), (Contraction), and (Commutative) axioms by Lemma 5.6.7, the (Product) axiom by Lemma 5.6.8, and the (Convergence) axiom by Lemma 5.6.9.

We prove that Ψ_A satisfies the (Image) axiom. Denote by $\mathcal{A} \subset \mathcal{L}^c(H)$ the smallest C* subalgebra that contains A and is closed under strong convergence (i.e. if $(A_i)_{i\in\mathbb{N}}$ is a sequence in $\mathcal{A}$ and $A \in \mathcal{L}^c(H)$ is an operator satisfying $Ax = \lim_{i\to\infty} A_i x$ for all $x \in H$, then $A \in \mathcal{A}$). Since the image of the operator $\Psi_A : B(\Sigma) \to \mathcal{L}^c(H)$ is such a C* subalgebra of $\mathcal{L}^c(H)$, by the (Product), (Conjugation), (Normalization), and (Convergence) axioms, it must contain $\mathcal{A}$. To prove the converse inclusion, consider the set

$$\mathcal{F} := \{f \in B(\Sigma) \mid \Psi_A(f) \in \mathcal{A}\}.$$

This is a complex unital subalgebra of $B(\Sigma)$ because $\mathcal{A} \subset \mathcal{L}^c(H)$ is a complex unital subalgebra and the map $\Psi_A : B(\Sigma) \to \mathcal{L}^c(H)$ is a unital algebra homomorphism by Lemma 5.6.8. It contains the identity map because Ψ_A satisfies the (Normalization) axiom, and it is invariant under complex conjugation because Ψ_A satisfies the (Conjugation) axiom. Moreover, $\mathcal{F}$ is closed under pointwise convergence of bounded sequences by Lemma 5.6.9. Hence $\mathcal{F}$ satisfies the requirements of Lemma 5.6.10 and therefore $\mathcal{F} = B(\Sigma)$. This shows that Ψ_A satisfies the (Image) axiom.

We prove that Ψ_A satisfies the (Eigenvector) axiom. Fix a real number $\lambda \in \mathrm{P}\sigma(A) \subset \Sigma$ and vector $x \in H$ such that $Ax = \lambda x$. Define

$$\mathcal{F} := \{f \in B(\Sigma) \mid \Psi_A(f)x = f(\lambda)x\}.$$

This set is a complex unital subalgebra of $B(\Sigma)$ that contains the identity and is invariant under complex conjugation, because Ψ_A satisfies the (Product), (Normalization), and (Conjugation) axioms. Moreover, if $f_i \in \mathcal{F}$ is a bounded sequence that converges pointwise to a function $f : \Sigma \to \mathbb{C}$, then $f \in \mathcal{F}$ by Lemma 5.6.9. Hence $\mathcal{F} = B(\Sigma)$ by Lemma 5.6.10. This shows that Ψ_A satisfies the (Eigenvector) axiom.

We prove that Ψ_A satisfies the (Spectrum) axiom. Let $f \in B(\Sigma)$ and let $\mu \in \mathbb{C} \setminus \overline{f(\Sigma)}$. Define the function $g : \Sigma \to \mathbb{C}$ by

$$g(\lambda) := (\mu - f(\lambda))^{-1} \qquad \text{for } \lambda \in \Sigma.$$

Then g is measurable and bounded and $g(\mu - f) = (\mu - f)g = 1$. Hence

$$\Psi_A(g)\,(\mu\mathbb{1} - \Psi_A(f)) = (\mu\mathbb{1} - \Psi_A(f))\,\Psi_A(g) = \mathbb{1}$$

by Lemma 5.6.8. Thus $\mu\mathbb{1} - \Psi_A(f)$ is bijective and so $\mu \notin \sigma(\Psi_A(f))$. Hence the spectrum of the operator $\Psi_A(f)$ is contained in the closure of the set $f(\Sigma)$. This shows that Ψ_A satisfies the (Spectrum) axiom.

We prove uniqueness. Thus assume that

$$\Psi : B(\Sigma) \to \mathcal{L}^c(H)$$

is any complex linear operator that satisfies the (Product), (Conjugation), (Normalization), and (Convergence) axioms. Then

$$\Psi(f) = \Psi_A(f)$$

for every polynomial $f = p|_\Sigma : \Sigma \to \mathbb{C}$ in z and $\overline{z}$ by the (Product), (Conjugation), and (Normalization) axioms. Define

$$\mathcal{F} := \{ f \in B(\Sigma) \,|\, \Psi(f) = \Psi_A(f) \} .$$

This set is a complex subalgebra of $B(\Sigma)$ and contains the polynomials in z and $\overline{z}$ by what we have just observed. It is invariant under complex conjugation because both Ψ and Ψ_A satisfy the (Conjugation) axiom. Moreover, if $(f_i)_{i\in\mathbb{N}}$ is a bounded sequence in $\mathcal{F}$ that converges pointwise to a function $f : \Sigma \to \mathbb{C}$, then f is a bounded Borel measurable function and

$$\Psi(f)x = \lim_{i\to\infty} \Psi(f_i)x = \lim_{i\to\infty} \Psi_A(f_i)x = \Psi_A(f)x \qquad \text{for all } x \in H,$$

by the (Convergence) axiom for Ψ and by Lemma 5.6.9 for Ψ_A, and so $f \in \mathcal{F}$. Thus the set $\mathcal{F}$ satisfies the requirements of Lemma 5.6.10 and so

$$\mathcal{F} = B(\Sigma).$$

This proves uniqueness.

We prove the (Composition) axiom. Fix a continuous function $f : \Sigma \to \mathbb{R}$ and define the set

$$\mathcal{G} := \{ g \in B(f(\Sigma)) \,|\, (g \circ f)(A) = g(f(A)) \} .$$

This set is a complex unital subalgebra of $B(f(\Sigma))$ because the maps

$$B(f(\Sigma)) \to \mathcal{L}^c(H) : g \mapsto (g \circ f)(A)$$

and

$$B(f(\Sigma)) \to \mathcal{L}^c(H) : g \mapsto g(f(A))$$

are both unital C* algebra homomorphisms. Second, the set $\mathcal{G}$ contains the identity map by definition and is invariant under complex conjugation by the (Conjugation) axiom. Third, the subspace $\mathcal{G}$ is closed under pointwise convergence of bounded sequences by the (Convergence) axiom. Hence

$$\mathcal{G} = B(f(\Sigma))$$

by Lemma 5.6.10. This proves Theorem 5.6.5. □

The final theorem of this subsection establishes some useful additional properties of the spectral measure and the measurable functional calculus of a normal operator.

Theorem 5.6.11 (Spectral Projections for Normal Operators). *Let H be a nonzero complex Hilbert space and let $A \in \mathcal{L}^c(H)$ be a normal operator. Denote its spectrum by $\Sigma := \sigma(A) \subset \mathbb{C}$.*

(i) *Let $\Omega \subset \Sigma$ be a nonempty Borel set and let $\chi_\Omega : \Sigma \to \{0, 1\}$ be the characteristic function of Ω. Then*

$$P_\Omega := \chi_\Omega(A)$$

is an orthogonal projection, its image

$$E_\Omega := \mathrm{im}(P_\Omega)$$

is an A-invariant subspace of H, and

$$\Sigma \setminus \overline{\Sigma \setminus \Omega} \subset \sigma(A|_{E_\Omega}) \subset \overline{\Omega}. \tag{5.6.27}$$

(ii) *Let $f \in B(\Sigma)$ and let $\lambda \in \Sigma$. If f is continuous at λ, then*

$$f(\lambda) \in \sigma(f(A)).$$

(iii) *Let $\lambda \in \Sigma$ and define $P_\lambda := P_{\{\lambda\}} \in \mathcal{L}^c(H)$. Then*

$$P_\lambda = P_\lambda^2 = P_\lambda^*, \qquad \mathrm{im}(P_\lambda) = \ker(\lambda \mathbb{1} - A). \tag{5.6.28}$$

Proof. We prove (i). When $\Omega = \Sigma$ or $\Omega = \emptyset$ there is nothing to prove. (The zero operator on the zero vector space has an empty spectrum.) Thus assume $\Omega \neq \Sigma$ and $\Omega \neq \emptyset$. Since $\chi_\Omega = \chi_\Omega^2 = \overline{\chi}_\Omega$, the operator P_Ω is an orthogonal projection. It commutes with A and hence its image $E_\Omega := \mathrm{im}(P_\Omega)$ is invariant under A.

For $c \in \mathbb{C}$ define $f_c : \sigma(A) \to \mathbb{C}$ by

$$f_c(\lambda) := \begin{cases} \lambda, & \text{for } \lambda \in \Omega, \\ c, & \text{for } \lambda \in \sigma(A) \setminus \Omega. \end{cases}$$

Then $f_c = \chi_\Omega \mathrm{id} + c\chi_{\sigma(A)\setminus\Omega}$, hence

$$f_c(A) = AP_\Omega + c(\mathbb{1} - P_\Omega),$$

and hence

$$\sigma(AP_\Omega + c(\mathbb{1} - P_\Omega)) \subset \overline{\Omega} \cup \{c\} \qquad \text{for all } c \in \mathbb{C},$$

by the (Spectrum) axiom in Theorem 5.6.5. If $\lambda \in \mathbb{C} \setminus \overline{\Omega}$ and $c \neq \lambda$, it follows that the operator

$$\lambda \mathbb{1} - f_c(A) = (\lambda \mathbb{1} - A)P_\Omega + (\lambda - c)(\mathbb{1} - P_\Omega)$$

is invertible and hence so is the operator $\lambda \mathbb{1} - A|_{E_\Omega} : E_\Omega \to E_\Omega$. This shows that $\sigma(A|_{E_\Omega}) \subset \overline{\Omega}$. Now let $\lambda \in \Sigma \setminus \overline{\Sigma \setminus \Omega}$. Then

$$\lambda \notin \sigma(A|_{E_{\Sigma\setminus\Omega}}) = \sigma(A|_{E_\Omega^\perp})$$

by what we have just proved and hence $\lambda \in \sigma(A|_{E_\Omega})$. This proves part (i).

We prove (ii). Suppose, by contradiction, that the operator

$$f(\lambda)\mathbb{1} - f(A)$$

is invertible and define

$$\varepsilon := \|(f(\lambda)\mathbb{1} - f(A))^{-1}\|^{-1}.$$

Then Corollary 1.5.7 asserts that the operator $\mu\mathbb{1} - f(A)$ is invertible for every $\mu \in \mathbb{C}$ with $|\mu - f(\lambda)| < \varepsilon$. Hence

$$\sigma(f(A)) \cap B_\varepsilon(f(\lambda)) = \emptyset. \tag{5.6.29}$$

Since f is continuous at λ, there exists a $\delta > 0$ such that every $\lambda' \in \Sigma$ satisfies

$$|\lambda - \lambda'| \le \delta \qquad \Longrightarrow \qquad |f(\lambda) - f(\lambda')| \le \frac{\varepsilon}{2}. \tag{5.6.30}$$

Define

$$\Omega := B_\delta(\lambda) \cap \Sigma.$$

Then $\overline{\Omega} \subset \overline{B}_\delta(\lambda) \cap \Sigma$, hence $f(\overline{\Omega}) \subset \overline{B}_{\varepsilon/2}(f(\lambda))$ by (5.6.30), and so

$$\overline{f(\overline{\Omega})} \subset \overline{B}_{\varepsilon/2}(f(\lambda)) \subset B_\varepsilon(f(\lambda)) \subset \mathbb{C} \setminus \sigma(f(A)). \tag{5.6.31}$$

Here the last step follows from (5.6.29). Moreover, $\Sigma \setminus \Omega = \Sigma \setminus B_\delta(\lambda)$ is a closed subset of $\mathbb{C}$ and so

$$\Sigma \setminus \overline{\Sigma \setminus \Omega} = \Omega. \tag{5.6.32}$$

Now let

$$P_\Omega := \chi_\Omega(A), \qquad E_\Omega := \mathrm{im}(P_\Omega)$$

as in (i) and define

$$A_\Omega := A|_{E_\Omega}.$$

Then $\Omega \subset \sigma(A_\Omega) \subset \overline{\Omega}$ by (5.6.27) and (5.6.32). This implies

$$\sigma(A_\Omega) = \overline{\Omega} \subset \Sigma,$$

because $\sigma(A_\Omega)$ is closed. Moreover, $\lambda \in \Omega \subset \sigma(A_\Omega)$ and so

$$E_\Omega \ne \{0\}. \tag{5.6.33}$$

For $g \in B(\Sigma)$ define

$$g_\Omega := g|_{\overline{\Omega}}.$$

Then the operator $P_\Omega = \chi_\Omega(A)$ commutes with $g(A)$ and so the subspace E_Ω is invariant under $g(A)$ for all $g \in B(\Sigma)$. We claim that

$$g_\Omega(A_\Omega) = g(A)|_{E_\Omega} : E_\Omega \to E_\Omega \qquad \text{for all } g \in B(\Sigma). \tag{5.6.34}$$

This formula clearly holds when g is a polynomial in z and $\overline{z}$, hence it holds for every continuous function $g : \Sigma \to \mathbb{C}$ by the Stone–Weierstraß Theorem 5.4.5, and hence it holds for all $g \in B(\Sigma)$ by Lemma 5.6.10. In particular, equation (5.6.34) holds for our fixed function $g = f$.

It follows from (5.6.31), (5.6.34), and the (Spectrum) axiom in Theorem 5.6.5 that

$$\begin{aligned}\sigma(f(A)|_{E_\Omega}) &= \sigma(f_\Omega(A_\Omega)) \\ &\subset \overline{f(\sigma(A_\Omega))} \\ &= \overline{f(\overline{\Omega})} \\ &\subset \mathbb{C} \setminus \sigma(f(A)).\end{aligned}$$

Since $\sigma(f(A)|_{E_\Omega}) \subset \sigma(f(A))$, this implies

$$\sigma(f(A)|_{E_\Omega}) = \emptyset,$$

in contradiction to the fact that $E_\Omega \neq \{0\}$ by (5.6.33). This proves part (ii).

We prove (iii). Write

$$\chi_\lambda := \chi_{\{\lambda\}}.$$

If $x \in H$ satisfies $Ax = \lambda x$, then

$$P_\lambda x = \chi_\lambda(A)x = \chi_\lambda(\lambda)x = x$$

by the (Eigenvector) axiom in Theorem 5.6.5. Thus

$$\ker(\lambda \mathbb{1} - A) \subset \operatorname{im}(P_\lambda).$$

Conversely, let $x \in \operatorname{im}(P_\lambda)$ and consider the map

$$g := \mathrm{id} : \Sigma \to \Sigma \subset \mathbb{C}.$$

Then

$$x = P_\lambda x, \qquad g\chi_\lambda = \lambda\chi_\lambda$$

and hence

$$\begin{aligned}Ax &= AP_\lambda x \\ &= g(A)\chi_\lambda(A)x \\ &= (g\chi_\lambda)(A)x \\ &= \lambda\chi_\lambda(A)x \\ &= \lambda P_\lambda x \\ &= \lambda x.\end{aligned}$$

This shows that

$$\operatorname{im}(P_\lambda) \subset \ker(\lambda \mathbb{1} - A).$$

Hence $\operatorname{im}(P_\lambda) = \ker(\lambda\mathbb{1} - A)$. This proves part (iii) and Theorem 5.6.11. □

5.7. Cyclic Vectors

The spectral measure can be used to identify a self-adjoint operator on a real or complex Hilbert space with a multiplication operator. This is the content of the next theorem, as formulated in [**72**, p. 227].

Theorem 5.7.1 (Spectral Theorem). *Let H be a nonzero complex Hilbert space and let*

$$A = A^* \in \mathcal{L}^c(H)$$

be a self-adjoint complex linear operator. Then there exists a collection of compact sets

$$\Sigma_i \subset \sigma(A),$$

each equipped with a Borel measure μ_i, indexed by $i \in I$, and an isomorphism

$$U : H \to \bigoplus_{i \in I} L^2(\Sigma_i, \mu_i) := \left\{ \psi = (\psi_i)_{i \in I} \;\middle|\; \begin{array}{l} \psi_i \in L^2(\Sigma_i, \mu_i) \text{ for all } i \in I \\ \text{and } \sum_{i \in I} \|\psi_i\|^2_{L^2(\Sigma_i, \mu_i)} < \infty \end{array} \right\}$$

such that the operator UAU^{-1} sends a tuple

$$\psi = (\psi_i)_{i \in I} \in \bigoplus_{i \in I} L^2(\Sigma_i, \mu_i)$$

to the tuple

$$UAU^{-1}\psi = ((UAU^{-1}\psi)_i)_{i \in I} \in \bigoplus_{i \in I} L^2(\Sigma_i, \mu_i)$$

given by

$$(UAU^{-1}\psi)_i(\lambda) = \lambda \psi_i(\lambda) \qquad \text{for } i \in I \text{ and } \lambda \in \Sigma_i.$$

Moreover, $\mu_i(\Omega) > 0$ for every $i \in I$ and every nonempty relatively open subset $\Omega \subset \Sigma_i$. If H is separable, then the index set I can be chosen countable.

Proof. See page 285. □

Theorem 5.7.1 can be viewed as a *diagonalization* of the operator A, extending the notion of diagonalization of a symmetric matrix. The proof is based on the notion of a cyclic vector.

Definition 5.7.2 (Cyclic Vector). Let H be a nonzero complex Hilbert space and let $A = A^* \in \mathcal{L}^c(H)$ be a self-adjoint complex linear operator. A vector $x \in H$ is called **cyclic for** A if

$$H = \overline{\operatorname{span}\{A^n x \mid n = 0, 1, 2, \dots\}}.$$

If such a cyclic vector exists, the Hilbert space H is necessarily separable.

Theorem 5.7.3 (Cyclic Vectors and Multiplication Operators). *Let H be a nonzero complex Hilbert space, let $A = A^* \in \mathcal{L}^c(H)$ be a self-adjoint complex linear operator, let $\Sigma := \sigma(A) \subset \mathbb{R}$ be the spectrum of A, and let $\mathcal{B} \subset 2^\Sigma$ be the Borel σ-algebra. Let $x \in H$ be a cyclic vector for A, let $\mu_x : \mathcal{B} \to [0, \infty)$ be the unique Borel measure that satisfies (5.6.2), and denote by $L^2(\Sigma, \mu_x)$ the complex L^2 space of μ_x. Then the following hold.*

(i) *There is a unique Hilbert space isometry $U : H \to L^2(\Sigma, \mu_x)$ such that*

$$U^{-1}\psi = \psi(A)x \qquad \text{for all } \psi \in C(\Sigma). \tag{5.7.1}$$

(ii) *Let $f : \Sigma \to \mathbb{C}$ be a bounded Borel measurable function. Then*

$$Uf(A)U^{-1}\psi = f\psi \tag{5.7.2}$$

for all $\psi \in L^2(\Sigma, \mu_x)$.

(iii) *The operator U in part (i) satisfies*

$$(UAU^{-1}\psi)(\lambda) = \lambda\psi(\lambda) \tag{5.7.3}$$

for all $\psi \in L^2(\Sigma, \mu_x)$ and all $\lambda \in \Sigma$.

(iv) *If $\Omega \subset \Sigma$ is a nonempty (relatively) open subset, then $\mu_x(\Omega) > 0$.*

Proof. We prove part (i). Define the map $T : C(\Sigma) \to H$ by

$$T\psi := \psi(A)x \qquad \text{for } \psi \in C(\Sigma). \tag{5.7.4}$$

Here $\psi(A) \in \mathcal{L}^c(H)$ is the operator in Theorem 5.4.7. The operator T is complex linear and it satisfies

$$\begin{aligned} \|T\psi\|_H^2 &= \langle \psi(A)x, \psi(A)x \rangle \\ &= \langle x, \psi(A)^*\psi(A)x \rangle \\ &= \langle x, \overline{\psi}(A)\psi(A)x \rangle \\ &= \langle x, |\psi|^2(A)x \rangle \\ &= \int_\Sigma |\psi|^2 \, d\mu_x \\ &= \|\psi\|_{L^2}^2 \end{aligned} \tag{5.7.5}$$

for all $\psi \in C(\Sigma)$. Here the penultimate step follows from the definition of the Borel measure μ_x on Σ in (5.6.2). Equation (5.7.5) shows the operator $T : C(\Sigma) \to H$ is an isometric embedding with respect to the L^2 norm on $C(\Sigma)$. By a standard result in measure theory, $C(\Sigma)$ is a dense subset of $L^2(\Sigma, \mu_x)$ (see for example [**75**, Thm. 4.15]). More precisely, the obvious map from $C(\Sigma)$ to $L^2(\Sigma, \mu_x)$ has a dense image. Hence the usual approximation argument shows that T extends to an isometric embedding of $L^2(\Sigma, \mu_x)$

into H which will still be denoted by

$$T : L^2(\Sigma, \mu_x) \to H. \tag{5.7.6}$$

(Given $\psi \in L^2(\Sigma, \mu_x)$, choose a sequence $\psi_n \in C(\Sigma)$ that L^2 converges to f; then $(T\psi_n)_{n\in\mathbb{N}}$ is a Cauchy sequence in H by (5.7.5); so $(T\psi_n)_{n\in\mathbb{N}}$ converges; the limit is independent of the choice of the sequence ψ_n that L^2 converges to ψ and is by definition the image $T\psi := \lim_{n\to\infty} T\psi_n$ of ψ under T.) Since the extended operator (5.7.6) is an isometric embedding it is, in particular, injective and has a closed image.

We prove that it is surjective. To see this, consider the sequence of continuous functions $\psi_n : \Sigma \to \mathbb{R}$ defined by

$$\psi_n(\lambda) := \lambda^n$$

for $n \in \mathbb{N}$ and $\lambda \in \Sigma$. Then $\psi_n(A) = A^n$ by the (Normalization) and (Product) axioms in Theorem 5.4.7. By definition of T in (5.7.4) this implies that the vector $A^n x = \psi_n(A)x = T\psi_n$ belongs to the image of T for all $n \in \mathbb{N}$. Since T is complex linear it follows that $\operatorname{span}\{x, Ax, A^2x, \dots\} \subset \operatorname{im}(T)$. Since x is a cyclic vector for A, this implies

$$H = \overline{\operatorname{span}\{x, Ax, A^2x, \dots\}} \subset \overline{\operatorname{im}(T)} = \operatorname{im}(T).$$

Thus the extended operator $T : L^2(\Sigma, \mu_x) \to H$ is an isometric isomorphism by (5.7.5). Its inverse $U := T^{-1} : H \to L^2(\Sigma, \mu_x)$ satisfies equation (5.7.1) by definition and is uniquely determined by this condition in view of the above extension argument. This proves part (i).

We prove part (ii). Since $C(\Sigma)$ is dense in $L^2(\Sigma, \mu_x)$, it suffices to prove the identity (5.7.2) for $\psi \in C(\Sigma)$. Assume first that $f \in C(\Sigma)$. If $\psi \in C(\Sigma)$, then it follows from (5.7.1) and the (Product) axiom in Theorem 5.6.5 that

$$f(A)U^{-1}\psi = f(A)\psi(A)x = (f\psi)(A)x = U^{-1}(f\psi)$$

and hence

$$Uf(A)U^{-1}\psi = f\psi.$$

Thus (5.7.2) holds for all $\psi \in C(\Sigma)$ and therefore, by continuity, for all $\psi \in L^2(\Sigma, \mu_x)$. Define

$$\mathcal{F} := \left\{ f \in B(\Sigma) \,|\, Uf(A)U^{-1}\psi = f\psi \text{ for all } \psi \in L^2(\Sigma, \mu_x) \right\}.$$

This set is a subalgebra of $B(\Sigma)$ by definition and $C(\Sigma) \subset \mathcal{F}$ by what we have just proved. Moreover, $\mathcal{F}$ is closed under pointwise convergence of bounded functions by the (Convergence) axiom in Theorem 5.6.5. Hence $\mathcal{F} = B(\Sigma)$ by Lemma 5.6.10 and this proves part (ii).

Part (iii) follows from part (ii) by taking $f = \mathrm{id} : \Sigma \to \Sigma \subset \mathbb{C}$.

We prove part (iv). Let $\Omega \subset \Sigma$ be a nonempty relatively open subset and suppose, by contradiction, that $\mu_x(\Omega) = 0$. Fix an element $\lambda_0 \in \Omega$ and define the functions $f, g : \Sigma \to \mathbb{C}$ by

$$f(\lambda) := \begin{cases} \frac{1}{\lambda_0 - \lambda}, & \text{for } \lambda \in \Sigma \setminus \Omega, \\ 0, & \text{for } \lambda \in \Omega, \end{cases} \qquad g(\lambda) := \begin{cases} 1, & \text{for } \lambda \in \Sigma \setminus \Omega, \\ 0, & \text{for } \lambda \in \Omega. \end{cases}$$

Then f is a bounded measurable function because Ω is open, and $g \overset{a.e.}{=} 1$ because $\mu_x(\Omega) = 0$. Moreover,

$$f(\lambda_0 - \mathrm{id}) = (\lambda_0 - \mathrm{id}) f = g$$

and hence it follows from parts (ii) and (iii) that

$$\begin{aligned} (U^{-1} f(A)(\lambda_0 \mathbb{1} - A) U)\psi &= (U^{-1}(\lambda_0 \mathbb{1} - A) f(A) U)\psi \\ &= (U^{-1} g(A) U)\psi \\ &= g\psi \overset{a.e.}{=} \psi \end{aligned}$$

for all $\psi \in \mathcal{L}^2(\Sigma, \mu_x)$. Thus the operator $\lambda_0 \mathbb{1} - A$ is bijective and therefore $\lambda_0 \in \Sigma \setminus \sigma(A)$, a contradiction. This proves Theorem 5.7.3. □

The essential hypothesis in Theorem 5.7.3 is the existence of a cyclic vector and not every self-adjoint operator admits a cyclic vector. However, given a self-adjoint operator $A = A^* \in \mathcal{L}^c(H)$ and any nonzero vector $x \in H$ one can restrict A to the smallest closed A-invariant subspace of H that contains x and apply Theorem 5.7.3 to the restriction of A to this subspace.

Corollary 5.7.4. *Let H be a complex Hilbert space, let $x \in H \setminus \{0\}$, and let $A = A^* \in \mathcal{L}^c(H)$. Then*

$$H_x := \overline{\mathrm{span}\{x, Ax, A^2x, \dots\}} \tag{5.7.7}$$

is the smallest closed A-invariant linear subspace of H that contains x. Define $A_x := A|_{H_x} : H_x \to H_x$, let $\Sigma_x := \sigma(A_x)$, let $\mathcal{B}_x \subset 2^{\Sigma_x}$ be the Borel σ-algebra, and let $\mu_x : \mathcal{B}_x \to [0, \infty)$ be the unique Borel measure that satisfies (5.6.2) for all $f \in C(\Sigma_x)$. Then there exists a unique Hilbert space isometry $U_x : H_x \to L^2(\Sigma_x, \mu_x)$ such that

$$U_x^{-1}\psi = \psi(A_x)x \qquad \text{for all } \psi \in C(\Sigma_x). \tag{5.7.8}$$

This operator satisfies

$$U_x f(A_x) U_x^{-1} \psi = f\psi \tag{5.7.9}$$

for all $f \in B(\Sigma_x)$ and all $\psi \in L^2(\Sigma_x, \mu_x)$. Moreover, $\mu_x(\Omega) > 0$ for every nonempty relatively open subset $\Omega \subset \Sigma_x$.

Proof. This follows directly from Theorem 5.7.3. □

Proof of Theorem 5.7.1. Here is a reformulation of the assertion.

Let H be a nonzero complex Hilbert space and let

$$A = A^* \in \mathcal{L}^c(H).$$

Then there exists a nonempty collection of nontrivial pairwise orthogonal closed A-invariant complex linear subspaces $H_i \subset H$ for $i \in I$ such that

$$A_i := A|_{H_i} : H_i \to H_i$$

admits a cyclic vector for each $i \in I$ and

$$H = \bigoplus_{i \in I} H_i.$$

Thus there is a collection of nonempty compact subsets $\Sigma_i \subset \sigma(A)$, Borel measures μ_i on Σ_i, and Hilbert space isometries

$$U_i : H_i \to L^2(\Sigma_i, \mu_i)$$

for $i \in I$, such that $\mu_i(\Omega) > 0$ for all $i \in I$ and all nonempty relatively open subsets $\Omega \subset \Sigma_i$ and

$$(U_i A_i U_i^{-1} \psi_i)(\lambda) = \lambda \psi_i(\lambda) \tag{5.7.10}$$

for all $i \in I$, all $\psi_i \in L^2(\Sigma_i, \mu_i)$, and all $\lambda \in \Sigma_i$.

Call a subset $S \subset H$ **A-orthonormal** if it satisfies the condition

$$\langle x, A^k y \rangle = \begin{cases} 1, & \text{if } x = y,\ k = 0, \\ 0, & \text{if } x \neq y, \end{cases} \qquad \text{for all } x, y \in S \text{ and } k \in \mathbb{N}_0. \tag{5.7.11}$$

The collection $\mathscr{S} := \{S \subset H \mid S \text{ satisfies (5.7.11)}\}$ of all A-orthonormal subsets of H is nonempty because $\{x\} \in \mathscr{S}$ for every unit vector $x \in H$. Moreover, $\mathscr{S}$ is partially ordered by inclusion and every nonempty chain in $\mathscr{S}$ has a supremum. Hence it follows from the Lemma of Zorn that $\mathscr{S}$ contains a maximal element $S \in \mathscr{S}$. If $S \in \mathscr{S}$ is a maximal element, then Corollary 5.7.4 implies that the collection $\{H_x\}_{x \in S}$ defined by (5.7.7) satisfies the requirements of Theorem 5.7.1 as formulated above. □

Exercise 5.7.5. Let $\Sigma \subset \mathbb{R}$ be a nonempty compact set and let μ be a Borel measure on Σ such that every nonempty relatively open subset of Σ has positive measure. Define the operator $A : L^2(\Sigma, \mu) \to L^2(\Sigma, \mu)$ by

$$(A\psi)(\lambda) := \lambda \psi(\lambda) \qquad \text{for } \psi \in L^2(\Sigma, \mu) \text{ and } \lambda \in \Sigma. \tag{5.7.12}$$

Prove that A is self-adjoint and $\sigma(A) = \Sigma$. Find a cyclic vector for A. Theorem 5.7.1 shows that every self-adjoint operator on a complex Hilbert space is a direct sum of operators of the form (5.7.12).

Exercise 5.7.6. Let H be a nonzero complex Hilbert space and let A be a compact self-adjoint operator on H. Prove that A admits a cyclic vector if and only if A is injective and $E_\lambda := \ker(\lambda 1\!\!1 - A)$ has dimension one for every $\lambda \in \mathrm{P}\sigma(A)$.

Exercise 5.7.7. Let

$$A = A^* \in \mathbb{C}^{n\times n}$$

be a Hermitian matrix and let $e_1, \dots, e_n$ be an orthonormal basis of eigenvectors, so $Ae_i = \lambda_i e_i$ for $i = 1, \dots, n$ with $\lambda_i \in \mathbb{R}$. Thus

$$\Sigma := \sigma(A) = \{\lambda_1, \dots, \lambda_n\}.$$

Assume $\lambda_i \neq \lambda_j$ for $i \neq j$.

(i) Prove that $f(A)x = \sum_{i=1}^n f(\lambda_i)\langle e_i, x\rangle e_i$ for $x \in \mathbb{C}^n$ and $f : \Sigma \to \mathbb{C}$.

(ii) Prove that $x := \sum_{i=1}^n e_i$ is a cyclic vector and that $\mu_x = \sum_{i=1}^n \delta_{\lambda_i}$ is the sum of the Dirac measures, so

$$\int_\Sigma f\,d\mu_x = \sum_{i=1}^n f(\lambda_i)$$

for $f : \Sigma \to \mathbb{C}$.

(iii) Let $U : \mathbb{C}^n \to L^2(\Sigma, \mu_x)$ be the isometry in Theorem 5.7.3. Prove that $(Ux)(\lambda_i) = \langle e_i, x\rangle$ for $x \in \mathbb{C}^n$ and $U^{-1}\psi = \sum_{i=1}^n \psi(\lambda_i)e_i$ for $\psi \in L^2(\Sigma, \mu_x)$.

Exercise 5.7.8. Let H be an infinite-dimensional separable complex Hilbert space and let $A = A^* \in \mathcal{L}^c(H)$ be a self-adjoint operator. Assume that $(e_i)_{i\in\mathbb{N}}$ is an orthonormal basis of eigenvectors of A so that $Ae_i = \lambda_i e_i$ for all $i \in \mathbb{N}$, where $\lambda_i \in \mathbb{R}$. Thus $\sup_{i\in\mathbb{N}} |\lambda_i| < \infty$ and $\Sigma = \overline{\{\lambda_i \,|\, i \in \mathbb{N}\}}$. Assume $\lambda_i \neq \lambda_j$ for $i \neq j$.

(i) Prove that

$$f(A)x = \sum_{i=1}^\infty f(\lambda_i)\langle e_i, x\rangle e_i$$

for every $x \in H$ and every bounded function $f : \Sigma \to \mathbb{C}$.

(ii) Choose a sequence $\varepsilon_i > 0$ such that $\sum_{i=1}^\infty \varepsilon_i^2 < \infty$. Prove that the vector $x := \sum_{i=1}^\infty \varepsilon_i e_i$ is cyclic for A and that $\mu_x = \sum_{i=1}^\infty \varepsilon_i^2 \delta_{\lambda_i}$.

(iii) Prove that the map $\psi \mapsto (\psi(\lambda_i))_{i\in\mathbb{N}}$ defines an isomorphism

$$L^2(\Sigma, \mu_x) \cong \left\{\eta = (\eta_i)_{i=1}^\infty \in \mathbb{C}^{\mathbb{N}} \,\middle|\, \sum_{i=1}^\infty \varepsilon_i^2 |\eta_i|^2 < \infty\right\}.$$

Prove that the operator $U : H \to L^2(\Sigma, \mu_x)$ in Theorem 5.7.3 is given by $(U^{-1}\psi) = \sum_{i=1}^\infty \varepsilon_i \psi(\lambda_i) e_i$. Prove that the operator $\Lambda := UAU^{-1}$ on the Hilbert space $L^2(\Sigma, \mu_x)$ is given by $\eta \mapsto (\lambda_i \eta_i)_{i\in\mathbb{N}}$.

Exercise 5.7.9. Here is an example with a rather different flavor. Consider the Hilbert space

$$H := \ell^2(\mathbb{Z}, \mathbb{C}) = \left\{ x = (x_n)_{n\in\mathbb{Z}} \in \mathbb{C}^{\mathbb{Z}} \,\middle|\, \sum_{n=-\infty}^{\infty} |x_n|^2 < \infty \right\}$$

and define the operator $A : H \to H$ by

$$Ax := (x_{n-1} + x_{n+1})_{n\in\mathbb{Z}} \qquad \text{for } x = (x_n)_{n\in\mathbb{Z}} \in H.$$

Thus $A = L + L^*$, where $L : H \to H$ is given by $Lx = (x_{n+1})_{n\in\mathbb{Z}}$. The vectors $e_i = (\delta_{in})_{n\in\mathbb{Z}}$ for $i \in \mathbb{Z}$ form an orthonormal basis of H.

(i) Define $a^{\mathrm{ev}} := e_0$ and $a^{\mathrm{odd}} := e_1 - e_{-1}$. Then

$$\begin{aligned} H^{\mathrm{ev}} &:= \overline{\mathrm{span}\{A^k a^{\mathrm{ev}} \,|\, k = 0, 1, 2, \dots\}} \\ &= \{x = (x_n)_{n\in\mathbb{Z}} \in H \,|\, x_n - x_{-n} = 0 \text{ for all } n \in \mathbb{Z}\}, \\ H^{\mathrm{odd}} &:= \overline{\mathrm{span}\{A^k a^{\mathrm{odd}} \,|\, k = 0, 1, 2, \dots\}} \\ &= \{x = (x_n)_{n\in\mathbb{Z}} \in H \,|\, x_n + x_{-n} = 0 \text{ for all } n \in \mathbb{Z}\}, \\ H &= H^{\mathrm{ev}} \oplus H^{\mathrm{odd}}. \end{aligned}$$

(ii) Define the operator $\Phi : H \to L^2([0,1])$ by $(\Phi x)(t) := \sum_{n\in\mathbb{Z}} e^{2\pi \mathbf{i} n t} x_n$ for $x \in H$ and $t \in [0,1]$. Then Φ is an isometric isomorphism and

$$(\Phi A \Phi^{-1} f)(t) = 2\cos(2\pi t) f(t) \qquad \text{for } f \in L^2([0,1]) \text{ and } 0 \le t \le 1.$$

Find a formula for $\Phi g(A) \Phi^{-1}$ for every continuous function $g : [-2,2] \to \mathbb{C}$.

(iii) $\mathrm{P}\sigma(A) = \emptyset$ and $\Sigma := \sigma(A) = [-2,2]$.

(iv) Let μ^{ev}, respectively μ^{odd}, be the Borel measure on $[-2,2]$ determined by equation (5.6.2) with x replaced by a^{ev}, respectively a^{odd}. Then

$$\mu^{\mathrm{ev}} = \frac{1}{\pi\sqrt{4-\lambda^2}}\, d\lambda, \qquad \mu^{\mathrm{odd}} = \frac{\sqrt{4-\lambda^2}}{\pi}\, d\lambda.$$

Hint: Use parts (ii) and (iii) with $(\Phi a^{\mathrm{ev}})(t) = 1$, $(\Phi a^{\mathrm{odd}})(t) = 2\mathbf{i}\sin(2\pi t)$.

(v) There exists a unique isomorphism $U^{\mathrm{ev}} : H^{\mathrm{ev}} \to L^2([-2,2], \mu^{\mathrm{ev}})$ such that $U^{\mathrm{ev}} f(A) a^{\mathrm{ev}} = f$ for all $f \in C([-2,2])$. It satisfies

$$(U^{\mathrm{ev}} A (U^{\mathrm{ev}})^{-1} \psi)(\lambda) = \lambda \psi(\lambda)$$

for $\psi \in L^2([-2,2], \mu^{\mathrm{ev}})$ and $\lambda \in [-2,2]$.

(vi) There exists a unique isomorphism $U^{\mathrm{odd}} : H^{\mathrm{odd}} \to L^2([-2,2], \mu^{\mathrm{odd}})$ such that $U^{\mathrm{odd}} f(A) a^{\mathrm{odd}} = f$ for all $f \in C([-2,2])$. It satisfies

$$(U^{\mathrm{odd}} A (U^{\mathrm{odd}})^{-1} \psi)(\lambda) = \lambda \psi(\lambda)$$

for $\psi \in L^2([-2,2], \mu^{\mathrm{odd}})$ and $\lambda \in [-2,2]$.

5.8. Problems

Exercise 5.8.1 (Invariant Subspaces). Let H be a complex Hilbert space and $A \in \mathcal{L}^c(H)$. Let $E \subset H$ be a closed complex linear subspace of H. The subspace E is called **invariant under** A if, for all $x \in H$,

$$x \in E \qquad \Longrightarrow \qquad Ax \in E.$$

Prove that E is invariant under A if and only if $E^\perp$ is invariant under A^*.

Exercise 5.8.2 (The Spectrum of $A + A^*$). Let $A : H \to H$ be a normal operator on a nonzero complex Hilbert space H.

(a) Prove that

$$\text{(5.8.1)} \quad \mathrm{Re}\lambda \geq 0 \text{ for all } \lambda \in \sigma(A) \qquad \Longleftrightarrow \qquad \mathrm{Re}\langle x, Ax\rangle \geq 0 \text{ for all } x \in H.$$

Hint: If $\mathrm{Re}\langle x, Ax\rangle \geq 0$ for all $x \in H$ use the Cauchy–Schwarz inequality for $\mathrm{Re}\langle x, Ax - \lambda x\rangle$ with $\mathrm{Re}\lambda < 0$. If $\mathrm{Re}\lambda \geq 0$ for all $\lambda \in \sigma(A)$ prove that $\|e^{-tA}\| \leq 1$ for all $t \geq 0$ and differentiate the function $t \mapsto \|e^{-tA}x\|^2$.

(b) Prove that

$$\text{(5.8.2)} \qquad \begin{aligned} \sup_{\|x\|=1} \mathrm{Re}\langle x, Ax\rangle &= \sup_{\lambda\in\sigma(A)} \mathrm{Re}\lambda, \\ \inf_{\|x\|=1} \mathrm{Re}\langle x, Ax\rangle &= \inf_{\lambda\in\sigma(A)} \mathrm{Re}\lambda. \end{aligned}$$

(c) Prove that

$$\text{(5.8.3)} \qquad \sigma(A) \cap \mathbf{i}\mathbb{R} = \emptyset \qquad \Longleftrightarrow \qquad A + A^* \quad \text{is bijective.}$$

Hint 1: If $A + A^*$ is bijective, use the Open Mapping Theorem 2.2.1 and Lemma 5.3.14 to deduce that A is bijective. Then replace A with $A + \mathbf{i}\lambda\mathbb{1}$.

Hint 2: If $\sigma(A) \cap \mathbf{i}\mathbb{R} = \emptyset$, use Theorem 5.2.12 to find an A-invariant direct sum decomposition $H = H^- \oplus H^+$ such that $\pm\mathrm{Re}\lambda > 0$ for all $\lambda \in \sigma(A|_{H^\pm})$. Prove that $H^\pm$ is invariant under A^* and use part (b) for $A|_{H^\pm}$.

(d) Prove that

$$\text{(5.8.4)} \qquad \sigma(A + A^*) = \left\{\lambda + \overline{\lambda} \,|\, \lambda \in \sigma(A)\right\}.$$

Hint: Apply part (c) to the operator $A - \mu\mathbb{1}$ for $\mu \in \mathbb{R}$.

(e) Prove that the hypothesis that A is normal cannot be removed in (a)–(d).

Hint: Find a matrix $A \in \mathbb{R}^{2\times2}$ and a vector $x \in \mathbb{R}^2$ such that $\sigma(A) = \{0\}$ and $\langle x, Ax\rangle > 0$.

Exercise 5.8.3 (The Spectrum of $p(A)$). Let $A : X \to X$ be a bounded complex linear operator on a nonzero complex Banach space X and let $p(z) = \sum_{k=0}^n a_k z^k$ be a polynomial with complex coefficients. Prove directly, without using Theorem 5.2.12, that the operator

$$p(A) := \sum_{k=0}^n a_k A^k$$

satisfies

$$\sigma(p(A)) = p(\sigma(A)). \tag{5.8.5}$$

Hint: To prove that $p(\sigma(A)) \subset \sigma(p(A))$ fix an element $\lambda \in \sigma(A)$ and use the fact that there exists a polynomial q with complex coefficients such that $p(z) - p(\lambda) = (z - \lambda)q(z)$ for all $z \in \mathbb{C}$. To prove the converse inclusion, assume $a := a_n \neq 0$, fix an element $\mu \in \sigma(p(A))$, and let $\lambda_1, \dots, \lambda_n$ be the zeros of the polynomial $p - \mu$ so that $p(z) - \mu = a \prod_{i=1}^n (z - \lambda_i)$ for all $z \in \mathbb{C}$. Show that $A - \lambda_i \mathbb{1}$ is not bijective for some i.

Exercise 5.8.4 (Stone–Weierstraß Theorem (real)). Here is another proof of the Stone–Weierstraß Theorem 5.4.5 for real valued functions.

Let M be a compact Hausdorff space and let $\mathcal{A} \subset C(M)$ be a subalgebra of the algebra of real valued continuous functions on M. Assume that $\mathcal{A}$ contains the constant functions and separates points (i.e. for all $x, y \in M$ there exists an $f \in \mathcal{A}$ such that $f(x) \neq f(y)$). Then $\mathcal{A}$ is dense in $C(M)$.

(a) The proof is by contradiction. Assume $\mathcal{A}$ is not dense in $C(M)$ and choose an element $f \in C(M)$ such that $d(f, \mathcal{A}) := \inf_{g \in \mathcal{A}} \|f - g\| = 1$.

(b) For a closed subset $K \subset M$ define $\|g\|_K := \sup_{x \in K} |g(x)|$ for $g \in C(M)$ and $d_K(f, \mathcal{A}) := \inf_{g \in \mathcal{A}} \|f - g\|_K$. Prove that there exists a smallest closed subset $K \subset M$ such that $d_K(f, \mathcal{A}) = 1$. **Hint:** Zorn's Lemma.

(c) Prove that K contains more than one point. Deduce that there exists a function $h \in \mathcal{A}$ such that $\min_K h = 0$ and $\max_K h = 1$.

(d) Define

$$K_0 := \{x \in K \,|\, h(x) \leq 2/3\}, \qquad K_1 := \{x \in K \,|\, h(x) \geq 1/3\}.$$

Find functions $g_0, g_1 \in \mathcal{A}$ such that $\|f - g_0\|_{K_0} < 1$ and $\|f - g_1\|_{K_1} < 1$.

(e) For $n \in \mathbb{N}$ define $h_n := (1 - h^n)^{2^n} \in \mathcal{A}$. Prove that

$$\|f - h_n g_0 - (1 - h_n) g_1\|_K < 1$$

for n sufficiently large and this contradicts the definition of K.

Hint: Use Bernoulli's inequality $(1 + t)^n \geq 1 + nt$ for $t \geq -1$ and the inequality $(1 - t) \leq (1 + t)^{-1}$ for $0 \leq t \leq 1$ to show that h_n converges uniformly to one on $K_0 \setminus K_1$ and converges uniformly to zero on $K_1 \setminus K_0$.

Exercise 5.8.5 (Stone–Weierstraß Theorem (complex)).

Let M be a compact Hausdorff space and let $\mathcal{A} \subset C(M, \mathbb{C})$ be a complex subalgebra of the algebra of complex valued continuous functions on M. Assume that $\mathcal{A}$ contains the constant functions, separates points, and is invariant under complex conjugation. Then $\mathcal{A}$ is dense in $C(M, \mathbb{C})$.

(a) Deduce the complex Stone–Weierstraß Theorem from the real Stone–Weierstraß Theorem.

(b) Find an example which shows that the hypothesis that $\mathcal{A}$ is invariant under complex conjugation cannot be removed in the complex Stone–Weierstraß Theorem. **Hint:** See Example 5.4.6.

Exercise 5.8.6 (Trigonometric Polynomials). Trigonometric polynomials are the elements of the smallest algebra $\mathcal{A} \subset C(\mathbb{R}/2\pi\mathbb{Z})$ that contains the functions sin and cos.

(a) Every element $p \in \mathcal{A}$ has the form

$$p(t) = \sum_{k=0}^{n} \bigl(a_k \cos(kt) + b_k \sin(kt)\bigr) \qquad \text{for } t \in \mathbb{R},$$

where $a_k, b_k \in \mathbb{R}$.

(b) The trigonometric polynomials form a dense subalgebra of the space $C(\mathbb{R}/2\pi\mathbb{Z})$ of continuous 2π-periodic real valued functions on $\mathbb{R}$.

(c) Why does this not contradict the fact that there exist continuous real valued 2π-periodic functions on the real axis whose Fourier series do not converge uniformly? (See Exercise 2.5.5.)

Exercise 5.8.7 (The Spectrum in a Banach Algebra). Let $\mathcal{A}$ be a complex unital Banach algebra. Define the **spectrum** of an element $a \in \mathcal{A}$ by

$$\sigma(a) := \bigl\{\lambda \in \mathbb{C} \,\big|\, \lambda\mathbb{1} - a \text{ is not invertible}\bigr\}.$$

Prove the following.

(a) The spectrum $\sigma(a)$ is a nonempty compact subset of $\mathbb{C}$ for every $a \in \mathcal{A}$.

(b) The Gelfand–Mazur Theorem. *If every nonzero element of $\mathcal{A}$ is invertible then $\mathcal{A}$ is isomorphic to $\mathbb{C}$.* **Hint:** See the proof of Theorem 5.5.2.

(c) Every nonzero quaternion is invertible. Why does the Gelfand–Mazur Theorem not apply?

(d) $\sigma(ab) \cup \{0\} = \sigma(ba) \cup \{0\}$ for all $a, b \in \mathcal{A}$.

Exercise 5.8.8 (Cayley–Hamilton). Let A be a complex $n \times n$-matrix with spectral radius r_A. Prove the Cauchy integral formula

$$p(A) = \frac{1}{2\pi\mathbf{i}} \int_{|z|=r} p(z)\big(z\mathbb{1} - A\big)^{-1} dz \tag{5.8.6}$$

for every $r > r_A$ and every polynomial $p(z) \in \mathbb{C}[z]$. Deduce that $p_A(A) = 0$, where

$$p_A(z) := \det(z\mathbb{1} - A)$$

is the characteristic polynomial of A.

Exercise 5.8.9 (Volterra Operator). Let $H := L^2([0,1])$ and define the operator $T : H \to H$ by

$$(Tf)(t) := \int_0^t f(s)\, ds$$

for $f \in L^2([0,1])$.

(a) Verify the formula

$$(T^n f)(t) := \frac{1}{(n-1)!} \int_0^t (t-s)^n f(s)\, ds$$

for $n \in \mathbb{N}$, $0 \leq t \leq 1$, and $f \in L^2([0,1])$.

(b) Determine the spectrum and the spectral radius of T.

(c) Prove that T is compact and injective. **Hint:** Arzelà–Ascoli.

(d) Compute the adjoint operator T^*.

(e) Is T self-adjoint? Is T normal?

(f) Prove that the operator

$$P := T + T^*$$

is an orthogonal projection, i.e. it satisfies

$$P^2 = P = P^*.$$

What is its image?

(g) Compute the eigenvalues and eigenvectors, the spectral radius, and the norm of the operator T^*T. **Hint:** Differentiate $T^*Tf = \lambda f$ twice.

(h) Prove that

$$\|T\| = \frac{2}{\pi}.$$

Hint: Compute the largest eigenvalue of T^*T and use equation (5.3.10).

Exercise 5.8.10 (Exponential Function and Logarithm). Let $\mathcal{A}$ be a unital Banach algebra.

(a) For $a \in \mathcal{A}$ define

$$\exp(a) := \sum_{n=0}^{\infty} \frac{a^n}{n!}.$$

If $a, b \in \mathcal{A}$ commute, prove that

$$\exp(a+b) = \exp(a)\exp(b).$$

Prove that $\exp(a)$ is invertible for every $a \in \mathcal{A}$.

(b) Let $a \in \mathcal{A}$ and suppose that the spectrum of a is contained in the open unit disc in $\mathbb{C}$. Show that the element

$$\log(a) := -\sum_{n=1}^{\infty} \frac{(1-a)^n}{n}$$

is a well-defined element of $\mathcal{A}$ and satisfies

$$\exp(\log(a)) = a.$$

(c) Show that $\exp(\mathcal{A})$ contains an open neighborhood of the unit $\mathbb{1}$.

(d) Let $\mathcal{G} \subset \mathcal{A}$ denote the group of invertible elements of $\mathcal{A}$. Recall that $\mathcal{G}$ is an open subset of $\mathcal{A}$ and denote by $\mathcal{G}_0$ the identity component of $\mathcal{G}$. Show that $\mathcal{G}_0$ is an open and closed normal subgroup of $\mathcal{G}$. Show that $\mathcal{G}_0$ is the smallest subgroup of $\mathcal{G}$ that contains the set $\exp(\mathcal{A})$. Show that every element of $\mathcal{G}_0$ is a composition of finitely many elements of $\exp(\mathcal{A})$.

(e) Suppose $\mathcal{A}$ is commutative. Prove that

$$\mathcal{G}_0 = \exp(\mathcal{A}).$$

Deduce that $\mathcal{G}/\mathcal{G}_0$ is torsion free. **Hint:** Let $g \in \mathcal{G}$ and assume

$$g^n \in \exp(\mathcal{A}).$$

Choose an element $a \in \mathcal{A}$ with

$$g^n = \exp(a)$$

and define

$$h := g\exp\left(-\frac{a}{n}\right).$$

Then $h^n = \mathbb{1}$. Use this to prove that the set

$$\{\lambda \in \mathbb{C} \mid (1-\lambda)\mathbb{1} + \lambda h \notin \mathcal{G}\}$$

is finite. Deduce that $h \in \mathcal{G}_0$ and so $g \in \mathcal{G}_0$.

Exercise 5.8.11 (The Gelfand Spectrum). This exercise expands the discussion in Subsection 5.5.1, with an emphasis on the complex valued unital algebra homomorphisms rather than the maximal ideals. Let $\mathcal{A}$ be a complex commutative unital Banach algebra with $\|\mathbb{1}\| = 1$.

(a) Show that every maximal ideal $\mathcal{J} \subset \mathcal{A}$ is closed and satisfies $\mathcal{A}/\mathcal{J} \cong \mathbb{C}$. Show that every noninvertible element of $\mathcal{A}$ is contained in a maximal ideal.

(b) Let $\Lambda : \mathcal{A} \to \mathbb{C}$ be a unital algebra homomorphism, i.e. it is linear and

$$\Lambda(ab) = \Lambda(a)\Lambda(b), \qquad \Lambda(\mathbb{1}) = 1$$

for all $a, b \in \mathcal{A}$. Show that Λ is surjective and

$$\|\Lambda\| = 1.$$

(c) Show that the set of unital algebra homomorphisms $\Lambda : \mathcal{A} \to \mathbb{R}$ is a weak* closed subset $\widehat{\mathcal{A}}$ of the unit ball in the complex dual space $\mathcal{A}^*$ of $\mathcal{A}$. Show that there is a one-to-one correspondence between the elements $\Lambda \in \widehat{\mathcal{A}}$ and the maximal ideals $\mathcal{J} \subset \mathcal{A}$. Thus the set $\widehat{\mathcal{A}}$ can be identified with the **Gelfand spectrum** $\mathrm{Spec}(\mathcal{A})$ (the set of maximal ideals in $\mathcal{A}$).

(d) Show that the spectrum of an element $a \in \mathcal{A}$ is determined by $\widehat{\mathcal{A}}$, i.e.

$$\sigma(a) = \left\{\Lambda(a) \,|\, \Lambda \in \widehat{\mathcal{A}}\right\}.$$

(e) Let $\mathcal{A} := C([0,1])$ be the space of complex valued continuous functions on the unit interval. Show that there is a homeomorphism

$$\mathrm{ev} : [0,1] \to \widehat{\mathcal{A}}$$

that assigns to each element $x \in [0,1]$ the evaluation map at x.

(f) The **Gelfand transform** is the map

$$\Gamma : \mathcal{A} \to C(\widehat{\mathcal{A}})$$

that assigns to each $a \in \mathcal{A}$ the evaluation map $\Gamma_a : \widehat{\mathcal{A}} \to \mathbb{C}$ given by

$$\Gamma_a(\Lambda) := \Lambda(a)$$

for $\Lambda \in \widehat{\mathcal{A}}$. Show that the Gelfand transform is a norm decreasing algebra homomorphism. Show that the functions in $\Gamma(\mathcal{A}) := \{\Gamma_a \,|\, a \in \mathcal{A}\} \subset C(\widehat{\mathcal{A}})$ separate the points in $\widehat{\mathcal{A}}$.

(g) In the case $\mathcal{A} = C([0,1])$ show that the Gelfand transform is an isometric isomorphism. Extend this result to the case where the unit interval is replaced by any compact metric space. (More generally, by Theorem 5.5.8, the Gelfand transform is an isometric isomorphism whenever $\mathcal{A}$ admits the structure of a C* algebra such that $\|a^*a\| = \|a\|^2$ for all $a \in \mathcal{A}$.)

Chapter 6

Unbounded Operators

This chapter is devoted to the spectral theory of unbounded linear operators on a Banach space X. The domain of an unbounded operator A is a linear subspace of X denoted by $\mathrm{dom}(A)$. In most of the relevant examples this subspace is dense and the linear operator $A : \mathrm{dom}(A) \to X$ has a closed graph. Section 6.1 examines the basic definition, discusses several examples, and examines the spectrum of an unbounded operator. Section 6.2 introduces the dual of an unbounded operator. Section 6.3 deals with unbounded operators on Hilbert spaces. It introduces the adjoint of an unbounded operator and examines the spectra of unbounded normal and self-adjoint operators. Section 6.4 extends the functional calculus and the spectral measure to unbounded self-adjoint operators.

6.1. Unbounded Operators on Banach Spaces

6.1.1. Definition and Examples.

Definition 6.1.1 (Unbounded Operator). Let X and Y be real or complex Banach spaces. An **unbounded (complex) linear operator from X to Y** is a pair $(A, \mathrm{dom}(A))$, where $\mathrm{dom}(A) \subset X$ is a (complex) linear subspace and

$$A : \mathrm{dom}(A) \to Y$$

is a (complex) linear map. An unbounded operator $A : \mathrm{dom}(A) \to Y$ is called **densely defined** if its domain is a dense subspace of X. It is called **closed** if its **graph**, defined by

$$\mathrm{graph}(A) := \{(x, Ax) \mid x \in \mathrm{dom}(A)\},$$

is a closed linear subspace of $X \times Y$ with respect to the product topology.

We have already encountered unbounded operators in Definition 2.2.11. Recall that the domain $\mathrm{dom}(A) \subset X$ of an unbounded operator

$$A : \mathrm{dom}(A) \to Y$$

is a normed vector space with the **graph norm of** A, defined in (2.2.8) by

$$\|x\|_A := \|x\|_X + \|Ax\|_Y \qquad \text{for } x \in \mathrm{dom}(A).$$

Thus an unbounded operator can also be viewed as a bounded operator from its domain, equipped with the graph norm, to its target space. By Exercise 2.2.12 an unbounded operator $A : \mathrm{dom}(A) \to Y$ has a closed graph if and only if its domain is a Banach space with respect to the graph norm. By Lemma 2.2.19 an unbounded operator $A : \mathrm{dom}(A) \to Y$ is closeable, i.e. it extends to an unbounded operator with a closed graph, if and only if every sequence $(x_n)_{n\in\mathbb{N}}$ in $\mathrm{dom}(A)$ such that $\lim_{n\to\infty} \|x_n\|_X = 0$ and $(Ax_n)_{n\in\mathbb{N}}$ is a Cauchy sequence in Y satisfies $\lim_{n\to\infty} \|Ax_n\|_Y = 0$. We emphasize that the case $\mathrm{dom}(A) = X$ is not excluded in Definition 6.1.1. Thus bounded operators are examples of unbounded operators. The Closed Graph Theorem 2.2.13 asserts in the case $\mathrm{dom}(A) = X$ that A has a closed graph if and only if A is bounded. The emphasis in the present chapter is on unbounded operators $A : \mathrm{dom}(A) \to Y$ whose domains are proper linear subspaces of X and whose graphs are closed.

Example 6.1.2. Let $X := C([0,1])$ be the Banach space of continuous real valued functions on $[0,1]$ with the supremum norm. Then the formula

$$\mathrm{dom}(A) := C^1([0,1]), \qquad Af := f', \tag{6.1.1}$$

defines an unbounded operator on $C([0,1])$ with a dense domain and a closed graph. The graph norm of A is the standard C^1 norm on $C^1([0,1])$. (See Example 2.2.10 and equation (2.2.9).)

Example 6.1.3. Let H be a separable complex Hilbert space, let $(e_i)_{i\in\mathbb{N}}$ be a complex orthonormal basis, and let $(\lambda_i)_{i\in\mathbb{N}}$ be a sequence of complex numbers. Define the operator

$$A_\lambda : \mathrm{dom}(A_\lambda) \to H$$

by

$$\begin{aligned} \mathrm{dom}(A_\lambda) &:= \left\{ x \in H \,\middle|\, \sum_{i=1}^\infty |\lambda_i \langle e_i, x\rangle|^2 < \infty \right\}, \\ A_\lambda x &:= \sum_{i=1}^\infty \lambda_i \langle e_i, x\rangle e_i \qquad \text{for } x \in \mathrm{dom}(A). \end{aligned} \tag{6.1.2}$$

This is an unbounded operator with a dense domain and a closed graph. It is bounded if and only if the sequence $(\lambda_i)_{i\in\mathbb{N}}$ is bounded.

Example 6.1.4 (Vector Fields). Here is an example for readers who are familiar with some basic notions of differential topology (smooth manifolds, tangent bundles, and vector fields). Let M be a compact smooth manifold and let $v : M \to TM$ be a smooth vector field. Consider the Banach space

$$X := C(M)$$

of continuous functions $f : M \to \mathbb{R}$ equipped with the supremum norm. Define the operator $D_v : \mathrm{dom}(D_v) \to C(M)$ by

$$\begin{aligned} \mathrm{dom}(D_v) &:= \left\{ f \in C(M) \;\middle|\; \begin{array}{l} \text{the partial derivative of } f \\ \text{in the direction } v(p) \\ \text{exists for every } p \in M \\ \text{and depends continuously on } p \end{array} \right\}, \\ (D_v f)(p) &:= \left.\frac{d}{dt}\right|_{t=0} f(\gamma(t)), \quad \gamma : \mathbb{R} \to M, \;\; \gamma(0) = p, \;\; \dot\gamma(0) = v(p). \end{aligned} \tag{6.1.3}$$

Here $\gamma : \mathbb{R} \to M$ is chosen as any smooth curve in M that passes through p at $t = 0$ and whose derivative at $t = 0$ is the tangent vector $v(p) \in T_pM$. The operator D_v has a dense domain and a closed graph. With the appropriate modifications this discussion carries over to manifolds with boundary. Then Example 6.1.2 is the special case $M = [0, 1]$ and $v = \partial/\partial t$.

Example 6.1.5 (Derivative). Fix a constant $1 \le p \le \infty$ and consider the Banach space $X := L^p(\mathbb{R}, \mathbb{C})$. Define the operator $A : \mathrm{dom}(A) \to X$ by

$$\begin{aligned} \mathrm{dom}(A) &:= W^{1,p}(\mathbb{R}, \mathbb{C}) \\ &:= \left\{ f \in L^p(\mathbb{R}, \mathbb{C}) \;\middle|\; \begin{array}{l} f \text{ is absolutely continuous} \\ \text{and } \frac{df}{ds} \in L^p(\mathbb{R}, \mathbb{C}) \end{array} \right\}, \\ Af &:= \frac{df}{ds} \qquad \text{for } f \in W^{1,p}(\mathbb{R}, \mathbb{C}). \end{aligned} \tag{6.1.4}$$

Here s is the variable in $\mathbb{R}$. Recall that an absolutely continuous function is almost everywhere differentiable, that its derivative is locally integrable, and that it can be written as the integral of its derivative, i.e. the fundamental theorem of calculus holds in this setting (see [**75**, Thm. 6.19]). The operator (6.1.4) has a closed graph and, for $1 \le p < \infty$, it has a dense domain. For $p = \infty$ its domain is the space $W^{1,\infty}(\mathbb{R}, \mathbb{C})$ of bounded globally Lipschitz continuous functions $f : \mathbb{R} \to \mathbb{C}$. These are the bounded absolutely continuous functions with bounded derivative and do not form a dense subspace of $L^\infty(\mathbb{R}, \mathbb{C})$. The closure of the subspace $W^{1,\infty}(\mathbb{R}, \mathbb{C})$ in $L^\infty(\mathbb{R}, \mathbb{C})$ is the space of bounded uniformly continuous functions $f : \mathbb{R} \to \mathbb{C}$.

Example 6.1.6 (Laplace Operator). Fix an integer $n \in \mathbb{N}$ and a real number $1 < p < \infty$. Consider the **Laplace operator**

$$\Delta := \sum_{i=1}^{n} \frac{\partial^2}{\partial x_i^2} : W^{2,p}(\mathbb{R}^n) \to L^p(\mathbb{R}^n). \tag{6.1.5}$$

Its domain is the Sobolev space $W^{2,p}(\mathbb{R}^n)$ of equivalence classes, up to equality almost everywhere, of real valued L^p functions on $\mathbb{R}^n$ whose distributional derivatives up to order two can be represented by L^p functions. This subspace contains the compactly supported smooth functions and so is dense in $L^p(\mathbb{R}^n)$. The proof that this operator has a closed graph requires elliptic regularity and the Calderón–Zygmund inequality (see [**75**, Thm. 7.43]).

Example 6.1.7 (Schrödinger Operator). Define the unbounded linear operator A on the Hilbert space $H := L^2(\mathbb{R}, \mathbb{C})$ by

$$\begin{aligned} \operatorname{dom}(A) &:= \left\{ \psi \in L^2(\mathbb{R}, \mathbb{C}) \;\middle|\; \begin{array}{l} \psi \text{ is absolutely continuous,} \\ \frac{d\psi}{dx} \text{ is absolutely continuous,} \\ \text{and } \frac{d^2\psi}{dx^2} \in L^2(\mathbb{R}, \mathbb{C}) \end{array} \right\}, \\ A\psi &:= \mathbf{i}\hbar \frac{d^2\psi}{dx^2} \qquad \text{for } \psi \in \operatorname{dom}(A) = W^{2,2}(\mathbb{R}, \mathbb{C}). \end{aligned} \tag{6.1.6}$$

Here $\hbar$ is a positive real number and x is the variable in $\mathbb{R}$. Another variant of the Schrödinger operator on $L^2(\mathbb{R}, \mathbb{C})$ is given by

$$\begin{aligned} \operatorname{dom}(A) &:= \left\{ \psi \in L^2(\mathbb{R}, \mathbb{C}) \;\middle|\; \begin{array}{l} \psi \text{ is absolutely continuous and} \\ \frac{d\psi}{dx} \text{ is absolutely continuous, and} \\ \int_{-\infty}^{\infty} \left| -\hbar^2 \frac{d^2\psi}{dx^2} + x^2\psi \right|^2 dx < \infty \end{array} \right\}, \\ (A\psi)(x) &:= \mathbf{i}\hbar \frac{d^2\psi}{dx^2}(x) + \frac{x^2}{\mathbf{i}\hbar}\psi(x) \quad \text{for } \psi \in \operatorname{dom}(A) \text{ and } x \in \mathbb{R}. \end{aligned} \tag{6.1.7}$$

The operators (6.1.6) and (6.1.7) are both densely defined and closed.

Example 6.1.8 (Multiplication Operator). Let $(M, \mathcal{A}, \mu)$ be a measure space and let $f : M \to \mathbb{R}$ be a measurable function. Let $1 \leq p < \infty$ and define the operator $A_f : \operatorname{dom}(A_f) \to L^p(\mu)$ by

$$\begin{aligned} \operatorname{dom}(A_f) &:= \{ \psi \in L^p(\mu) \mid f\psi \in L^p(\mu) \}, \\ A_f \psi &:= f\psi \qquad \text{for } \psi \in \operatorname{dom}(A_f). \end{aligned} \tag{6.1.8}$$

This operator has a dense domain and a closed graph.

There are many other interesting examples of unbounded operators that play important roles in differential geometry and topology and other fields of mathematics. Their study goes beyond the scope of the present book, whose purpose is merely to provide the necessary functional analytic background.

6.1.2. The Spectrum of an Unbounded Operator. The following definition is the natural analogue of the definition of the spectrum of a bounded complex linear operator in Definition 5.2.1. We restrict the discussion to operators with closed graphs.

Definition 6.1.9 (Spectrum). Let X be a complex Banach space and let $A : \mathrm{dom}(A) \to X$ be an unbounded complex linear operator with a closed graph (whose domain is a complex linear subspace of X). The **spectrum of** A is the set

$$\begin{aligned}\sigma(A) &:= \left\{ \lambda \in \mathbb{C} \,\middle|\, \begin{array}{l} \text{the operator } \lambda\mathbb{1} - A : \mathrm{dom}(A) \to X \\ \text{is not bijective} \end{array} \right\} \\ &= \mathrm{P}\sigma(A) \cup \mathrm{R}\sigma(A) \cup \mathrm{C}\sigma(A). \end{aligned} \tag{6.1.9}$$

Here $\mathrm{P}\sigma(A)$ is the **point spectrum**, $\mathrm{R}\sigma(A)$ is the **residual spectrum**, and $\mathrm{C}\sigma(A)$ is the **continuous spectrum**. These are defined by

$$\begin{aligned} \mathrm{P}\sigma(A) &:= \left\{ \lambda \in \mathbb{C} \,\middle|\, \text{the operator } \lambda\mathbb{1} - A \text{ is not injective} \right\}, \\ \mathrm{R}\sigma(A) &:= \left\{ \lambda \in \mathbb{C} \,\middle|\, \begin{array}{l} \text{the operator } \lambda\mathbb{1} - A \text{ is injective} \\ \text{and its image is not dense} \end{array} \right\}, \\ \mathrm{C}\sigma(A) &:= \left\{ \lambda \in \mathbb{C} \,\middle|\, \begin{array}{l} \text{the operator } \lambda\mathbb{1} - A \text{ is injective} \\ \text{and its image is dense,} \\ \text{but it is not surjective} \end{array} \right\}. \end{aligned} \tag{6.1.10}$$

The **resolvent set of** A is the complement of the spectrum, denoted by

$$\rho(A) := \mathbb{C} \setminus \sigma(A) = \left\{ \lambda \in \mathbb{C} \,\middle|\, \begin{array}{l} \text{the operator} \\ \lambda\mathbb{1} - A : \mathrm{dom}(A) \to X \\ \text{is bijective} \end{array} \right\}. \tag{6.1.11}$$

For $\lambda \in \rho(A)$ the linear operator $R_\lambda(A) := (\lambda\mathbb{1} - A)^{-1} : X \to X$ is called the **resolvent operator** of A associated to λ. A complex number λ belongs to the point spectrum $\mathrm{P}\sigma(A)$ if and only if there exists a nonzero vector $x \in \mathrm{dom}(A)$ such that $Ax = \lambda x$. The elements $\lambda \in \mathrm{P}\sigma(A)$ are called **eigenvalues** of A and the nonzero vectors $x \in \ker(\lambda\mathbb{1} - A)$ are called **eigenvectors**.

The first observation about this definition is that, for every $\lambda \in \rho(A)$, the resolvent operator $R_\lambda(A) := (\lambda\mathbb{1} - A)^{-1}$ has a closed graph because A does, and hence is bounded by the Closed Graph Theorem 2.2.13 (see Exercise 6.5.2). The resolvent set may actually be empty for unbounded operators with closed graphs or it may be the entire complex plane as we will see below. The second observation is that the resolvent set is always an open subset of the complex plane and the map $\rho(A) \to \mathcal{L}^c(X) : \lambda \mapsto R_\lambda(A)$ is holomorphic. This is the content of the next lemma.

Lemma 6.1.10 (Resolvent Operator). *Let X be a complex Banach space and let $A : \mathrm{dom}(A) \to X$ be an unbounded complex linear operator with a closed graph. Let $\mu \in \rho(A)$ and let $\lambda \in \mathbb{C}$ such that*

$$|\lambda - \mu| \, \|(\mu\mathbb{1} - A)^{-1}\| < 1. \tag{6.1.12}$$

Then $\lambda \in \rho(A)$ and

$$(\lambda\mathbb{1} - A)^{-1} = \sum_{k=0}^{\infty} (\mu - \lambda)^k (\mu\mathbb{1} - A)^{-k-1}. \tag{6.1.13}$$

Proof. Define the bounded linear operator $T_\lambda \in \mathcal{L}(X)$ by

$$T_\lambda x := x - (\mu - \lambda)(\mu\mathbb{1} - A)^{-1}x$$

for $x \in X$. By (6.1.12) and Corollary 1.5.7 this operator is bijective and

$$T_\lambda^{-1} = \sum_{k=0}^{\infty} (\mu - \lambda)^k (\mu\mathbb{1} - A)^{-k}.$$

Moreover, for all $x \in \mathrm{dom}(A)$,

$$T_\lambda(\mu\mathbb{1} - A)x = (\mu\mathbb{1} - A)x - (\mu - \lambda)x = (\lambda\mathbb{1} - A)x.$$

Hence the operator $\lambda\mathbb{1} - A : \mathrm{dom}(A) \to X$ is bijective and

$$(\lambda\mathbb{1} - A)^{-1} = (\mu\mathbb{1} - A)^{-1}T_\lambda^{-1} = \sum_{k=0}^{\infty} (\mu - \lambda)^k (\mu\mathbb{1} - A)^{-k-1}.$$

This proves (6.1.13) and Lemma 6.1.10. □

The third observation is that the resolvent identity of Lemma 5.2.6 continues to hold for unbounded operators.

Lemma 6.1.11 (Resolvent Identity). *Let X be a complex Banach space, let $A : \mathrm{dom}(A) \to X$ be an unbounded complex linear operator with a closed graph, and let $\lambda, \mu \in \rho(A)$. Then the resolvent operators*

$$R_\lambda(A) := (\lambda\mathbb{1} - A)^{-1}, \qquad R_\mu(A) := (\mu\mathbb{1} - A)^{-1}$$

commute and

$$R_\lambda(A) - R_\mu(A) = (\mu - \lambda)R_\lambda(A)R_\mu(A). \tag{6.1.14}$$

Proof. Let $x \in X$. Then

$$\begin{aligned}(\lambda\mathbb{1} - A)\left(R_\lambda(A)x - R_\mu(A)x\right) &= x - (\mu\mathbb{1} - A)R_\mu(A)x + (\mu - \lambda)R_\mu(A)x \\ &= (\mu - \lambda)R_\mu(A)x\end{aligned}$$

and hence $R_\lambda(A)x - R_\mu(A)x = (\mu - \lambda)R_\lambda(A)R_\mu(A)x$. This proves (6.1.14). Interchange the roles of λ and μ to obtain that $R_\lambda(A)$ and $R_\mu(A)$ commute. This proves Lemma 6.1.11. □

The fourth observation is that the spectrum of an unbounded operator with a nonempty resolvent set is related to the spectrum of its resolvent operator as follows.

Lemma 6.1.12 (Spectrum and Resolvent Operator). *Let X be a complex Banach space and let $A : \mathrm{dom}(A) \to X$ be an unbounded complex linear operator with a closed graph such that $\mathrm{dom}(A) \subsetneq X$. Let $\mu \in \rho(A)$. Then*

$$
\begin{aligned}
\mathrm{P}\sigma(R_\mu(A)) &= \left\{ \frac{1}{\mu - \lambda} \,\middle|\, \lambda \in \mathrm{P}\sigma(A) \right\}, \\
\mathrm{R}\sigma(R_\mu(A)) \setminus \{0\} &= \left\{ \frac{1}{\mu - \lambda} \,\middle|\, \lambda \in \mathrm{R}\sigma(A) \right\}, \\
\mathrm{C}\sigma(R_\mu(A)) \setminus \{0\} &= \left\{ \frac{1}{\mu - \lambda} \,\middle|\, \lambda \in \mathrm{C}\sigma(A) \right\}, \qquad (6.1.15) \\
\sigma(R_\mu(A)) &= \left\{ \frac{1}{\mu - \lambda} \,\middle|\, \lambda \in \sigma(A) \right\} \cup \{0\}, \\
\rho(R_\mu(A)) &= \left\{ \frac{1}{\mu - \lambda} \,\middle|\, \lambda \in \rho(A) \setminus \{\mu\} \right\}.
\end{aligned}
$$

Moreover, if $\lambda \in \rho(A) \setminus \{\mu\}$, then

$$
R_{(\mu-\lambda)^{-1}}(R_\mu(A)) = (\mu - \lambda)(\mu\mathbb{1} - A)R_\lambda(A), \qquad (6.1.16)
$$

and if $\lambda \in \mathrm{P}\sigma(A)$ and $k \in \mathbb{N}$, then

$$
\ker((\mu - \lambda)^{-1}\mathbb{1} - R_\mu(A))^k = \ker(\lambda\mathbb{1} - A)^k.
$$

Proof. First observe that $R_\mu(A)$ is injective and

$$
\mathrm{im}(R_\mu(A)) = \mathrm{dom}(A) \subsetneq X.
$$

Hence

$$
0 \in \mathrm{R}\sigma(R_\mu(A)) \cup \mathrm{C}\sigma(R_\mu(A)).
$$

Second, if $\lambda \in \mathbb{C} \setminus \{\mu\}$, then

$$
\frac{1}{\mu - \lambda}\mathbb{1} - R_\mu(A) = \frac{1}{\mu - \lambda}(\lambda\mathbb{1} - A)R_\mu(A) \in \mathcal{L}^c(X). \qquad (6.1.17)
$$

The left-hand side is injective if and only if $\lambda\mathbb{1} - A$ is injective, has a dense image if and only if $\lambda\mathbb{1} - A$ has a dense image, and is surjective if and only if $\lambda\mathbb{1} - A$ is surjective. This proves (6.1.15) and (6.1.16). Now let $\lambda \in \mathrm{P}\sigma(A)$ and $k \in \mathbb{N}$ and consider the linear subspace

$$
E_k := \ker(\lambda\mathbb{1} - A)^k = \left\{ x \in \mathrm{dom}(A^\infty) \,|\, (\lambda\mathbb{1} - A)^k x = 0 \right\}.
$$

This subspace is invariant under the operator $R_\mu(A)$ and so under $R_\mu(A)^k$. Thus it follows from (6.1.17) that

$$
E_k \subset \ker\left((\mu - \lambda)^{-1}\mathbb{1} - R_\mu(A)\right)^k.
$$

To prove the converse inclusion, we proceed by induction on k. Suppose first that $x \in \ker\left((\mu-\lambda)^{-1}1\!\!1 - R_\mu(A)\right)$. Then

$$x = (\mu-\lambda)R_\mu(A)x \in \mathrm{dom}(A)$$

and hence

$$\begin{aligned} Ax &= (\mu-\lambda)AR_\mu(A)x \\ &= (\mu-\lambda)(\mu R_\mu(A)x - x) \\ &= \mu(\mu-\lambda)R_\mu(A)x + (\lambda-\mu)x \\ &= \lambda x. \end{aligned}$$

This implies $x \in E_1$. Now let $k \geq 2$, assume

$$E_{k-1} = \ker((\mu-\lambda)^{-1}1\!\!1 - R_\mu(A))^{k-1},$$

and fix an element

$$x \in \ker\left((\mu-\lambda)^{-1}1\!\!1 - R_\mu(A)\right)^k.$$

Then

$$x - (\mu-\lambda)R_\mu(A)x \in E_{k-1} \subset \mathrm{dom}(A^\infty)$$

by the induction hypothesis. This implies $x \in \mathrm{dom}(A)$ and

$$R_\mu(A)(\lambda x - Ax) = x - (\mu-\lambda)R_\mu(A)x \in E_{k-1}$$

by (6.1.17). Hence $\lambda x - Ax \in E_{k-1}$, because E_{k-1} is invariant under $\mu 1\!\!1 - A$, and hence $x \in E_k$. This proves Lemma 6.1.12. □

Lemma 6.1.12 allows us to carry over the results about the spectra of bounded linear operators to unbounded operators. An important special case concerns operators with compact resolvent.

Definition 6.1.13 (Operator with Compact Resolvent). An unbounded operator $A : \mathrm{dom}(A) \to X$ on a complex Banach space X with a closed graph and $\mathrm{dom}(A) \subset X$ is said to have a **compact resolvent** if $\rho(A) \neq \emptyset$ and the resolvent operator $R_\lambda(A) = (\lambda 1\!\!1 - A)^{-1} \in \mathcal{L}^c(X)$ is compact for all $\lambda \in \rho(A)$.

Exercise 6.1.14. Let $A : \mathrm{dom}(A) \to X$ be an unbounded operator on a complex Banach space X with a closed graph and $\mathrm{dom}(A) \subset X$.

(i) Prove that $R_\lambda(A)$ is compact for some $\lambda \in \rho(A)$ if and only if it is compact for all $\lambda \in \rho(A)$.

(ii) Let $\lambda \in \mathrm{P}\sigma(A)$ and define $E_k := \ker(\lambda 1\!\!1 - A)^k$ for $k \in \mathbb{N}$. Assume that $E_m = E_{m+1}$. Prove that $E_m = E_k$ for every integer $k \geq m$.

Theorem 6.1.15 (Spectrum and Compact Resolvent). *Let X be a complex Banach space and let $A : \mathrm{dom}(A) \to X$ be an unbounded complex linear operator on X with compact resolvent. Then $\sigma(A) = \mathrm{P}\sigma(A)$ is a discrete subset of $\mathbb{C}$ and the subspace $E_\lambda := \bigcup_{k=1}^{\infty} \ker(\lambda\mathbb{1} - A)^k$ is finite-dimensional for all $\lambda \in \mathrm{P}\sigma(A)$.*

Proof. Let $\mu \in \rho(A)$. Then zero is not an eigenvalue of $R_\mu(A)$. Since the operator $R_\mu(A)$ is compact, it follows from Theorem 5.2.8 that

$$\sigma(R_\mu(A)) \setminus \{0\} = \mathrm{P}\sigma(R_\mu(A))$$

is a discrete subset of $\mathbb{C} \setminus \{0\}$ and that the generalized eigenspace of $R_\mu(A)$ associated to every eigenvalue $z = (\mu - \lambda)^{-1}$ is finite-dimensional. Hence Lemma 6.1.12 asserts that

$$\begin{aligned} \sigma(A) &= \left\{ \mu - \frac{1}{z} \,\middle|\, z \in \sigma(R_\mu(A)) \setminus \{0\} \right\} \\ &= \left\{ \mu - \frac{1}{z} \,\middle|\, z \in \mathrm{P}\sigma(R_\mu(A)) \right\} \\ &= \mathrm{P}\sigma(A) \end{aligned}$$

is a discrete subset of $\mathbb{C}$ and that $\dim E_\lambda < \infty$ for all $\lambda \in \sigma(A)$. This proves Theorem 6.1.15. □

Example 6.1.16. Consider the complex Hilbert space $H := \ell^2(\mathbb{N}, \mathbb{C})$ (see part (ii) of Exercise 5.3.5). Let $(\lambda_i)_{i\in\mathbb{N}}$ be a sequence of complex numbers and define the unbounded operator $A_\lambda : \mathrm{dom}(A_\lambda) \to H$ by

$$\mathrm{dom}(A_\lambda) := \left\{ x = (x_i)_{i\in\mathbb{N}} \in \mathbb{C}^{\mathbb{N}} \,\middle|\, \sum_{i=1}^{\infty} |x_i|^2 < \infty,\ \sum_{i=1}^{\infty} |\lambda_i x_i|^2 < \infty \right\}$$

and

$$A_\lambda x := (\lambda_i x_i)_{i\in\mathbb{N}} \qquad \text{for } x = (x_i)_{i\in\mathbb{N}} \in \mathrm{dom}(A_\lambda).$$

This operator has a dense domain and a closed graph by Example 6.1.3 and its spectrum is given by $\mathrm{R}\sigma(A_\lambda) = \emptyset$ and

$$\mathrm{P}\sigma(A_\lambda) = \{\lambda_i \,|\, i \in \mathbb{N}\}, \qquad \sigma(A_\lambda) = \overline{\{\lambda_i \,|\, i \in \mathbb{N}\}}.$$

Here the overline denotes the closure (and not complex conjugation). Thus the resolvent set $\rho(A_\lambda)$ is empty if and only if the sequence $(\lambda_i)_{i\in\mathbb{N}}$ is dense in $\mathbb{C}$. The operator A_λ has a compact resolvent if and only if the sequence $|\lambda_i|$ diverges to infinity as i tends to ∞.

Example 6.1.16 shows that the spectrum of an unbounded densely defined closed operator on a separable Hilbert space can be any nonempty closed subset of the complex plane. The next example shows that the spectrum can also be empty.

Example 6.1.17. Consider the complex Hilbert space $H := L^2([0,1],\mathbb{C})$ (Exercise 5.3.5) and let $W^{1,2}([0,1],\mathbb{C})$ be the space of all absolutely continuous functions $u : [0,1] \to \mathbb{C}$ with square integrable derivatives. Let t be the variable in the unit interval $[0,1]$.

(i) Define the operator $A_0 : \mathrm{dom}(A_0) \to H$ by

$$\mathrm{dom}(A_0) := \left\{u \in W^{1,2}([0,1],\mathbb{C}) \,\middle|\, u(0) = 0\right\}, \qquad A_0 u := \dot{u}.$$

Let $f \in L^2([0,1],\mathbb{C})$ and $\lambda \in \mathbb{C}$. A function $u \in \mathrm{dom}(A_0)$ satisfies the equation $\lambda u - A_0 u = f$ if and only if $u \in W^{1,2}([0,1],\mathbb{C})$ and

$$\dot{u} = \lambda u - f, \qquad u(0) = 0.$$

This equation has a unique solution given by

$$u(t) = -\int_0^t e^{\lambda(t-s)} f(s)\, ds \qquad \text{for } 0 \le t \le 1.$$

Hence $\lambda\mathbb{1} - A_0$ is invertible for all $\lambda \in \mathbb{C}$ and so $\sigma(A_0) = \emptyset$.

(ii) Define $A_P : \mathrm{dom}(A_P) \to H$ (periodic boundary conditions) by

$$\mathrm{dom}(A_P) := \left\{u \in W^{1,2}([0,1],\mathbb{C}) \,\middle|\, u(0) = u(1)\right\}, \qquad A_P u := \mathbf{i}\dot{u}.$$

Let $f \in L^2([0,1],\mathbb{C})$ and $\lambda \in \mathbb{C}$. A function $u \in \mathrm{dom}(A_P)$ satisfies the equation $\lambda u - A_P u = f$ if and only if $u \in W^{1,2}([0,1],\mathbb{C})$ and

$$\dot{u} = -\mathbf{i}\lambda u + \mathbf{i}f, \qquad u(0) = u(1).$$

This equation has a unique solution if and only if $e^{-\mathbf{i}\lambda} \ne 1$, and in this case the solution is given by

$$u(t) = e^{-\mathbf{i}\lambda t}u_0 + \mathbf{i}\int_0^t e^{-\mathbf{i}\lambda(t-s)} f(s)\, ds, \qquad u_0 := \int_0^1 \frac{\mathbf{i}e^{-\mathbf{i}\lambda(1-s)}f(s)}{1 - e^{-\mathbf{i}\lambda}}\, ds.$$

Thus $\sigma(A_P) = \mathrm{P}\sigma(A_P) = 2\pi\mathbb{Z}$.

(iii) Define $A_L : \mathrm{dom}(A_L) \to H \times H$ (Lagrangian boundary conditions) by

$$\mathrm{dom}(A_L) := \left\{(u,v) \in W^{1,2}([0,1],\mathbb{C}^2) \,\middle|\, v(0) = v(1) = 0\right\}$$

and $A_L(u,v) := (-\dot{v}, \dot{u})$. **Exercise:** Show that $\sigma(A_L) = \mathrm{P}\sigma(A_L) = \pi\mathbb{Z}$.

(iv) The operators in (i), (ii), and (iii) have compact resolvent. Removing the boundary conditions one obtains the operator

$$A = \frac{d}{dt} : \mathrm{dom}(A) = W^{1,2}([0,1],\mathbb{C}) \to L^2([0,1],\mathbb{C})$$

with $\sigma(A) = \mathrm{P}\sigma(A) = \mathbb{C}$, which has no resolvent at all.

6.1.3. Spectral Projections. The holomorphic functional calculus in Section 5.2.4 does not carry over to unbounded operators unless one imposes rather stringent conditions on the asymptotic behavior of the holomorphic functions in question. However, the basic construction can be used to define certain spectral projections.

Definition 6.1.18 (Dunford Integral). Let $A : \mathrm{dom}(A) \to X$ be an unbounded complex linear operator with a closed graph on a complex Banach space X and let $\Sigma \subset \sigma(A)$ be a compact set. Call Σ **isolated** if $\sigma(A) \setminus \Sigma$ is a closed subset of $\mathbb{C}$. Call an open set $U \subset \mathbb{C}$ an **isolating neighborhood** of Σ if $\sigma(A) \cap U = \Sigma$. Assume U is an isolating neighborhood of Σ and let γ be a cycle in $U \setminus \Sigma$ such that

$$\mathrm{w}(\gamma, \lambda) := \frac{1}{2\pi\mathbf{i}} \int_\gamma \frac{dz}{z - \lambda} = \begin{cases} 1, & \text{for } \lambda \in \Sigma, \\ 0, & \text{for } \lambda \in \mathbb{C} \setminus U. \end{cases} \tag{6.1.18}$$

(See Figure 5.2.1.) The operator $\Phi_{\Sigma,A}(f) \in \mathcal{L}^c(X)$ is defined by

$$\Phi_{\Sigma,A}(f) := \frac{1}{2\pi\mathbf{i}} \int_\gamma f(z)(z\mathbb{1} - A)^{-1}\, dz. \tag{6.1.19}$$

Theorem 6.1.19 (Spectral Projection). *Let X, A, Σ, U be as in Definition 6.1.18. Then the following hold.*

(i) *The operator $\Phi_{A,\Sigma}(f)$ is independent of the choice of the cycle γ in $U \setminus \Sigma$ satisfying (6.1.18) that is used to define it.*

(ii) *Let $f, g : U \to \mathbb{C}$ be holomorphic. Then $\Phi_{A,\Sigma}(f+g) = \Phi_{A,\Sigma}(f) + \Phi_{A,\Sigma}(g)$ and $\Phi_{A,\Sigma}(fg) = \Phi_{A,\Sigma}(f)\Phi_{A,\Sigma}(g)$.*

(iii) *Let $f : U \to \mathbb{C}$ be holomorphic. Then $\sigma(\Phi_{A,\Sigma}(f)) = f(\Sigma)$.*

(iv) *Let $V \subset \mathbb{C}$ be an open set and let $f : U \to V$ and $g : V \to \mathbb{C}$ be holomorphic functions. Then $g(\Phi_{A,\Sigma}(f)) = \Phi_{A,\Sigma}(g \circ f)$.*

(v) *Let γ be a cycle in $U \setminus \Sigma$ satisfying (6.1.18) and define*

$$P_\Sigma := \Phi_{A,\Sigma}(1) = \frac{1}{2\pi\mathbf{i}} \int_\gamma (z\mathbb{1} - A)^{-1}\, dz. \tag{6.1.20}$$

Then P_Σ is a projection, its image $X_\Sigma := \mathrm{im}(P_\Sigma) \subset \mathrm{dom}(A)$ is A-invariant, the operator $A_\Sigma := A|_{X_\Sigma} : X_\Sigma \to X_\Sigma$ is bounded, its spectrum is $\sigma(A_\Sigma) = \Sigma$, and the unbounded operator $A|_{Y_\Sigma \cap \mathrm{dom}(A)} : Y_\Sigma \cap \mathrm{dom}(A) \to Y_\Sigma := \ker(P_\Sigma)$ has the spectrum $\sigma(A) \setminus \Sigma$.

Proof. The proof of Theorem 6.1.19 is verbatim the same as that of Theorem 5.2.12 and will be omitted. □

6.2. The Dual of an Unbounded Operator

Definition 6.2.1 (Dual Operator). Let X and Y be real or complex Banach spaces and let

$$A : \mathrm{dom}(A) \to Y$$

be an unbounded operator with a dense domain $\mathrm{dom}(A) \subset X$. The **dual operator of** A is the linear operator

$$A^* : \mathrm{dom}(A^*) \to X^*, \qquad \mathrm{dom}(A^*) \subset Y^*,$$

defined as follows. Its domain is the linear subspace

$$\mathrm{dom}(A^*) := \left\{ y^* \in Y^* \;\middle|\; \begin{array}{l} \text{there exists a constant } c \geq 0 \text{ such that} \\ |\langle y^*, Ax\rangle| \leq c\,\|x\| \text{ for all } x \in \mathrm{dom}(A) \end{array} \right\}$$

and, for $y^* \in \mathrm{dom}(A^*)$, the element $A^*y^* \in X^*$ is the unique bounded linear functional on X that satisfies

$$\langle A^*y^*, x\rangle = \langle y^*, Ax\rangle \qquad \text{for all } x \in \mathrm{dom}(A).$$

Thus the graph of the linear operator A^* is the linear subspace of $Y^* \times X^*$ that is characterized by the condition

$$\begin{array}{l} y^* \in \mathrm{dom}(A^*) \\ \text{and } x^* = A^*y^* \end{array} \iff \begin{array}{l} \langle x^*, x\rangle = \langle y^*, Ax\rangle \\ \text{for all } x \in \mathrm{dom}(A). \end{array} \tag{6.2.1}$$

The next theorem summarizes some fundamental correspondences between the domains, kernels, and images of an unbounded linear operator and its dual. It is the analogue of Theorem 4.1.8 for unbounded operators.

Theorem 6.2.2 (Duality). *Let X and Y be Banach spaces and suppose that $A : \mathrm{dom}(A) \to Y$ is a linear operator with a dense domain $\mathrm{dom}(A) \subset X$. Then the following hold.*

(i) *The dual operator $A^* : \mathrm{dom}(A^*) \to X^*$ is closed.*

(ii) *Let $x \in X$ and $y \in Y$. Then*

$$(x, y) \in \overline{\mathrm{graph}(A)} \iff \begin{array}{l} \langle y^*, y\rangle = \langle A^*y^*, x\rangle \\ \textit{for all } y^* \in \mathrm{dom}(A^*). \end{array} \tag{6.2.2}$$

(iii) *A is closeable if and only if $\mathrm{dom}(A^*)$ is weak* dense in Y^*.*

(iv) $\mathrm{im}(A)^\perp = \ker(A^*)$ *and, if A has a closed graph, then* ${}^\perp\mathrm{im}(A^*) = \ker(A)$.

(v) *The operator A has a dense image if and only if A^* is injective.*

(vi) *Assume A has a closed graph. Then A is injective if and only if A^* has a weak* dense image.*

Proof. Part (i) follows directly from (6.2.1).

We prove part (ii). Let $x \in X$ and $y \in Y$. By Corollary 2.3.25, we have $(x, y) \in \overline{\mathrm{graph}(A)}$ if and only if, for all $(x^*, y^*) \in X^* \times Y^*$,

$$\langle x^*, \xi \rangle + \langle y^*, A\xi \rangle = 0 \text{ for all } \xi \in \mathrm{dom}(A) \quad \Longrightarrow \quad \langle x^*, x \rangle + \langle y^*, y \rangle = 0.$$

By (6.2.1) the equation $\langle x^*, \xi \rangle + \langle y^*, A\xi \rangle = 0$ holds for all $\xi \in \mathrm{dom}(A)$ if and only if

$$y^* \in \mathrm{dom}(A^*), \qquad A^*y^* = -x^*.$$

Thus $(x, y) \in \overline{\mathrm{graph}(A)}$ if and only if

$$\langle y^*, y \rangle = \langle A^*y^*, x \rangle \,. \quad \text{for all } y^* \in \mathrm{dom}(A^*).$$

This proves part (ii).

We prove part (iii). Fix an element $y \in Y$. Then it follows from (6.2.2) in part (ii) that $(0, y) \in \overline{\mathrm{graph}(A)}$ if and only if $\langle y^*, y \rangle = 0$ for all $y^* \in \mathrm{dom}(A^*)$, and this means that $y \in {}^\perp\mathrm{dom}(A^*)$. Thus

$$y \in {}^\perp\mathrm{dom}(A^*) \qquad \Longleftrightarrow \qquad (0, y) \in \overline{\mathrm{graph}(A)}. \tag{6.2.3}$$

Now Lemma 2.2.19 asserts that the operator A is closeable if and only if the projection $\overline{\mathrm{graph}(A)} \to X$ is injective, i.e. for all $y \in Y$,

$$(0, y) \in \overline{\mathrm{graph}(A)} \qquad \Longrightarrow \qquad y = 0.$$

By (6.2.3) this shows that A is closeable if and only if

$${}^\perp\mathrm{dom}(A^*) = \{0\},$$

and, by Corollary 3.1.26, this condition holds if and only if the domain of A^* is a weak* dense subspace of Y^*. This proves part (iii).

We prove part (iv). Note that

$$\begin{aligned} y^* \in \ker(A^*) &\iff y^* \circ A = 0 \\ &\iff y^* \in \mathrm{im}(A)^\perp \end{aligned}$$

and, if A is closed, then

$$\begin{aligned} x \in {}^\perp\mathrm{im}(A^*) &\iff \langle A^*y^*, x \rangle = 0 \text{ for all } y^* \in \mathrm{dom}(A^*) \\ &\iff (x, 0) \in \overline{\mathrm{graph}(A)} \\ &\iff x \in \mathrm{dom}(A) \text{ and } Ax = 0. \end{aligned}$$

Here the second step follows from (6.2.2) in part (ii) and the last step holds because A has a closed graph. This proves part (iv).

Part (v) follows from part (iv) and Corollary 2.3.25 and part (vi) follows from part (iv) and Corollary 3.1.26. This proves Theorem 6.2.2. □

The next result extends the Closed Image Theorem 4.1.16 to unbounded operators. In this form it was proved by Stefan Banach in 1932.

Theorem 6.2.3 (Closed Image Theorem). *Let X and Y be Banach spaces and let $A : \mathrm{dom}(A) \to Y$ be a linear operator with a dense domain $\mathrm{dom}(A) \subset X$ and a closed graph. Then the following are equivalent.*

(i) $\mathrm{im}(A) = {}^{\perp}\ker(A^*)$.

(ii) *The image of A is a closed subspace of Y.*

(iii) *There exists a constant $c > 0$ such that*

$$\inf_{A\xi=0} \|x + \xi\|_X \le c\,\|Ax\|_Y \qquad \text{for all } x \in \mathrm{dom}(A). \tag{6.2.4}$$

Here the infimum runs over all $\xi \in \mathrm{dom}(A)$ that satisfy $A\xi = 0$.

(iv) $\mathrm{im}(A^*) = \ker(A)^{\perp}$.

(v) *The image of A^* is a weak* closed subspace of X^*.*

(vi) *The image of A^* is a closed subspace of X^*.*

(vii) *There exists a constant $c > 0$ such*

$$\inf_{A^*\eta^*=0} \|y^* + \eta^*\|_{Y^*} \le c\,\|A^*y^*\|_{X^*} \qquad \text{for all } y^* \in \mathrm{dom}(A^*). \tag{6.2.5}$$

Here the infimum runs over all $\eta^* \in \mathrm{dom}(A^*)$ that satisfy $A^*\eta^* = 0$.

Proof. We prove that (i) is equivalent to (ii). By Corollary 3.1.18 and part (iv) of Theorem 6.2.2, we have

$$\overline{\mathrm{im}(A)} = {}^{\perp}(\mathrm{im}(A)^{\perp}) = {}^{\perp}\ker(A^*).$$

Hence (i) is equivalent to (ii).

We prove that (ii) is equivalent to (iii). By Exercise 2.2.12, the domain of A is a Banach space with the graph norm $\|x\|_A := \|x\|_X + \|Ax\|_Y$ for $x \in \mathrm{dom}(A)$. Thus A is also a bounded linear operator from the Banach spaces $\mathrm{dom}(A)$ to the Banach space Y. Hence it follows from the equivalence of (ii) and (iii) in Theorem 4.1.16 that A has a closed image if and only if there exists a constant $c > 0$ such that

$$\inf_{A\xi=0} \|x + \xi\|_A \le c\,\|Ax\|_Y \qquad \text{for all } x \in \mathrm{dom}(A).$$

Since $\|x + \xi\|_A = \|x + \xi\|_X + \|Ax\|_Y$ for $x \in \mathrm{dom}(A)$ and $\xi \in \ker(A)$, this is equivalent to part (iii). This shows that (ii) is equivalent to (iii).

We prove that (iii) implies (iv) by the same argument as in the proof of Theorem 4.1.16. The inclusion $\mathrm{im}(A^*) \subset \ker(A)^{\perp}$ follows directly from the definition of the dual operator. To prove the converse inclusion, fix an element $x^* \in \ker(A)^{\perp}$ so that

$$\langle x^*, \xi\rangle = 0 \qquad \text{for all } \xi \in \ker(A).$$

Then, for all $x \in \mathrm{dom}(A)$ and all $\xi \in \ker(A)$, we have

$$|\langle x^*, x\rangle| = |\langle x^*, x+\xi\rangle| \le \|x^*\|_{X^*} \|x+\xi\|_X .$$

Take the infimum over all ξ to obtain the estimate

$$|\langle x^*, x\rangle| \le \|x^*\|_{X^*} \inf_{A\xi=0} \|x+\xi\|_X \le c\,\|x^*\|_{X^*} \|Ax\|_Y$$

for all $x \in \mathrm{dom}(A)$. Here the second step follows from (6.2.4). This implies that there exists a unique bounded linear functional Λ on $\mathrm{im}(A) \subset Y$ such that $\Lambda \circ A = x^*$. The functional Λ extends to an element $y^* \in Y^*$ by the Hahn–Banach Theorem (Corollaries 2.3.4 and 2.3.5). The extended functional satisfies $y^* \circ A = x^*$. Hence $y^* \in \mathrm{dom}(A^*)$ and $x^* = A^*y^*$ by definition of the dual operator, and so $x^* \in \mathrm{im}(A^*)$. This shows that (iii) implies (iv).

That (iv) implies (v) and (v) implies (vi) follows directly from the definition of the weak* topology. That (vi) is equivalent to (vii) follows from the fact that (ii) is equivalent to (iii) (already proved).

We prove that (vi) implies (ii), following [**88**, pp. 205–206]. Assume A^* has a closed image. Consider the product space $X \times Y$ with the norm

$$\|(x,y)\|_{X\times Y} := \|x\|_X + \|y\|_Y \qquad \text{for } (x,y) \in X \times Y.$$

The dual space of $X \times Y$ is the product space $X^* \times Y^*$ with the norm

$$\|(x^*,y^*)\|_{X^*\times Y^*} := \max\Big\{ \|x^*\|_{X^*}, \|y^*\|_{Y^*} \Big\}$$

for $(x^*,y^*) \in X^* \times Y^*$. The graph of A is the closed subspace

$$\Gamma := \Big\{ (x,y) \in X \times Y \,\Big|\, x \in \mathrm{dom}(A),\ y = Ax \Big\} \subset X \times Y$$

and the projection $B : \Gamma \to Y$ onto the second factor is given by

$$B(x,y) := y = Ax \qquad \text{for } (x,y) \in \Gamma.$$

This is a bounded linear operator with $\mathrm{im}(B) = \mathrm{im}(A)$. We prove in four steps that A has a closed image.

Step 1. *The annihilator of* Γ *is given by*

$$\Gamma^\perp = \Big\{ (x^*,y^*) \in X^* \times Y^* \,\Big|\, y^* \in \mathrm{dom}(A^*),\ x^* = -A^*y^* \Big\}.$$

Thus $\Gamma^\perp \subset \mathrm{im}(A^*) \times Y^*$.

Fix a pair $(x^*,y^*) \in X^* \times Y^*$. Then we have $(x^*,y^*) \in \Gamma^\perp$ if and only if

$$\langle x^*, x\rangle + \langle y^*, Ax\rangle = 0 \qquad \text{for all } x \in \mathrm{dom}(A),$$

and this is equivalent to the conditions

$$y^* \in \mathrm{dom}(A^*), \qquad x^* = -A^*y^*$$

by (6.2.1). This proves Step 1.

Step 2. *Define the map* $X^* \times Y^* \to \Gamma^* : (x^*, y^*) \mapsto \Lambda_{x^*,y^*}$ *by*

$$\Lambda_{x^*,y^*}(x, Ax) := \langle x^*, x\rangle + \langle y^*, Ax\rangle \qquad \textit{for } x \in \mathrm{dom}(A).$$

This map induces an isometric isomorphism from $X^* \times Y^*/\Gamma^\perp$ *to* Γ^* *and so*

$$\|\Lambda_{x^*,y^*}\| = \inf_{\eta^* \in \mathrm{dom}(A^*)} \max\Big\{ \|x^* - A^*\eta^*\|_{X^*}, \|y^* + \eta^*\|_{Y^*} \Big\}$$

for all $(x^*, y^*) \in X^* \times Y^*$.

This follows from Step 1 and Corollary 2.3.26, respectively Corollary 2.4.2.

Step 3. *The image of the dual operator* $B^* : Y^* \to \Gamma^*$ *is given by*

$$\mathrm{im}(B^*) = \{\Lambda_{x^*,y^*} \,|\, x^* \in \mathrm{im}(A^*), y^* \in Y^*\}.$$

If $y^* \in Y^*$, then

$$B^*y^* = y^* \circ B = \Lambda_{0,y^*}.$$

Conversely, let $(x^*, y^*) \in \mathrm{im}(A^*) \times Y^*$ and choose $\eta^* \in \mathrm{dom}(A^*)$ such that

$$A^*\eta^* = x^*.$$

Then $\Lambda_{-x^*,\eta^*} = 0$ by Step 1 and so

$$\Lambda_{x^*,y^*} = \Lambda_{0,y^*+\eta^*} = B^*(y^* + \eta^*) \in \mathrm{im}(B^*).$$

This proves Step 3.

Step 4. B^* *has a closed image.*

Let $\Lambda_i \in \mathrm{im}(B^*) \subset \Gamma^*$ be a sequence that converges to $\Lambda \in \Gamma^*$ in the norm topology. Choose $(x^*, y^*) \in X^* \times Y^*$ such that $\Lambda = \Lambda_{x^*,y^*}$ and, by Step 3, choose a sequence $(x_i^*, y_i^*) \in X^* \times Y^*$ such that

$$\Lambda_i = \Lambda_{x_i^*,y^*}, \qquad x_i^* \in \mathrm{im}(A^*) \qquad \text{for all } i \in \mathbb{N}.$$

Then, by Step 2, there exists a sequence $\eta_i^* \in \mathrm{dom}(A^*)$ such that

$$\max\Big\{ \|x^* - x_i^* - A^*\eta_i^*\|_{X^*}, \|y^* - y_i^* + \eta_i^*\|_{Y^*} \Big\} \le \|\Lambda - \Lambda_i\| + 2^{-i}$$

for all i and so

$$\lim_{i\to\infty} \|x^* - x_i^* - A^*\eta_i^*\|_{X^*} = 0.$$

Thus

$$x^* = \lim_{i\to\infty} (x_i^* + A^*\eta_i^*) \in \mathrm{im}(A^*)$$

because A^* has a closed image by assumption. Hence $\Lambda = \Lambda_{x^*,y^*} \in \mathrm{im}(B^*)$ by Step 3 and this proves Step 4.

It follows from Step 4 and Theorem 4.1.16 that B has a closed image. Hence so does A because $\mathrm{im}(A) = \mathrm{im}(B)$. This shows that (vi) implies (ii) and completes the proof of Theorem 6.2.3. □

Corollary 6.2.4. *Let X and Y be Banach spaces and let $A : \mathrm{dom}(A) \to Y$ be a linear operator with a dense domain $\mathrm{dom}(A) \subset X$ and a closed graph. Then A is bijective if and only if its dual operator A^* is bijective. If this holds, then $A^{-1} : Y \to X$ is a bounded linear operator and $(A^*)^{-1} = (A^{-1})^*$.*

Proof. Assume A is bijective and recall from Exercise 2.2.12 that $\mathrm{dom}(A)$ is a Banach space with the graph norm because A has a closed graph. Thus $A : \mathrm{dom}(A) \to Y$ is a bijective bounded linear operator between Banach spaces. Hence $A^{-1} : Y \to \mathrm{dom}(A)$ is bounded by the Open Mapping Theorem 2.2.1 and so is $A^{-1} : Y \to X$ (same notation, different target space). Now let $x^* \in X^*$ and $y^* \in Y^*$. We prove that

$$\begin{array}{l} y^* \in \mathrm{dom}(A^*) \\ x^* = A^*y^* \end{array} \quad \iff \quad (A^{-1})^*x^* = y^*. \tag{6.2.6}$$

By (6.2.1), $y^* \in \mathrm{dom}(A^*)$ and $A^*y^* = x^*$ if and only if $\langle x^*, x\rangle = \langle y^*, Ax\rangle$ for all $x \in \mathrm{dom}(A)$, and this is equivalent to the condition $\langle x^*, A^{-1}y\rangle = \langle y^*, y\rangle$ for all $y \in Y$, because A is bijective. This is equivalent to $(A^{-1})^*x^* = y^*$, and this proves (6.2.6). By (6.2.6), we have $\mathrm{im}(A^{-1})^* = \mathrm{dom}(A^*)$ and

$$A^*(A^{-1})^* = \mathrm{id} : X^* \to X^*, \qquad (A^{-1})^*A^* = \mathrm{id} : \mathrm{dom}(A^*) \to \mathrm{dom}(A^*).$$

Thus A^* is bijective and $(A^*)^{-1} = (A^{-1})^*$. Conversely, if A^* is bijective, then A is injective by part (vi) of Theorem 6.2.2, has a dense image by part (v) of Theorem 6.2.2, and has a closed image by Theorem 6.2.3, and hence is bijective. This proves Corollary 6.2.4. □

Example 6.2.5. This example shows that the domain of the dual operator of a closed densely defined operator need not be dense (see part (iii) of Theorem 6.2.2). Consider the real Banach space $X = \ell^1$ and define the unbounded operator $A : \mathrm{dom}(A) \to \ell^1$ by

$$\mathrm{dom}(A) := \left\{ x = (x_i)_{i\in\mathbb{N}} \in \ell^1 \,\middle|\, \sum_{i=1}^{\infty} i|x_i| < \infty \right\},$$
$$Ax := (ix_i)_{i\in\mathbb{N}} \qquad \text{for } x = (x_i)_{i\in\mathbb{N}} \in \mathrm{dom}(A).$$

This operator has a dense domain. Moreover, it is bijective and has a bounded inverse, given by $A^{-1}y = (i^{-1}y_i)_{i\in\mathbb{N}}$ for $y = (y_i)_{i\in\mathbb{N}} \in \ell^1$. Hence A has a closed graph. Identify the dual space X^* with ℓ^∞ in the canonical way. Then the dual operator $A^* : \mathrm{dom}(A^*) \to \ell^\infty$ is given by

$$\mathrm{dom}(A^*) := \left\{ y = (y_i)_{i\in\mathbb{N}} \in \ell^\infty \,\middle|\, \sup_{i\in\mathbb{N}} i|y_i| < \infty \right\},$$
$$A^*y := (iy_i)_{i\in\mathbb{N}} \qquad \text{for } y = (y_i)_{i\in\mathbb{N}} \in \mathrm{dom}(A^*).$$

This operator is again bijective. However, its domain is contained in the proper closed subspace $c_0 \subset \ell^\infty$ of sequences of real numbers that converge to zero. It is weak* dense in $X^* = \ell^\infty$ but not dense.

The next lemma shows that the relation between the spectrum of a bounded linear operator and that of the dual operator in Lemma 5.2.5 carries over verbatim to densely defined unbounded operators with closed graphs.

Lemma 6.2.6 (Spectrum of A and A^*). *Let X be a complex Banach space, let $A : \mathrm{dom}(A) \to X$ be an unbounded complex linear operator with a closed graph and a dense domain $\mathrm{dom}(A) \subset X$, and denote by*

$$A^* : \mathrm{dom}(A^*) \to X^*$$

the dual operator. Then the following hold.

(i) $\sigma(A^*) = \sigma(A)$.

(ii) *The point, residual, and continuous spectra of A and A^* are related by*

$$\begin{aligned}
&\mathrm{P}\sigma(A^*) \subset \mathrm{P}\sigma(A) \cup \mathrm{R}\sigma(A), && \mathrm{P}\sigma(A) \subset \mathrm{P}\sigma(A^*) \cup \mathrm{R}\sigma(A^*),\\
&\mathrm{R}\sigma(A^*) \subset \mathrm{P}\sigma(A) \cup \mathrm{C}\sigma(A), && \mathrm{R}\sigma(A) \subset \mathrm{P}\sigma(A^*),\\
&\mathrm{C}\sigma(A^*) \subset \mathrm{C}\sigma(A), && \mathrm{C}\sigma(A) \subset \mathrm{R}\sigma(A^*) \cup \mathrm{C}\sigma(A^*).
\end{aligned}$$

(iii) *If X is reflexive, then $\mathrm{C}\sigma(A^*) = \mathrm{C}\sigma(A)$ and*

$$\begin{aligned}
&\mathrm{P}\sigma(A^*) \subset \mathrm{P}\sigma(A) \cup \mathrm{R}\sigma(A), && \mathrm{P}\sigma(A) \subset \mathrm{P}\sigma(A^*) \cup \mathrm{R}\sigma(A^*),\\
&\mathrm{R}\sigma(A^*) \subset \mathrm{P}\sigma(A), && \mathrm{R}\sigma(A) \subset \mathrm{P}\sigma(A^*).
\end{aligned}$$

Proof. Part (i) follows from the identity

$$(\lambda 1\!\!1_X - A)^* = \lambda 1\!\!1_{X^*} - A^*$$

and Corollary 6.2.4.

Part (ii) follows from the same arguments as part (iii) of Lemma 5.2.5, with Theorem 4.1.8 replaced by Theorem 6.2.2. If $\lambda \in \mathrm{P}\sigma(A^*)$, then $\lambda 1\!\!1 - A^*$ is not injective, hence $\lambda 1\!\!1 - A$ does not have a dense image by part (v) of Theorem 6.2.2, and therefore $\lambda \in \mathrm{P}\sigma(A) \cup \mathrm{R}\sigma(A)$. If $\lambda \in \mathrm{R}\sigma(A^*)$, then $\lambda 1\!\!1 - A^*$ is injective, hence $\lambda 1\!\!1 - A$ has a dense image, and so $\lambda \in \mathrm{P}\sigma(A) \cup \mathrm{C}\sigma(A)$. Third, if $\lambda \in \mathrm{C}\sigma(A^*)$, then $\lambda 1\!\!1 - A^*$ is injective and has a dense image and therefore also has a weak* dense image, thus it follows from parts (v) and (vi) of Theorem 6.2.2 that $\lambda 1\!\!1 - A$ is injective and has a dense image, and therefore $\lambda \in \mathrm{C}\sigma(A)$. This proves part (ii).

Part (iii) follows from part (ii) and the fact that

$$\mathrm{C}\sigma(A) = \mathrm{C}\sigma(A^*)$$

in the reflexive case, again by parts (v) and (vi) of Theorem 6.2.2. This proves Lemma 6.2.6. □

6.3. Unbounded Operators on Hilbert Spaces

The dual operator of an unbounded operator between Banach spaces was introduced in Definition 6.2.1. For Hilbert spaces this leads to the notion of the adjoint of an unbounded densely defined operator which we explain next. As in Example 4.1.6 and Definition 5.3.7, the idea is to replace the dual space of a Hilbert space by the original Hilbert space via the isomorphism of Theorem 1.4.4, respectively Theorem 5.3.6 in the complex case.

6.3.1. The Adjoint of an Unbounded Operator.

Definition 6.3.1 (Adjoint Operator). Let X and Y be complex Hilbert spaces and let $A : \mathrm{dom}(A) \to Y$ be an unbounded operator with a dense domain $\mathrm{dom}(A) \subset X$. The **adjoint operator**

$$A^* : \mathrm{dom}(A^*) \to X, \qquad \mathrm{dom}(A^*) \subset Y,$$

of A is defined as follows. Its domain is the linear subspace

$$\mathrm{dom}(A^*) := \left\{ y \in Y \,\middle|\, \begin{array}{l} \text{there exists a constant } c \geq 0 \text{ such that} \\ |\langle y, A\xi\rangle_Y| \leq c \|\xi\|_X \text{ for all } \xi \in \mathrm{dom}(A) \end{array} \right\}$$

and, for $y \in \mathrm{dom}(A^*)$, the element $A^*y \in X$ is the unique element of X that satisfies the equation

$$\langle A^*y, \xi\rangle_X = \langle y, A\xi\rangle_Y \qquad \text{for all } \xi \in \mathrm{dom}(A).$$

Thus the graph of the adjoint operator is characterized by the condition

$$\begin{array}{l} y \in \mathrm{dom}(A^*) \\ \text{and } x = A^*y \end{array} \quad \Longleftrightarrow \quad \begin{array}{l} \langle A^*y, \xi\rangle_X = \langle y, A\xi\rangle_Y \\ \text{for all } \xi \in \mathrm{dom}(A). \end{array} \tag{6.3.1}$$

The operator A is called **self-adjoint** if $X = Y$ and $A = A^*$.

Observe that an element $y \in Y$ belongs to the domain of A^* if and only if the complex linear functional $\mathrm{dom}(A) \to \mathbb{C} : \xi \mapsto \langle y, A\xi\rangle$ is bounded. In this case, the linear functional extends uniquely to a bounded complex linear functional on all of X, because $\mathrm{dom}(A)$ is a dense subspace of X, and Theorem 5.3.6 asserts that this extended complex linear functional is uniquely represented by an element of X. The reader may verify that $\mathrm{dom}(A^*)$ is a complex subspace of Y and that the operator $A^* : \mathrm{dom}(A^*) \to X$ is complex linear. Throughout the remainder of this chapter the symbol A^* will always denote the adjoint of an unbounded operator between Hilbert spaces as in Definition 6.3.1. The dual operator of Definition 6.2.1 is no longer used.

The next lemma summarizes the basic properties of the adjoint operator. Recall that in the Hilbert space setting the notation

$$S^\perp := \{y \in H \,|\, \langle x, y\rangle = 0 \text{ for all } x \in S\}$$

refers to the (complex) orthogonal complement of a subset $S \subset H$.

Lemma 6.3.2 (Properties of the Adjoint Operator). *Let X and Y be complex Hilbert spaces and let $A : \mathrm{dom}(A) \to Y$ be a linear operator with a dense domain $\mathrm{dom}(A) \subset X$. Then the following hold.*

(i) *If $P \in \mathcal{L}^c(X, Y)$ and $\lambda \in \mathbb{C}$, then $(A + P)^* = A^* + P^*$ and $(\lambda A)^* = \overline{\lambda} A^*$.*

(ii) *A is closeable if and only if $\mathrm{dom}(A^*)$ is a dense subspace of Y.*

(iii) *If A is closed, then $A^{**} = A$.*

(iv) *$\mathrm{im}(A)^\perp = \ker(A^*)$ and, if A is closed, then $\ker(A) = \mathrm{im}(A^*)^\perp$.*

(v) *A has a dense image if and only if A^* is injective.*

(vi) *Assume A is closed. Then A has a closed image if and only if A^* has a closed image if and only if $\mathrm{im}(A^*) = \ker(A)^\perp$.*

(vii) *If A is bijective, then so is A^* and $(A^{-1})^* = (A^*)^{-1}$.*

(viii) *If $X = Y = H$ and A is closed, then $\sigma(A^*) = \left\{\overline{\lambda} \,|\, \lambda \in \sigma(A)\right\}$ and*

$$
\begin{aligned}
\mathrm{P}\sigma(A^*) &\subset \left\{\overline{\lambda} \,|\, \lambda \in \mathrm{P}\sigma(A) \cup \mathrm{R}\sigma(A)\right\}, \\
\mathrm{R}\sigma(A^*) &\subset \left\{\overline{\lambda} \,|\, \lambda \in \mathrm{P}\sigma(A)\right\}, \\
\mathrm{C}\sigma(A^*) &= \left\{\overline{\lambda} \,|\, \lambda \in \mathrm{C}\sigma(A)\right\}.
\end{aligned}
$$

Proof. These assertions are proved by carrying over Theorem 6.2.2, Theorem 6.2.3, Corollary 6.2.4, and Lemma 6.2.6 to the Hilbert space setting. The details are left to the reader. □

6.3.2. Unbounded Self-Adjoint Operators. By definition, every self-adjoint operator on a Hilbert space $H = X = Y$ is symmetric, i.e. it satisfies

$$
\langle x, Ay \rangle = \langle Ax, y \rangle \qquad \text{for all } x, y \in \mathrm{dom}(A).
$$

However, the converse does not hold, even for operators with dense domains and closed graphs. (By Example 2.2.23 every symmetric operator is closeable.) Exercise 6.3.3 below illustrates the difference between symmetric and self-adjoint operators and shows how one can construct self-adjoint extensions of symmetric operators.

A skew-symmetric bilinear form

$$
\omega : V \times V \to \mathbb{R}
$$

on a real vector space V is called **symplectic** if it is nondegenerate, i.e. for every nonzero vector $v \in V$ there exists a vector $u \in V$ such that $\omega(u, v) \neq 0$. Assume $\omega : V \times V \to \mathbb{R}$ is a symplectic form. A linear subspace $\Lambda \subset V$ is called a **Lagrangian subspace** if $\omega(u, v) = 0$ for all $u, v \in \Lambda$ and if, for every $v \in V \setminus \Lambda$, there exists a vector $u \in \Lambda$ such that $\omega(u, v) \neq 0$.

Exercise 6.3.3 (Gelfand–Robbin Quotient). Let H be a real Hilbert space and let $A : \mathrm{dom}(A) \to H$ be a densely defined symmetric operator.

(i) Prove that $\mathrm{dom}(A) \subset \mathrm{dom}(A^*)$ and $A^*|_{\mathrm{dom}(A)} = A$.

(ii) Let $V := \mathrm{dom}(A^*)/\mathrm{dom}(A)$ and define the map $\omega : V \times V \to \mathbb{R}$ by

$$\omega(u, v) := \langle A^*x, y\rangle - \langle x, A^*y\rangle \tag{6.3.2}$$

for $x, y \in \mathrm{dom}(A^*)$, where $u := [x] \in V$ and $v := [y] \in V$. Prove that ω is a well-defined skew-symmetric bilinear form. Prove that ω is nondegenerate if and only if the operator A has a closed graph.

(iii) Assume A has a closed graph. For a subspace $\Lambda \subset V$ define the operator

$$A_\Lambda : \mathrm{dom}(A_\Lambda) \to H$$

by

$$\mathrm{dom}(A_\Lambda) := \{x \in \mathrm{dom}(A^*) \mid [x] \in \Lambda\}, \qquad A_\Lambda := A^*|_{\mathrm{dom}(A_\Lambda)}. \tag{6.3.3}$$

Prove that A_Λ is self-adjoint if and only if Λ is a Lagrangian subspace of V.

(iv) Prove that A admits a self-adjoint extension. **Hint:** The Lemma of Zorn.

(v) Prove that $\Lambda_0 := (\ker(A^*) + \mathrm{dom}(A))/\mathrm{dom}(A)$ is a Lagrangian subspace of V whenever A has a closed graph and a closed image.

Exercise 6.3.4. This example illustrates how the Gelfand–Robbin quotient gives rise to symplectic forms on the spaces of boundary data for symmetric differential operators. Let $n \in \mathbb{N}$ and consider the matrix

$$J := \begin{pmatrix} 0 & -\mathbb{1} \\ \mathbb{1} & 0 \end{pmatrix} \in \mathbb{R}^{2n \times 2n}.$$

Define the operator A on the Hilbert space $H := L^2([0,1], \mathbb{R}^{2n})$ by

$$\mathrm{dom}(A) := \left\{u \in W^{1,2}([0,1], \mathbb{R}^{2n}) \mid u(0) = u(1) = 0\right\}, \qquad Au := J\dot{u}.$$

Here $W^{1,2}([0,1], \mathbb{R}^{2n})$ denotes the space of all absolutely continuous functions $u : [0,1] \to \mathbb{R}^{2n}$ with square integrable derivatives. Prove the following.

(i) A is a symmetric operator with a closed graph.

(ii) $\mathrm{dom}(A^*) = W^{1,2}([0,1], \mathbb{R}^{2n})$ and $A^*u = J\dot{u}$ for all $u \in W^{1,2}([0,1], \mathbb{R}^{2n})$.

(iii) The map $W^{1,2}([0,1], \mathbb{R}^{2n}) \to \mathbb{R}^{2n} \times \mathbb{R}^{2n} : u \mapsto (u(0), u(1))$ descends to an isomorphism from the quotient space $V = \mathrm{dom}(A^*)/\mathrm{dom}(A)$ to $\mathbb{R}^{2n} \times \mathbb{R}^{2n}$. The resulting symplectic form determined by (6.3.2) on $\mathbb{R}^{2n} \times \mathbb{R}^{2n}$ is

$$\omega((u_0, u_1), (v_0, v_1)) = \langle Ju_1, v_1\rangle - \langle Ju_0, v_0\rangle$$

for $(u_0, u_1), (v_0, v_1) \in \mathbb{R}^{2n} \times \mathbb{R}^{2n}$. Here $\langle \cdot, \cdot\rangle$ denotes the standard inner product on $\mathbb{R}^{2n}$.

Exercise 6.3.5. Let H be a separable complex Hilbert space, let $(e_i)_{i\in\mathbb{N}}$ be a complex orthonormal basis, let $(\lambda_i)_{i\in\mathbb{N}}$ be a sequence of complex numbers, and let $A_\lambda : \mathrm{dom}(A_\lambda) \to H$ be the operator in Example 6.1.3. Prove that its adjoint is the operator $A_\lambda^* = A_{\overline{\lambda}}$ associated to the sequence $(\overline{\lambda}_i)_{i\in\mathbb{N}}$. Deduce that A_λ is self-adjoint if and only if $\lambda_i \in \mathbb{R}$ for all i.

Exercise 6.3.6. Prove that the operator A_f in Example 6.1.8 is self-adjoint for $p = 2$ and every measurable function $f : M \to \mathbb{R}$.

Another example of an unbounded self-adjoint operator is the Laplace operator on $\Delta : W^{2,2}(\mathbb{R}^n, \mathbb{C}) \to L^2(\mathbb{R}^n, \mathbb{C})$ in Example 6.1.6. The proof that this operator is self-adjoint requires elliptic regularity and goes beyond the scope of this book. However, this example can be recast as a special case of a general abstract setup, which is useful for many applications and which we now explain.

Definition 6.3.7 (Gelfand Triple). A **Gelfand triple** consists of a real Hilbert space $(H, \langle\cdot,\cdot\rangle_H)$ and a dense subspace $V \subset H$, equipped with an inner product $\langle\cdot,\cdot\rangle_V$ which renders V into a Hilbert space in its own right and the inclusion $V \hookrightarrow H$ into a bounded linear operator. Thus there exists a constant $\kappa > 0$ such that

$$\|v\|_H \le \kappa \|v\|_V \qquad \text{for all } v \in V. \tag{6.3.4}$$

We identify H with its dual space H^* via the isomorphism of Theorem 1.4.4. However, we do not identify V with its own dual space. Thus

$$V \subset H \subset V^*, \tag{6.3.5}$$

where the inclusion $H \cong H^* \hookrightarrow V^*$ assigns to each $u \in H$ the bounded linear functional $V \to \mathbb{R} : v \mapsto \langle u, v\rangle_H$. This is the dual operator of the inclusion $V \hookrightarrow H$ and so is injective and has a dense image by Theorem 4.1.8.

Theorem 6.3.8 (Gelfand Triples). *Let $V \subset H \subset V^*$ be a Gelfand triple and let $B : V \times V \to \mathbb{R}$ be a symmetric bilinear form. Suppose that there exist positive constants δ, c, and C such that*

$$\delta \|v\|_V^2 - c\|v\|_H^2 \le B(v, v) \le C\|v\|_V^2 \qquad \textit{for all } v \in V. \tag{6.3.6}$$

Then the linear subspace

$$\mathrm{dom}(A) := \left\{ u \in V \,\middle|\, \sup_{v\in V\setminus\{0\}} \frac{|B(u, v)|}{\|v\|_H} < \infty \right\} \tag{6.3.7}$$

is dense in V, there is a unique linear operator $A : \mathrm{dom}(A) \to H$ such that

$$\langle Au, v\rangle_H = B(u, v) \qquad \textit{for all } u \in \mathrm{dom}(A) \textit{ and all } v \in V, \tag{6.3.8}$$

and this operator A is self-adjoint. If H is a complex Hilbert space and V is a complex subspace of H such that the complex structure preserves the inner product on V and the bilinear form B, then A is complex linear.

Proof. The existence and uniqueness of an operator $A : \mathrm{dom}(A) \to H$ that satisfies (6.3.7) and (6.3.8) follow directly from the definitions and Theorem 1.4.4. Namely, if $u \in \mathrm{dom}(A)$, then, since V is dense in H, there exists a unique bounded linear functional $\Lambda_u : H \to \mathbb{R}$ such that

$$\Lambda_u(v) = B(u, v) \qquad \text{for all } v \in V,$$

and so, by Theorem 1.4.4, there exists a unique element $Au \in H$ such that

$$\langle Au, f \rangle_H = \Lambda_u(f) \qquad \text{for all } f \in H.$$

Then $A : \mathrm{dom}(A) \to H$ is a symmetric linear operator that satisfies (6.3.8). We prove in seven steps that A is self-adjoint.

Step 1. *If $u, v \in V$, then $|B(u, v)| \le C \|u\|_V \|v\|_V$.*

By Theorem 1.4.4 there exists a unique linear operator $\mathscr{B} : V \to V$ such that $\langle u, \mathscr{B}v \rangle_V = B(u, v)$ for all $u, v \in V$. Since B is symmetric, so is $\mathscr{B}$. Hence $\mathscr{B}$ is bounded by the Hellinger–Toeplitz Theorem (Corollary 2.2.16). Moreover, $|\langle v, \mathscr{B}v \rangle_V| = |B(v, v)| \le C$ for all $v \in V$ with $\|v\|_V = 1$ by (6.3.6). Hence $\|\mathscr{B}\|_{\mathcal{L}(V)} \le C$ by part (iv) of Theorem 5.3.16 and so

$$|B(u, v)| = |\langle u, \mathscr{B}v \rangle_V| \le \|u\|_V \|\mathscr{B}v\|_V \le C \|u\|_V \|v\|_V$$

for all $u, v \in V$. This proves Step 1.

Step 2. *If $u \in \mathrm{dom}(A)$, then $\|u\|_V \le \delta^{-1}\kappa \|cu + Au\|_H$.*

By (6.3.4), (6.3.6), and (6.3.8), every $u \in \mathrm{dom}(A)$ satisfies

$$\begin{aligned} \delta \|u\|_V^2 &\le c \|u\|_H^2 + B(u, u) \\ &= \langle cu + Au, u \rangle_H \\ &\le \|cu + Au\|_H \|u\|_H \\ &\le \kappa \|cu + Au\|_H \|u\|_V \end{aligned}$$

and this proves Step 2. (**Exercise:** Use Step 2 to show that A is closed.)

Step 3. *The formula*

$$\text{(6.3.9)} \qquad \langle u, v \rangle_B := c \langle u, v \rangle_H + B(u, v) \qquad \text{for } u, v \in V$$

defines an inner product on V whose norm $V \to \mathbb{R} : v \mapsto \|v\|_B := \sqrt{\langle v, v \rangle_B}$ is compatible with $\|\cdot\|_V$. Thus $(V, \langle \cdot, \cdot \rangle_B)$ is a Hilbert space.

The bilinear form (6.3.9) is symmetric because B is symmetric and satisfies the inequality $\delta \|v\|_V^2 \le B(v, v) \le c \|v\|_H^2 + C \|v\|_V^2 \le \left(c\kappa^2 + C\right) \|v\|_V^2$ for all $v \in V$ by (6.3.4) and (6.3.6). This proves Step 3.

The next step is the heart of the proof. It can be viewed as an abstract variant of the Dirichlet principle.

Step 4. *The operator* $c1\!\!1_H + A : \mathrm{dom}(A) \to H$ *is bijective.*

The operator is injective by Step 2. To prove that it is surjective, fix an element $f \in H$ and define the bounded linear functional $\Lambda : V \to \mathbb{R}$ by

$$\Lambda(v) := \langle f, v \rangle_H \qquad \text{for } v \in V.$$

Then, by Step 3 and Theorem 1.4.4, there exists an element $u \in V$ that satisfies $\langle u, v \rangle_B = \Lambda(v)$ for all $v \in V$. This implies

$$c \langle u, v \rangle_H + B(u, v) = \langle f, v \rangle_H$$

for all $v \in V$ and hence

$$|B(u, v)| = |\langle f - cu, v \rangle_H| \leq \|f - cu\|_H \|v\|_H .$$

Thus $u \in \mathrm{dom}(A)$ and, for all $v \in V$, we have

$$\langle cu + Au - f, v \rangle_H = c \langle u, v \rangle_H + B(u, v) - \langle f, v \rangle_H = 0.$$

Since V is dense in H, it follows that $cu + Au = f$ and this proves Step 4.

Step 5. *The subspace* $\mathrm{dom}(A) \subset V$ *defined by* (6.3.7) *is dense in* V.

Let $\iota : V \to H$ denote the canonical inclusion and let $\iota^* : H \to V$ be its adjoint operator with respect to the inner products $\langle \cdot, \cdot \rangle_H$ on H and $\langle \cdot, \cdot \rangle_B$ on V (see Step 3). Then ι^* has a dense image by Theorem 4.1.8. Let $f \in H$ and define $u := (c1\!\!1_H + A)^{-1} f \in \mathrm{dom}(A)$ by Step 4. Then $cu + Au = f$ and

$$\langle \iota^*(f), v \rangle_B = \langle f, \iota(v) \rangle_H = \langle cu + Au, v \rangle_H = \langle u, v \rangle_B = \big\langle (c1\!\!1 + A)^{-1} f, v \big\rangle_B$$

for all $v \in V$. This shows that $\iota^* = (c1\!\!1_H - A)^{-1} : H \to V$ and hence the subspace $\mathrm{dom}(A) = \mathrm{im}(\iota^*)$ is dense in V. This proves Step 5.

Step 6. *Let* $v \in H$ *and suppose that there is a constant* $K \geq 0$ *such that*

$$|\langle v, Au \rangle_H| \leq K \|u\|_V \qquad \text{for all } u \in \mathrm{dom}(A). \tag{6.3.10}$$

Then $v \in V$.

By (6.3.4) and (6.3.10) we have $|\langle v, cu + Au \rangle_H| \leq (c\kappa \|v\|_H + K) \|u\|_V$ for all $u \in \mathrm{dom}(A)$. Since $\mathrm{dom}(A)$ is dense in V by Step 5, this implies that there exists a unique bounded linear functional $\Lambda : V \to \mathbb{R}$ such that

$$\Lambda(u) = \langle v, cu + Au \rangle_H \qquad \text{for all } u \in \mathrm{dom}(A). \tag{6.3.11}$$

Hence, by Step 3 and Theorem 1.4.4, there exists a $w \in V$ such that

$$\Lambda(u) = \langle w, u \rangle_B = c \langle w, u \rangle_H + B(w, u) \qquad \text{for all } u \in V. \tag{6.3.12}$$

Take $u \in \mathrm{dom}(A)$ in (6.3.12) to obtain $\Lambda(u) = \langle w, cu + Au \rangle_H$. Hence it follows from (6.3.11) that $\langle v - w, cu + Au \rangle_H = 0$ for all $u \in \mathrm{dom}(A)$. Since the operator $c1\!\!1_H + A : \mathrm{dom}(A) \to H$ is surjective by Step 4, we have $v = w \in V$. This proves Step 6.

Step 7. *The operator $A : \mathrm{dom}(A) \to H$ is self-adjoint.*

The operator A is symmetric by definition. Hence $\mathrm{dom}(A) \subset \mathrm{dom}(A^*)$ and $A^*|_{\mathrm{dom}(A)} = A$. It remains to prove that $\mathrm{dom}(A^*) \subset \mathrm{dom}(A)$. To see this, fix an element $v \in \mathrm{dom}(A^*)$. Then

$$|\langle v, Au\rangle_H| = |\langle A^*v, u\rangle_H| \le \|A^*v\|_H \|u\|_H \le \kappa \|A^*v\|_H \|u\|_V$$

for all $u \in \mathrm{dom}(A)$ by (6.3.4). Hence $v \in V$ by Step 6. This implies

$$|B(v,u)| = |\langle v, Au\rangle_H| = |\langle A^*v, u\rangle_H| \le \|A^*v\|_H \|u\|_H$$

for all $u \in \mathrm{dom}(A)$. Since $\mathrm{dom}(A)$ is dense in V by Step 5, and the functions $V \to \mathbb{R} : u \mapsto \|u\|_H$ and $V \to \mathbb{R} : u \mapsto B(v,u)$ are continuous by (6.3.4) and Step 1, this implies $|B(v,u)| \le \|A^*v\|_H \|u\|_H$ for all $u \in V$ and therefore $v \in \mathrm{dom}(A)$. This proves Step 7 and Theorem 6.3.8. □

The next corollary explains how every closed densely defined unbounded operator gives rise to a self-adjoint operator by composition with its adjoint. The composition of two unbounded linear operators $A : \mathrm{dom}(A) \to Y$ with $\mathrm{dom}(A) \subset X$ and $B : \mathrm{dom}(B) \to Z$ with $\mathrm{dom}(B) \subset Y$ is the operator $BA : \mathrm{dom}(BA) \to Z$ defined by

$$\begin{aligned} \mathrm{dom}(BA) &:= \left\{x \in \mathrm{dom}(A) \,\middle|\, Ax \in \mathrm{dom}(B)\right\}, \\ BAx &:= B(Ax) \qquad \text{for } x \in \mathrm{dom}(BA). \end{aligned} \tag{6.3.13}$$

The domain of BA can be trivial even if A and B are densely defined. In the next theorem X and Y can either be real Hilbert spaces or complex Hilbert spaces with Hermitian inner products. In the latter case we assume that D is an unbounded complex linear operator and so D^*D is also complex linear.

Corollary 6.3.9 (The Operator D^*D). *Let X and Y be Hilbert spaces and let $D : \mathrm{dom}(D) \to Y$ be a closed unbounded operator with a dense domain $\mathrm{dom}(D) \subset X$. Then the operator $D^*D : \mathrm{dom}(D^*D) \to X$ is self-adjoint and its domain is dense in $\mathrm{dom}(D)$ with respect to the graph norm.*

Proof. This is a Gelfand triple with

$$H := X, \qquad V := \mathrm{dom}(D), \qquad \langle u, v\rangle_V := \langle u, v\rangle_X + \langle Du, Dv\rangle_Y$$

for $u, v \in \mathrm{dom}(D)$, and the bilinear form $B : V \times V \to \mathbb{R}$ is given by

$$B(u,v) := \langle Du, Dv\rangle_Y \qquad \text{for } u, v \in \mathrm{dom}(D) \subset X.$$

These data satisfy the hypotheses of Theorem 6.3.8 with $\delta = c = C = 1$. In particular, $\|v\|_V^2 = \|v\|_X^2 + \|Dv\|_Y^2 = \|v\|_X^2 + B(v,v) = \|v\|_B^2$ for all $v \in V$. The condition $\sup_{v\in V\setminus\{0\}} \|v\|_X^{-1} |\langle Du, Dv\rangle_Y| < \infty$ for $u \in V = \mathrm{dom}(D)$ in equation (6.3.7) is equivalent to $Du \in \mathrm{dom}(D^*)$, so the operator A in (6.3.8) agrees with D^*D. Hence the operator D^*D is self-adjoint by Theorem 6.3.8. This proves Corollary 6.3.9. □

Example 6.3.10 (Dirichlet Problem). The archetypal example of the situation in Theorem 6.3.8 and Corollary 6.3.9 is the operator

$$\nabla = \left(\frac{\partial}{\partial x_1}, \dots, \frac{\partial}{\partial x_n}\right) : W_0^{1,2}(\Omega) \to L^2(\Omega, \mathbb{R}^n).$$

Here $\Omega \subset \mathbb{R}^n$ is a bounded open set with smooth boundary (i.e. $\partial\Omega$ is a smooth $(n-1)$-dimensional submanifold of $\mathbb{R}^n$ and $\Omega = \mathrm{int}(\overline{\Omega})$) and $W_0^{1,2}(\Omega)$ is the completion of the space $C_0^\infty(\Omega)$ of smooth functions $u : \Omega \to \mathbb{R}$ with compact support with respect to the norm

$$\|u\|_{W_0^{1,2}} := \sqrt{\int_\Omega \sum_{i=1}^n \left|\frac{\partial u}{\partial u_i}(x)\right|^2 dx} = \sqrt{\int_\Omega |\nabla u(x)|^2 \, dx}.$$

The Poincaré inequality asserts that this norm controls the L^2 norm of u. This example corresponds to the Gelfand triple with

$$H = X = L^2(\Omega), \qquad V = \mathrm{dom}(D) = W_0^{1,2}(\Omega),$$

where the bilinear form

$$B : W_0^{1,2}(\Omega) \times W_0^{1,2}(\Omega) \to \mathbb{R}$$

is given by

$$B(u, v) := \int_\Omega \langle \nabla u(x), \nabla v(x) \rangle \, dx$$

for $u, v \in W_0^{1,2}(\Omega)$. The operator $D = \nabla : \mathrm{dom}(D) \to Y$ takes values in the Hilbert space $Y = L^2(\Omega, \mathbb{R}^n)$, and $A = D^*D$ is the **Laplace operator**

$$\Delta = \sum_{i=1}^n \frac{\partial^2}{\partial x_i^2} : W^{2,2}(\Omega) \cap W_0^{1,2}(\Omega) \to L^2(\Omega). \tag{6.3.14}$$

Here $W^{2,2}(\Omega)$ denotes the space of equivalence classes, up to equality almost everywhere, of all L^2 functions $u : \Omega \to \mathbb{R}$ whose distributional derivatives up to order two can be represented by L^2 functions. The proof that

$$\mathrm{dom}(D^*D) = W^{2,2}(\Omega) \cap W_0^{1,2}(\Omega)$$

(for all domains $\Omega \subset \mathbb{R}^n$ with *"sufficiently nice boundary"*) requires elliptic regularity and goes beyond the scope of this book. Once this is established, Corollary 6.3.9 asserts that the Laplace operator (6.3.14) is self-adjoint. Moreover, this example satisfies condition (6.3.6) with $c = 0$. Hence it follows from Step 4 in the proof of Theorem 6.3.8 that the operator (6.3.14) is bijective. This translates into the observation that the Dirichlet problem

$$\begin{aligned} \Delta u &= f \quad \text{in } \Omega, \\ u &= 0 \quad \text{on } \partial\Omega \end{aligned} \tag{6.3.15}$$

has a unique solution $u \in W^{2,2}(\Omega) \cap W_0^{1,2}(\Omega)$ for every $f \in L^2(\Omega)$.

6.3.3. Unbounded Normal Operators. The next theorem introduces unbounded normal operators on Hilbert spaces.

Theorem 6.3.11 (Unbounded Normal Operator). *Let H be a complex Hilbert space and let $A : \mathrm{dom}(A) \to H$ be a closed unbounded complex linear operator with a dense domain $\mathrm{dom}(A) \subset H$. The following are equivalent.*

(i) $AA^* = A^*A$.

(ii) $\mathrm{dom}(A) = \mathrm{dom}(A^*)$ *and* $\|Ax\| = \|A^*x\|$ *for all* $x \in \mathrm{dom}(A)$.

(iii) *There exist complex linear self-adjoint operators* $A_i : \mathrm{dom}(A_i) \to H$ *for* $i = 1, 2$ *such that* $\mathrm{dom}(A) = \mathrm{dom}(A^*) = \mathrm{dom}(A_1) \cap \mathrm{dom}(A_2)$ *and*

$$Ax = A_1x + \mathbf{i}A_2x, \qquad A^*x = A_1x - \mathbf{i}A_2x, \qquad \|Ax\|^2 = \|A_1x\|^2 + \|A_2x\|^2$$

for all $x \in \mathrm{dom}(A)$.

Definition 6.3.12 (Unbounded Normal Operator). A closed unbounded complex linear operator $A : \mathrm{dom}(A) \to H$ on a Hilbert space H with a dense domain $\mathrm{dom}(A) \subset H$ is called **normal** if it satisfies the equivalent conditions of Theorem 6.3.11.

Proof. We prove that (i) implies (ii). Assume $AA^* = A^*A$. Then every element $x \in \mathrm{dom}(A^*A) = \mathrm{dom}(AA^*)$ satisfies $x \in \mathrm{dom}(A) \cap \mathrm{dom}(A^*)$ as well as $Ax \in \mathrm{dom}(A^*)$ and $A^*x \in \mathrm{dom}(A)$, and hence

$$\|Ax\|^2 = \langle Ax, Ax\rangle = \langle x, A^*Ax\rangle = \langle x, AA^*x\rangle = \langle A^*x, A^*x\rangle = \|A^*x\|^2 .$$

Next we prove that $\mathrm{dom}(A) \subset \mathrm{dom}(A^*)$. Let $x \in \mathrm{dom}(A)$. Then Corollary 6.3.9 asserts that there exists a sequence $x_i \in \mathrm{dom}(A^*A)$ such that

$$\lim_{i\to\infty} \|x - x_i\| = 0, \qquad \lim_{i\to\infty} \|Ax - Ax_i\| = 0.$$

Thus $(Ax_i)_{i\in\mathbb{N}}$ is a Cauchy sequence in H and so is the sequence $(A^*x_i)_{i\in\mathbb{N}}$ because $\|A^*x_i - A^*x_j\| = \|Ax_i - Ax_j\|$ for all $i, j \in \mathbb{N}$ by what we already proved. Hence $(A^*x_i)_{i\in\mathbb{N}}$ converges to some element $y := \lim_{i\to\infty} A^*x_i$. Since the sequence $(x_i)_{i\in\mathbb{N}}$ converges to x and $(A^*x_i)_{i\in\mathbb{N}}$ converges to y and A^* has a closed graph, it follows that $x \in \mathrm{dom}(A^*)$ and $A^*x = y$. Hence

$$\|A^*x\| = \|y\| = \lim_{i\to\infty} \|A^*x_i\| = \lim_{i\to\infty} \|Ax_i\| = \|Ax\| .$$

This shows that $\mathrm{dom}(A) \subset \mathrm{dom}(A^*)$ and $\|A^*x\| = \|Ax\|$ for all $x \in \mathrm{dom}(A)$. The converse inclusion $\mathrm{dom}(A^*) \subset \mathrm{dom}(A)$ follows by interchanging the roles of A and A^*. This shows that (i) implies (ii).

We prove that (ii) implies (i). Assume $\mathrm{dom}(A) = \mathrm{dom}(A^*)$ and

$$\|Ax\| = \|A^*x\| \qquad \text{for all } x \in \mathrm{dom}(A).$$

Then the same argument as in the proof of Lemma 5.3.14 shows that

$$\langle Ax, Ay\rangle = \langle A^*x, A^*y\rangle \qquad \text{for all } x, y \in \mathrm{dom}(A). \tag{6.3.16}$$

Now let $x \in \mathrm{dom}(A^*A)$. Then $x \in \mathrm{dom}(A)$ and $Ax \in \mathrm{dom}(A^*)$ and, by equation (6.3.16), we have

$$|\langle A^*x, A^*\xi\rangle| = |\langle Ax, A\xi\rangle| = |\langle A^*Ax, \xi\rangle| \leq \|A^*Ax\| \, \|\xi\|$$

for all $\xi \in \mathrm{dom}(A^*)$. This implies $A^*x \in \mathrm{dom}(A)$ and hence $x \in \mathrm{dom}(AA^*)$. Thus we have proved that $\mathrm{dom}(A^*A) \subset \mathrm{dom}(AA^*)$. The same argument, with the roles of A and A^* reversed, shows that

$$\mathrm{dom}(A^*A) = \mathrm{dom}(AA^*).$$

Now let $x \in \mathrm{dom}(A^*A) = \mathrm{dom}(AA^*)$. Then, by equation (6.3.16), we have

$$\langle A^*Ax, \xi\rangle = \langle Ax, A\xi\rangle = \langle A^*x, A^*\xi\rangle = \langle AA^*x, \xi\rangle$$

for all $\xi \in \mathrm{dom}(A) = \mathrm{dom}(A^*)$. Since $\mathrm{dom}(A)$ is dense in H, this implies $A^*Ax = AA^*x$. Thus we have proved that (ii) implies (i).

We prove that (ii) implies (iii). Assume $\mathrm{dom}(A) = \mathrm{dom}(A^*)$ and

$$\|Ax\| = \|A^*x\| \qquad \text{for all } x \in \mathrm{dom}(A).$$

Define the operators $B_1, B_2 : \mathrm{dom}(A) \to H$ by

$$B_1x := \frac{1}{2}(Ax + A^*x), \qquad B_2x := \frac{1}{2\mathbf{i}}(Ax - A^*x)$$

for $x \in \mathrm{dom}(A)$. These operators are symmetric and hence closeable by Example 2.2.23. Thus they admit self-adjoint extensions $A_i : \mathrm{dom}(A_i) \to H$ for $i = 1, 2$ by Exercise 6.3.3. Moreover, $\mathrm{dom}(A) \subset \mathrm{dom}(A_1) \cap \mathrm{dom}(A_2)$ and every element $x \in \mathrm{dom}(A) = \mathrm{dom}(A^*)$ satisfies

$$Ax = A_1x + \mathbf{i}A_2x, \qquad A^*x = A_1x - \mathbf{i}A_2x,$$

and

$$\begin{aligned}
\|Ax\|^2 &= \frac{1}{2}\Big(\|Ax\|^2 + \|A^*x\|^2\Big) \\
&= \frac{1}{4}\Big(\|Ax + A^*x\|^2 + \|Ax - A^*x\|^2\Big) \\
&= \|A_1x\|^2 + \|A_2x\|^2.
\end{aligned}$$

Now let $x \in \mathrm{dom}(A_1) \cap \mathrm{dom}(A_2)$. Then

$$\begin{aligned}
|\langle x, A\xi\rangle| &= |\langle x, A_1\xi + \mathbf{i}A_2\xi\rangle| \\
&= |\langle A_1x, \xi\rangle + \langle A_2x, \mathbf{i}\xi\rangle| \\
&\leq \Big(\|A_1x\| + \|A_2x\|\Big)\,\|\xi\|
\end{aligned}$$

for every $\xi \in \mathrm{dom}(A)$ and hence

$$x \in \mathrm{dom}(A^*) = \mathrm{dom}(A).$$

This shows that (ii) implies (iii).

We prove that (iii) implies (ii). Assume $A_i : \mathrm{dom}(A_i) \to H$ for $i = 1, 2$ are self-adjoint operators that satisfy the following four conditions.

(a) $\mathrm{dom}(A_1) \cap \mathrm{dom}(A_2)$ is a dense subspace of H.

(b) $\|A_1x + \mathbf{i}A_2x\|^2 = \|A_1x\|^2 + \|A_2x\|^2$ for all $x \in \mathrm{dom}(A_1) \cap \mathrm{dom}(A_2)$.

(c) Let $y \in H$ and $c > 0$ such that $|\langle y, A_1x + \mathbf{i}A_2x\rangle| \le c\|x\|$ for every element $x \in \mathrm{dom}(A_1) \cap \mathrm{dom}(A_2)$. Then $y \in \mathrm{dom}(A_1) \cap \mathrm{dom}(A_2)$.

(d) Let $x \in H$ and $c > 0$ such that $|\langle x, A_1y - \mathbf{i}A_2y\rangle| \le c\|y\|$ for every element $y \in \mathrm{dom}(A_1) \cap \mathrm{dom}(A_2)$. Then $x \in \mathrm{dom}(A_1) \cap \mathrm{dom}(A_2)$.

Define the operator $A : \mathrm{dom}(A) \to H$ by

$$\begin{aligned} \mathrm{dom}(A) &:= \mathrm{dom}(A_1) \cap \mathrm{dom}(A_2), \\ Ax &:= A_1x + \mathbf{i}A_2x \qquad \text{for } x \in \mathrm{dom}(A_1) \cap \mathrm{dom}(A_2). \end{aligned} \tag{6.3.17}$$

Its domain is dense by (a). We prove that its adjoint operator is given by

$$\begin{aligned} \mathrm{dom}(A^*) &= \mathrm{dom}(A_1) \cap \mathrm{dom}(A_2), \\ A^*y &= A_1y - \mathbf{i}A_2y \qquad \text{for } y \in \mathrm{dom}(A_1) \cap \mathrm{dom}(A_2). \end{aligned} \tag{6.3.18}$$

Let $y \in \mathrm{dom}(A^*)$. Then $\langle y, Ax\rangle = \langle A^*y, x\rangle$ for all $x \in \mathrm{dom}(A)$ and this implies $y \in \mathrm{dom}(A_1) \cap \mathrm{dom}(A_2)$ by (c). Hence

$$\langle A^*y, x\rangle = \langle y, A_1x + \mathbf{i}A_2x\rangle = \langle A_1y - \mathbf{i}A_2y, x\rangle$$

for all $x \in \mathrm{dom}(A_1) \cap \mathrm{dom}(A_2)$, and hence $A^*y = A_1y - \mathbf{i}A_2y$ by (a). The converse inclusion $\mathrm{dom}(A_1) \cap \mathrm{dom}(A_2) \subset \mathrm{dom}(A^*)$ follows directly from the assumptions. This shows that (6.3.18) is the adjoint operator of (6.3.17) and vice versa by the same argument, using (d) instead of (c). In particular, A has a closed graph. Moreover, it follows from (b) that $\|Ax\| = \|A^*x\|$ for all $x \in \mathrm{dom}(A) = \mathrm{dom}(A^*)$. This shows that (iii) implies (ii) and completes the proof of Theorem 6.3.11. □

Let H be a separable complex Hilbert space, equipped with an orthonormal basis $(e_i)_{i\in\mathbb{N}}$. Then the operator $A_\lambda : \mathrm{dom}(A_\lambda) \to H$ in Example 6.1.3 is normal for every sequence of complex numbers $(\lambda_i)_{i\in\mathbb{N}}$. The operator A_λ is bounded if and only if the sequence $(\lambda_i)_{i\in\mathbb{N}}$ is bounded, it is self-adjoint if and only if $\lambda_i \in \mathbb{R}$ for all i (Exercise 6.3.5), it is compact if and only if $\lim_{i\to\infty}|\lambda_i| = 0$ (Example 4.2.8), and it has a compact resolvent if and only if $\lim_{i\to\infty}|\lambda_i| = \infty$. This example shows that the domains of the self-adjoint operators $A_1 = A_{\mathrm{Re}\lambda}$ and $A_2 = A_{\mathrm{Im}\lambda}$ in Theorem 6.3.11 may differ dramatically from the domain of $A = A_\lambda$. It also shows that every nonempty closed subset of the complex plane can be the spectrum of an unbounded normal operator (Example 6.1.16). In particular, the resolvent set can be empty. The next theorem shows that every normal operator has a nonempty spectrum.

Theorem 6.3.13 (Spectrum of a Normal Operator). *Let H be a nonzero complex Hilbert space and let $A : \mathrm{dom}(A) \to H$ be an unbounded normal operator with $\mathrm{dom}(A) \subsetneq H$. Then the following hold.*

(i) *If $\lambda \in \mathbb{C}$, then $\lambda \mathbb{1} - A$ is normal and, if $\lambda \in \rho(A)$, then the resolvent operator $R_\lambda(A) = (\lambda \mathbb{1} - A)^{-1}$ is normal.*

(ii) $\sigma(A) \neq \emptyset$.

(iii) $\mathrm{R}\sigma(A) = \emptyset$ *and* $\mathrm{P}\sigma(A^*) = \{\overline{\lambda} \,|\, \lambda \in \mathrm{P}\sigma(A)\}$.

(iv) *If A has a compact resolvent, then the spectrum $\sigma(A) = \mathrm{P}\sigma(A)$ is discrete, for each $\lambda \in \mathrm{P}\sigma(A)$ the eigenspace $E_\lambda := \ker(\lambda \mathbb{1} - A)$ is finite-dimensional, and A admits an orthonormal basis of eigenvectors.*

(v) *If A is self-adjoint, then $\sigma(A) \subset \mathbb{R}$ and*

$$\begin{aligned} \sup \sigma(A) &= \sup \{\langle x, Ax \rangle \,|\, x \in \mathrm{dom}(A),\ \|x\| = 1\}, \\ \inf \sigma(A) &= \inf \{\langle x, Ax \rangle \,|\, x \in \mathrm{dom}(A),\ \|x\| = 1\}. \end{aligned} \tag{6.3.19}$$

Proof. We prove part (i). Let $\lambda \in \mathbb{C}$. Then $(\lambda \mathbb{1} - A)^* = \overline{\lambda}\mathbb{1} - A^*$ by part (i) of Lemma 6.3.2. Hence

$$\begin{aligned} \|\lambda x - Ax\|^2 &= |\lambda|^2 \|x\|^2 - 2\mathrm{Re}\langle \lambda x, Ax \rangle + \|Ax\|^2 \\ &= |\overline{\lambda}|^2 \|x\|^2 - 2\mathrm{Re}\langle A^* x, \overline{\lambda} x \rangle + \|A^* x\|^2 \\ &= \|\overline{\lambda} x - A^* x\|^2 \end{aligned}$$

for all $x \in \mathrm{dom}(A) = \mathrm{dom}(\lambda \mathbb{1} - A) = \mathrm{dom}(\overline{\lambda}\mathbb{1} - A^*)$. Thus $\lambda \mathbb{1} - A$ is normal. If A is invertible, then

$$\begin{aligned} A^{-1}(A^{-1})^* &= A^{-1}(A^*)^{-1} \\ &= (A^*A)^{-1} \\ &= (AA^*)^{-1} \\ &= (A^*)^{-1}A^{-1} \\ &= (A^{-1})^* A^{-1} \end{aligned}$$

by part (vii) of Lemma 6.3.2, and hence A^{-1} is normal. This proves part (i).

We prove part (ii). If $\rho(A) = \emptyset$, then $\sigma(A) = \mathbb{C} \neq \emptyset$. If $\rho(A) \neq \emptyset$ and $\mu \in \rho(A)$, then $R_\mu(A)$ is normal by part (i), hence

$$\sup_{z \in \sigma(R_\mu(A))} |z| = \|R_\mu(A)\| > 0$$

by Theorem 5.3.15, and hence

$$\sigma(A) = \{\mu - z^{-1} \,|\, z \in \sigma(R_\mu(A)) \setminus \{0\}\} \neq \emptyset$$

by Lemma 6.1.12. This proves part (ii).

We prove part (iii). Fix an element $\lambda \in \mathbb{C} \setminus (\mathrm{P}\sigma(A) \cup \mathrm{C}\sigma(A))$. Then the operator $\lambda \mathbb{1} - A$ is normal by part (i) and is injective because $\lambda \notin \mathrm{P}\sigma(A)$. Hence the adjoint operator $(\lambda \mathbb{1} - A)^* = \overline{\lambda}\mathbb{1} - A^*$ is injective by definition of a normal operator in Theorem 6.3.11. Thus $\lambda \mathbb{1} - A$ has a dense image by part (v) of Lemma 6.3.2 and so $\lambda \mathbb{1} - A$ is surjective because $\lambda \notin \mathrm{C}\sigma(A)$. Thus $\lambda \in \rho(A)$ and this proves part (iii).

We prove part (iv). By assumption $\rho(A) \neq \emptyset$ and the resolvent operator $R_\mu(A)$ is compact for all $\mu \in \rho(A)$. Fix an element $\mu \in \rho(A)$. Then Theorem 5.2.8 asserts that $\sigma(R_\mu(A)) \setminus \{0\} = \mathrm{P}\sigma(R_\mu(A))$, that the spectrum of $R_\mu(A)$ can only accumulate at the origin, and that the eigenspaces of $R_\mu(A)$ are all finite-dimensional. Moreover, Theorem 5.3.15 asserts that the operator $R_\mu(A)$ admits an orthonormal basis of eigenvectors. Hence part (iv) follows from Lemma 6.1.12.

We prove part (v). Assume A is self-adjoint and let $\lambda \in \mathbb{C} \setminus \mathbb{R}$. Then

$$\|\lambda x - Ax\|^2 = (\mathrm{Im}\lambda)^2 \|x\|^2 + \|(\mathrm{Re}\lambda)x - Ax\|^2 \geq (\mathrm{Im}\lambda)^2 \|x\|^2$$

for all $x \in \mathrm{dom}(A)$ as in the proof of Theorem 5.3.16. Hence $\lambda \mathbb{1} - A$ is injective and has a closed image by Theorem 6.2.3. Replace λ by $\overline{\lambda}$ to deduce that the adjoint operator $\overline{\lambda}\mathbb{1} - A^* = \overline{\lambda}\mathbb{1} - A$ is also injective, hence $\lambda\mathbb{1} - A$ has a dense image by part (iv) of Lemma 6.3.2, so $\lambda \mathbb{1} - A$ is bijective and $\lambda \in \rho(A)$.

Now let $\lambda \in \mathbb{R}$ and assume

$$\lambda > \sup_{x \in \mathrm{dom}(A),\, \|x\|=1} \langle x, Ax \rangle =: c.$$

Then

$$\|x\|\|\lambda x - Ax\| \geq \langle x, \lambda x - Ax \rangle \geq (\lambda - c)\|x\|^2 \qquad \text{for all } x \in \mathrm{dom}(A).$$

Hence $\lambda \mathbb{1} - A$ is injective and has a closed image by Theorem 6.2.3 and so is bijective by Lemma 6.3.2. This shows that $\sigma(A) \subset (-\infty, c]$.

Conversely, assume

$$c := \sup \sigma(A) < \infty.$$

We must prove that $\langle x, Ax \rangle \leq c$ for all $x \in \mathrm{dom}(A)$ with $\|x\| = 1$. Suppose, by contradiction, that there exists an element $x \in \mathrm{dom}(A)$ such that $\|x\| = 1$ and $\langle x, Ax \rangle > c$. Choose a real number μ such that $c < \mu < \langle x, Ax \rangle$ and define $\xi := \mu x - Ax$. Then $\mu \in \rho(A)$ by assumption and

$$\langle \xi, R_\mu(A)\xi \rangle = \langle \mu x - Ax, x \rangle = \mu - \langle Ax, x \rangle < 0.$$

However, by Lemma 6.1.12, we have

$$\sigma(R_\mu(A)) = \{(\mu - \lambda)^{-1} \,|\, \lambda \in \sigma(A)\} \cup \{0\} \subset [0, \infty)$$

in contradiction to Theorem 5.3.16. This proves Theorem 6.3.13. □

6.4. Functional Calculus and Spectral Measures

The purpose of the present section is to extend the measurable functional calculus and the spectral measure to unbounded self-adjoint operators.

6.4.1. Functional Calculus. For a topological space Σ let $B(\Sigma)$ denote the C* algebra of bounded Borel measurable functions $f : \Sigma \to \mathbb{C}$ with the supremum norm $\|f\| := \sup_{\lambda\in\Sigma}|f(\lambda)|$. Denote by $C_b(\Sigma) \subset B(\Sigma)$ the C* subalgebra of complex valued bounded continuous functions on Σ. The next theorem extends the functional calculus of Theorem 5.6.5 to unbounded self-adjoint operators.

Theorem 6.4.1 (Functional Calculus). *Let H be a nonzero complex Hilbert space, let $A : \mathrm{dom}(A) \to H$ be an unbounded self-adjoint operator, and let $\Sigma := \sigma(A) \subset \mathbb{R}$. Then there exists a C^* algebra homomorphism*

$$B(\Sigma) \to \mathcal{L}^c(H) : f \mapsto f(A) =: \Psi_A(f) \tag{6.4.1}$$

that satisfies the following axioms.

(Normalization) *Let $f_i \in B(\Sigma)$ be a sequence such that $\sup_{i\in\mathbb{N}} |f_i(\lambda)| \le |\lambda|$ and $\lim_{i\to\infty} f_i(\lambda) = \lambda$ for all $\lambda \in \Sigma$. Then*

$$\lim_{i\to\infty} f_i(A)x = Ax \qquad \text{for all } x \in \mathrm{dom}(A).$$

(Convergence) *Let $f_i \in B(\Sigma)$ be a sequence such that $\sup_{i\in\mathbb{N}} \|f_i\| < \infty$ and let $f \in B(\Sigma)$ such that $\lim_{i\to\infty} f_i(\lambda) = f(\lambda)$ for all $\lambda \in \Sigma$. Then*

$$\lim_{i\to\infty} f_i(A)x = f(A)x \qquad \text{for all } x \in H.$$

(Positive) *If $f \in B(\Sigma, \mathbb{R})$ and $f \ge 0$, then $f(A) = f(A)^* \ge 0$.*

(Contraction) *$\|f(A)\| \le \|f\|$ for all $f \in B(\Sigma)$ and $\|f(A)\| = \|f\|$ for all $f \in C_b(\Sigma)$.*

(Commutative) *If $B \in \mathcal{L}^c(H)$ satisfies $AB = BA$, then $f(A)B = Bf(A)$ for all $f \in B(\Sigma)$.*

(Eigenvector) *If $\lambda \in \Sigma$ and $x \in \mathrm{dom}(A)$ satisfy $Ax = \lambda x$, then every function $f \in B(\Sigma)$ satisfies $f(A)x = f(\lambda)x$.*

(Spectrum) *If $f \in B(\Sigma)$, then $f(A)$ is normal and $\sigma(f(A)) \subset \overline{f(\Sigma)}$. Moreover, $\sigma(f(A)) = \overline{f(\Sigma)}$ for all $f \in C_b(\Sigma)$.*

(Composition) *If $f \in C_b(\Sigma)$ and $g \in B(\overline{f(\Sigma)})$, then $(g \circ f)(A) = g(f(A))$.*

The C^* algebra homomorphism (6.4.1) is uniquely determined by the (Normalization) and (Convergence) axioms.

Proof. See page 329. □

Theorem 6.4.2 (Cayley Transform). *Let H be a complex Hilbert space.*

(i) *Let $A : \mathrm{dom}(A) \to H$ be a self-adjoint operator. Then the operator*

$$U := (A - \mathbf{i}\mathbb{1})(A + \mathbf{i}\mathbb{1})^{-1} : H \to H \tag{6.4.2}$$

is unitary, the operator $\mathbb{1} - U : H \to H$ is injective, and

$$\mathrm{dom}(A) = \mathrm{im}(\mathbb{1} - U), \qquad A = \mathbf{i}(\mathbb{1} + U)(\mathbb{1} - U)^{-1}. \tag{6.4.3}$$

The operator U is called the **Cayley transform** *of A.*

(ii) *Let $U \in \mathcal{L}^c(H)$ be a unitary operator such that $\mathbb{1} - U$ is injective. Then the operator*

$$A := \mathbf{i}(\mathbb{1} + U)(\mathbb{1} - U)^{-1} : \mathrm{dom}(A) \to H, \qquad \mathrm{dom}(A) := \mathrm{im}(\mathbb{1} - U),$$

is self-adjoint and U is the Cayley transform of A.

(iii) *Let $A : \mathrm{dom}(A) \to H$ be a self-adjoint operator and let $U \in \mathcal{L}^c(H)$ be its Cayley transform. Define the Möbius transformation $\phi : \mathbb{R} \to S^1 \setminus \{1\}$ by*

$$\phi(\lambda) := \frac{\lambda - \mathbf{i}}{\lambda + \mathbf{i}}, \qquad \phi^{-1}(\mu) = \mathbf{i}\frac{1 + \mu}{1 - \mu} \tag{6.4.4}$$

for $\lambda \in \mathbb{R}$ and $\mu \in S^1 \setminus \{1\}$. Then

$$\sigma(U) \setminus \{1\} = \phi(\sigma(A)), \qquad \mathrm{P}\sigma(U) = \phi(\mathrm{P}\sigma(A)), \tag{6.4.5}$$

and

$$\ker(\lambda\mathbb{1} - A) = \ker(\phi(\lambda)\mathbb{1} - U) \tag{6.4.6}$$

for all $\lambda \in \mathbb{R}$.

Proof. We prove (i). The operators

$$A \pm \mathbf{i}\mathbb{1} : \mathrm{dom}(A) \to H$$

are bijective and have bounded inverses by part (v) of Theorem 6.3.13 and they are normal by part (i) of Theorem 6.3.13. Hence

$$\|Ax - \mathbf{i}x\| = \|Ax + \mathbf{i}x\| \qquad \text{for all } x \in \mathrm{dom}(A)$$

and so the Cayley transform

$$U := (A - \mathbf{i}\mathbb{1})(A + \mathbf{i}\mathbb{1})^{-1}$$

in (6.4.2) is a unitary operator on H (Lemma 5.3.14). The operator U satisfies

$$\mathbb{1} - U = 2\mathbf{i}(A + \mathbf{i}\mathbb{1})^{-1}, \qquad \mathbb{1} + U = 2A(A + \mathbf{i}\mathbb{1})^{-1}.$$

Thus $\mathbb{1} - U$ is injective, $\mathrm{im}(\mathbb{1} - U) = \mathrm{dom}(A)$, and $\mathbf{i}^{-1}A(\mathbb{1} - U) = \mathbb{1} + U$, and hence A and U satisfy (6.4.3). This proves part (i).

We prove (ii). Assume $U \in \mathcal{L}^c(H)$ is a unitary operator such that $\mathbb{1} - U$ is injective. Then $1 \in \mathbb{C} \setminus \mathrm{P}\sigma(U)$ and hence the operator $\mathbb{1} - U$ has a dense image by Theorem 5.3.15. Define the operator $A : \mathrm{dom}(A) \to H$ by (6.4.3). We prove that A is self-adjoint. Thus let

$$x \in \mathrm{dom}(A^*), \qquad y := A^*x.$$

Then

$$\langle y, \zeta \rangle = \langle x, A\zeta \rangle = \langle x, \mathbf{i}(\mathbb{1} + U)(\mathbb{1} - U)^{-1}\zeta \rangle$$

for all $\zeta \in \mathrm{dom}(A) = \mathrm{im}(\mathbb{1} - U)$ and hence

$$\langle y, \xi - U\xi \rangle = \langle x, \mathbf{i}(\xi + U\xi) \rangle \qquad \text{for all } \xi \in H.$$

This implies $U^*y - y = \mathbf{i}(U^*x + x)$ and hence

$$y - Uy = \mathbf{i}(x + Ux). \tag{6.4.7}$$

Thus

$$x = \tfrac{1}{2}(x - Ux) + \tfrac{1}{2}(x + Ux) = \tfrac{1}{2}(\mathbb{1} - U)(x - \mathbf{i}y) \in \mathrm{im}(\mathbb{1} - U) = \mathrm{dom}(A),$$

hence

$$(\mathbb{1} - U)^{-1}x = \tfrac{1}{2}(x - \mathbf{i}y),$$

and therefore

$$Ax = \mathbf{i}(\mathbb{1} + U)(\mathbb{1} - U)^{-1}x = \tfrac{1}{2}(\mathbb{1} + U)(\mathbf{i}x + y) = y.$$

Here the last equation follows from (6.4.7). This shows that A is self-adjoint. Moreover, $A + \mathbf{i}\mathbb{1} = 2\mathbf{i}(\mathbb{1} - U)^{-1}$ and $\mathbb{A} - \mathbf{i}\mathbb{1} = 2\mathbf{i}U(\mathbb{1} - U)^{-1}$, and hence $U = (A - \mathbf{i}\mathbb{1})(A + \mathbf{i}\mathbb{1})^{-1}$ is the Cayley transform of A. This proves part (ii).

We prove (iii). Fix a real number λ. Then, by (6.4.2) and (6.4.4),

$$\begin{aligned}(\lambda + \mathbf{i})(\phi(\lambda)\mathbb{1} - U)(Ax + \mathbf{i}x) &= (\lambda - \mathbf{i})(Ax + \mathbf{i}x) - (\lambda + \mathbf{i})(Ax - \mathbf{i}x) \\ &= 2\mathbf{i}(\lambda x - Ax)\end{aligned}$$

for all $x \in \mathrm{dom}(A)$. Since the operator $A + \mathbf{i}\mathbb{1} : \mathrm{dom}(A) \to H$ is surjective, this implies that $\lambda\mathbb{1} - A : \mathrm{dom}(A) \to H$ is bijective if and only if $\phi(\lambda)\mathbb{1} - U$ is bijective. Moreover, if $x \in \mathrm{dom}(A)$ satisfies $Ax = \lambda x$, then

$$\begin{aligned}(\lambda + \mathbf{i})^2(\phi(\lambda)x - Ux) &= (\lambda + \mathbf{i})(\phi(\lambda)\mathbb{1} - U)(\lambda x + \mathbf{i}x) \\ &= (\lambda + \mathbf{i})(\phi(\lambda)\mathbb{1} - U)(Ax + \mathbf{i}x) \\ &= 2\mathbf{i}(\lambda x - Ax) \\ &= 0.\end{aligned}$$

Conversely, let $x \in H$ such that $Ux = \phi(\lambda)x$. Then $(1 - \phi(\lambda))x = x - Ux$ and so $x \in \mathrm{im}(\mathbb{1} - U) = \mathrm{dom}(A)$. Moreover, $\xi := (\mathbb{1} - U)^{-1}x = (1 - \phi(\lambda))^{-1}x$ and so

$$Ax = \mathbf{i}(\xi + U\xi) = \mathbf{i}\frac{x + Ux}{1 - \phi(\lambda)} = \mathbf{i}\frac{1 + \phi(\lambda)}{1 - \phi(\lambda)}x = \lambda x.$$

This proves part (iii) and Theorem 6.4.2. □

With these preparations we are now ready to establish the functional calculus for general unbounded self-adjoint operators. We give a proof of Theorem 6.4.1 which reduces the result to the functional calculus for bounded normal operators in Theorem 5.6.5 via the Cayley transform.

Proof of Theorem 6.4.1. Let $A : \mathrm{dom}(A) \to H$ be a self-adjoint operator with domain $\mathrm{dom}(A) \subsetneq H$ (so A is not bounded) and spectrum

$$\Sigma := \sigma(A) \subset \mathbb{R}.$$

Let

$$U := (A - \mathbf{i}1\!\!1)(A + \mathbf{i}1\!\!1)^{-1} \in \mathcal{L}^c(H)$$

be the Cayley transform of A. Then U is a unitary operator and $1\!\!1 - U$ is injective and not surjective, because $\mathrm{im}(1\!\!1 - U) = \mathrm{dom}(A) \neq H$, and so

$$1 \in \sigma(U).$$

Hence it follows from part (iii) of Theorem 6.4.2 that the spectrum of U is the (compact) set

$$\sigma(U) = \phi(\Sigma) \cup \{1\} \subset S^1. \tag{6.4.8}$$

Now denote by

$$B(\sigma(U)) \to \mathcal{L}^c(H) : g \mapsto g(U)$$

the C* algebra homomorphism in Theorem 5.6.5, and define the map

$$B(\Sigma) \to \mathcal{L}^c(H) : f \mapsto f(A)$$

by

$$f(A) := (f \circ \phi^{-1})(U) \qquad \text{for } f \in B(\Sigma). \tag{6.4.9}$$

Here the bounded measurable function $f \circ \phi^{-1} : S^1 \setminus \{1\} \to \mathbb{C}$ is extended to all of S^1 by setting $(f \circ \phi^{-1})(1) := 0$. We prove in seven steps that the map (6.4.9) satisfies the requirements of Theorem 6.4.1.

Step 1. *The map* (6.4.9) *is a C^* algebra homomorphism. In particular, it satisfies* $1(A) = 1\!\!1$.

Define $g_0 : \sigma(U) \to \mathbb{C}$ by $g_0(1) := 1$ and

$$g_0(\mu) := 0 \qquad \text{for } \mu \in \sigma(U) \setminus \{1\}.$$

Then the operator $g_0(U)$ is the orthogonal projection onto the kernel of the operator $1\!\!1 - U$ by part (iii) of Theorem 5.6.11, and so $g_0(U) = 0$ because $1\!\!1 - U$ is injective. This implies

$$1(A) = (1 \circ \phi^{-1})(U) = (1 - g_0)(U) = 1\!\!1.$$

That the map (6.4.9) is linear and preserves multiplication follows directly from the definition. This proves Step 1.

Step 2. *The map* (6.4.9) *satisfies the (Normalization) axiom.*

Let $f_i : \Sigma \to \mathbb{C}$ be a sequence of bounded measurable functions such that

$$\sup_{i\in\mathbb{N}}|f_i(\lambda)| \le |\lambda|, \qquad \lim_{i\to\infty} f_i(\lambda) = \lambda \qquad \text{for all } \lambda \in \Sigma.$$

For $i \in \mathbb{N}$ define the function $h_i : \sigma(U) \to \mathbb{C}$ by

$$h_i(\mu) := (f_i \circ \phi^{-1})(\mu)(1-\mu) \qquad \text{for } \mu \in \sigma(U),$$

so $h_i : \sigma(U) \to \mathbb{C}$ is a bounded measurable function and

$$h_i(U) = f_i(A)(\mathbb{1} - U). \tag{6.4.10}$$

Moreover, $\phi^{-1}(\mu) = \mathbf{i}(1+\mu)(1-\mu)^{-1}$ for $\mu \in \sigma(U) \setminus \{1\}$ and hence

$$\begin{aligned} |h_i(\mu)| &= \left| f_i\left(\mathbf{i}\frac{1+\mu}{1-\mu}\right)\right| |1-\mu| \\ &\le |1+\mu| \\ &\le 2 \end{aligned}$$

for all $\mu \in \sigma(U) \setminus \{1\}$. Since $h_i(1) = 0$ for all i, this implies

$$\sup_{i\in\mathbb{N}}|h_i(\mu)| \le 2, \qquad \lim_{i\to\infty} h_i(\mu) = \mathbf{i}(1+\mu-2g_0(\mu)) \qquad \text{for all } \mu \in \sigma(U), \tag{6.4.11}$$

where $g_0 : \sigma(U) \to \mathbb{C}$ is as in the proof of Step 1. Now let

$$x \in \mathrm{dom}(A) = \mathrm{im}(\mathbb{1} - U)$$

and define

$$\xi := (\mathbb{1} - U)^{-1}x.$$

Then it follows from (6.4.3), (6.4.10), (6.4.11), and the (Convergence) axiom in Theorem 5.6.5 that

$$\begin{aligned} \lim_{i\to\infty} f_i(A)x &= \lim_{i\to\infty} f_i(A)(\xi - U\xi) \\ &= \lim_{i\to\infty} h_i(U)\xi \\ &= \mathbf{i}(\xi + U\xi) \\ &= Ax. \end{aligned}$$

This proves Step 2.

Step 3. *The map* (6.4.9) *satisfies the (Convergence), (Positive), (Commutative), and (Eigenvector) axioms.*

The (Convergence) and (Positive) axioms follow directly from the definition and the corresponding axioms in Theorem 5.6.5. The (Commutative) axiom follows from the (Commutative) axiom in Theorem 5.6.5 and the fact that an operator $B \in \mathcal{L}^c(H)$ commutes with A if and only if it commutes with U (and hence also with $U^* = U^{-1}$). The (Eigenvector) axiom follows from equation (6.4.6) and the (Eigenvector) axiom in Theorem 5.6.5.

Step 4. *The map* (6.4.9) *satisfies the (Spectrum) axiom.*

Let $f \in B(\Sigma)$ and $\mu \in \mathbb{C} \setminus \overline{f(\Sigma)}$, and define the function $g : \Sigma \to \mathbb{C}$ by

$$g(\lambda) := \frac{1}{\mu - f(\lambda)} \qquad \text{for } \lambda \in \Sigma.$$

Then g is bounded and measurable and satisfies $g(\mu - f) = (\mu - f)g = 1$. Hence $g(A)(\mu\mathbb{1} - f(A)) = (\mu\mathbb{1} - f(A))g(A) = \mathbb{1}$ by Step 1, so $\mu\mathbb{1} - f(A)$ is invertible and thus $\mu \in \rho(f(A))$. This shows that $\sigma(f(A)) \subset \overline{f(\Sigma)}$.

Let $f \in C_b(\Sigma)$ and define the function $g : \sigma(U) \to \mathbb{C}$ by

$$g(z) := \begin{cases} f(\phi^{-1}(z)), & \text{for } z \in \sigma(U) \setminus \{1\}, \\ 0, & \text{for } z = 1. \end{cases}$$

Then g is continuous at every point $z \in \sigma(U) \setminus \{1\}$ and $f(A) = g(U)$. Hence

$$f(\lambda) = g(\phi(\lambda)) \in \sigma(g(U)) = \sigma(f(A)) \qquad \text{for all } \lambda \in \Sigma$$

by part (ii) of Theorem 5.6.11. Hence $\overline{f(\Sigma)} \subset \sigma(f(A))$ because the spectrum of $f(A)$ is a closed subset of $\mathbb{C}$. This proves Step 4.

Step 5. *The map* (6.4.9) *satisfies the (Contraction) axiom.*

This follows from Step 4 and the formula $\|f(A)\| = \sup_{\mu \in \sigma(f(A))} |\mu|$ in part (ii) of Theorem 5.3.15.

Step 6. *The map* (6.4.9) *satisfies the (Composition) axiom.*

Fix a function $f \in C_b(\Sigma)$ and define $A_f := f(A)$. Then $\Sigma_f := \sigma(A_f) = \overline{f(\Sigma)}$ by Step 4. Consider the map $B(\Sigma_f) \to \mathcal{L}^c(H) : g \mapsto g(A_f) := (g \circ f)(A)$. This map is a C* algebra homomorphism by Step 1, it is continuous by Step 5, it satisfies the (Normalization) axiom $\mathrm{id}(A_f) = A_f$ by definition, and it satisfies the (Convergence) axiom by Step 3. Hence Step 6 follows from uniqueness in Theorem 5.6.5.

Step 7. *The* C^* *algebra homomorphism* (6.4.9) *is uniquely determined by the (Normalization) and (Convergence) axioms.*

Let $B(\Sigma) \to \mathcal{L}^c(H) : f \mapsto f(A)$ be any C* algebra homomorphism that satisfies the (Normalization) and (Convergence) axioms and define $U := \phi(A)$. Then $U(A + \mathbf{i}\mathbb{1}) = A - \mathbf{i}\mathbb{1}$ by the (Normalization) axiom, so U is the Cayley transform of A. Define the map $B(\sigma(U)) \to \mathcal{L}^c(H) : g \mapsto g(U)$ by $g(U) := (g \circ \phi)(A)$ for $g \in B(\sigma(U))$. By definition, this map is a C* algebra homomorphism that satisfies the (Convergence) axiom. Moreover, it satisfies $\mathrm{id}(U) = \phi(A) = U$. Hence the map $g \mapsto g(U)$ agrees with the functional calculus in Theorem 5.6.5. This proves Step 7 and Theorem 6.4.1. □

6.4.2. Spectral Measures. Let $\mathcal{B} \subset 2^{\mathbb{R}}$ be the Borel σ-algebra. Theorem 6.4.1 allows us to assign to every unbounded self-adjoint operator on a complex Hilbert space a projection valued measure (see Definition 5.6.1).

Definition 6.4.3 (Spectral Measure). Let H be a nonzero complex Hilbert space, let $A : \mathrm{dom}(A) \to H$ be an unbounded self-adjoint operator with spectrum $\Sigma := \sigma(A) \subset \mathbb{R}$, and let

$$\Psi_A : B(\Sigma) \to \mathcal{L}^c(H)$$

be the C* algebra homomorphism of Theorem 6.4.1. For every $\Omega \in \mathcal{B}$ define the operator $P_\Omega \in \mathcal{L}^c(H)$ by

$$P_\Omega := \Psi_A(\chi_\Omega|_\Sigma). \tag{6.4.12}$$

By Theorem 6.4.1 these operators are orthogonal projections and the map

$$\mathcal{B} \to \mathcal{L}^c(H) : \Omega \mapsto P_\Omega \tag{6.4.13}$$

is a projection valued measure. It is called the **spectral measure of** A.

Conversely, every projection valued measure (6.4.12) on the real axis gives rise to a family of self-adjoint operators $A_f : \mathrm{dom}(A_f) \to H$, one for every Borel measurable function $f : \mathbb{R} \to \mathbb{R}$. If f is bounded, then this operator is bounded, so $\mathrm{dom}(A_f) = H$, and it is given by the formula $A_f := \Psi(f)$ in Theorem 5.6.2. For unbounded functions f the operator A_f will in general be unbounded.

Theorem 6.4.4 (The Operator A_f). *Let H be a nonzero complex Hilbert space and fix any projection valued measure $\mathcal{B} \to \mathcal{L}^c(H) : \Omega \mapsto P_\Omega$ on the real axis. Define the signed Borel measures $\mu_{y,x} : \mathcal{B} \to \mathbb{R}$ by*

$$\mu_{y,x}(\Omega) := \mathrm{Re}\langle y, P_\Omega x\rangle \qquad \text{for } x, y \in H \text{ and } \Omega \in \mathcal{B}. \tag{6.4.14}$$

Let $f : \mathbb{R} \to \mathbb{R}$ be a Borel measurable function. Then the formula

$$\begin{aligned} \mathrm{dom}(A_f) &:= \left\{ x \in H \,\middle|\, \int_{\mathbb{R}} f^2\, d\mu_{x,x} < \infty \right\}, \\ \mathrm{Re}\langle y, A_f x\rangle &:= \int_{\mathbb{R}} f\, d\mu_{y,x} \qquad \text{for } x \in \mathrm{dom}(A_f) \text{ and } y \in H, \end{aligned} \tag{6.4.15}$$

defines a self-adjoint operator $A_f : \mathrm{dom}(A_f) \to H$. This operator satisfies the equation

$$\|A_f x\|^2 = \int_{\mathbb{R}} f^2\, d\mu_{x,x}$$

for all $x \in \mathrm{dom}(A_f)$.

Proof. For $x, y \in H$ the function $\mu_{y,x} : \mathcal{B} \to \mathbb{R}$ is a signed Borel measure. Its **total variation** is the Borel measure $|\mu_{y,x}| : \mathcal{B} \to [0, \infty)$, defined by

$$|\mu_{y,x}|(\Omega) := \sup\left\{\mu_{y,x}(\Omega') - \mu_{y,x}(\Omega \setminus \Omega') \,\middle|\, \Omega' \in \mathcal{B},\, \Omega' \subset \Omega\right\}$$

for every Borel set $\Omega \subset \mathbb{R}$ (see [**75**, Thm. 5.12]). By definition, the total variation satisfies $|\mu_{y,x}(\Omega)| \leq |\mu_{y,x}|(\Omega)$ for all $\Omega \in \mathcal{B}$. The positive and negative parts of $\mu_{y,x}$ are the Borel measures $\mu^{\pm}_{y,x} : \mathcal{B} \to [0, \infty)$, defined by

$$\mu^{\pm}_{y,x}(\Omega) := \frac{|\mu_{y,x}|(\Omega) \pm \mu_{y,x}(\Omega)}{2} \qquad \text{for } \Omega \in \mathcal{B}.$$

They satisfy

$$\mu_{y,x} = \mu^{+}_{y,x} - \mu^{-}_{y,x}, \qquad |\mu_{y,x}| = \mu^{+}_{y,x} + \mu^{-}_{y,x}.$$

Let us now fix a Borel measurable function $f : \mathbb{R} \to \mathbb{R}$. Then two vectors $x, y \in H$ satisfy $\int_{\mathbb{R}} |f| d|\mu_{y,x}| < \infty$ if and only if $\int_{\mathbb{R}} |f| d\mu^{\pm}_{y,x} < \infty$, and if this holds, then the integral of f with respect to $\mu_{y,x}$ is defined by

$$\int_{\mathbb{R}} f\, d\mu_{y,x} := \int_{\mathbb{R}} f\, d\mu^{+}_{y,x} - \int_{\mathbb{R}} f\, d\mu^{-}_{y,x}.$$

With this understood, we prove in eight steps that the operator A_f is well defined and self-adjoint and satisfies $\|A_f x\|^2 = \int_{\mathbb{R}} f^2\, d\mu_{x,x}$ for all $x \in \mathrm{dom}(A_f)$.

Step 1. *The signed Borel measures $\mu_{y,x}$ in (6.4.14) satisfy the inequality*

$$|\mu_{y,x}|(\Omega) \leq \sqrt{\mu_{x,x}(\Omega)}\sqrt{\mu_{y,y}(\Omega)} \tag{6.4.16}$$

for all $x, y \in H$ and all $\Omega \in \mathcal{B}$.

Fix two elements $x, y \in H$. If $\Omega_1, \Omega_2 \in \mathcal{B}$ are disjoint and $\Omega_1 \cup \Omega_2 =: \Omega$, then

$$\begin{aligned} \|P_{\Omega_1} x\|^2 + \|P_{\Omega_1} x\|^2 &= \langle x, P_{\Omega_1} x\rangle + \langle x, P_{\Omega_2} x\rangle \\ &= \langle x, P_\Omega x\rangle \\ &= \mu_{x,x}(\Omega). \end{aligned}$$

By the Cauchy–Schwarz inequality, this implies

$$\begin{aligned} \mu_{y,x}(\Omega') - \mu_{y,x}(\Omega \setminus \Omega') &= \mathrm{Re}\langle P_{\Omega'} y, P_{\Omega'} x\rangle - \mathrm{Re}\langle P_{\Omega\setminus\Omega'} y, P_{\Omega\setminus\Omega'} x\rangle \\ &\leq \|P_{\Omega'} x\|\, \|P_{\Omega'} y\| + \|P_{\Omega\setminus\Omega'} x\|\, \|P_{\Omega\setminus\Omega'} y\| \\ &\leq \sqrt{\|P_{\Omega'} x\|^2 + \|P_{\Omega\setminus\Omega'} x\|^2}\sqrt{\|P_{\Omega'} y\|^2 + \|P_{\Omega\setminus\Omega'} y\|^2} \\ &= \sqrt{\mu_{x,x}(\Omega)}\sqrt{\mu_{y,y}(\Omega)} \end{aligned}$$

for every pair of Borel sets $\Omega' \subset \Omega \subset \mathbb{R}$. Fix a Borel set $\Omega \subset \mathbb{R}$ and take the supremum over all Borel sets $\Omega' \subset \Omega$ to obtain (6.4.16). This proves Step 1.

Step 2. *Let $g : \mathbb{R} \to \mathbb{R}$ be a Borel measurable function. Then*

$$\int_{\mathbb{R}} |g|\, d|\mu_{y,x}| \le \|y\| \sqrt{\int_{\mathbb{R}} g^2\, d\mu_{x,x}} \tag{6.4.17}$$

for all $x, y \in H$.

For every finite collection of pairwise disjoint Borel sets $\Omega_1, \dots, \Omega_n \subset \mathbb{R}$ and every finite collection of positive real numbers $a_1, \dots, a_n$, we have

$$\sum_{i=1}^n a_i |\mu_{y,x}|(\Omega_i) \le \left(\sum_{i=1}^n a_i^2 \mu_{x,x}(\Omega_i)\right)^{1/2} \left(\sum_{i=1}^n \mu_{y,y}(\Omega_i)\right)^{1/2}$$

by Step 1 and the Cauchy–Schwarz inequality. Moreover,

$$\sum_{i=1}^n \mu_{y,y}(\Omega_i) = \mu_{y,y}\left(\bigcup_{i=1}^n \Omega_i\right) \le \|y\|^2.$$

This proves (6.4.17) for the Borel measurable step function $g := \sum_{i=1}^n a_i \chi_{\Omega_i}$. Since every nonnegative Borel measurable function can be approximated pointwise from below by a sequence of Borel measurable step functions (see for example [**75**, Thm. 1.26]), Step 2 follows from the Lebesgue Monotone Convergence Theorem.

Step 3. *The operator $A_f : \mathrm{dom}(A_f) \to H$ in (6.4.15) is well defined. More precisely, fix an element $x \in \mathrm{dom}(A_f)$. Then the function $|f| : \mathbb{R} \to [0, \infty)$ is integrable with respect to the Borel measure $|\mu_{y,x}|$ for every $y \in H$, and there exists a unique element $A_f x \in H$ such that*

$$\mathrm{Re}\langle y, A_f x\rangle = \int_{\mathbb{R}} f\, d\mu_{y,x}$$

for all $y \in H$. Moreover, $\|A_f x\|^2 \le \int_{\mathbb{R}} f^2\, d\mu_{x,x}$.

Fix an element $x \in \mathrm{dom}(A_f)$ and define $c := \left(\int_{\mathbb{R}} f^2\, d\mu_{x,x}\right)^{1/2} < \infty$. Then Step 2 asserts that $\int_{\mathbb{R}} |f| d|\mu_{y,x}| \le c\,\|y\| < \infty$ and so the integral $\int_{\mathbb{R}} f d\mu_{y,x}$ is well defined for all $y \in H$. Now define the map $\Lambda_x : H \to \mathbb{R}$ by

$$\Lambda_x(y) := \int_{\mathbb{R}} f\, d\mu_{y,x} \qquad \text{for } y \in H.$$

This map is real linear and satisfies the inequality

$$|\Lambda_x(y)| \le \int_{\mathbb{R}} |f|\, d|\mu_{y,x}| \le \left(\int_{\mathbb{R}} f^2\, d\mu_{x,x}\right)^{1/2} \|y\| = c\,\|y\|$$

for all $y \in H$ by Step 2. Hence, by Theorem 1.4.4, there exists a unique element $A_f x \in H$ such that $\mathrm{Re}\langle y, A_f x\rangle = \int_{\mathbb{R}} f\, d\mu_{y,x}$ for all $y \in H$. Moreover $\|A_f x\| = \|\Lambda_x\| \le c$ and this proves Step 3.

Step 4. *The set* $\mathrm{dom}(A_f) \subset H$ *is a complex linear subspace and the operator* A_f *in* (6.4.15) *is complex linear and symmetric.*

Let $x, x' \in \mathrm{dom}(A_f)$. Then

$$\begin{aligned}\mu_{x+x',x+x'}(\Omega) &= \langle x + x', P_\Omega x + P_\Omega x' \rangle \\ &= \|P_\Omega x\|^2 + 2\mathrm{Re}\langle P_\Omega x', P_\Omega x\rangle + \|P_\Omega x'\|^2 \\ &\le 2\|P_\Omega x\|^2 + 2\|P_\Omega x'\|^2 \\ &= 2\mu_{x,x}(\Omega) + 2\mu_{x',x'}(\Omega)\end{aligned}$$

for all $\Omega \in \mathcal{B}$ and this implies $x+x' \in \mathrm{dom}(A_f)$. Moreover, $\mu_{\lambda x,\lambda x} = |\lambda|^2\mu_{x,x}$, so $\lambda x \in \mathrm{dom}(A_f)$ for all $\lambda \in \mathbb{C}$. Thus $\mathrm{dom}(A_f)$ is a complex subspace of H. Since $\mu_{y,x+x'} = \mu_{y,x} + \mu_{y,x'}$ and $\mu_{y,\lambda x} = \lambda\mu_{y,x}$ for all $x, x' \in \mathrm{dom}(A_f)$ and all $\lambda \in \mathbb{R}$, the operator A_f is real linear. To prove that it is complex linear, let $x \in \mathrm{dom}(A_f)$ and $y \in H$. Then $\mu_{y,\mathbf{i}x} = -\mu_{\mathbf{i}y,x}$ and hence

$$\mathrm{Re}\langle y, A_f\mathbf{i}x\rangle = \int_{\mathbb{R}} f\,d\mu_{y,\mathbf{i}x} = -\int_{\mathbb{R}} f\,d\mu_{\mathbf{i}y,x} = -\mathrm{Re}\langle \mathbf{i}y, A_f x\rangle = \mathrm{Re}\langle y, \mathbf{i}A_f x\rangle.$$

This shows that $A_f\mathbf{i}x = \mathbf{i}A_f x$ for all $x \in \mathrm{dom}(A_f)$, so A_f is complex linear. Moreover, A_f is symmetric because the bilinear map

$$\mathrm{dom}(A_f) \times \mathrm{dom}(A_f) \to \mathcal{M}(\mathbb{R}) : (x, y) \mapsto \mu_{x,y}$$

is symmetric. This proves Step 4.

Step 5. *The operator* $A_f : \mathrm{dom}(A_f) \to H$ *in* (6.4.15) *has a dense domain.*

For $n \in \mathbb{N}$ define $\Omega_n := \{\lambda \in \mathbb{R}\,|\,|f(\lambda)| \le n\}$. Then $\mathbb{R} = \bigcup_{n=1}^\infty \Omega_n$. Hence it follows from the (σ-Additive) and (Normalization) axioms in Definition 5.6.1 that $\lim_{n\to\infty} P_{\Omega_n} x = x$ for all $x \in H$. Now let $x \in H$ and define $x_n := P_{\Omega_n} x$. Then $\mu_{x_n,x_n}(\Omega) = \mu_{x,x}(\Omega \cap \Omega_n)$ for all $\Omega \in \mathcal{B}$ by the (Intersection) axiom in Definition 5.6.1. Hence

$$\int_{\mathbb{R}} f^2\,d\mu_{x_n,x_n} = \int_{\Omega_n} f^2\,d\mu_{x,x} \le n^2\|x\|^2$$

and so $x_n \in \mathrm{dom}(A_f)$ for all $n \in \mathbb{N}$. This proves Step 5.

Step 6. *Let* $x \in \mathrm{dom}(A_f)$ *and* $\Omega \in \mathcal{B}$. *Then* $P_\Omega x \in \mathrm{dom}(A_f)$ *and*

$$A_f P_\Omega x = P_\Omega A_f x.$$

The estimate $\int_{\mathbb{R}} f^2\,d\mu_{P_\Omega x,P_\Omega x} = \int_\Omega f^2\,d\mu_{x,x} < \infty$ implies $P_\Omega x \in \mathrm{dom}(A_f)$. Moreover,

$$\mathrm{Re}\langle y, A_f P_\Omega x\rangle = \int_{\mathbb{R}} f\,d\mu_{y,P_\Omega x} = \int_{\mathbb{R}} f\,d\mu_{P_\Omega y,x} = \mathrm{Re}\langle P_\Omega y, A_f x\rangle$$

for all $y \in H$ and this proves Step 6.

Step 7. *Let $x \in \mathrm{dom}(A_f)$. Then f is integrable with respect to the Borel measure $|\mu_{x,A_fx}|$ and*

$$\int_{\mathbb{R}} f^2 \, d\mu_{x,x} = \int_{\mathbb{R}} f \, d\mu_{x,A_fx} = \|A_fx\|^2 . \tag{6.4.18}$$

That f is integrable with respect to $|\mu_{x,A_fx}| = |\mu_{A_fx,x}|$ was proved in Step 3. Next we observe that

$$\int_{\mathbb{R}} \chi_\Omega \, d\mu_{x,A_fx} = \mu_{x,A_fx}(\Omega) = \mathrm{Re}\langle P_\Omega x, A_fx\rangle = \int_{\mathbb{R}} f \, d\mu_{P_\Omega x,x} = \int_{\mathbb{R}} \chi_\Omega f d\mu_{x,x}$$

for every Borel set $\Omega \subset \mathbb{R}$. This shows that $\int_{\mathbb{R}} g \, d\mu_{x,A_fx} = \int_{\mathbb{R}} gf d\mu_{x,x}$ for every Borel measurable step function $g : \mathbb{R} \to \mathbb{R}$. Now approximate f pointwise by a sequence of Borel measurable step functions $g_n : \mathbb{R} \to \mathbb{R}$ such that $|g_n(\lambda)| \le |f(\lambda)|$ for all $n \in \mathbb{N}$ and all $\lambda \in \mathbb{R}$ (see [**75**, Thm. 1.26]). Then the Lebesgue Dominated Convergence Theorem asserts that

$$\int_{\mathbb{R}} f \, d\mu_{x,A_fx} = \lim_{n\to\infty} \int_{\mathbb{R}} g_n \, d\mu_{x,A_fx} = \lim_{n\to\infty} \int_{\mathbb{R}} g_n f \, d\mu_{x,x} = \int_{\mathbb{R}} f^2 d\mu_{x,x}.$$

This proves the first equality in (6.4.18). The second equality follows from Step 3 and this proves Step 7.

Step 8. *The operator $A_f : \mathrm{dom}(A_f) \to H$ in (6.4.15) is self-adjoint.*

By Step 4 it suffices to prove that $\mathrm{dom}(A_f^*) \subset \mathrm{dom}(A_f)$. Let $x \in \mathrm{dom}(A_f^*)$ and define $y := A_f^*x$. Then, for all $\xi \in \mathrm{dom}(A_f)$, we have

$$\int_{\mathbb{R}} f d\mu_{x,\xi} = \mathrm{Re}\langle x, A_f\xi\rangle = \mathrm{Re}\langle A_f^*x, \xi\rangle = \mathrm{Re}\langle y, \xi\rangle \tag{6.4.19}$$

by Step 3. For $n \in \mathbb{N}$ let $\Omega_n := \{\lambda \in \mathbb{R} \,|\, |f(\lambda)| \le n\}$ and $x_n := P_{\Omega_n}x$ as in the proof of Step 5. Then

$$\int_{\mathbb{R}} f^2 \, d\mu_{x,x} = \lim_{n\to\infty} \int_{\Omega_n} f^2 \, d\mu_{x,x} = \lim_{n\to\infty} \int_{\mathbb{R}} f^2 \, d\mu_{x_n,x_n}$$

by the Lebesgue Monotone Convergence Theorem. Moreover, it follows from Steps 5 and 6 that $A_fx_n = A_fP_{\Omega_n}x_n = P_{\Omega_n}A_fx_n \in \mathrm{dom}(A_f)$. Hence

$$\int_{\mathbb{R}} f^2 \, d\mu_{x_n,x_n} = \int_{\mathbb{R}} f \, d\mu_{x,A_fx_n} = \mathrm{Re}\langle y, A_fx_n\rangle \le \|y\| \sqrt{\int_{\mathbb{R}} f^2 \, d\mu_{x_n,x_n}}.$$

Here the first equality uses Step 7 and the fact that the signed Borel measures μ_{x_n,A_fx_n} and μ_{x,A_fx_n} agree, the second equality follows from (6.4.19) with $\xi := A_fx_n \in \mathrm{dom}(A_f)$, and the inequality follows from Step 3. Thus

$$\int_{\mathbb{R}} f^2 \, d\mu_{x,x} = \lim_{n\to\infty} \int_{\mathbb{R}} f^2 \, d\mu_{x_n,x_n} \le \|y\|^2$$

and so $x \in \mathrm{dom}(A_f)$. This proves Step 8 and Theorem 6.4.4. □

Remark 6.4.5. **(i)** Theorem 6.4.4 can be used to extend the functional calculus for self-adjoint operators to unbounded functions $f : \mathbb{R} \to \mathbb{R}$, starting from a projection valued measure as in Theorem 5.6.2. This functional calculus can then be used to prove that the operator $A_f + \mathbf{i}\mathbb{1}$ is invertible and thus gives rise to an alternative proof that A_f is self-adjoint. This approach is used in Kato [**44**, p. 355]. Steps 6 and 7 in the above proof of Theorem 6.4.4 can be understood as a special case of this functional calculus, using one unbounded function f and the bounded functions χ_Ω for $\Omega \in \mathcal{B}$.

(ii) There is an entirely different approach to the measurable functional calculus for unbounded self-adjoint operators. One can start by assigning to an unbounded self-adjoint operator A its spectral measure and use Theorem 5.6.2 to construct the C* algebra homomorphism $\Psi_A : B(\Sigma) \to \mathcal{L}^c(H)$. For the construction of the spectral measure one can proceed as follows. First show that every self-adjoint operator $A : \mathrm{dom}(A) \to H$ can be written as a difference $A = A^+ - A^-$ of two positive semidefinite self-adjoint operators $A^\pm : \mathrm{dom}(A^\pm) \to H$ with $\mathrm{dom}(A^+) \cap \mathrm{dom}(A^-) = \mathrm{dom}(A)$. Then the operators $\mathbb{1} + A^\pm$ are invertible by Theorem 6.3.13 and one can use the spectral measures of their inverses in Theorem 5.6.3 to find the spectral measure for A. This approach is taken in Kato [**44**, pp. 353–361]. It does not require the functional calculus for normal operators in Section 5.5.

(iii) Suppose the projection valued measure is supported on a closed subset $\Sigma \subset \mathbb{R}$. Then the functional calculus for unbounded functions can be used as in Step 5 of the proof of Theorem 5.6.2 to show that $\sigma(A_f) \subset \overline{f(\Sigma)}$.

(iv) The functional calculus extends to unbounded normal operators. It can be reduced to the self-adjoint case by writing an unbounded normal operator as $A = A_1 + \mathbf{i}A_2$ where A_1 and A_2 are self-adjoint (Theorem 6.3.11). For bounded normal operators this approach is outlined in [**72**, pp. 245–247].

The next theorem shows that (6.4.13), (6.4.14), (6.4.15) give rise to a one-to-one correspondence between projection valued measures on the real axis with values in $\mathcal{L}^c(H)$ and unbounded self-adjoint operators on H.

Theorem 6.4.6 (Spectral Measures). *Let H be a nonzero complex Hilbert space and let $\mathcal{B} \subset 2^{\mathbb{R}}$ be the Borel σ-algebra.*

(i) *Let $A : \mathrm{dom}(A) \to H$ be a self-adjoint operator and let $\{P_\Omega\}_{\Omega\in\mathcal{B}}$ be the spectral measure of A in Definition* 6.4.3. *Then $A = A_{\mathrm{id}}$ is the operator in Theorem* 6.4.4 *with $f = \mathrm{id} : \mathbb{R} \to \mathbb{R}$.*

(ii) *Let $\mathcal{B} \to \mathcal{L}^c(H) : \Omega \mapsto P_\Omega$ be a projection valued measure and let A_{id} be the operator in Theorem* 6.4.4 *with $f = \mathrm{id} : \mathbb{R} \to \mathbb{R}$. Then $\{P_\Omega\}_{\Omega\in\mathcal{B}}$ is the spectral measure of A_{id} in Definition* 6.4.3.

Proof. See page 338. □

Corollary 6.4.7 (Characterization of the Spectral Measure). *Let $A : \mathrm{dom}(A) \to H$ be a self-adjoint operator on a nonzero complex Hilbert space H. Then there exists a unique projection valued measure $\{P_\Omega\}_{\Omega\in\mathcal{B}}$ on the real axis such that*

$$(6.4.20)\qquad \begin{aligned} \mathrm{dom}(A) &= \left\{ x \in H \,\middle|\, \int_{\mathbb{R}} \lambda^2 \, d\mu_{x,x}(\lambda) < \infty \right\}, \\ \mathrm{Re}\langle y, Ax\rangle &= \int_{\mathbb{R}} \lambda \, d\mu_{y,x}(\lambda) \qquad \textit{for } x \in \mathrm{dom}(A) \textit{ and } y \in H, \end{aligned}$$

where $\{\mu_{y,x}\}_{x,y\in H}$ is the collection of signed Borel measures on the real axis defined by $\mu_{y,x}(\Omega) := \mathrm{Re}\,\langle y, P_\Omega x\rangle$ for all $x, y \in H$ and all Borel sets $\Omega \subset \mathbb{R}$. It agrees with the spectral measure of Definition 6.4.3.

Proof. Uniqueness follows from part (ii) of Theorem 6.4.6, and existence follows from Theorem 6.4.4 and part (i) of Theorem 6.4.6. □

Proof of Theorem 6.4.6. We prove part (i). Let $A : \mathrm{dom}(A) \to H$ be an unbounded self-adjoint operator with spectrum

$$\Sigma := \sigma(A)$$

and take $\{P_\Omega\}_{\Omega\in\mathcal{B}}$ to be the projection valued measure in Definition 6.4.3, associated to the C* algebra homomorphism $\Psi_A : B(\Sigma) \to \mathcal{L}^c(H)$ in Theorem 6.4.1. For $i \in \mathbb{N}$ define the function $f_i : \mathbb{R} \to \mathbb{R}$ by

$$f_i(\lambda) := \begin{cases} \lambda, & \text{if } |\lambda| \le i, \\ 0, & \text{if } |\lambda| > i. \end{cases}$$

Then the (Normalization) axiom in Theorem 6.4.1 asserts that

$$(6.4.21)\qquad \lim_{i\to\infty} \Psi_A(f_i|_\Sigma)x = Ax \qquad \text{for all } x \in \mathrm{dom}(A).$$

Moreover, by definition of P_Ω in (6.4.12) and of $\mu_{y,x}$ in (6.4.14), we have

$$\mu_{y,x}(\Omega) = \mathrm{Re}\langle y, P_\Omega x\rangle = \mathrm{Re}\langle y, \Psi_A(\chi_{\Sigma\cap\Omega})x\rangle$$

for all $x, y \in H$ and all $\Omega \in \mathcal{B}$. Hence the (Convergence) axiom implies

$$\int_{\mathbb{R}} f \, d\mu_{y,x} = \mathrm{Re}\langle y, \Psi_A(f|_\Sigma)x\rangle$$

for all $x, y \in H$ and all bounded Borel measurable functions $f : \mathbb{R} \to \mathbb{R}$. In particular,

$$(6.4.22)\qquad \int_{\mathbb{R}} f_i \, d\mu_{y,x} = \mathrm{Re}\langle y, \Psi_A(f_i|_\Sigma)x\rangle, \qquad \int_{\mathbb{R}} f_i^2 \, d\mu_{x,x} = \|\Psi_A(f_i|_\Sigma)x\|^2$$

for all $i \in \mathbb{N}$ and all $x, y \in H$.

Now let $x \in \mathrm{dom}(A)$. Then, by equations (6.4.21) and (6.4.22) and the Lebesgue Monotone Convergence Theorem, we have

$$\int_{\mathbb{R}} \lambda^2 \, d\mu_{x,x}(\lambda) = \lim_{i\to\infty} \int_{\mathbb{R}} f_i^2 \, d\mu_{x,x} = \lim_{i\to\infty} \|\Psi_A(f_i|_\Sigma)x\|^2 = \|Ax\|^2 .$$

This implies $x \in \mathrm{dom}(A_{\mathrm{id}})$ and hence, by equations (6.4.21) and (6.4.22) and the Lebesgue Dominated Convergence Theorem,

$$\begin{aligned}\mathrm{Re}\langle y, A_{\mathrm{id}}x\rangle &= \int_{\mathbb{R}} \lambda \, d\mu_{y,x} \\ &= \lim_{i\to\infty} \int_{\mathbb{R}} f_i \, d\mu_{y,x} \\ &= \lim_{i\to\infty} \mathrm{Re}\langle y, \Psi_A(f_i|_\Sigma)x\rangle \\ &= \mathrm{Re}\langle y, Ax\rangle\end{aligned}$$

for all $y \in H$. Thus $\mathrm{dom}(A) \subset \mathrm{dom}(A_{\mathrm{id}})$ and $A_{\mathrm{id}}|_{\mathrm{dom}(A)} = A$. This implies $A_{\mathrm{id}} = A$ by Exercise 6.5.4 and proves part (i).

We prove part (ii). Thus let $\mathcal{B} \to \mathcal{L}^c(H) : \Omega \mapsto P_\Omega$ be a projection valued measure on the real axis, let

$$A := A_{\mathrm{id}}$$

be the operator in Theorem 6.4.4 with $f = \mathrm{id}$, and let $\Psi : B(\mathbb{R}) \to \mathcal{L}^c(H)$ be the C* algebra homomorphism in Theorem 5.6.2 associated to $\{P_\Omega\}_{\Omega\in\mathcal{B}}$. Then Ψ satisfies the (Convergence) axiom in Theorem 6.4.1 by definition. We prove that

$$P_{\mathbb{R}\setminus\Sigma} = 0, \qquad \Sigma := \sigma(A_{\mathrm{id}}). \tag{6.4.23}$$

Suppose, by contradiction, that $P_{\mathbb{R}\setminus\Sigma} \neq 0$, choose a vector $x \in X$ such that $P_{\mathbb{R}\setminus\Sigma}x \neq 0$, and consider the Borel measure $\mu_x : \mathcal{B} \to [0,\infty)$ defined by $\mu_x(\Omega) := \langle x, P_\Omega x\rangle$ for $\Omega \in \mathcal{B}$. Then $\mu_x(\mathbb{R}\setminus\Sigma) > 0$ and so, since every Borel measure on $\mathbb{R}$ is inner regular by [**75**, Thm. 3.18], there exists a compact set $K \subset \mathbb{R} \setminus \Sigma$ such that $\mu_x(K) > 0$. Hence $P_K \neq 0$ and so

$$E_K := \mathrm{im}(P_K)$$

is a nonzero closed subspace of H. Since the identity function $f = \mathrm{id} : \mathbb{R} \to \mathbb{R}$ is bounded on K, it follows from the definition of the operator $A = A_{\mathrm{id}}$ in (6.4.15) that $E_K \subset \mathrm{dom}(A)$ and E_K is invariant under A. Since $E_K \neq \{0\}$ and the operator $A_K := A|_{E_K} : E_K \to E_K$ is self-adjoint, its spectrum is nonempty. Since $\mu_{y,x}(\Omega) = \mu_{y,x}(\Omega \cap K)$ for $x, y \in E_K$ and $\Omega \in \mathcal{B}$, we have

$$\mathrm{Re}\langle y, A_K x\rangle = \int_{\mathbb{R}} \lambda \, d\mu_{y,x}(\lambda) = \int_K \lambda \, d\mu_{y,x}(\lambda) \qquad \text{for all } x, y \in E_K.$$

Hence $\sigma(A_K) \subset K$ by Theorem 5.6.2 and so $\emptyset \neq \sigma(A_K) \subset \sigma(A) \cap K = \emptyset$, a contradiction. This proves (6.4.23).

Since $P_{\mathbb{R}\setminus\Sigma} = 0$, the C* algebra homomorphism Ψ of Theorem 5.6.2 descends to a unique C* algebra homomorphism

$$\Psi_\Sigma : B(\Sigma) \to \mathcal{L}^c(H)$$

such that $\Psi(f) = \Psi_\Sigma(f|_\Sigma)$ for all $f \in B(\mathbb{R})$. We prove that Ψ_Σ satisfies the (Normalization) axiom in Theorem 6.4.1 with $A = A_{\mathrm{id}}$. To see this, choose a sequence of bounded Borel measurable functions $f_i : \mathbb{R} \to \mathbb{R}$ that satisfy

$$\sup_{i\in\mathbb{N}} |f_i(\lambda)| \le |\lambda|, \qquad \lim_{i\to\infty} f_i(\lambda) = \lambda \qquad \text{for all } \lambda \in \mathbb{R}.$$

Fix an element $x \in \mathrm{dom}(A_{\mathrm{id}})$. Then the identity function $\mathrm{id} : \mathbb{R} \to \mathbb{R}$ is square integrable with respect to the Borel measure $\mu_{x,x}$, and is integrable with respect to the Borel measure $|\mu_{y,x}|$ for every $y \in H$. Hence it follows from the Lebesgue Dominated Convergence Theorem and the Hahn Decomposition Theorem that

$$\begin{aligned}\langle y, A_{\mathrm{id}}x\rangle &= \int_{\mathbb{R}} \lambda \, d\mu_{y,x}(\lambda) \\ &= \lim_{i\to\infty} \int_{\mathbb{R}} f_i \, d\mu_{y,x} \\ &= \lim_{i\to\infty} \langle y, \Psi_\Sigma(f_i|_\Sigma)x\rangle\end{aligned}$$

for all $y \in H$ and

$$\begin{aligned}\|A_{\mathrm{id}}x\|^2 &= \int_{\mathbb{R}} \lambda^2 \, d\mu_{x,x}(\lambda) \\ &= \lim_{i\to\infty} \int_{\mathbb{R}} f_i^2 \, d\mu_{x,x} \\ &= \lim_{i\to\infty} \|\Psi_\Sigma(f_i|_\Sigma)x\|^2 .\end{aligned}$$

Hence the sequence $\Psi_\Sigma(f_i|_\Sigma)x$ converges weakly to $A_{\mathrm{id}}x$ and its norm converges to that of $A_{\mathrm{id}}x$. By Exercise 3.7.1 this implies

$$\lim_{i\to\infty} \|A_{\mathrm{id}}x - \Psi_\Sigma(f_i|_\Sigma)x\| = 0.$$

Thus the reduced C* algebra homomorphism $\Psi_\Sigma : B(\Sigma) \to \mathcal{L}^c(H)$ satisfies the (Normalization) axiom in Theorem 6.4.1 with $A = A_{\mathrm{id}}$. Hence it follows from uniqueness in Theorem 6.4.1 that $\Psi_\Sigma = \Psi_{A_{\mathrm{id}}}$ is the functional calculus associated to the self-adjoint operator A_{id}. Hence

$$P_\Omega = \Psi(\chi_\Omega) = \Psi_\Sigma(\chi_\Omega|_\Sigma) = \Psi_{A_{\mathrm{id}}}(\chi_\Omega|_\Sigma)$$

for every Borel set $\Omega \subset \mathbb{R}$. Here the first equality holds by definition of the C* algebra homomorphism $\Psi : B(\mathbb{R}) \to \mathcal{L}^c(H)$ in Theorem 5.6.2. Hence the projection valued measure $\{P_\Omega\}_{\Omega\in\mathcal{B}}$ is the spectral measure of A_{id} as introduced in Definition 6.4.3. This proves part (ii) and Theorem 6.4.6. □

Example 6.4.8. Let $A : \mathrm{dom}(A) \to H$ be a self-adjoint operator on a nonzero complex Hilbert space H.

(i) Consider the operator family

$$\mathbb{R} \to \mathcal{L}(H) : t \mapsto U(t)$$

associated to the functions $\lambda \mapsto e^{\mathbf{i}\lambda t}$ via the functional calculus of Theorem 6.4.1. In terms of the spectral measure the operators $U(t)$ are determined by the formula

$$\langle y, U(t)x\rangle := \int_{-\infty}^{\infty} e^{\mathbf{i}\lambda t}\, d\langle y, P_\lambda x\rangle \qquad \text{for all } x, y \in H \text{ and all } t \in \mathbb{R}.$$

Here the expression $\int_{\mathbb{R}} f(\lambda)\, d\langle y, P_\lambda x\rangle$ denotes the integral of a Borel measurable function $f : \mathbb{R} \to \mathbb{C}$ with respect to the complex valued Borel measure

$$\mathcal{B} \to \mathbb{C} : \Omega \mapsto \langle y, P_\Omega x\rangle$$

on the real axis. The operator family $\mathbb{R} \to \mathcal{L}^c(H) : t \mapsto U(t)$ is strongly continuous, by the (Convergence) axiom, and satisfies

$$U(s+t) = U(s)U(t), \qquad U(0) = \mathbb{1}$$

for all $s, t \in \mathbb{R}$. This means that U is a *strongly continuous group* of (unitary) operators. Such groups play an important role in quantum mechanics. For example, they appear as solutions of the Schrödinger equation.

(ii) Assume, in addition, that

$$\langle x, Ax\rangle \le 0 \qquad \text{for all } x \in \mathrm{dom}(A).$$

Then $\sigma(A) \subset (-\infty, 0]$ and a similar construction leads to an operator family

$$[0, \infty) \to \mathcal{L}^c(H) : t \mapsto S(t)$$

associated to the functions $\lambda \mapsto e^{\lambda t}$ on the negative real axis. In terms of the spectral measure the operators $S(t)$ are determined by the formula

$$\langle y, S(t)x\rangle := \int_{-\infty}^{0} e^{\lambda t}\, d\langle y, P_\lambda x\rangle \qquad \text{for all } x, y \in H \text{ and all } t \ge 0.$$

The restriction $t \ge 0$ is needed to obtain bounded functions on the negative real axis and bounded linear operators $S(t)$. These operators form a *strongly continuous semigroup* of operators on H. For example, the solutions of the heat equation on $\mathbb{R}^n$ can be expressed in this form with A the Laplace operator. The study of strongly continuous semigroups is the subject of the next and final chapter of this book.

6.5. Problems

Exercise 6.5.1 (Unbounded Operators and their Inverses). Let X and Y be Banach spaces and let $A : \mathrm{dom}(A) \to Y$ be an unbounded operator with a dense domain $\mathrm{dom}(A) \subset X$. Assume A is injective and let

$$A^{-1} : \mathrm{dom}(A^{-1}) \to X$$

be its inverse with the domain

$$\mathrm{dom}(A^{-1}) := \mathrm{im}(A) = \{Ax \mid x \in \mathrm{dom}(A)\} \subset Y.$$

(a) Prove that A has a closed graph if and only if A^{-1} has a closed graph.

(b) Assume A is surjective. Prove that A^{-1} is bounded if and only if A has a closed graph.

(c) Assume that A is surjective, $\mathrm{dom}(A)$ is a dense subspace of X, and A^{-1} is a compact operator. Prove that X is separable. (See Exercise 4.5.2.)

(d) Assume $Y = X$ and A has a closed graph. Prove that

$$\sigma(A^{-1}) \setminus \{0\} = \{\lambda^{-1} \mid \lambda \in \sigma(A) \setminus \{0\}\}.$$

(e) Assume $X = Y$ and A has a closed graph. Prove that $0 \notin \sigma(A) \cup \sigma(A^{-1})$ if and only if $\mathrm{dom}(A) = X$ and $A : X \to X$ is bijective and bounded.

(f) Find an example of an injective unbounded operator $A : \mathrm{dom}(A) \to X$ with a closed graph such that $0 \in \sigma(A) \cap \sigma(A^{-1})$.

Exercise 6.5.2 (Closed Graphs and Inverses). Let X be a complex Banach space and let $A : \mathrm{dom}(A) \to X$ be an unbounded complex linear operator. Let $\lambda \in \mathbb{C}$ and suppose that the operator $\lambda\mathbb{1} - A : \mathrm{dom}(A) \to X$ is bijective. Prove that the following are equivalent.

(i) The operator $(\lambda\mathbb{1} - A)^{-1} : X \to X$ is bounded.

(ii) A has a closed graph.

Hint: Show that $\lambda\mathbb{1} - A$ has a closed graph if and only if A has a closed graph. Use Exercise 2.2.12 and the Open Mapping Theorem 2.2.1.

Exercise 6.5.3 (Symmetric and Surjective Implies Self-Adjoint). Let H be a complex Hilbert space and let $A : \mathrm{dom}(A) \to H$ be an unbounded symmetric complex linear operator with a dense domain. Prove that the following are equivalent.

(i) There exists a $\lambda \in \mathbb{C}$ such that $\lambda\mathbb{1} - A : \mathrm{dom}(A) \to H$ is surjective.

(ii) A is self-adjoint.

Exercise 6.5.4 (Uniqueness of Self-Adjoint Operators). Let H be a complex Hilbert space and let A, B be unbounded self-adjoint operators on H such that

$$\mathrm{dom}(A) \subset \mathrm{dom}(B), \qquad B|_{\mathrm{dom}(A)} = A.$$

Then $B = A$.

Exercise 6.5.5 (Bounded Self-Adjoint Operators). Let H be a complex Hilbert space and let $A : \mathrm{dom}(A) \to H$ be a self-adjoint operator on H. Prove that $\mathrm{dom}(A) = H$ if and only if $\sigma(A)$ is a bounded subset of $\mathbb{R}$. **Hint:** Theorem 6.4.1 with $f = \mathrm{id}$.

Exercise 6.5.6 (The Unbounded Open Mapping Theorem).

(a) Let X and Y be Banach spaces and let $A : \mathrm{dom}(A) \to Y$ be a closed unbounded operator with a dense domain $\mathrm{dom}(A) \subset X$. Let $\delta > 0$ and assume

$$\{y \in Y \mid \|y\| \le \delta\} \subset \overline{\{Ax \mid x \in \mathrm{dom}(A),\ \|x\|_X < 1\}}. \tag{6.5.1}$$

Prove that

$$\{y \in Y \mid \|y\| < \delta\} \subset \{Ax \mid x \in \mathrm{dom}(A),\ \|x\|_X < 1\}. \tag{6.5.2}$$

Hint: The proof of Lemma 2.2.3 carries over almost verbatim to operators with dense domains and closed graphs.

(b) Prove that (vii) implies (i) in Theorem 6.2.3 by carrying over the proof of the corresponding statement in Theorem 4.1.16 to unbounded operators. **Hint:** Use part (a).

Exercise 6.5.7 (Spectral Projection). Let $A : \mathrm{dom}(A) \to X$ be an operator on a complex Banach space X with a compact resolvent (see Definition 6.1.13).

(a) If $\mathrm{dom}(A) = X$ prove that $\dim X < \infty$.

(b) Let $\lambda \in \sigma(A)$ and define $P_\lambda \in \mathcal{L}^c(X)$ by (6.1.20) with $\Sigma := \{\lambda\}$, i.e.

$$P_\lambda := \frac{1}{2\pi\mathbf{i}} \int_\gamma (z\mathbb{1} - A)^{-1}\, dz, \tag{6.5.3}$$

where $\gamma(t) := \lambda + re^{2\pi\mathbf{i}t}$ for $0 \le t \le 1$ and $r > 0$ sufficiently small. Prove that P_λ is the unique projection that commutes with A and whose image is the generalized eigenspace

$$\mathrm{im}(P_\lambda) = E_\lambda := \bigcup_{k=1}^{\infty} \ker(\lambda\mathbb{1} - A)^k. \tag{6.5.4}$$

Exercise 6.5.8 (Square Root). Let H be a complex Hilbert space.

(a) Call an unbounded self-adjoint operator $A : \mathrm{dom}(A) \to H$ **positive semidefinite** if it satisfies

$$\langle x, Ax \rangle \geq 0 \qquad \text{for all } x \in \mathrm{dom}(A).$$

Assume $A : \mathrm{dom}(A) \to H$ is a positive semidefinite operator. Prove that there exists a unique self-adjoint operator $B : \mathrm{dom}(B) \to H$ such that

$$B^2 = A, \qquad \langle x, Bx \rangle \geq 0 \qquad \text{for all } x \in \mathrm{dom}(B).$$

The operator B is called the **square root of** A and is denoted by

$$B =: \sqrt{A} =: A^{1/2}.$$

Hint: Theorem 6.4.4 with $f(\lambda) := \sqrt{\lambda}$.

(b) Let $A : \mathrm{dom}(A) \to H$ be an unbounded self-adjoint operator. Prove that the positive semidefinite operator

$$|A| := \sqrt{A^2}$$

has the same domain as A and satisfies

$$0 \leq |\langle x, Ax \rangle| \leq \langle x, |A|x \rangle \qquad \text{for all } x \in \mathrm{dom}(A).$$

Let $A^{\pm}$ be self-adjoint extensions of the symmetric operators $\frac{1}{2}(|A| \pm A)$. Show that $A^{\pm}$ are positive semidefinite and satisfy

$$\mathrm{dom}(A) = \mathrm{dom}(A^+) \cap \mathrm{dom}(A^-)$$

and

$$A = A^+ - A^-, \qquad |A| = A^+ + A^-.$$

Hint: Theorem 6.4.4 with $f(\lambda) = |\lambda|$.

Exercise 6.5.9 (Densely Defined Operators and their Adjoints). Let X and Y be real Hilbert spaces and let $A : \mathrm{dom}(A) \to Y$ be an unbounded operator with a dense domain $\mathrm{dom}(A) \subset X$.

(a) The graph of the adjoint operator $A^* : \mathrm{dom}(A^*) \to X$ is the orthogonal complement of the subspace $\{(y, x) \in Y \times X \mid x \in \mathrm{dom}(A),\, y = -Ax\}$. Thus A^* has a closed graph.

(b) The operator A is closeable if and only if $\mathrm{dom}(A^*)$ is a dense subspace of Y. **Hint:** Carry over the proof of part (iii) of Theorem 6.2.2 to the Hilbert space setting.

(c) If A is closed, then $A^{**} = A$. **Hint:** Use (a).

Exercise 6.5.10 (Symplectic Vector Spaces). Let (V, ω) be a **symplectic vector space**, i.e. V is a real vector space and $\omega : V \times V \to \mathbb{R}$ is a **nondegenerate** skew-symmetric bilinear form, so for every nonzero vector $v \in V$, there is a vector $w \in V$ such that $\omega(v, w) \neq 0$. The **symplectic complement** of a linear subspace $W \subset V$ is the linear subspace

$$W^\omega := \{v \in V \,|\, \omega(v, w) = 0 \text{ for all } w \in W\}.$$

A linear subspace $W \subset V$ is called **ω-reflexive** if $W^{\omega\omega} = W$. An ω-reflexive subspace $W \subset V$ is called **isotropic** if $W \subset W^\omega$, **coistropic** if $W^\omega \subset W$, and **Lagrangian** if $W = W^\omega$. A **complex structure** on V is a linear operator $J : V \to V$ such that $J^2 = -\mathbb{1}$. A complex structure is called **compatible with** ω if the formula

$$\langle u, v\rangle := \omega(u, Jv) \qquad \text{for } u, v \in V \tag{6.5.5}$$

defines an inner product on V.

(a) Let J be an ω-compatible complex structure on V and let $W \subset V$ be a linear subspace. Prove that the orthogonal and symplectic complements of W are related by

$$W^\omega = JW^\perp, \qquad W^{\omega\omega} = W^{\perp\perp}.$$

Deduce that W is ω-reflexive if and only if it is closed with respect to the inner product (6.5.5).

(b) Let $W \subset V$ be an isotropic subspace and define $\overline{V} := W^\omega/W$. Prove that the formula $\overline{\omega}([u], [v]) := \omega(u, v)$ for $u, v \in W^\omega$ defines a symplectic form on $\overline{V}$. This construction is called **symplectic reduction**.

(c) Let H be a real Hilbert space. Show that the formulas

$$\omega(z, \zeta) := \langle x, \eta\rangle_H - \langle y, \xi\rangle_H, \qquad J(x, y) := (-y, x)$$

for $z = (x, y)$, $\zeta = (\xi, \eta) \in H \times H$ define a symplectic form ω and an ω-compatible complex structure J on the Hilbert space $H \times H$ that induce the standard inner product.

(d) Let H be a real Hilbert space, let $A : \mathrm{dom}(A) \to H$ be a densely defined unbounded operator on H, and let ω be the symplectic form on $H \times H$ in (c). Define $\mathrm{graph}(A) := \{(Ax, x) \,|\, x \in \mathrm{dom}(A)\}$. Show that

$$\mathrm{graph}(A^*) = \mathrm{graph}(A)^\omega.$$

Deduce that A is closed if and only if its graph is an ω-reflexive subspace of $H \times H$, and that A is self-adjoint if and only if its graph is a Lagrangian subspace. If A is closed and symmetric, show that the reduced space $\mathrm{graph}(A^*)/\mathrm{graph}(A)$ in (b) is naturally isomorphic to the Gelfand–Robbin quotient $\mathrm{dom}(A^*)/\mathrm{dom}(A)$ in Exercise 6.3.3.

Exercise 6.5.11 (The Gelfand–Robbin Quotient). The purpose of this exercise is to introduce a natural inner product on the Gelfand–Robbin quotient and to examine its properties. Assume throughout that H is a real Hilbert space and that $A : \mathrm{dom}(A) \to H$ is a densely defined symmetric operator with a closed graph.

(a) Prove that the domain of A^* is a Hilbert space with the inner product

$$\langle x, y\rangle_{A^*} := \langle x, y\rangle_H + \langle A^*x, A^*y\rangle_H \qquad \text{for } x, y \in \mathrm{dom}(A^*) \tag{6.5.6}$$

and that $\mathrm{dom}(A)$ is a closed subspace of $\mathrm{dom}(A^*)$. Let $V \subset \mathrm{dom}(A^*)$ be the orthogonal complement of $\mathrm{dom}(A)$. Prove that

$$V = \{x \in \mathrm{dom}(A^*) \mid A^*x \in \mathrm{dom}(A^*),\ A^*A^*x + x = 0\}. \tag{6.5.7}$$

Thus V is canonically isomorphic to the Gelfand–Robbin quotient in Exercise 6.3.3 and the inner product (6.5.6) renders V into a Hilbert space.

(b) Prove that the linear map $A^*|_V : V \to V$ is a complex structure on V and that it is compatible with the symplectic form

$$\omega(x, y) = \langle A^*x, y\rangle_H - \langle x, A^*y\rangle_H \qquad \text{for } x, y \in V. \tag{6.5.8}$$

Prove that $\omega(x, A^*y) = \langle x, y\rangle_{A^*}$ for all $x, y \in V$. Prove that every Lagrangian subspace of V is closed.

(c) Assume A has a closed image. Prove that

$$\Lambda_0 := \{x \in V \mid A^*x \in \mathrm{im}(A)\} \tag{6.5.9}$$

is a Lagrangian subspace of V.

Hint 1: If $x, y \in \Lambda_0$ and $\xi, \eta \in \mathrm{dom}(A)$ satisfy that $A\xi = A^*x$ and $A\eta = A^*y$, then $\langle A^*x, y\rangle = \langle \xi, A^*y\rangle = \langle \xi, A\eta\rangle = \langle A\xi, \eta\rangle = \langle A^*x, \eta\rangle = \langle x, A^*y\rangle$.

Hint 2: Let $x \in V$ such that $\omega(x, y) = 0$ for all $y \in \Lambda_0$. Prove that

$$\langle A^*x, y\rangle = 0 \qquad \text{for all } y \in \ker(A^*)$$

and so $A^*x \in \mathrm{im}(A)$ by Lemma 6.3.2. To see this, let $y \in \ker(A^*)$ and choose $\eta \in \mathrm{dom}(A)$ such that $\langle y - \eta, \xi\rangle_{A^*} = 0$ for all $\xi \in \mathrm{dom}(A)$. Deduce that $A\eta \in \mathrm{dom}(A^*)$ and $A^*A\eta = y - \eta$. This implies $y - \eta \in \Lambda_0$ and hence

$$0 = \langle A^*x, y - \eta\rangle_{A^*} = \langle A^*x, y\rangle.$$

(d) Assume A has a closed image. Prove that the orthogonal complement of Λ_0 with respect to the inner product (6.5.6) is the Lagrangian subspace

$$\Lambda_0^\perp = A^*\Lambda_0 = V \cap \mathrm{im}(A). \tag{6.5.10}$$

Hint: The first equation is a general fact about symplectic vector spaces with compatible complex structures (see part (a) of Exercise 6.5.10).

(e) Assume A has a closed image and let $\Lambda_0 \subset V$ be as in (6.5.9). Prove that

$$\operatorname{dom}(A) \oplus \Lambda_0 = \operatorname{dom}(A) + \ker(A^*), \qquad \operatorname{im}(A) \oplus \Lambda_0 = \operatorname{im}(A^*). \tag{6.5.11}$$

Hint: To prove the inclusion $\ker(A^*) \subset \operatorname{dom}(A) \oplus \Lambda_0$ use the argument in Hint 2 for part (c). That $\operatorname{im}(A) \cap \Lambda_0 = \{0\}$ follows from (6.5.10). To prove the inclusion $\operatorname{im}(A^*) \subset \operatorname{im}(A) \oplus \Lambda_0$ use the fact that, by (6.5.10),

$$\begin{aligned}\operatorname{dom}(A^*) &= \operatorname{dom}(A) \oplus \Lambda_0 \oplus (V \cap \operatorname{im}(A)) \\ &= (\operatorname{dom}(A) + \ker(A^*)) \oplus (V \cap \operatorname{im}(A)).\end{aligned}$$

This implies that, for every $y \in \operatorname{im}(A^*)$, there exist elements $\xi \in \operatorname{dom}(A)$ and $x \in V \cap \operatorname{im}(A)$ such that $y = A\xi + A^*x$. Thus we have $A^*x \in V$ and $A^*A^*x = -x \in \operatorname{im}(A)$, and so $A^*x \in \Lambda_0$.

(f) Assume A^* is surjective and let $\Lambda \subset V$ be a Lagrangian subspace of V. Denote by

$$A_\Lambda : \operatorname{dom}(A_\Lambda) = \operatorname{dom}(A) \oplus \Lambda \to H$$

the corresponding self-adjoint extension as in part (iii) of Exercise 6.3.3. Prove that A_Λ is a Fredholm operator if and only if (V, Λ_0, Λ) is a Fredholm triple with respect to the inner product (6.5.6) (see Exercise 4.5.15).

Hint: The domain of A_Λ is a closed subspace of the domain of A^* with respect to the graph norm and $A_\Lambda : \operatorname{dom}(A_\Lambda) \to H$ is a bounded linear operator with respect to the graph norm of A^* on its domain. Moreover,

$$\operatorname{im}(A_\Lambda) = \operatorname{im}(A) + A^*\Lambda. \tag{6.5.12}$$

Use this to prove that

$$H = \operatorname{im}(A_\Lambda) \oplus (\Lambda_0 \cap \Lambda). \tag{6.5.13}$$

Let $y \in \operatorname{im}(A_\Lambda) \cap \Lambda_0 \cap \Lambda$, $\xi \in \operatorname{dom}(A)$, and $x \in \Lambda$ such that $y = A\xi + A^*x$. Then $A\xi \in V \cap \operatorname{im}(A) = A^*\Lambda_0$, hence $y \in A^*(\Lambda_0 + \Lambda) \cap \Lambda_0 \cap \Lambda$, and so $y = 0$. Next, let $y \in H$ and use (6.5.11) to find $\xi \in \operatorname{dom}(A)$, $y_0 \in \Lambda_0 \cap \Lambda$, $x_1 \in \Lambda$ such that $y_1 := A^*x_1 \in \Lambda_0 \cap A^*\Lambda$ and $y = A\xi + y_0 + y_1 = A_\Lambda(\xi + x_1) + y_0$. If (V, Λ_0, Λ) is a Fredholm triple, then $\dim \operatorname{coker}(A_\Lambda) < \infty$ by (6.5.13), and hence A_Λ has a closed image by Lemma 4.3.2.

Chapter 7

Semigroups of Operators

Strongly continuous semigroups play an important role in the study of many linear partial differential equations such as the heat equation, the wave equation, and the Schrödinger equation. The finite-dimensional model of a strongly continuous semigroup is the exponential matrix associated to a first order linear ordinary differential equation. The concept of the exponential operator carries over naturally to infinite-dimensional Banach spaces X and can be used to find a solution of the **Cauchy problem**

$$\dot{x} = Ax, \qquad x(0) = x_0$$

for every bounded linear operator $A \in \mathcal{L}(X)$ and every initial value $x_0 \in X$. The unique solution $x : \mathbb{R} \to X$ of this equation is given by

$$x(t) = e^{tA}x_0 = \sum_{k=0}^{\infty} \frac{t^k}{k!} A^k x_0 \qquad \text{for } t \in \mathbb{R}.$$

(See Exercise 5.2.13.) The aforementioned partial differential equations can be expressed in the same form, however, with the caveat that the operator A is unbounded with a dense domain and that the solutions may only exist in forward time. In such situations it is convenient to use the solutions, rather than the equation, as the starting point. This leads to the notion of a strongly continuous semigroup, introduced in Section 7.1 along with several examples. That section also derives some of their basic properties and discusses the infinitesimal generator. The main result is the Hille–Yosida–Phillips Theorem in Section 7.2 which characterizes infinitesimal generators of strongly continuous semigroups. The dual semigroup is the subject of Section 7.3 and analytic semigroups are discussed in Section 7.4. A preparatory Section 7.5 is devoted to Banach space valued measurable functions, and inhomogeneous equations are examined in Section 7.6.

7.1. Strongly Continuous Semigroups

7.1.1. Definition and Examples. The existence and uniqueness theorem for solutions of a time-independent ordinary differential equation implies that the solutions define a flow. This means that the value of the solution with initial condition x_0 at time $s+t$ agrees with the value at time s of the solution whose initial condition is taken to be the value of the original solution at time t. For linear differential equations on Banach spaces this translates into a semigroup condition on the family of linear operators, parametrized by a nonnegative real variable t, that assign to a given initial condition the solution of the respective linear differential equation at time t. Continuous dependence on time translates into strong continuity of the semigroup of operators and continuous dependence on the initial condition translates into boundedness of the operators.

Definition 7.1.1 (Strongly Continuous Semigroup). Let X be a real Banach space. A **one-parameter semigroup (of operators on** X**)** is a map $S:[0,\infty)\to\mathcal{L}(X)$ that satisfies

$$S(0)=\mathbb{1},\qquad S(s+t)=S(s)S(t) \tag{7.1.1}$$

for all $s,t\geq 0$. A **one-parameter group (of operators on** X**)** is a map $S:\mathbb{R}\to\mathcal{L}(X)$ that satisfies (7.1.1) for all $s,t\in\mathbb{R}$. A **strongly continuous semigroup (of operators on** X**)** is a map $S:[0,\infty)\to\mathcal{L}(X)$ that satisfies (7.1.1) for all $s,t\geq 0$ and satisfies

$$\lim_{t\to 0}\|S(t)x-x\|=0 \tag{7.1.2}$$

for all $x\in X$. A **strongly continuous group (of operators on** X**)** is a map $S:\mathbb{R}\to\mathcal{L}(X)$ that satisfies (7.1.1) for all $s,t\in\mathbb{R}$ and satisfies (7.1.2) for all $x\in X$.

Example 7.1.2 (Groups Generated by Bounded Operators). Let X be a real Banach space and let $A:X\to X$ be a bounded linear operator. Then the operators

$$S(t):=e^{tA}=\sum_{k=0}^{\infty}\frac{t^kA^k}{k!} \tag{7.1.3}$$

for $t\in\mathbb{R}$ form a strongly continuous group of operators on X. In this example the map $\mathbb{R}\to\mathcal{L}(X):t\mapsto S(t)$ is continuous with respect to the norm topology on $\mathcal{L}(X)$ (see Exercise 5.2.13).

Example 7.1.3 (Semigroups and Orthonormal Bases). Let H be a separable complex Hilbert space, let $(e_i)_{i\in\mathbb{N}}$ be a complex orthonormal basis, and let $(\lambda_i)_{i\in\mathbb{N}}$ be a sequence of complex numbers such that

$$\sup_{i\in\mathbb{N}} \mathrm{Re}\lambda_i < \infty.$$

Define the map $S : [0,\infty) \to \mathcal{L}^c(H)$ by

$$S(t)x := \sum_{i=1}^{\infty} e^{\lambda_i t}\langle e_i, x\rangle e_i \tag{7.1.4}$$

for $x \in H$ and $t \geq 0$. **Exercise:** Show that this is a strongly continuous semigroup of operators on H. Show that it extends to a strongly continuous group $S : \mathbb{R} \to \mathcal{L}^c(H)$ if and only if

$$\sup_{i\in\mathbb{N}} |\mathrm{Re}\lambda_i| < \infty.$$

Example 7.1.4 (Shift Semigroups). Fix a constant $1 \leq p < \infty$ and let $X = L^p([0,\infty))$ be the Banach space of real valued L^p-functions on $[0,\infty)$ with respect to the Lebesgue measure.

(i) Define the map $L : [0,\infty) \to \mathcal{L}(X)$ by

$$(L(t)f)(s) := f(s+t) \tag{7.1.5}$$

for $f \in L^p([0,\infty))$ and $s,t \geq 0$. **Exercise:** Show that this is a strongly continuous semigroup of operators. Show that this example extends to the space of continuous functions on $[0,\infty)$ that converge to zero at infinity. Show that strong continuity fails when $L^p([0,\infty))$ is replaced by $L^\infty([0,\infty))$ or by the space of bounded continuous real valued functions on $[0,\infty)$. Show that the formula (7.1.5) defines a group on $L^p(\mathbb{R})$ for $1 \leq p < \infty$.

(ii) Define the map $R : [0,\infty) \to \mathcal{L}(X)$ by

$$(R(t)f)(s) := \begin{cases} 0, & \text{if } s < t, \\ f(s-t), & \text{if } s \geq t, \end{cases} \tag{7.1.6}$$

for $f \in L^p([0,\infty))$ and $s,t \geq 0$. **Exercise:** Show that this is a strongly continuous semigroup of isometric embeddings. Show that this example extends to the space of continuous functions $f : [0,\infty) \to \mathbb{R}$ that vanish at the origin and converge to zero at infinity.

(iii) Define the map $S : [0,\infty) \to \mathcal{L}(L^p([0,1]))$ by

$$(S(t)f)(s) := \begin{cases} f(s+t), & \text{if } s+t \leq 1, \\ 0, & \text{if } s+t > 1, \end{cases} \tag{7.1.7}$$

for $f \in L^p([0,1])$, $s \in [0,1]$, and $t \geq 0$. **Exercise:** Show that this is a strongly continuous semigroup of operators such that $S(t) = 0$ for $t \geq 1$.

Example 7.1.5 (Flows). Let (M, d) be a compact metric space and suppose that the map

$$\mathbb{R} \times M \to M : (t, p) \mapsto \phi_t(p)$$

is a **continuous flow**, i.e. it is continuous and satisfies

$$\phi_0 = \mathrm{id}, \qquad \phi_{s+t} = \phi_s \circ \phi_t$$

for all $s, t \in \mathbb{R}$. Let $X := C(M)$ be the Banach space of continuous real valued functions on M equipped with the supremum norm. Define

$$S(t)f := f \circ \phi_t \qquad \text{for } t \in \mathbb{R} \text{ and } f \in C(M). \tag{7.1.8}$$

Then $S : \mathbb{R} \to \mathcal{L}(C(M))$ is a strongly continuous group of operators.

Example 7.1.6 (Heat Equation). Fix a positive integer n and a real number $1 \leq p < \infty$. Define the **heat kernel** $K_t : \mathbb{R}^n \to \mathbb{R}$ by

$$K_t(x) := \frac{1}{(4\pi t)^{n/2}} e^{-|x|^2/4t} \qquad \text{for } x \in \mathbb{R}^n \text{ and } t > 0. \tag{7.1.9}$$

Here $|x| := \sqrt{\sum_{i=1}^n x_i^2}$ denotes the Euclidean norm of $x = (x_1, \dots, x_n) \in \mathbb{R}^n$. These functions are nonnegative and Lebesgue integrable and satisfy

$$\int_{\mathbb{R}^n} K_t(\xi)\, d\xi = 1, \qquad K_{s+t} = K_s * K_t \tag{7.1.10}$$

for all $s, t > 0$, where $(f*g)(x) := \int_{\mathbb{R}^n} f(x-\xi)g(\xi)\, d\xi$ denotes the convolution of two Lebesgue integrable functions $f, g : \mathbb{R}^n \to \mathbb{R}$. Equation (7.1.10) implies that the operators $S(t) : L^p(\mathbb{R}^n) \to L^p(\mathbb{R}^n)$, defined by

$$S(t)f := \begin{cases} K_t * f, & \text{for } t > 0, \\ f, & \text{for } t = 0, \end{cases} \tag{7.1.11}$$

define a semigroup of operators. Since $\lim_{t\to 0} \sup_{|x|\geq\delta} K_t(x) = 0$ for all $\delta > 0$ and $\int_{\mathbb{R}^n} K_t = 1$ for all $t > 0$, the functions $S(t)f = K_t * f$ converge uniformly to f for every continuous function $f : \mathbb{R}^n \to \mathbb{R}$ with compact support. The convergence is also in $L^p(\mathbb{R}^n)$. Since $C_c(\mathbb{R}^n)$ is dense in $L^p(\mathbb{R}^n)$ by [**75**, Thm. 4.15] and $\|S(t)\| \leq 1$ for all $t \geq 0$ by Young's inequality, it follows from Theorem 2.1.5 that $\lim_{t\to 0} \|S(t)f - f\|_{L^p} = 0$ for all $f \in L^p(\mathbb{R}^n)$. Thus the semigroup (7.1.11) is strongly continuous. Moreover, for each $f \in L^p(\mathbb{R}^n)$, the function $u : (0, \infty) \times \mathbb{R}^n \to \mathbb{R}$, defined by $u(t, x) := (K_t * f)(x)$ for $t > 0$ and $x \in \mathbb{R}^n$, is smooth and satisfies the **heat equation**

$$\frac{\partial u}{\partial t} = \sum_{i=1}^n \frac{\partial^2 u}{\partial x_i^2}, \qquad \lim_{t\to 0} \int_{\mathbb{R}^n} |u(t, x) - f(x)|^p\, dx = 0. \tag{7.1.12}$$

Exercise: Fill in the details.

Example 7.1.7 (Wave Equation). Let $L^2(\mathbb{R})$ be the space of square integrable real valued functions on $\mathbb{R}$ with respect to the Lebesgue measure, modulo equality almost everywhere, and let $W^{1,2}(\mathbb{R})$ denote the space of absolutely continuous functions $f : \mathbb{R} \to \mathbb{R}$ such that f and f' are square integrable. Then $H := W^{1,2}(\mathbb{R}) \times L^2(\mathbb{R})$ is a Hilbert space with the norm

$$\|(f,g)\|_H := \sqrt{\int_{-\infty}^{\infty} \left(|f(x)|^2 + \left|\frac{df}{dx}(x)\right|^2 + |g(x)|^2 \right) dx}$$

for $f \in W^{1,2}(\mathbb{R})$ and $g \in L^2(\mathbb{R})$. Given a pair $(f,g) \in H$, define the function $u : \mathbb{R}^2 \to \mathbb{R}$ by

$$u(t,x) = \frac{f(x+t) + f(x-t)}{2} + \frac{1}{2}\int_{x-t}^{x+t} g(s)\,ds \tag{7.1.13}$$

for $t, x \in \mathbb{R}$. Then $u(t,\cdot) \in W^{1,2}(\mathbb{R})$ and $\partial_t u(t,\cdot) \in L^2(\mathbb{R})$ for all $t \in \mathbb{R}$, and the linear operators $S(t) : H \to H$ given by $S(t)(f,g) := (u(t,\cdot), \partial_t u(t,\cdot))$ for $(f,g) \in H$ and $t \in \mathbb{R}$ define a strongly continuous group of operators on H. If f and g are smooth, then the function (7.1.13) is the unique solution of the one-dimensional **wave equation**

$$\frac{\partial^2 u}{\partial t^2} = \frac{\partial^2 u}{\partial x^2}, \qquad u(0,x) = f(x), \qquad \frac{\partial u}{\partial t}(0,x) = g(x). \tag{7.1.14}$$

The energy identity asserts that the function

$$E(t) := \frac{1}{2}\int_{-\infty}^{\infty} \left(\left|\frac{\partial u}{\partial x}(t,x)\right|^2 + \left|\frac{\partial u}{\partial t}(t,x)\right|^2 \right) dx$$

is constant for every solution of (7.1.14). Thus the operators $S(t) \in \mathcal{L}(H)$ extend to isometries of the completion $\mathcal{H}$ of H with respect to the norm

$$\|(f,g)\|_{\mathcal{H}} := \sqrt{\int_{-\infty}^{\infty} \left(\left|\frac{df}{dx}(x)\right|^2 + |g(x)|^2 \right) dx}.$$

The completion can be identified with the quotient of the space of all pairs (f,g), where $g \in L^2(\mathbb{R})$ and $f : \mathbb{R} \to \mathbb{R}$ is absolutely continuous with square integrable derivative, under the equivalence relation $(f_1,g_1) \sim (f_2,g_2)$ iff $g_1 = g_2$ and $f_1 - f_2$ is constant (Exercise 7.7.5). If one identifies $\mathcal{H}$ with $\mathscr{H} := L^2(\mathbb{R},\mathbb{R}^2)$ via the isomorphism $\mathcal{H} \to \mathscr{H} : (f,g) \mapsto (f',g)$, one obtains the strongly continuous group $\mathscr{S} : \mathbb{R} \to \mathcal{L}(\mathscr{H})$ of isometries, given by $\mathscr{S}(t)(f,g) := (u(t,\cdot), v(t,\cdot))$ for $t \in \mathbb{R}$ and $f, g \in L^2(\mathbb{R})$, where

$$\begin{aligned} u(t,x) &:= \frac{f(x+t)+f(x-t)}{2} + \frac{g(x+t)-g(x-t)}{2}, \\ v(t,x) &:= \frac{f(x+t)-f(x-t)}{2} + \frac{g(x+t)+g(x-t)}{2}. \end{aligned} \tag{7.1.15}$$

7.1.2. Basic Properties. The next two lemmas examine some of the elementary properties of strongly continuous semigroups.

Lemma 7.1.8. *Let X be a real Banach space and let $S : [0,\infty) \to \mathcal{L}(X)$ be a strongly continuous semigroup. Then the following hold.*

(i) $\sup_{0\le t\le T} \|S(t)\| < \infty$ *for all* $T > 0$.

(ii) *The function* $[0,\infty) \to X : t \mapsto S(t)x$ *is continuous for all* $x \in X$.

(iii) *The function* $t^{-1}\log\|S(t)\|$ *converges in* $\mathbb{R}\cup\{-\infty\}$ *as* t *tends to infinity and*

$$\lim_{t\to\infty} t^{-1}\log\|S(t)\| = \inf_{t>0} t^{-1}\log\|S(t)\| =: \omega_0. \tag{7.1.16}$$

(iv) *Let* ω_0 *be as in (iii) and fix a real number* $\omega > \omega_0$. *Then there exists a constant* $M \ge 1$ *such that*

$$\|S(t)\| \le Me^{\omega t} \qquad \textit{for all } t \ge 0. \tag{7.1.17}$$

Proof. To prove (i) we show first that there exist constants $\delta > 0$ and $M \ge 1$ such that, for all $t \in \mathbb{R}$,

$$0 \le t \le \delta \qquad \Longrightarrow \qquad \|S(t)\| \le M. \tag{7.1.18}$$

Suppose by contradiction that there do not exist such constants. Then

$$\sup_{0\le t\le\delta} \|S(t)\| = \infty$$

for all $\delta > 0$. Hence there exists a sequence of real numbers $t_n > 0$ such that $\lim_{n\to\infty} t_n = 0$ and the sequence $\|S(t_n)\|$ is unbounded. By the Uniform Boundedness Theorem 2.1.1 this implies that there exists an element $x \in X$ such that the sequence $\|S(t_n)x\|$ is unbounded. This contradicts the fact that $\lim_{n\to\infty} \|S(t_n)x - x\| = 0$. Thus we have proved (7.1.18).

Now fix a number $T > 0$ and choose $N \in \mathbb{N}$ such that $N\delta > T$. Fix an element $t \in [0,T]$. Then there exists a unique integer $k \in \{0,1,\dots,N-1\}$ such that $k\delta \le t < (k+1)\delta$ and hence, by (7.1.18),

$$\|S(t)\| = \|S(\delta)^k S(t-k\delta)\| \le \|S(\delta)\|^k \|S(t-k\delta)\| \le M^{k+1} \le M^N.$$

This proves part (i).

Part (ii) follows from part (i) and the inequalities

$$\|S(t+h)x - S(t)x\| \le \|S(t)\| \, \|S(h)x - x\|$$

and

$$\|S(t-h)x - S(t)x\| \le \|S(t-h)\| \, \|x - S(h)x\|$$

for $0 \le h \le t$.

We prove part (iii). Equation (7.1.16) holds obviously with $\omega_0 = -\infty$ whenever $S(t) = 0$ for some $t > 0$. Hence assume $S(t) \neq 0$ for all $t > 0$. Then for every $t > 0$ there is a constant $c \geq 1$ such that $c^{-1} \leq \|S(s)\| \leq c$ for $0 \leq s \leq t$. Define the function $g : [0, \infty) \to \mathbb{R}$ by

$$g(t) := \log \|S(t)\| \qquad \text{for } t \geq 0.$$

Then it follows from the semigroup property and part (i) that

$$g(0) = 0, \qquad g(s+t) \leq g(s) + g(t), \qquad M(t) := \sup_{0 \leq s \leq t} |g(s)| < \infty$$

for all $s, t \geq 0$. Fix a real number $t_0 > 0$ and let $t > 0$ be any positive real number. Then there exists an integer $k \geq 0$ and a real number s such that

$$t = kt_0 + s, \qquad 0 \leq s < t_0.$$

Hence

$$\frac{g(t)}{t} \leq \frac{kg(t_0) + g(s)}{t} = \frac{g(t_0)}{t_0} - \frac{sg(t_0)}{t_0 t} + \frac{g(s)}{t} \leq \frac{g(t_0)}{t_0} + \frac{2M(t_0)}{t}.$$

This implies

$$\limsup_{t \to \infty} \frac{g(t)}{t} \leq \frac{g(t_0)}{t_0}.$$

Since this holds for all $t_0 > 0$, we have $\limsup_{t\to\infty} t^{-1} g(t) \leq \inf_{t>0} t^{-1} g(t)$ and this proves part (iii).

We prove part (iv). Fix a real number $\omega > \omega_0$. By part (iii) there exists a constant $T > 0$ such that

$$\frac{\log \|S(t)\|}{t} \leq \omega \qquad \text{for all } t \geq T.$$

Thus $\log \|S(t)\| \leq \omega t$ and so $\|S(t)\| \leq e^{\omega t}$ for all $t \geq T$. Define

$$M := \sup_{0 \leq t \leq T} \|S(t)\| \, e^{-\omega t}.$$

Then $\|S(t)\| \leq M e^{\omega t}$ for all $t \geq 0$ and this proves Lemma 7.1.8. □

Lemma 7.1.9. *Let X be a real Banach space and let $S : [0, \infty) \to \mathcal{L}(X)$ be a strongly continuous semigroup. Then the following hold.*

(i) *The operator $S(t)$ is injective for some $t > 0$ if and only if it is injective for all $t > 0$.*

(ii) *The operator $S(t)$ is surjective for some $t > 0$ if and only if it is surjective for all $t > 0$.*

(iii) *The operator $S(t)$ has a dense image for some $t > 0$ if and only if it has a dense image for all $t > 0$.*

(iv) *Assume $S(t)$ is injective for all $t > 0$. Then $S(t)$ has a closed image for some $t > 0$ if and only if it has a closed image for all $t > 0$.*

Proof. We prove part (i). Assume that there exists a real number $t_0 > 0$ such that $S(t_0)$ is injective. Let $t > 0$ and choose an integer $k > 0$ such that $kt_0 \geq t$. If $x \in X$ satisfies $S(t)x = 0$, then $S(t_0)^k x = S(kt_0 - t)S(t)x = 0$ and hence $x = 0$. Thus $S(t)$ is injective for all $t > 0$.

We prove part (ii). Assume that there exists a real number $t_0 > 0$ such that $S(t_0)$ is surjective. Let $t > 0$ and choose an integer $k > 0$ such that $kt_0 \geq t$. Then $S(kt_0) = S(t_0)^k$ is surjective and this implies that $\mathrm{im}(S(t)) \supset \mathrm{im}(S(t)S(kt_0 - t)) = \mathrm{im}(S(kt_0)) = X$. Thus $S(t)$ is surjective for all $t > 0$.

We prove part (iii). Assume that there exists a real number $t_0 > 0$ such that $S(t_0)$ has a dense image. Let $t > 0$ and choose an integer $k > 0$ such that $kt_0 \geq t$. Then the operator $S(kt_0) = S(t_0)^k$ has a dense image. Since $\mathrm{im}(S(t)) \supset \mathrm{im}(S(t)S(kt_0 - t)) = \mathrm{im}(S(kt_0))$ this implies that $S(t)$ has a dense image.

We prove part (iv). Thus assume $S(t)$ is injective for all $t > 0$ and that there exists a real number $t_0 > 0$ such that $S(t_0)$ has a closed image. Then it follows from part (ii) of Corollary 4.1.17 that there exists a constant $\delta > 0$ such that $\delta \|x\| \leq \|S(t_0)x\|$ for all $x \in X$. By induction this implies $\delta^k \|x\| \leq \|S(kt_0)x\|$ for all $x \in X$ and all $k \in \mathbb{N}$. Let $t > 0$ and choose an integer $k > 0$ such that $kt_0 \geq t$. Then

$$\|S(kt_0 - t)\| \, \|S(t)x\| \geq \|S(kt_0)x\| \geq \delta^k \|x\|$$

and so $\|S(t)x\| \geq \|S(kt_0 - t)\|^{-1} \delta^k \|x\|$ for all $x \in X$. Hence $S(t)$ has a closed image by Theorem 4.1.16 and this proves Lemma 7.1.9. □

Example 7.1.10. This example shows that the hypothesis that $S(t)$ is injective for all $t > 0$ cannot be removed in part (iv) of Lemma 7.1.9. Consider the real Banach space

$$X := \left\{ f \in \mathcal{L}^2([0,1]) \,\middle|\, f \text{ is continuous on } [0, \tfrac{1}{2}] \text{ and } f(0) = 0 \right\} \Big/ {\sim} .$$

Here the equivalence relation is defined by $f \sim g$ if and only if $f - g$ vanishes almost everywhere on the interval $[\frac{1}{2}, 1]$, and the norm is defined by

$$\|f\|_X := \sup_{0 \leq s \leq \frac{1}{2}} |f(s)| + \sqrt{\int_{\frac{1}{2}}^1 f(s)^2 \, ds}$$

for $f \in X$. Then the formula

$$(S(t)f)(s) := \begin{cases} f(s - t), & \text{if } s \geq t, \\ 0, & \text{if } s < t, \end{cases}$$

for $f \in X$, $t \geq 0$, and $0 \leq s \leq 1$ defines a strongly continuous semigroup on X. The operator $S(t)$ has a nontrivial kernel for all $t > 0$, does not have a closed image for $0 < t < 1$, and vanishes for all $t \geq 1$.

7.1.3. The Infinitesimal Generator. The starting point of the present section was to introduce strongly continuous semigroups of operators as a generalization of the space of solutions of a linear differential equation. Given such a space of *"solutions"* it is then a natural question to ask whether there is actually a linear differential equation that a given strongly continuous semigroup provides the solutions of. The quest for such an equation leads to the following definition.

Definition 7.1.11 (Infinitesimal Generator). Let X be a real Banach space and let $S : [0, \infty) \to \mathcal{L}(X)$ be a strongly continuous semigroup. The **infinitesimal generator of** S is the linear operator $A : \mathrm{dom}(A) \to X$, whose domain is the linear subspace $\mathrm{dom}(A) \subset X$ defined by

$$\mathrm{dom}(A) := \left\{ x \in X \,\middle|\, \text{the limit } \lim_{h \searrow 0} \frac{S(h)x - x}{h} \text{ exists} \right\}, \tag{7.1.19}$$

and which is given by

$$Ax := \lim_{h \searrow 0} \frac{S(h)x - x}{h} \qquad \text{for } x \in \mathrm{dom}(A). \tag{7.1.20}$$

Example 7.1.12. Let H be a separable complex Hilbert space, let $(e_i)_{i \in \mathbb{N}}$ be a complex orthonormal basis, and let $(\lambda_i)_{i \in \mathbb{N}}$ be a sequence of complex numbers such that

$$\sup_{i \in \mathbb{N}} \mathrm{Re}\lambda_i < \infty.$$

Let $S : [0, \infty) \to \mathcal{L}^c(H)$ be the strongly continuous semigroup in Example 7.1.3, i.e.

$$S(t)x = \sum_{i=1}^{\infty} e^{\lambda_i t} \langle e_i, x \rangle e_i$$

for $x \in H$ and $t \geq 0$. Then the infinitesimal generator of S is the linear operator

$$A : \mathrm{dom}(A) \to H$$

in Example 6.1.3, given by

$$\mathrm{dom}(A) = \left\{ x \in H \,\middle|\, \sum_{i=1}^{\infty} |\lambda_i \langle e_i, x \rangle|^2 < \infty \right\} \tag{7.1.21}$$

and

$$Ax = \sum_{i=1}^{\infty} \lambda_i \langle e_i, x \rangle e_i \qquad \text{for } x \in \mathrm{dom}(A). \tag{7.1.22}$$

Exercise: Prove this.

Lemma 7.1.13. *Let X be a real Banach space and let $S : [0, \infty) \to \mathcal{L}(X)$ be a strongly continuous semigroup with infinitesimal generator*

$$A : \mathrm{dom}(A) \to X.$$

Let $x \in X$. Then the following are equivalent.

(i) $x \in \mathrm{dom}(A)$.

(ii) *The function $[0, \infty) \to X : t \mapsto S(t)x$ is continuously differentiable, takes values in the domain of A, and satisfies the differential equation*

$$\frac{d}{dt} S(t)x = AS(t)x = S(t)Ax \qquad \text{for all } t \geq 0. \tag{7.1.23}$$

Proof. That (ii) implies (i) follows directly from the definitions. To prove the converse, fix an element $x \in \mathrm{dom}(A)$. Then, for $t \geq 0$, we have

$$S(t)Ax = \lim_{h \searrow 0} S(t) \frac{S(h)x - x}{h} = \lim_{h \searrow 0} \frac{S(t+h)x - S(t)x}{h}$$

and, for $t > 0$,

$$S(t)Ax = \lim_{h \searrow 0} S(t-h) \frac{S(h)x - x}{h} = \lim_{h \searrow 0} \frac{S(t-h)x - S(t)x}{-h}.$$

This shows that the function $[0, \infty) \to X : t \mapsto S(t)x$ is continuously differentiable and that its derivative at $t \geq 0$ is $S(t)Ax$. Moreover,

$$\lim_{h \searrow 0} \frac{S(h)S(t)x - S(t)x}{h} = \lim_{h \searrow 0} S(t) \frac{S(h)x - x}{h} = S(t)Ax.$$

Thus $S(t)x \in \mathrm{dom}(A)$ and

$$AS(t)x = S(t)Ax.$$

This proves Lemma 7.1.13. □

Lemma 7.1.14 (Variation of Constants). *Let X be a real Banach space and let $S : [0, \infty) \to \mathcal{L}(X)$ be a strongly continuous semigroup with infinitesimal generator $A : \mathrm{dom}(A) \to X$. Let $f : [0, \infty) \to X$ be a continuously differentiable function and define the function $x : [0, \infty) \to X$ by*

$$x(t) := \int_0^t S(t-s) f(s)\, ds \qquad \text{for } t \geq 0. \tag{7.1.24}$$

Then x is continuously differentiable, $x(t) \in \mathrm{dom}(A)$ for all $t \geq 0$, and

$$\dot{x}(t) = Ax(t) + f(t) = S(t)f(0) + \int_0^t S(t-s)\dot{f}(s)\, ds \tag{7.1.25}$$

for all $t \geq 0$.

Proof. Fix a constant $t \geq 0$ and let $h > 0$. Then

$$\begin{aligned}
\frac{S(h)x(t) - x(t)}{h} &= \frac{S(h) - \mathbb{1}}{h} \int_0^t S(s)f(t-s)\,ds \\
&= \frac{1}{h}\int_0^t S(s+h)f(t-s)\,ds - \frac{1}{h}\int_0^t S(s)f(t-s)\,ds \\
&= \frac{1}{h}\int_h^{t+h} S(s)f(t+h-s)\,ds - \frac{1}{h}\int_0^t S(s)f(t-s)\,ds \\
&= \frac{1}{h}\int_t^{t+h} S(s)f(t+h-s)\,ds - \frac{1}{h}\int_0^h S(s)f(t+h-s)\,ds \\
&\quad + \int_0^t S(s)\frac{f(t+h-s) - f(t-s)}{h}\,ds.
\end{aligned}$$

Take the limit $h \to 0$ to obtain $x(t) \in \mathrm{dom}(A)$ and

$$Ax(t) = S(t)f(0) - f(t) + \int_0^t S(t-s)\dot{f}(s)\,ds. \tag{7.1.26}$$

This proves the second equation in (7.1.25) and shows that Ax is continuous. Next observe that

$$\begin{aligned}
\frac{x(t+h) - x(t)}{h} &= \frac{1}{h}\int_0^{t+h} S(t+h-s)f(s)\,ds - \frac{1}{h}\int_0^t S(t-s)f(s)\,ds \\
&= \frac{S(h)x(t) - x(t)}{h} + \frac{1}{h}\int_t^{t+h} S(t+h-s)f(s)\,ds
\end{aligned}$$

for all $h > 0$. Take the limit $h \to 0$ to obtain that x is right differentiable and $\frac{d}{dt^+}x(t) = Ax(t) + f(t)$. Third, observe that

$$\begin{aligned}
\frac{x(t) - x(t-h)}{h} &= \frac{1}{h}\int_0^t S(t-s)f(s)\,ds - \frac{1}{h}\int_0^{t-h} S(t-h-s)f(s)\,ds \\
&= \frac{1}{h}\int_0^t S(t-s)f(s)\,ds - \frac{1}{h}\int_h^t S(t-s)f(s-h)\,ds \\
&= \frac{1}{h}\int_0^h S(t-s)f(s)\,ds + \int_h^t S(t-s)\frac{f(s) - f(s-h)}{h}\,ds
\end{aligned}$$

for $0 < h < t$. Take the limit $h \to 0$ to obtain that x is left differentiable and $\frac{d}{dt^-}x(t) = S(t)f(0) + \int_0^t S(t-s)\dot{f}(s)\,ds = Ax(t) + f(t)$. Here the last equation follows from (7.1.26). This proves Lemma 7.1.14. □

Example 7.1.15. Let $x \in X$ and take $f(t) = x$ in Lemma 7.1.14. Then

$$\int_0^t S(s)x\,ds \in \mathrm{dom}(A), \qquad A\int_0^t S(s)x\,ds = S(t)x - x$$

for all $t > 0$.

Lemma 7.1.16. *Let X be a real Banach space and let $S : [0, \infty) \to \mathcal{L}(X)$ be a strongly continuous semigroup with infinitesimal generator*

$$A : \mathrm{dom}(A) \to X.$$

For $n \in \mathbb{N}$ define the linear subspaces $\mathrm{dom}(A^n) \subset X$ recursively by

$$\mathrm{dom}(A^1) := \mathrm{dom}(A), \qquad \mathrm{dom}(A^n) := \{x \in \mathrm{dom}(A) \mid Ax \in \mathrm{dom}(A^{n-1})\}$$

for $n \geq 2$. Then the linear subspace $\mathrm{dom}(A^\infty) := \bigcap_{n\in\mathbb{N}} \mathrm{dom}(A^n)$ is dense in X and A has a closed graph.

Proof. The proof has three steps.

Step 1. *Let $x \in X$ and let $\phi : \mathbb{R} \to X$ be a smooth function with compact support contained in the interval $[\delta, \delta^{-1}]$ for some constant $0 < \delta < 1$. Then, for every $n \in \mathbb{N}$, we have $\int_0^\infty \phi(t)S(t)x\,dt \in \mathrm{dom}(A^n)$ and*

$$A^n \int_0^\infty \phi(t)S(t)x\,dt = (-1)^n \int_0^\infty \phi^{(n)}(t)S(t)x\,dt.$$

For $n = 1$ this follows from Lemma 7.1.14 with $t > \delta^{-1}$ and $f(s) := \phi(t-s)x$ for $s \geq 0$. For $n \geq 2$ the assertion follows by induction.

Step 2. $\mathrm{dom}(A^\infty)$ *is dense in X.*

Let $x \in X$ and choose a smooth function $\phi : \mathbb{R} \to [0, \infty)$, with compact support in the interval $[1/2, 1]$, such that $\int_0^1 \phi(t)\,dt = 1$. Define

$$x_n := n \int_0^\infty \phi(nt)S(t)x\,dt \qquad \text{for } n \in \mathbb{N}.$$

Then $x_n \in \mathrm{dom}(A^\infty)$ by Step 1 and

$$\|x_n - x\| = \left\| n \int_0^{1/n} \phi(nt)(S(t)x - x)\,dt \right\| \leq \sup_{0\leq t\leq 1/n} \|S(t)x - x\|.$$

Hence $\lim_{n\to\infty}\|x_n - x\| = 0$ and this proves Step 2.

Step 3. *A has a closed graph.*

Choose a sequence $x_n \in \mathrm{dom}(A)$ and $x, y \in X$ such that

$$\lim_{n\to\infty} \|x_n - x\| = 0, \qquad \lim_{n\to\infty} \|Ax_n - y\| = 0.$$

Then, by Lemma 7.1.13,

$$S(t)x - x = \lim_{n\to\infty} \big(S(t)x_n - x_n\big) = \lim_{n\to\infty} \int_0^t S(s)Ax_n\,ds = \int_0^t S(s)y\,ds$$

for all $t > 0$. Hence $y = \lim_{t\searrow 0} t^{-1}(S(t)x - x)$ and this implies $x \in \mathrm{dom}(A)$ and $Ax = y$. This proves Step 3 and Lemma 7.1.16. □

Recall from Exercise 2.2.12 that the domain of a closed densely defined operator $A : \mathrm{dom}(A) \to X$ is a Banach space with the **graph norm**

$$\|x\|_A := \|x\|_X + \|Ax\|_X \qquad \text{for } x \in \mathrm{dom}(A).$$

Moreover, the operator A can be viewed as a bounded operator from $\mathrm{dom}(A)$ to X rather than as an unbounded densely defined operator from X to itself.

Lemma 7.1.17. *Let X be a real Banach space and let $S : [0, \infty) \to \mathcal{L}(X)$ be a strongly continuous semigroup. Let $A : \mathrm{dom}(A) \to X$ be a linear operator with a dense domain $\mathrm{dom}(A) \subset X$ and a closed graph. Then the following are equivalent.*

(i) *The operator A is the infinitesimal generator of the semigroup S.*

(ii) *If $x \in \mathrm{dom}(A)$ and $t > 0$, then $S(t)x \in \mathrm{dom}(A)$, $AS(t)x = S(t)Ax$, and $S(t)x - x = \int_0^t S(s)Ax\,ds$.*

(iii) *If $x_0 \in \mathrm{dom}(A)$, then the function $[0, \infty) \to X : t \mapsto x(t) := S(t)x_0$ is continuously differentiable, takes values in $\mathrm{dom}(A)$, and satisfies the differential equation $\dot{x}(t) = Ax(t)$ for all $t \geq 0$.*

Proof. That (i) implies (ii) follows directly from Lemma 7.1.13. That (ii) implies (iii) follows directly from part (vii) of Lemma 5.1.10. We prove in three steps that (iii) implies (i). Assume A satisfies (iii).

Step 1. *Let $x \in \mathrm{dom}(A)$ and $t > 0$. Then*

$$\int_0^t S(s)x\,ds \in \mathrm{dom}(A), \qquad A\int_0^t S(s)x\,ds = S(t)x - x. \tag{7.1.27}$$

By part (iii) the function $\xi : [0, t] \to X$ defined by $\xi(s) := S(s)x$ for $0 \leq s \leq t$ takes values in $\mathrm{dom}(A)$ and the function $A\xi = \dot{\xi} : [0, t] \to X$ is continuous. Hence the function $\xi : [0, t] \to \mathrm{dom}(A)$ is continuous with respect to the graph norm. Thus it follows from part (iii) of Lemma 5.1.10 that

$$\int_0^t \xi(s)\,ds \in \mathrm{dom}(A)$$

and

$$A\int_0^t \xi(s)\,ds = \int_0^t A\xi(s)\,ds = \xi(t) - \xi(0) = S(t)x - x.$$

This proves Step 1.

Step 2. *If $x \in X$ and $t > 0$, then* (7.1.27) *holds.*

Let $x \in X$ and $t > 0$. Choose a sequence $x_i \in \mathrm{dom}(A)$ that converges to x. Then $\xi_i := \int_0^t S(s)x_i\,ds \in \mathrm{dom}(A)$ and $A\xi_i = S(t)x_i - x_i$ by Step 1. Since A has a closed graph, ξ_i converges to $\int_0^t S(s)x\,ds$, and $A\xi_i$ converges to $S(t)x - x$, it follows that x and t satisfy (7.1.27). This proves Step 2.

Step 3. *Let $x, y \in X$. Then*

$$\lim_{h\to 0} \frac{S(h)x - x}{h} = y \qquad \Longleftrightarrow \qquad x \in \mathrm{dom}(A), \quad Ax = y. \tag{7.1.28}$$

If $x \in \mathrm{dom}(A)$ and $y = Ax$, then $\lim_{h\to 0} h^{-1}(S(h)x - x) = y$ by part (iii). Conversely, suppose that $\lim_{h\to 0} h^{-1}(S(h)x - x) = y$. For each $h > 0$ define $x_h := h^{-1}\int_0^h S(s)x\,ds$. Then $\lim_{h\to 0} x_h = x$ and by Step 2 $x_h \in \mathrm{dom}(A)$ and $Ax_h = h^{-1}(S(h)x - x)$. Hence $\lim_{h\to 0} Ax_h = y$. Since A has a closed graph this implies $x \in \mathrm{dom}(A)$ and $Ax = y$. This proves Lemma 7.1.17. □

Lemma 7.1.18. *Let X be a real Banach space and let $S : [0, \infty) \to \mathcal{L}(X)$ be a strongly continuous semigroup with infinitesimal generator A. Then the following are equivalent.*

(i) $\mathrm{dom}(A) = X$.

(ii) *A is bounded.*

(iii) *The semigroup S is continuous in the norm topology on $\mathcal{L}(X)$.*

Proof. The Closed Graph Theorem 2.2.13 asserts that (i) and (ii) are equivalent. That (ii) implies (iii) follows from Exercise 1.5.4 and Corollary 7.2.3 below. We prove that (iii) implies (i), following [**26**, p. 615]. Assume that $S : [0, \infty) \to \mathcal{L}(X)$ is continuous with respect to the norm topology on $\mathcal{L}(X)$. Then $\lim_{t\to 0} \|S(t) - \mathbb{1}\| = 0$. Hence there exists a $\delta > 0$ such that $\sup_{0\le t\le\delta} \|S(t) - \mathbb{1}\| < 1$. For $0 \le t \le \delta$ define

$$B(t) := \sum_{n=1}^{\infty} \frac{(-1)^{n-1}}{n} (S(t) - \mathbb{1})^n.$$

Then the following hold.

(I) The function $B : [0, \delta] \to \mathcal{L}(X)$ is norm-continuous.

(II) $e^{B(t)} = S(t)$ for $0 \le t \le \delta$.

(III) If $k \in \mathbb{N}$ and $0 \le t \le \delta/k$, then $B(kt) = kB(t)$.

Part (II) uses the fact that the power series $f(z) := \sum_{n=1}^{\infty}(-1)^{n-1}(z-1)^n/n$ satisfies $\exp(f(z)) = z$ for all $z \in \mathbb{C}$ with $|z - 1| < 1$. Part (III) follows from the fact that $f(z^k) = kf(z)$ whenever $|z^j - 1| < 1$ for $j = 1, 2, \dots, k$.

By (III), $B(\delta) = \ell B(\delta/\ell)$ and so $B(k\delta/\ell) = kB(\delta/\ell) = (k/\ell)B(\delta)$ for all integers $0 \le k \le \ell$. Since B is continuous by (I), this implies

$$B(t) = t\delta^{-1}B(\delta) \qquad \text{for } 0 \le t \le \delta.$$

(Approximate $t\delta^{-1}$ by a sequence of rational numbers in $[0, 1]$.) Now define the operator $A := \delta^{-1}B(\delta) \in \mathcal{L}(X)$. Then by (II) we have $S(t) = e^{B(t)} = e^{tA}$ for $0 \le t \le \delta$. So $S(t) = e^{tA}$ for all $t \ge 0$ and this proves Lemma 7.1.18. □

7.2. The Hille–Yosida–Phillips Theorem

7.2.1. Well-Posed Cauchy Problems. Let us now change the point of view and suppose that $A : \mathrm{dom}(A) \to X$ is a linear operator on a Banach space X whose domain is a linear subspace $\mathrm{dom}(A) \subset X$. Consider the **Cauchy problem**

$$\dot{x} = Ax, \qquad x(0) = x_0. \tag{7.2.1}$$

Definition 7.2.1. (i) Let $I \subset [0, \infty)$ be a closed interval with $0 \in I$. A continuously differentiable function $x : I \to X$ is called a **solution** of (7.2.1) if it takes values in $\mathrm{dom}(A)$ and $x(0) = x_0$ and $\dot{x}(t) = Ax(t)$ for all $t \in I$.

(ii) The Cauchy problem (7.2.1) is called **well-posed** if it satisfies the following axioms.

(Existence) For each $x_0 \in \mathrm{dom}(A)$ there is a solution of (7.2.1) on $[0, \infty)$.

(Uniqueness) Let $x_0 \in \mathrm{dom}(A)$ and $T > 0$. If $x, y : [0, T] \to X$ are solutions of (7.2.1), then $x(t) = y(t)$ for all $t \in [0, T]$.

(Continuous Dependence) Define the map $\phi : [0, \infty) \times \mathrm{dom}(A) \to X$ by $\phi(t, x_0) := x(t)$ for $t \geq 0$ and $x_0 \in \mathrm{dom}(A)$, where $x : [0, \infty) \to X$ is the unique solution of (7.2.1). Then, for every $T > 0$, there is a constant $M \geq 1$ such that $\|\phi(t, x_0)\| \leq M\|x_0\|$ for all $t \in [0, T]$ and all $x_0 \in \mathrm{dom}(A)$.

The next theorem characterizes well-posed Cauchy problems and was proved by Ralph S. Phillips [**68**] in 1954.

Theorem 7.2.2 (Phillips). *Let $A : \mathrm{dom}(A) \to X$ be a linear operator with a dense domain $\mathrm{dom}(A) \subset X$ and a closed graph. The following are equivalent.*

(i) *A is the infinitesimal generator of a strongly continuous semigroup.*

(ii) *The Cauchy problem (7.2.1) is well-posed.*

Proof. We prove that (i) implies (ii). Thus assume that A is the infinitesimal generator of a strongly continuous semigroup $S : [0, \infty) \to \mathcal{L}(X)$ and fix an element $x_0 \in \mathrm{dom}(A)$. Then the function $[0, \infty) \to X : t \mapsto S(t)x_0$ is a solution of equation (7.2.1) by Lemma 7.1.13. To prove uniqueness, assume that $x : [0, \infty) \to X$ is any solution of (7.2.1). Fix a constant $t > 0$. We will prove that the function $[0, t] \to X : s \mapsto S(t - s)x(s)$ is constant. To see this, note that $x(s) \in \mathrm{dom}(A)$ and so

$$\lim_{\substack{h \to 0 \\ h \leq t-s}} \frac{S(t - s - h)x(s) - S(t - s)x(s)}{-h} = S(t - s)Ax(s) \qquad \text{for } 0 \leq s \leq t.$$

This implies

$$\begin{aligned}
&\lim_{h\to 0} \frac{S(t-s-h)x(s+h) - S(t-s)x(s)}{h} \\
&= \lim_{h\to 0} S(t-s-h)\left(\frac{x(s+h)-x(s)}{h} - Ax(s)\right) \\
&\quad + \lim_{h\to 0}\left(\frac{S(t-s-h)x(s) - S(t-s)x(s)}{h} + S(t-s)Ax(s)\right) \\
&\quad + \lim_{h\to 0}\big(S(t-s-h)Ax(s) - S(t-s)Ax(s)\big) \\
&= 0.
\end{aligned}$$

Hence the function $[0,t] \to X : s \mapsto S(t-s)x(s)$ is everywhere differentiable and its derivative vanishes. Thus it is constant and hence $x(t) = S(t)x_0$. Since $t > 0$ was chosen arbitrarily this proves uniqueness. Continuous dependence follows from the estimate $\|S(t)\| \le Me^{\omega t}$ in Lemma 7.1.8. This shows that (i) implies (ii).

We prove that (ii) implies (i). Assume the Cauchy problem (7.2.1) is well-posed and let

$$\phi : [0,\infty) \times \mathrm{dom}(A) \to \mathrm{dom}(A)$$

be the map that assigns to each element $x_0 \in \mathrm{dom}(A)$ the unique solution $[0,\infty) \to X : t \mapsto \phi(t, x_0)$ of (7.2.1). We claim that, for each $t \ge 0$, there is a unique bounded linear operator $S(t) : X \to X$ such that

$$S(t)x_0 = \phi(t, x_0) \qquad \text{for all } x_0 \in \mathrm{dom}(A). \tag{7.2.2}$$

To see this, note first that the space of solutions $x : [0,\infty) \to X$ of (7.2.1) is a linear subspace of the space of all functions from $[0,\infty)$ to X. Hence it follows from uniqueness that the map $\mathrm{dom}(A) \to X : x_0 \mapsto \phi(t, x_0)$ is linear. Second, it follows from continuous dependence that the linear operator $\mathrm{dom}(A) \to X : x_0 \mapsto \phi(t, x_0)$ is bounded. Since $\mathrm{dom}(A)$ is a dense linear subspace of X it follows that this operator extends uniquely to a bounded linear operator $S(t) \in \mathcal{L}(X)$. More precisely, fix an element $x \in X$. Then there exists a sequence $x_n \in \mathrm{dom}(A)$ that converges to x. Hence $(x_n)_{n\in\mathbb{N}}$ is a Cauchy sequence in X and so is the sequence $(\phi(t, x_n))_{n\in\mathbb{N}}$ by continuous dependence. Hence it converges and the limit

$$S(t)x := \lim_{n\to\infty} \phi(t, x_n)$$

is independent of the choice of the sequence $x_n \in \mathrm{dom}(A)$ used to define it. This proves the existence of a bounded linear operator $S(t)$ that satisfies (7.2.2).

We prove that these operators form a one-parameter semigroup. Fix a real number $t \geq 0$ and an element $x_0 \in \mathrm{dom}(A)$. Then

$$S(t)x_0 = \phi(t, x_0) \in \mathrm{dom}(A)$$

and the function $[0, \infty) \to X : s \mapsto S(s+t)x_0 = \phi(s+t, x_0)$ is a solution of the Cauchy problem (7.2.1) with x_0 replaced by $S(t)x_0 = \phi(t, x_0)$. Hence

$$S(s+t, x_0) = \phi(s, S(t)x_0) = S(s)S(t)x_0.$$

Since this holds for all $x_0 \in \mathrm{dom}(A)$, the set $\mathrm{dom}(A)$ is dense in X, and the operators $S(s+t)$ and $S(s)S(t)$ are both continuous maps, it follows that $S(s+t) = S(s)S(t)$ for all $s \geq 0$. This shows that $S : [0, \infty) \to \mathcal{L}(X)$ is a one-parameter semigroup.

We prove that S is strongly continuous. To see this, fix an element $x \in X$ and a constant $\varepsilon > 0$. By continuous dependence there exists an $M \geq 1$ such that $\sup_{0 \leq t \leq 1} \|\phi(t, x_0)\| \leq M \|x_0\|$ for all $x_0 \in \mathrm{dom}(A)$. This shows that $\sup_{0 \leq t \leq 1} \|S(t)\| \leq M$. Choose an element $y \in \mathrm{dom}(A)$ such that

$$\|x - y\| \leq \frac{\varepsilon}{2(M+1)}.$$

Next choose a constant $0 < \delta < 1$ such that, for all $t \in \mathbb{R}$,

$$0 \leq t < \delta \qquad \Longrightarrow \qquad \|\phi(t, y) - y\| < \frac{\varepsilon}{2}.$$

Fix a real number $0 \leq t < \delta$. Then

$$\begin{aligned} \|S(t)x - x\| &\leq \|S(t)x - S(t)y\| + \|S(t)y - y\| + \|y - x\| \\ &\leq (M+1)\|x - y\| + \|\phi(t, y) - y\| < \frac{\varepsilon}{2} + \frac{\varepsilon}{2} = \varepsilon. \end{aligned}$$

This shows that S is strongly continuous.

We prove that A is the infinitesimal generator of S. Let $x_0 \in \mathrm{dom}(A)$ and define the function $x : [0, \infty) \to X$ by $x(t) := S(t)x_0 = \phi(t, x_0)$. It is continuously differentiable, takes values in $\mathrm{dom}(A)$, and satisfies the equation $\dot{x}(t) = Ax(t)$ for all $t \geq 0$. Thus A and S satisfy condition (iii) in Lemma 7.1.17, so A is the infinitesimal generator of S. This proves Theorem 7.2.2. □

Corollary 7.2.3 (Uniqueness). *A linear operator on a Banach space is the infinitesimal generator of at most one strongly continuous semigroup.*

Proof. Let A be the infinitesimal generator of two strongly continuous semigroups $S, T : [0, \infty) \to \mathcal{L}(X)$. Let $x_0 \in \mathrm{dom}(A)$. Then the functions $x(t) := S(t)x_0$ and $y(t) := T(t)x_0$ both satisfy (7.2.1) and hence agree by Theorem 7.2.2. Since $\mathrm{dom}(A)$ is dense in X by Lemma 7.1.16, it follows that $S(t)x = T(t)x$ for all $x \in X$ and all $t \geq 0$. □

Theorem 7.2.4 (Strongly Continuous Groups). *Let X be a real Banach space, let $S : [0,\infty) \to \mathcal{L}(X)$ be a strongly continuous semigroup, and let $A : \mathrm{dom}(A) \to X$ be the infinitesimal generator of S. Then the following are equivalent.*

(i) *The semigroup S extends to a strongly continuous group $S : \mathbb{R} \to \mathcal{L}(X)$.*

(ii) *$-A$ is the infinitesimal generator of a strongly continuous semigroup.*

(iii) *The operator $S(t)$ is bijective for all $t > 0$.*

Proof. We prove that (i) implies (ii). Thus assume that S extends to a strongly continuous group $S : \mathbb{R} \to \mathcal{L}(X)$. Then

$$S(t)S(-t) = S(-t)S(t) = \mathbb{1}$$

for all $t > 0$ by definition of a one-parameter group of operators. This implies that $S(t)$ is bijective and

$$S(t)^{-1} = S(-t)$$

for all $t > 0$. Define the map $T : [0,\infty) \to \mathcal{L}(X)$ by

$$T(t) := S(-t) = S(t)^{-1} \qquad \text{for } t \geq 0.$$

Then T is a strongly continuous semigroup by definition. Denote its infinitesimal generator by $B : \mathrm{dom}(B) \to X$. We must prove that $B = -A$. To see this, choose a constant $M \geq 1$ such that

$$\|S(t)\| \leq M \quad \text{and} \quad \|T(t)\| \leq M \qquad \text{for } 0 \leq t \leq 1.$$

Now let $x \in \mathrm{dom}(A)$. Then

$$\begin{aligned} \left\| \frac{T(h)x - x}{h} + Ax \right\| &\leq \left\| T(h) \left(\frac{x - S(h)x}{h} + Ax \right) \right\| + \|Ax - T(h)Ax\| \\ &\leq M \left\| \frac{x - S(h)x}{h} + Ax \right\| + \|Ax - T(h)Ax\| \end{aligned}$$

for $0 < h < 1$. Since the right-hand side converges to zero it follows that

$$x \in \mathrm{dom}(B), \qquad Bx = -Ax.$$

Thus we have proved that

$$\mathrm{dom}(A) \subset \mathrm{dom}(B), \qquad B|_{\mathrm{dom}(A)} = -A.$$

Interchange the roles of S and T to obtain

$$\mathrm{dom}(B) = \mathrm{dom}(A), \qquad B = -A.$$

This shows that (i) implies (ii).

We prove that (ii) implies (iii). Let $T : [0, \infty) \to \mathcal{L}(X)$ be the strongly continuous semigroup generated by $-A$. We prove that $S(t)$ is bijective and $T(t) = S(t)^{-1}$ for all $t > 0$. To see this, fix an element $x \in \mathrm{dom}(A)$ and a real number $t > 0$. Define the functions $y, z : [0, t] \to X$ by

$$y(s) := S(t-s)x, \qquad z(s) := T(t-s)x \qquad \text{for } 0 \le s \le t.$$

Then y and z are continuously differentiable, take values in the domain of A, and satisfy the Cauchy problems

$$\dot{y}(s) = -Ay(s) \quad \text{for } 0 \le s \le t, \qquad y(0) = S(t)x,$$

and

$$\dot{z}(s) = Az(s) \quad \text{for } 0 \le s \le t, \qquad z(0) = T(t)x.$$

By Theorem 7.2.2 this implies

$$y(s) = T(s)S(t)x, \qquad z(s) = S(s)T(t)x \qquad \text{for } 0 \le s \le t.$$

Take $s = t$ to obtain $T(t)S(t)x = y(t) = x$ and $S(t)T(t)x = z(t) = x$. Thus we have proved that $S(t)T(t)x = T(t)S(t)x = x$ for all $t > 0$ and all $x \in \mathrm{dom}(A)$. Since the domain of A is dense in X this implies

$$S(t)T(t) = T(t)S(t) = \mathbb{1} \qquad \text{for all } t > 0.$$

Hence $S(t)$ is bijective for all $t > 0$. This shows that (ii) implies (iii).

We prove that (iii) implies (i). Thus assume that $S(t)$ is bijective for all $t > 0$. Then $S(t)^{-1} : X \to X$ is a bounded linear operator for every $t > 0$ by the Open Mapping Theorem 2.2.1. Define

$$S(-t) := S(t)^{-1} \qquad \text{for } t > 0.$$

We prove that the extended function $S : \mathbb{R} \to \mathcal{L}(X)$ is a one-parameter group. The formula $S(t+s) = S(t)S(s)$ holds by definition whenever $s, t \ge 0$ or $s, t \le 0$. Moreover, if $0 \le s < t$, then $S(t-s)S(s) = S(t)$ and hence

$$S(t-s) = S(t)S(s)^{-1} = S(t)S(-s).$$

This implies that, for $0 \le t < s$, we have $S(s-t) = S(s)S(-t)$ and hence

$$S(t-s) = S(s-t)^{-1} = S(-t)^{-1}S(s)^{-1} = S(t)S(-s).$$

This shows that S is a one-parameter group. Strong continuity at $t = 0$ follows from the equation

$$S(-h)x - x = S(h)^{-1}(x - S(h)x)$$

for $h > 0$. Strong continuity at $-t < 0$ follows from the equation

$$S(-t+h)x - S(-t)x = S(t)^{-1}\big(S(h)x - x\big)$$

for $h \in \mathbb{R}$. This proves Theorem 7.2.4. $\square$

7.2.2. The Hille–Yosida–Phillips Theorem. The following theorem is the main result of this chapter. For the special case $M = 1$ it was discovered by Hille [**35**] and Yosida [**87**] independently in 1948. It was extended to the case $M > 1$ by Phillips [**67**] in 1952.

Theorem 7.2.5 (Hille–Yosida–Phillips). *Let X be a real Banach space and $A : \mathrm{dom}(A) \to X$ be a linear operator with a dense domain $\mathrm{dom}(A) \subset X$. Fix real numbers ω and $M \geq 1$. Then the following are equivalent.*

(i) *The operator A is the infinitesimal generator of a strongly continuous semigroup $S : [0, \infty) \to \mathcal{L}(X)$ that satisfies*

$$\|S(t)\| \leq Me^{\omega t} \qquad \text{for all } t \geq 0. \tag{7.2.3}$$

(ii) *For every real number $\lambda > \omega$ the operator $\lambda \mathbb{1} - A : \mathrm{dom}(A) \to X$ is invertible and*

$$\|(\lambda \mathbb{1} - A)^{-k}\| \leq \frac{M}{(\lambda - \omega)^k} \qquad \text{for all } \lambda > \omega \text{ and all } k \in \mathbb{N}. \tag{7.2.4}$$

Proof. See page 371. □

The necessity of the condition (7.2.4) is a straightforward consequence of Lemma 7.2.6 below which expresses the resolvent operator $(\lambda \mathbb{1} - A)^{-1}$ in terms of the semigroup. At this point it is convenient to allow for λ to be a complex number and therefore to extend the discussion to complex Banach spaces. When X is a real Banach space we will tacitly assume that X has been complexified so as to make sense of the operator $\lambda \mathbb{1} - A : \mathrm{dom}(A) \to X$ for complex numbers λ (see Exercise 5.1.5).

Lemma 7.2.6 (Resolvent Identity for Semigroups). *Let X be a complex Banach space and let*

$$A : \mathrm{dom}(A) \to X$$

be the infinitesimal generator of a strongly continuous semigroup

$$S : [0, \infty) \to \mathcal{L}^c(X).$$

Let $\lambda \in \mathbb{C}$ such that

$$\mathrm{Re}\lambda > \omega_0 := \lim_{t \to \infty} \frac{\log\|S(t)\|}{t}. \tag{7.2.5}$$

Then $\lambda \in \rho(A)$ and

$$(\lambda \mathbb{1} - A)^{-k} x = \frac{1}{(k-1)!} \int_0^\infty t^{k-1} e^{-\lambda t} S(t) x \, dt \tag{7.2.6}$$

for all $x \in X$ and all $k \in \mathbb{N}$.

Proof. We first prove the assertion for $k = 1$. Fix a complex number λ such that $\mathrm{Re}\lambda > \omega_0$ and choose a real number ω such that $\omega_0 < \omega < \mathrm{Re}\lambda$. By Lemma 7.1.8, there exists a constant $M \geq 1$ such that $\|S(t)\| \leq Me^{\omega t}$ for all $t \geq 0$. Hence $\|e^{-\lambda t}S(t)x\| \leq Me^{(\omega-\mathrm{Re}\lambda)t}\|x\|$ for all $x \in X$ and all $t \geq 0$. This implies that the formula

$$R_\lambda x := \int_0^\infty e^{-\lambda t}S(t)x\,dt = \lim_{T\to\infty}\int_0^T e^{-\lambda t}S(t)x\,dt \quad \text{for } x \in X$$

defines a bounded linear operator $R_\lambda \in \mathcal{L}^c(X)$. We prove the following.

Claim 1. *If $x \in X$ and $T > 0$, then $\xi_T := \int_0^T e^{-\lambda t}S(t)x\,dt \in \mathrm{dom}(A)$ and*

$$A\xi_T = e^{-\lambda T}S(T)x - x + \lambda\int_0^T e^{-\lambda t}S(t)x\,dt =: \eta_T.$$

Claim 2. *If $x \in \mathrm{dom}(A)$ and $T > 0$, then $\int_0^T e^{-\lambda t}S(t)Ax\,ds = \eta_T$.*

Claim 1 follows from Lemma 7.1.14 with $t = T$ and $f(t) := e^{-\lambda(T-t)}x$. Claim 2 follows from integration by parts with $\frac{d}{dt}S(t)x = S(t)Ax$. Now

$$A\xi_T = \eta_T, \qquad \lim_{T\to\infty}\xi_T = R_\lambda x, \qquad \lim_{T\to\infty}\eta_T = \lambda R_\lambda x - x$$

by Claim 1. Since A has a closed graph this implies

$$R_\lambda x \in \mathrm{dom}(A), \qquad AR_\lambda x = \lambda R_\lambda x - x \qquad \text{for all } x \in X.$$

If $x \in \mathrm{dom}(A)$ it follows from Claim 2 that

$$R_\lambda Ax = \lim_{T\to\infty}\int_0^T e^{-\lambda t}S(t)Ax\,dt = \lambda R_\lambda x - x.$$

Thus $(\lambda\mathbb{1}-A)R_\lambda x = x$ for all $x \in X$ and $R_\lambda(\lambda\mathbb{1}-A)x = x$ for all $x \in \mathrm{dom}(A)$. Hence $\lambda\mathbb{1}-A$ is bijective and $(\lambda\mathbb{1}-A)^{-1} = R_\lambda$. This proves (7.2.6) for $k = 1$. To prove the equation for $k \geq 2$ observe that the function

$$\rho(A) \to X : \lambda \mapsto (\lambda\mathbb{1} - A)^{-1}x$$

is holomorphic by Lemma 6.1.10 and satisfies

$$\begin{aligned}
(\lambda\mathbb{1} - A)^{-k}x &= \frac{(-1)^{k-1}}{(k-1)!}\frac{d^{k-1}}{d\lambda^{k-1}}(\lambda\mathbb{1} - A)^{-1}x \\
&= \frac{(-1)^{k-1}}{(k-1)!}\frac{d^{k-1}}{d\lambda^{k-1}}\int_0^\infty e^{-\lambda t}S(t)x\,dt \\
&= \frac{1}{(k-1)!}\int_0^\infty t^{k-1}e^{-\lambda t}S(t)x\,dt
\end{aligned}$$

for all $x \in X$ and all $\lambda \in \mathbb{C}$ with $\mathrm{Re}\lambda > \omega_0$. This proves Lemma 7.2.6. $\square$

It follows from Lemma 7.2.6 that

$$\sup_{\lambda\in\sigma(A)} \mathrm{Re}\lambda \le \omega_0 = \lim_{t\to\infty} \frac{\log\|S(t)\|}{t} \tag{7.2.7}$$

for every strongly continuous semigroup S with infinitesimal generator A. The following example by Einar Hille shows that the inequality in (7.2.7) can be strict.

Example 7.2.7. Fix a real number $\omega > 0$ and consider the Banach space

$$X := \left\{ f : [0,\infty) \to \mathbb{C} \,\middle|\, \begin{array}{l} f \text{ is continuous and bounded} \\ \text{and } \int_0^\infty e^{\omega s}|f(s)|\,ds < \infty \end{array} \right\},$$

equipped with the norm

$$\|f\| := \sup_{s\ge 0}|f(s)| + \int_0^\infty e^{\omega s}|f(s)|\,ds \qquad \text{for } f \in X.$$

The formula

$$(S(t)f)(s) := f(s+t) \qquad \text{for } f \in X \text{ and } s,t \ge 0$$

defines a strongly continuous semigroup on X and its infinitesimal generator is the operator $A : \mathrm{dom}(A) \to X$ given by

$$\mathrm{dom}(A) = \left\{ u : [0,\infty) \to \mathbb{C} \,\middle|\, \begin{array}{l} u \text{ is continuously differentiable} \\ \text{and } u, \dot{u} \in X \end{array} \right\},$$
$$Au = \dot{u}.$$

The operator $S(t)$ satisfies $\|S(t)\| = 1$ for all $t \ge 0$ and so

$$\omega_0 = \lim_{t\to\infty} \frac{\log\|S(t)\|}{t} = 0.$$

Now let $\lambda \in \mathbb{C}$ with $\mathrm{Re}\lambda > -\omega$ and let $f \in X$. Then, for $u \in \mathrm{dom}(A)$,

$$\lambda u - Au = f \qquad \iff \qquad \dot{u} = \lambda u - f.$$

This equation has a unique solution $u \in \mathrm{dom}(A)$ given by

$$u(s) = \int_s^\infty e^{\lambda(s-t)} f(t)\,dt \qquad \text{for } s \ge 0.$$

Thus the operator $\lambda\mathbb{1} - A$ is bijective for all $\lambda \in \mathbb{C}$ with $\mathrm{Re}\lambda > -\omega$. It has a one-dimensional kernel for all $\lambda \in \mathbb{C}$ with $\mathrm{Re}\lambda < -\omega$. Thus

$$\sup_{\lambda\in\sigma(A)} \mathrm{Re}\lambda = -\omega < 0 = \lim_{t\to\infty} \frac{\log\|S(t)\|}{t}.$$

Exercise: For $t > 0$ the spectrum of $S(t)$ is the closed unit disc and the point spectrum of $S(t)$ is the open disc of radius $e^{-\omega t}$ centered at the origin.

Proof of Theorem 7.2.5. We prove that (i) implies (ii). Thus assume that $A : \mathrm{dom}(A) \to X$ is the infinitesimal generator of a strongly continuous semigroup $S : [0, \infty) \to \mathcal{L}(X)$ that satisfies (7.2.3). Fix a real number

$$\lambda > \omega$$

and a positive integer k. Then

$$(\lambda \mathbb{1} - A)^{-k} x = \frac{1}{(k-1)!} \int_0^\infty t^{k-1} e^{-\lambda t} S(t) x \, dt$$

for all $x \in X$ by Lemma 7.2.6 and hence

$$\begin{aligned} \|(\lambda \mathbb{1} - A)^{-k} x\| &\le \frac{1}{(k-1)!} \int_0^\infty t^{k-1} e^{-\lambda t} \|S(t)x\| \, dt \\ &\le \frac{M \|x\|}{(k-1)!} \int_0^\infty t^{k-1} e^{-(\lambda-\omega)t} \, dt \\ &= \frac{M\|x\|}{(\lambda-\omega)^k}. \end{aligned}$$

Hence the operator A satisfies (ii).

We prove that (ii) implies (i). Thus assume that $A : \mathrm{dom}(A) \to X$ is a linear operator with a dense domain such that

$$\lambda \mathbb{1} - A : \mathrm{dom}(A) \to X$$

is bijective and satisfies the estimate (7.2.4) for $\lambda > \omega$. We prove in five steps that A is the infinitesimal generator of a strongly continuous semigroup that satisfies the estimate (7.2.3).

Step 1. $x = \lim_{\lambda\to\infty} \lambda(\lambda \mathbb{1} - A)^{-1} x$ *for all* $x \in X$.

If $x \in \mathrm{dom}(A)$, then

$$\lambda(\lambda \mathbb{1} - A)^{-1} x - x = A(\lambda \mathbb{1} - A)^{-1} x = (\lambda \mathbb{1} - A)^{-1} A x$$

for all $\lambda > \omega$ and so it follows from (7.2.4) that

$$\left\|\lambda(\lambda \mathbb{1} - A)^{-1} x - x\right\| \le \frac{M}{\lambda - \omega} \|Ax\|.$$

Thus

$$x = \lim_{\lambda\to\infty} \lambda(\lambda \mathbb{1} - A)^{-1} x$$

for all $x \in \mathrm{dom}(A)$. Moreover

$$\left\|\lambda(\lambda \mathbb{1} - A)^{-1}\right\| \le \frac{M\lambda}{\lambda - \omega} \le 2M \qquad \text{for all } \lambda > 2\omega.$$

Hence Step 1 follows from Theorem 2.1.5.

Step 2. *For $\lambda > \omega$ and $t \geq 0$ define*

$$A_\lambda := \lambda A(\lambda\mathbb{1} - A)^{-1}, \qquad S_\lambda(t) := e^{tA_\lambda} = \sum_{k=0}^{\infty} \frac{t^k A_\lambda^k}{k!}.$$

Then

$$\|S_\lambda(t)\| \leq M e^{\frac{\lambda\omega t}{\lambda-\omega}}$$

for all $\lambda > \omega$ and all $t \geq 0$.

The operator A_λ can be written as

$$A_\lambda = \lambda^2(\lambda\mathbb{1} - A)^{-1} - \lambda\mathbb{1}.$$

Hence

$$\begin{aligned}
\|S_\lambda(t)\| &= e^{-\lambda t}\left\|e^{t\lambda^2(\lambda\mathbb{1}-A)^{-1}}\right\| \\
&\leq e^{-\lambda t}\sum_{k=0}^{\infty}\frac{t^k\lambda^{2k}}{k!}\left\|(\lambda\mathbb{1} - A)^{-k}\right\| \\
&\leq e^{-\lambda t}\sum_{k=0}^{\infty}\frac{t^k\lambda^{2k}}{k!}\frac{M}{(\lambda-\omega)^k} \\
&= Me^{-\lambda t}e^{\frac{\lambda^2 t}{\lambda-\omega}} = Me^{\frac{\lambda\omega t}{\lambda-\omega}}
\end{aligned}$$

for all $\lambda > \omega$ and all $t \geq 0$. This proves Step 2.

Step 3. *Fix real numbers $\lambda > \mu > \omega$. Then*

$$\|S_\lambda(t)x - S_\mu(t)x\| \leq M^2 e^{\frac{\mu\omega t}{\mu-\omega}} t\, \|A_\lambda x - A_\mu x\|$$

for all $x \in X$ and all $t \geq 0$.

Since $A_\lambda A_\mu = A_\mu A_\lambda$, we have $A_\lambda S_\mu(t) = S_\mu(t)A_\lambda$ and so

$$\begin{aligned}
S_\lambda(t)x - S_\mu(t)x &= \int_0^t \frac{d}{ds}S_\mu(t-s)S_\lambda(s)x\,ds \\
&= \int_0^t S_\mu(t-s)S_\lambda(s)(A_\lambda x - A_\mu x)\,ds
\end{aligned}$$

for all $x \in X$ and all $t \geq 0$. Hence

$$\begin{aligned}
\|S_\lambda(t)x - S_\mu(t)x\| &\leq \int_0^t \|S_\mu(t-s)\|\,\|S_\lambda(s)\|\,ds\,\|A_\lambda x - A_\mu x\| \\
&\leq M^2 e^{\frac{\mu\omega t}{\mu-\omega}}\int_0^t e^{-\frac{\mu\omega s}{\mu-\omega}}e^{\frac{\lambda\omega s}{\lambda-\omega}}\,ds\,\|A_\lambda x - A_\mu x\| \\
&\leq M^2 e^{\frac{\mu\omega t}{\mu-\omega}} t\,\|A_\lambda x - A_\mu x\|.
\end{aligned}$$

Here the last step uses the inequality $\frac{\lambda\omega}{\lambda-\omega} \leq \frac{\mu\omega}{\mu-\omega}$. This proves Step 3.

Step 4. *The limit*

$$S(t)x := \lim_{\lambda\to\infty} S_\lambda(t)x \tag{7.2.8}$$

exists for all $x \in X$ and all $t \geq 0$. The resulting map $S : [0,\infty) \to \mathcal{L}(X)$ is a strongly continuous semigroup that satisfies (7.2.3).

Assume first that $x \in \mathrm{dom}(A)$. Then $\lim_{\lambda\to\infty} A_\lambda x = Ax$ by Step 1. Hence the limit (7.2.8) exists for all $t \geq 0$ by Step 3 and the convergence is uniform on every compact interval $[0,T]$. Since the operator family $\{S_\lambda(t)\}_{\lambda\geq 2\omega}$ is bounded by Step 2 it follows from Theorem 2.1.5 that the limit (7.2.8) exists for all $x \in X$ and that $S(t) \in \mathcal{L}(X)$ for all $t \geq 0$. Apply Theorem 2.1.5 to the operator family $X \to C([0,T],X) : x \mapsto S_\lambda(\cdot)x$ to deduce that the map $[0,T] \to X : t \mapsto S(t)x$ is continuous for all $x \in X$ and all $T > 0$. Moreover,

$$S(s)S(t)x = \lim_{\lambda\to\infty} S_\lambda(s)S_\lambda(t)x = \lim_{\lambda\to\infty} S_\lambda(s+t)x = S(s+t)x$$

for all $s,t \geq 0$ and all $x \in X$ and $S(0)x = \lim_{\lambda\to\infty} S_\lambda(t)x = x$ for all $x \in X$. Thus S is a strongly continuous semigroup. By Step 2 it satisfies the estimate

$$\|S(t)x\| = \lim_{\lambda\to\infty} \|S_\lambda(t)x\| \leq \lim_{\lambda\to\infty} Me^{\frac{\lambda\omega t}{\lambda-\omega}} \|x\| = Me^{\omega t}\|x\|$$

and this proves Step 4.

Step 5. *The operator A is the infinitesimal generator of S.*

Let B be the infinitesimal generator of S and let $x \in \mathrm{dom}(A)$. Then

$$\|S_\lambda(t)A_\lambda x - S(t)Ax\| \leq \|S_\lambda(t)\| \, \|A_\lambda x - Ax\| + \|S_\lambda(t)Ax - S(t)Ax\|$$

for all $t \geq 0$. Hence it follows from Step 1 and Step 2 that the functions $S_\lambda(\cdot)A_\lambda x : [0,h] \to X$ converge uniformly to $S(\cdot)Ax$ as λ tends to infinity. This implies

$$\int_0^h S(t)Ax\,dt = \lim_{\lambda\to\infty}\int_0^h S_\lambda(t)A_\lambda x\,dt = \lim_{\lambda\to\infty} S_\lambda(h)x - x = S(h)x - x$$

for all $h > 0$ and so

$$\lim_{h\to 0}\frac{S(h)x - x}{h} = \lim_{h\to 0}\frac{1}{h}\int_0^h S(t)Ax\,dt = Ax.$$

This shows that $\mathrm{dom}(A) \subset \mathrm{dom}(B)$ and $B|_{\mathrm{dom}(A)} = A$. Now let $y \in \mathrm{dom}(B)$ and $\lambda > \omega$. Define $x := (\lambda\mathbb{1} - A)^{-1}(\lambda y - By)$. Then $x \in \mathrm{dom}(A) \subset \mathrm{dom}(B)$ and $\lambda x - Bx = \lambda x - Ax = \lambda y - By$. Since $\lambda\mathbb{1} - B : \mathrm{dom}(B) \to X$ is injective by Lemma 7.2.6, this implies $y = x \in \mathrm{dom}(A)$. Thus $\mathrm{dom}(B) \subset \mathrm{dom}(A)$ and so $\mathrm{dom}(B) = \mathrm{dom}(A)$. This proves Step 5 and Theorem 7.2.5. □

Corollary 7.2.8. *Let X be a complex Banach space and let*

$$A : \mathrm{dom}(A) \to X$$

be a complex linear operator with a dense domain $\mathrm{dom}(A) \subset X$. *Fix two real numbers $M \geq 1$ and ω. Then the following are equivalent.*

(i) *The operator A is the infinitesimal generator of a strongly continuous semigroup $S : [0, \infty) \to \mathcal{L}^c(X)$ that satisfies the estimate* (7.2.3).

(ii) *For every real number $\lambda > \omega$ the operator $\lambda 1\!\!1 - A : \mathrm{dom}(A) \to X$ is bijective and satisfies the estimate* (7.2.4).

(iii) *For every $\lambda \in \mathbb{C}$ with $\mathrm{Re}\lambda > \omega$ the operator $\lambda 1\!\!1 - A : \mathrm{dom}(A) \to X$ is bijective and satisfies the estimate*

$$\|(\lambda 1\!\!1 - A)^{-k}\| \leq \frac{M}{(\mathrm{Re}\lambda - \omega)^k} \qquad \text{for all } k \in \mathbb{N}. \tag{7.2.9}$$

Proof. That (i) implies (iii) follows from Lemma 7.2.6 by the same argument that was used in the proof of Theorem 7.2.5. That (iii) implies (ii) is obvious and that (ii) implies (i) follows from Theorem 7.2.5 and the fact that the operators $S_\lambda(t)$ in the proof of Theorem 7.2.5 are complex linear whenever A is complex linear. This proves Corollary 7.2.8. □

7.2.3. Contraction Semigroups. The archetypal example of a contraction semigroup is the heat flow in Example 7.1.6. Here is the general definition.

Definition 7.2.9 (Contraction Semigroup). Let X be a real Banach space. A **contraction semigroup on** X is a strongly continuous semigroup $S : [0, \infty) \to \mathcal{L}(X)$ that satisfies the inequality

$$\|S(t)\| \leq 1 \tag{7.2.10}$$

for all $t \geq 0$.

Definition 7.2.10 (Dissipative Operator). Let X be a complex Banach space. A complex linear operator $A : \mathrm{dom}(A) \to X$ with a dense domain $\mathrm{dom}(A) \subset X$ is called **dissipative** if, for every $x \in \mathrm{dom}(A)$, there exists an element $x^* \in X^*$ such that

$$\|x^*\|^2 = \|x\|^2 = \langle x^*, x \rangle, \qquad \mathrm{Re}\langle x^*, Ax \rangle \leq 0. \tag{7.2.11}$$

When $X = H$ is a complex Hilbert space, a linear operator $A : \mathrm{dom}(A) \to H$ with a dense domain $\mathrm{dom}(A) \subset H$ is dissipative if and only if

$$\mathrm{Re}\langle x, Ax \rangle \leq 0 \tag{7.2.12}$$

for all $x \in \mathrm{dom}(A)$.

The next theorem characterizes the infinitesimal generators of contraction semigroups. It was proved by Lumer–Phillips [**58**] in 1961. They also introduced the notion of a dissipative operator.

Theorem 7.2.11 (Lumer–Phillips). *Let X be a complex Banach space and let $A : \mathrm{dom}(A) \to X$ be a complex linear operator with a dense domain $\mathrm{dom}(A) \subset X$. Then the following are equivalent.*

(i) *The operator A is the infinitesimal generator of a contraction semigroup.*

(ii) *For every real number $\lambda > 0$ the operator $\lambda 1\!\!1 - A : \mathrm{dom}(A) \to X$ is bijective and satisfies the estimate*

$$\left\|(\lambda 1\!\!1 - A)^{-1}\right\| \le \frac{1}{\lambda}. \tag{7.2.13}$$

(iii) *For every $\lambda \in \mathbb{C}$ with $\mathrm{Re}\lambda > 0$ the operator $\lambda 1\!\!1 - A : \mathrm{dom}(A) \to X$ is bijective and satisfies the estimate*

$$\left\|(\lambda 1\!\!1 - A)^{-1}\right\| \le \frac{1}{\mathrm{Re}\lambda}. \tag{7.2.14}$$

(iv) *The operator A is dissipative and there exists a $\lambda > 0$ such that the operator $\lambda 1\!\!1 - A : \mathrm{dom}(A) \to X$ has a dense image.*

Proof. The equivalence of (i), (ii), and (iii) follows from Corollary 7.2.8 with $M = 1$ and $\omega = 0$. We prove the remaining implications in three steps.

Step 1. *If A is dissipative, then*

$$\|\lambda x - Ax\| \ge \mathrm{Re}\lambda\, \|x\| \tag{7.2.15}$$

for all $x \in \mathrm{dom}(A)$ and all $\lambda \in \mathbb{C}$ with $\mathrm{Re}\lambda > 0$.

Let $x \in \mathrm{dom}(A)$ and $\lambda \in \mathbb{C}$ such that $\mathrm{Re}\lambda > 0$. Since A is dissipative, there exists an element $x^* \in X^*$ such that (7.2.11) holds. This implies

$$\begin{aligned}\|x\|\,\|\lambda x - Ax\| &= \|x^*\|\,\|\lambda x - Ax\| \\ &\ge \mathrm{Re}\langle x^*, \lambda x - Ax\rangle \\ &= \mathrm{Re}\lambda\langle x^*, x\rangle - \mathrm{Re}\langle x^*, Ax\rangle \\ &\ge \mathrm{Re}\lambda\, \|x\|^2.\end{aligned}$$

Hence

$$\|\lambda x - Ax\| \ge \mathrm{Re}\lambda\, \|x\|$$

and this proves Step 1.

Step 2. *We prove that (iv) implies (iii).*

Assume A satisfies (iv) and define the set

$$\Omega = \{\lambda \in \mathbb{C} \mid \mathrm{Re}\lambda > 0 \text{ and } \lambda\mathbb{1} - A \text{ has a dense image}\}.$$

This set is nonempty by (iv). Moreover, it follows from Step 1 that the operator $\lambda\mathbb{1} - A : \mathrm{dom}(A) \to X$ is injective and has a closed image for every $\lambda \in \mathbb{C}$ with $\mathrm{Re}\lambda > 0$. Hence $\Omega \subset \rho(A)$ and

$$\left\|(\lambda\mathbb{1} - A)^{-1}\right\| \le \frac{1}{\mathrm{Re}\lambda} \qquad \text{for all } \lambda \in \Omega \subset \rho(A). \tag{7.2.16}$$

If $\lambda \in \Omega$ and $|\mu - \lambda| < \mathrm{Re}\lambda$, then $\mathrm{Re}\mu > 0$ and $|\mu - \lambda|\|(\lambda\mathbb{1} - A)^{-1}\| < 1$, hence $\mu \in \rho(A)$ by Lemma 6.1.10, and hence $\mu \in \Omega$. Thus

$$\lambda \in \Omega \text{ and } |\mu - \lambda| < \mathrm{Re}\lambda \qquad \Longrightarrow \qquad \mu \in \Omega. \tag{7.2.17}$$

Fix an element $\lambda \in \Omega$. Then it follows from (7.2.17) that

$$\{\mu \in \mathbb{C} \mid \mathrm{Im}\mu = \mathrm{Im}\lambda,\ 0 < \mathrm{Re}\mu < 2\mathrm{Re}\lambda\} \subset \Omega.$$

Thus an induction argument shows that

$$\{\mu \in \mathbb{C} \mid \mathrm{Im}\mu = \mathrm{Im}\lambda,\ \mathrm{Re}\mu > 0\} \subset \Omega.$$

Hence it follows from (7.2.17) that $B_{\mathrm{Re}\mu}(\mu) \subset \Omega$ for every $\mu \in \mathbb{C}$ such that $\mathrm{Im}\mu = \mathrm{Im}\lambda$ and $\mathrm{Re}\mu > 0$. The union of these open discs is the entire positive half-plane in $\mathbb{C}$. Thus $\{z \in \mathbb{C} \mid \mathrm{Re}z > 0\} = \Omega \subset \rho(A)$ and hence it follows from (7.2.16) that A satisfies (iii). This proves Step 2.

Step 3. *We prove that (i) implies (iv).*

Assume that $A : \mathrm{dom}(A) \to X$ is the infinitesimal generator of a contraction semigroup $S : [0, \infty) \to \mathcal{L}^c(X)$. Let $x \in \mathrm{dom}(A)$. By the Hahn–Banach Theorem (Corollary 2.3.23) there exists an element $x^* \in X^*$ such that

$$\|x^*\|^2 = \|x\|^2 = \langle x^*, x\rangle.$$

Since S is a contraction semigroup this implies

$$\mathrm{Re}\langle x^*, S(h)x - x\rangle \le \|x^*\|\|S(h)x\| - \|x\|^2 \le 0$$

for all $h > 0$ and hence

$$\mathrm{Re}\langle x^*, Ax\rangle = \lim_{h\to 0} \frac{\mathrm{Re}\langle x^*, S(h)x - x\rangle}{h} \le 0.$$

This proves Step 3 and Theorem 7.2.11. □

7.3. The Dual Semigroup

When $S : [0, \infty) \to \mathcal{L}(X)$ is a strongly continuous semigroup on a real Banach space X the dual operators define a semigroup

$$S^* : [0, \infty) \to \mathcal{L}(X^*),$$

called the **dual semigroup**. One might expect that the dual semigroup is again strongly continuous, however, an elementary example shows that this need not always be the case (see Example 7.3.3 below). The failure of strong continuity of the dual semigroup is related to the fact that the Banach space X in Example 7.3.3 is not reflexive. On a reflexive Banach space it turns out that the dual semigroup is always strongly continuous and this is the content of Corollary 7.3.2 below, which will be derived as a consequence of the main theorem about the dual semigroup. The other subsections deal with self-adjoint semigroups and with unitary groups.

7.3.1. The Dual Semigroup and its Infinitesimal Generator. The following theorem is the main result of the present section. It was proved in 1955 by R. S. Phillips [**69**].

Theorem 7.3.1 (Phillips). *Let $S : [0, \infty) \to \mathcal{L}(X)$ be a strongly continuous semigroup on a real Banach space X and let $A : \mathrm{dom}(A) \to X$ be its infinitesimal generator. Denote by*

$$[0, \infty) \to \mathcal{L}(X^*) : t \mapsto S^*(t) := S(t)^*$$

the dual semigroup and by

$$E := \left\{ x^* \in X^* \,\middle|\, \begin{array}{l} \textit{there exists a sequence } x_i^* \in \mathrm{dom}(A^*) \\ \textit{such that } \lim_{i\to\infty} \|x_i^* - x^*\| = 0 \end{array} \right\} \tag{7.3.1}$$

the strong closure of the domain of the dual operator $A^* : \mathrm{dom}(A^*) \to X^*$. Then the following hold.

(i) *Let $x^* \in X^*$. Then $x^* \in E$ if and only if $\lim_{t\to 0} \|S^*(t)x^* - x^*\| = 0$.*

(ii) *The closed subspace $E \subset X^*$ is invariant under the operator $S^*(t)$ for every $t \geq 0$ and the map $T : [0, \infty) \to \mathcal{L}(E)$, defined by*

$$T(t) := S^*(t)|_E \qquad \textit{for } t \geq 0,$$

is a strongly continuous semigroup.

(iii) *The infinitesimal generator of the strongly continuous semigroup T in part (ii) is the operator $B : \mathrm{dom}(B) \to E$ with*

$$\mathrm{dom}(B) = \{x^* \in \mathrm{dom}(A^*) \mid A^*x^* \in E\}$$

and $Bx^* = A^*x^*$ for $x^* \in \mathrm{dom}(B)$.

Proof. It follows directly from Lemma 4.1.3 that S^* is a one-parameter semigroup. The remaining assertions are proved in eight steps.

Step 1. *Let $x^* \in X^*$ and $h > 0$ and define the element $x_h^* \in X^*$ by*

$$\langle x_h^*, x\rangle = \frac{1}{h}\int_0^h \langle x^*, S(t)x\rangle\, dt \tag{7.3.2}$$

for $x \in X$. Then $x_h^* \in \mathrm{dom}(A^*)$ and $A^*x_h^* = h^{-1}(S^*(h)x^* - x^*)$.

Let $M := \sup_{0\le t\le h}\|S(t)\|$. The functional $X \to \mathbb{R} : x \mapsto \frac{1}{h}\int_0^h \langle x^*, S(t)x\rangle dt$ is linear and satisfies the inequality

$$\begin{aligned}\left|\frac{1}{h}\int_0^h \langle x^*, S(t)x\rangle\, dt\right| &\le \frac{1}{h}\int_0^h |\langle x^*, S(t)x\rangle|\, dt\\ &\le \frac{1}{h}\int_0^h \|x^*\|\,\|S(t)x\|\, dt\\ &\le M\,\|x^*\|\,\|x\|\end{aligned}$$

for all $x \in X$. Hence (7.3.2) defines an element $x_h^* \in X^*$. For $x \in \mathrm{dom}(A)$ this element satisfies the equation

$$\begin{aligned}\langle x_h^*, Ax\rangle &= \left\langle x^*, \int_0^h \frac{S(t)Ax}{h} dt\right\rangle\\ &= \left\langle x^*, \frac{S(h)x - x}{h}\right\rangle\\ &= \left\langle \frac{S^*(h)x^* - x^*}{h}, x\right\rangle.\end{aligned}$$

Here the second step follows from Lemma 7.1.13. This implies $x_h^* \in \mathrm{dom}(A^*)$ and $A^*x_h^* = h^{-1}(S^*(h)x^* - x^*)$. This proves Step 1.

Step 2. *Let $x^* \in \mathrm{dom}(A^*)$ and $t > 0$. Then $S^*(t)x^* \in \mathrm{dom}(A^*)$ and*

$$A^*S^*(t)x^* = S^*(t)A^*x^*.$$

If $x\in\mathrm{dom}(A)$, then $S(t)x\in\mathrm{dom}(A)$ and $S(t)Ax=AS(t)x$ by Lemma 7.1.13, and hence $\langle S^*(t)A^*x^*, x\rangle = \langle A^*x^*, S(t)x\rangle = \langle x^*, AS(t)x\rangle = \langle S^*(t)x^*, Ax\rangle$. By definition of the dual operator, this implies that $S^*(t)x^* \in \mathrm{dom}(A^*)$ and $A^*S^*(t)x^* = S^*(t)A^*x^*$. This proves Step 2.

Step 3. *Let $x^* \in E$ and $t \ge 0$. Then $S^*(t)x^* \in E$.*

Choose a sequence $x_i^* \in \mathrm{dom}(A^*)$ such that $\lim_{i\to\infty}\|x_i^* - x^*\| = 0$. Then it follows from Step 2 that $S^*(t)x_i^* \in \mathrm{dom}(A^*)$. Since $S^*(t) : X^* \to X^*$ is a bounded linear operator, we also have $\lim_{i\to\infty}\|S^*(t)x_i^* - S^*(t)x^*\| = 0$, and hence $S^*(t)x^* \in E$. This proves Step 3.

Step 4. *Let* $x^* \in \operatorname{dom}(A^*)$ *and* $x \in X$. *Then*

$$\langle S^*(t)x^* - x^*, x\rangle = \int_0^t \langle S^*(s)A^*x^*, x\rangle\, ds.$$

By Example 7.1.15 we have

$$\int_0^t S(s)x\, ds \in \operatorname{dom}(A), \qquad A\int_0^t S(s)x\, ds = S(t)x - x,$$

and hence

$$\begin{aligned}\langle S^*(t)x^* - x^*, x\rangle &= \langle x^*, S(t)x - x\rangle \\ &= \langle x^*, A\int_0^t S(s)x\, ds\rangle \\ &= \langle A^*x^*, \int_0^t S(s)x\, ds\rangle \\ &= \int_0^t \langle A^*x^*, S(s)x\rangle\, ds \\ &= \int_0^t \langle S^*(s)A^*x^*, x\rangle\, ds.\end{aligned}$$

Here the fourth equality follows from Lemma 5.1.8. This proves Step 4.

Step 5. *If* $x^* \in E$, *then* $\lim_{t\to 0}\|S^*(t)x^* - x^*\| = 0$.

Define $M := \sup_{0\le t\le 1}\|S(t)\|$ and let $x^* \in \operatorname{dom}(A^*)$. Then, by Step 4,

$$\begin{aligned}\langle S^*(t)x^* - x^*, x\rangle &= \int_0^t \langle A^*x^*, S(s)x\rangle\, ds \\ &\le \|A^*x^*\| \int_0^t \|S(s)x\|\, ds \\ &\le tM\,\|A^*x^*\|\,\|x\|\end{aligned}$$

for $0 \le t \le 1$. This implies

$$\|S^*(t)x^* - x^*\| = \sup_{x\in X\setminus\{0\}} \frac{\langle S^*(t)x^* - x^*, x\rangle}{\|x\|} \le tM\,\|A^*x^*\|$$

for $0 \le t \le 1$ and so $\lim_{t\to 0}\|S^*(t)x^* - x^*\| = 0$. Since $\operatorname{dom}(A^*)$ is dense in E and $\|S^*(t)\| = \|S(t)\| \le M$ for $0 \le t \le 1$, it follows from the Banach–Steinhaus Theorem 2.1.5 that

$$\lim_{t\to 0}\|S^*(t)x^* - x^*\| = 0 \qquad \text{for all } x^* \in E.$$

This proves Step 5.

Step 6. *Let $x^* \in X^*$ such that* $\lim_{t\to 0}\|S^*(t)x^* - x^*\| = 0$. *Then $x^* \in E$.*

For $h > 0$ let $x_h^* \in X^*$ be as in Step 1. Then $x_h^* \in \mathrm{dom}(A^*)$ and

$$\langle x_h^* - x^*, x\rangle = \frac{1}{h}\int_0^h \langle x^*, S(t)x - x\rangle\, dt.$$

Now fix a constant $\varepsilon > 0$ and choose $\delta > 0$ such that

$$0 \le t < \delta \qquad \Longrightarrow \qquad \|S^*(t)x^* - x^*\| < \varepsilon.$$

Let $0 < h < \delta$. Then

$$\begin{aligned}\frac{\langle x^*, S(t)x - x\rangle}{\|x\|} &= \frac{\langle S^*(t)x^* - x^*, x\rangle}{\|x\|} \\ &\le \|S^*(t)x^* - x^*\| \\ &\le \varepsilon\end{aligned}$$

for $0 \le t \le h$ and $x \in X \setminus \{0\}$, and hence

$$\frac{\langle x_h^* - x^*, x\rangle}{\|x\|} = \frac{1}{h}\int_0^h \frac{\langle x^*, S(t)x - x\rangle}{\|x\|}\, dt \le \varepsilon.$$

Take the supremum over all $x \in X \setminus \{0\}$ to obtain the inequality

$$\|x_h^* - x^*\| = \sup_{x\in X\setminus\{0\}} \frac{\langle x_h^* - x^*, x\rangle}{\|x\|} \le \varepsilon$$

for $0 < h < \delta$. Thus we have proved that

$$\lim_{h\to 0}\|x_h^* - x^*\| = 0,$$

and hence $x^* \in E$. This proves Step 6.

Step 7. *Let $x^* \in \mathrm{dom}(A^*)$ such that $y^* := A^*x^* \in E$. Then*

$$\lim_{t\to 0}\left\|\frac{S^*(t)x^* - x^*}{t} - y^*\right\| = 0.$$

By Step 3 and Step 5, S^* restricts to a strongly continuous semigroup on the subspace E. Thus the function $[0,\infty) \to E : t \mapsto S^*(t)y^* = S^*(t)A^*x^*$ is continuous and so

$$S^*(t)x^* - x^* = \int_0^t S^*(s)y^*\, ds$$

for all $t > 0$ by Step 4. Hence

$$\left\|\frac{S^*(t)x^* - x^*}{t} - y^*\right\| = \left\|\frac{1}{t}\int_0^t \Big(S^*(s)y^* - y^*\Big)\, ds\right\| \le \sup_{0\le s\le t}\|S^*(s)y^* - y^*\|$$

and this proves Step 7.

Step 8. *Let $x^*, y^* \in X^*$ such that $\lim_{h\to 0}\left\|\frac{S^*(t)x^*-x^*}{t} - y^*\right\| = 0$. Then*

$$x^* \in \mathrm{dom}(A^*), \qquad y^* = A^*x^* \in E.$$

It follows from the assumptions of Step 8 that $\lim_{t\to 0}\|S^*(t)x^* - x^*\| = 0$ and hence $x^* \in E$ by Step 6. This implies $t^{-1}(S^*(t)x^* - x^*) \in E$ by Step 3, and so $y^* \in E$ because E is a closed subspace of X^*. Since the function $[0,h] \to E : t \mapsto S^*(t)x^*$ is continuous by Step 3 and Step 5, the element $x_h^* \in X^*$ in Step 1 is given by

$$x_h^* = \frac{1}{h}\int_0^h S^*(t)x^*\,dt$$

and converges to x^* as h tends to zero. Moreover, by Step 1, we have that $x_h^* \in \mathrm{dom}(A^*)$ and $A^*x_h^* = h^{-1}(S^*(h)x^* - x^*)$ converges to y^* as h tends to zero. Since A^* is a closed operator, this implies $x^* \in \mathrm{dom}(A^*)$ and $A^*x^* = y^* \in E$. This proves Step 8.

Part (i) follows from Steps 5 and 6, part (ii) from Steps 3 and 5, and part (iii) from Steps 7 and 8. This proves Theorem 7.3.1. □

Corollary 7.3.2. *Let X be a real reflexive Banach space and let S be a strongly continuous semigroup on X with the infinitesimal generator A. Then the dual semigroup $S^* : [0,\infty) \to \mathcal{L}(X^*)$ is strongly continuous and its infinitesimal generator is the dual operator $A^* : \mathrm{dom}(A^*) \to X^*$.*

Proof. The domain of the dual operator A^* is weak* dense in X^* by part (iii) of Theorem 6.2.2, and so it is dense because X is reflexive. Hence the result follows from Theorem 7.3.1 with $E = X^*$. □

The shift group in the following example shows that Corollary 7.3.2 does not extend to nonreflexive Banach spaces. In Example 7.3.3 the subspace E is not invariant under A^* although it is invariant under $S^*(t)$ for all t.

Example 7.3.3. Let $X := L^1(\mathbb{R})$ and, for $t \in \mathbb{R}$, define the linear operator $S(t) : L^1(\mathbb{R}) \to L^1(\mathbb{R})$ for $t \in \mathbb{R}$ by

$$(S(t)f)(s) := f(s+t) \qquad \text{for } f \in L^1(\mathbb{R}) \text{ and } s,t \in \mathbb{R}.$$

Then $X^* \cong L^\infty(\mathbb{R})$ and under this identification the dual group is given by

$$(S^*(t)g)(s) := g(s-t) \qquad \text{for } g \in L^\infty(\mathbb{R}) \text{ and } s,t \in \mathbb{R}.$$

For a general element $g \in L^\infty(\mathbb{R})$ the function $\mathbb{R} \to L^\infty(\mathbb{R}) : t \mapsto S^*(t)g$ is weak* continuous but not continuous. In this example the domain of A^* is the space of bounded Lipschitz continuous functions on $\mathbb{R}$. This space is weak* dense in $L^\infty(\mathbb{R})$ but not dense. Its closure is the space $E \subset L^\infty(\mathbb{R})$ of bounded uniformly continuous functions on $\mathbb{R}$.

7.3.2. Self-Adjoint Semigroups. The next theorem characterizes the infinitesimal generators of self-adjoint semigroups.

Theorem 7.3.4 (Self-Adjoint Semigroups). *Let H be a real Hilbert space and let $A : \mathrm{dom}(A) \to H$ be a linear operator with a dense domain $\mathrm{dom}(A) \subset H$. Then the following are equivalent.*

(i) *The operator A is the infinitesimal generator of a strongly continuous semigroup $S : [0, \infty) \to \mathcal{L}(H)$ such that $S(t) = S(t)^*$ for all $t \geq 0$.*

(ii) *The operator A is self-adjoint and*

$$\sup_{x \in \mathrm{dom}(A) \setminus \{0\}} \frac{\langle x, Ax \rangle}{\|x\|^2} < \infty.$$

If these equivalent conditions are satisfied, then

$$\frac{\log\|S(t)\|}{t} = \sup_{\substack{x \in \mathrm{dom}(A) \\ x \neq 0}} \frac{\langle x, Ax \rangle}{\|x\|^2} \tag{7.3.3}$$

for all $t > 0$.

Proof. We prove that (i) implies (ii) and

$$\sup_{\substack{x \in \mathrm{dom}(A) \\ x \neq 0}} \frac{\langle x, Ax \rangle}{\|x\|^2} \leq \frac{\log\|S(t)\|}{t} = \lim_{s \to \infty} \frac{\log\|S(s)\|}{s} \qquad \text{for all } t > 0. \tag{7.3.4}$$

For Hilbert spaces Theorem 7.3.1 asserts that the adjoint A^* of the infinitesimal generator A of a semigroup S is the infinitesimal generator of the adjoint semigroup S^*. Since $S(t)^* = S(t)$ for all $t \geq 0$ in the case at hand, it follows that the infinitesimal generator A is self-adjoint. Moreover,

$$\|S(t)\| = \|S(t)^n\|^{1/n} = \|S(nt)\|^{1/n}$$

by part (i) of Theorem 5.3.15 and hence

$$\frac{\log\|S(t)\|}{t} = \frac{\log\|S(nt)\|}{nt} \qquad \text{for all } t > 0 \text{ and all } n \in \mathbb{N}.$$

Take the limit $n \to \infty$ and use Lemma 7.1.8 to obtain

$$\frac{\log\|S(t)\|}{t} = \omega_0 := \lim_{s \to \infty} \frac{\log\|S(s)\|}{s} \qquad \text{for all } t > 0.$$

This implies $\log\|S(t)\| = t\omega_0$ and so $\|S(t)\| = e^{t\omega_0}$ for all $t > 0$. Thus

$$\langle x, S(t)x \rangle \leq e^{t\omega_0}\|x\|^2 \qquad \text{for all } x \in H \text{ and all } t \geq 0.$$

Differentiate this inequality at $t = 0$ to obtain $\langle x, Ax \rangle \leq \omega_0\|x\|^2$ for every $x \in \mathrm{dom}(A)$. This shows that (i) implies (ii) and (7.3.4).

We prove that (ii) implies (i). Thus assume A is self-adjoint and

$$\omega := \sup_{\substack{x\in\mathrm{dom}(A)\\ x\neq 0}} \frac{\langle x, Ax\rangle}{\|x\|^2} < \infty.$$

We prove in five steps that A generates a self-adjoint semigroup.

Step 1. *If $\lambda > \omega$ and $x \in \mathrm{dom}(A)$, then $\|\lambda x - Ax\| \geq (\lambda - \omega)\|x\|$.*

Let $x \in \mathrm{dom}(A)$ and $\lambda > \omega$. Then $\langle x, Ax\rangle \leq \omega\,\|x\|^2$ and so

$$\|x\|\|\lambda x - Ax\| \geq \langle x, \lambda x - Ax\rangle \geq (\lambda - \omega)\|x\|^2.$$

This proves Step 1.

Step 2. *If $\lambda > \omega$, then $\lambda\mathbb{1} - A$ is injective and has a closed image.*

Let $\lambda > \omega$. Assume x_n is a sequence in $\mathrm{dom}(A)$ such that $y_n := \lambda x_n - Ax_n$ converges to y. Then x_n is a Cauchy sequence by Step 1 and so converges to some element $x \in H$. Hence $Ax_n = \lambda x_n - y_n$ converges to $\lambda x - y$. Since A has a closed graph by Theorem 6.2.2, this implies $x \in \mathrm{dom}(A)$ and $Ax = \lambda x - y$. Thus $y = \lambda x - Ax \in \mathrm{im}(\lambda\mathbb{1} - A)$, and so $\lambda\mathbb{1} - A$ has a closed image. That it is injective follows directly from the estimate in Step 1. This proves Step 2.

Step 3. *If $\lambda > \omega$, then $\lambda\mathbb{1} - A$ is surjective.*

Let $\lambda > \omega$ and suppose $y \in H$ is orthogonal to the image of $\lambda\mathbb{1} - A$. Then $\langle y, \lambda x\rangle = \langle y, Ax\rangle$ for all $x \in \mathrm{dom}(A)$. Hence $y \in \mathrm{dom}(A^*) = \mathrm{dom}(A)$ and $Ay = A^*y = \lambda y$. Thus $y = 0$ by Step 2. This shows that $\lambda\mathbb{1} - A$ has a dense image. Hence it is surjective by Step 2. This proves Step 3.

Step 4. *The operator A is the infinitesimal generator of a strongly continuous semigroup $S : [0,\infty) \to \mathcal{L}(H)$ such that $\|S(t)\| \leq e^{\omega t}$ for all $t \geq 0$.*

Let $\lambda > \omega$. Then $\lambda\mathbb{1} - A : \mathrm{dom}(A) \to H$ is bijective by Step 2 and Step 3 and $\|(\lambda\mathbb{1} - A)^{-1}\| \leq (\lambda - \omega)^{-1}$ by Step 1. Hence Step 4 follows from the Hille–Yosida–Phillips Theorem 7.2.5 with $M = 1$.

Step 5. *The semigroup S in Step 4 is self-adjoint and satisfies* (7.3.3).

The operator $A = A^*$ is the infinitesimal generator of S by Step 4 and of the adjoint semigroup S^* by Theorem 7.3.1. Hence Corollary 7.2.3 asserts that $S(t) = S^*(t)$ for all $t \geq 0$. This implies that A and S satisfy (7.3.4). By (7.3.4), we have $\omega \leq t^{-1}\log\|S(t)\|$ and by Step 4 we have $\|S(t)\| \leq e^{\omega t}$ and hence $t^{-1}\log\|S(t)\| \leq \omega$ for all $t > 0$. Thus equality holds in (7.3.4). This proves (7.3.3) and Theorem 7.3.4. □

7.3.3. Unitary Groups. On complex Hilbert spaces it is interesting to examine the infinitesimal generators of strongly continuous unitary groups. This is the content of Theorem 7.3.6 below which was proved in 1932 by M. H. Stone [**81**].

Definition 7.3.5. Let H be a complex Hilbert space. A strongly continuous group $S : \mathbb{R} \to \mathcal{L}^c(H)$ is called **unitary** if $\|S(t)x\| = \|x\|$ for all $t \in \mathbb{R}$ and all $x \in H$ or, equivalently,

$$S^*(t) = S(t)^{-1} = S(-t)$$

for all $t \in \mathbb{R}$, where $S^*(t) = S(t)^*$ denotes the adjoint operator of $S(t)$.

Theorem 7.3.6 (Stone). *Let H be a complex Hilbert space and suppose that $A : \mathrm{dom}(A) \to H$ is a linear operator with a dense domain $\mathrm{dom}(A) \subset H$. Then the following are equivalent.*

(i) *A is the infinitesimal generator of a unitary group.*

(ii) *The operator $\mathbf{i}A : \mathrm{dom}(A) \to H$ is self-adjoint.*

Proof. We prove that (i) implies (ii). Thus assume that A is the infinitesimal generator of a unitary group $S : \mathbb{R} \to \mathcal{L}^c(H)$. Then

$$S^*(t) = S(t)^{-1} = S(-t) \qquad \text{for all } t \in \mathbb{R}.$$

The operator $-A : \mathrm{dom}(A) \to H$ is the infinitesimal generator of the group $\mathbb{R} \to \mathcal{L}^c(H) : t \mapsto S(-t)$ by Theorem 7.2.4 and $A^* : \mathrm{dom}(A^*) \to H$ is the infinitesimal generator of the group $\mathbb{R} \to \mathcal{L}^c(H) : t \mapsto S^*(t)$ by Theorem 7.3.1. Hence

$$A^* = -A$$

and so

$$(\mathbf{i}A)^* = -\mathbf{i}A^* = \mathbf{i}A.$$

Thus $\mathbf{i}A$ is self-adjoint.

We prove that (ii) implies (i). Suppose that

$$A = \mathbf{i}B,$$

where $B : \mathrm{dom}(B) \to H$ is a complex linear self-adjoint operator. Then A has a dense domain $\mathrm{dom}(A) = \mathrm{dom}(B)$ and a closed graph. Moreover,

$$A^* = (\mathbf{i}B)^* = -\mathbf{i}B^* = -\mathbf{i}B = -A.$$

This implies

$$\mathrm{Re}\langle x, Ax\rangle = \frac{\langle x, Ax\rangle + \langle Ax, x\rangle}{2} = \frac{\langle x, (A + A^*)x\rangle}{2} = 0 \tag{7.3.5}$$

for all $x \in \mathrm{dom}(A)$.

We prove that the operator $\mathbb{1} - A : \mathrm{dom}(A) \to H$ has a dense image. Assume that $y \in H$ is orthogonal to the image of $\mathbb{1} - A$. Then

$$0 = \langle y, x - Ax \rangle = \langle y, x \rangle - \langle y, Ax \rangle \qquad \text{for all } x \in \mathrm{dom}(A).$$

Hence it follows from the definition of the adjoint operator that

$$y \in \mathrm{dom}(A^*) = \mathrm{dom}(A), \qquad y = A^*y = -Ay.$$

This implies $\|y\|^2 = -\langle y, Ay \rangle = -\langle A^*y, y \rangle = -\|y\|^2$ and so $y = 0$. Hence the operator $\mathbb{1} - A$ has a dense image by the Hahn–Banach Theorem (Corollary 2.3.25).

Since $\mathbb{1} - A$ has a dense image it follows from (7.3.5) and the Lumer–Phillips Theorem 7.2.11 that A is the infinitesimal generator of a contraction semigroup $S : [0, \infty) \to \mathcal{L}^c(H)$. The adjoint semigroup $S^* : [0, \infty) \to \mathcal{L}^c(H)$ is also a contraction semigroup and is generated by the operator A^* by Theorem 7.3.1. Hence $-A = A^*$ is the infinitesimal generator of the semigroup S^* and so S extends to a strongly continuous group $S : \mathbb{R} \to \mathcal{L}^c(H)$ by Theorem 7.2.4. Since S^* is the group generated by $-A = A^*$ it follows that $S(t)^{-1} = S(-t) = S^*(t)$ for all $t \in \mathbb{R}$ and this proves Theorem 7.3.6. $\square$

Example 7.3.7 (Shift Group). Consider the Hilbert space

$$H := L^2(\mathbb{R}, \mathbb{C})$$

and define the operator $A : \mathrm{dom}(A) \to H$ by

$$\begin{aligned} \mathrm{dom}(A) &:= W^{1,2}(\mathbb{R}, \mathbb{C}) \\ &:= \left\{ f : \mathbb{R} \to \mathbb{C} \;\middle|\; \begin{array}{l} f \text{ is absolutely continuous} \\ \text{and } f, \frac{df}{ds} \in L^2(\mathbb{R}, \mathbb{C}) \end{array} \right\}, \\ Af &:= \frac{df}{ds} \qquad \text{for } f \in W^{1,2}(\mathbb{R}, \mathbb{C}). \end{aligned} \tag{7.3.6}$$

Here s is the variable in $\mathbb{R}$. Recall that an absolutely continuous function is almost everywhere differentiable, that its derivative is locally integrable, and that it can be written as the integral of its derivative, i.e. the fundamental theorem of calculus holds in this setting (see [**75**, Thm. 6.19]). The operator

$$\mathbf{i}A = \mathbf{i}\frac{d}{ds} : W^{1,2}(\mathbb{R}, \mathbb{C}) \to L^2(\mathbb{R}, \mathbb{C})$$

is self-adjoint and hence A generates a unitary group $U : \mathbb{R} \to \mathcal{L}^c(L^2(\mathbb{R}, \mathbb{C}))$. This group is in fact the shift group in Example 7.1.4 given by

$$(U(t)f)(s) = f(s + t) \qquad \text{for } f \in L^2(\mathbb{R}, \mathbb{C}) \text{ and } s, t \in \mathbb{R}.$$

(See also Example 7.3.3 and Exercise 7.7.3.) **Exercise:** Verify the details.

Example 7.3.8 (Schrödinger Equation). (i) Define the unbounded linear operator A on the Hilbert space $H := L^2(\mathbb{R}, \mathbb{C})$ by

$$
\begin{aligned}
\operatorname{dom}(A) &:= W^{2,2}(\mathbb{R}, \mathbb{C}) \\
&:= \left\{ f : \mathbb{R} \to \mathbb{C} \;\middle|\; \begin{array}{l} f \text{ is absolutely continuous and} \\ \frac{df}{dx} \text{ is absolutely continuous and} \\ \int_{-\infty}^{\infty} (|f|^2 + |\frac{df}{dx}|^2 + |\frac{d^2f}{dx^2}|^2)\, dx < \infty \end{array} \right\}, \qquad (7.3.7) \\
Af &:= \mathbf{i}\hbar \frac{d^2 f}{dx^2} \qquad \text{for } f \in W^{2,2}(\mathbb{R}, \mathbb{C}).
\end{aligned}
$$

(See Example 6.1.7.) Here $\hbar$ is a positive real number (**Planck's constant**) and x is the variable in $\mathbb{R}$. The operator

$$\mathbf{i}A = -\hbar \frac{d^2}{dx^2} : W^{2,2}(\mathbb{R}, \mathbb{C}) \to L^2(\mathbb{R}, \mathbb{C})$$

is self-adjoint and hence A generates a unitary group $U : \mathbb{R} \to \mathcal{L}^c(L^2(\mathbb{R}, \mathbb{C}))$. If $f : \mathbb{R} \to \mathbb{C}$ is a smooth function with compact support and $u : \mathbb{R}^2 \to \mathbb{C}$ is defined by $u(t,x) := (U(t)f)(x)$, then u satisfies the **Schrödinger equation**

$$\mathbf{i}\hbar \frac{\partial u}{\partial t} = -\hbar^2 \frac{\partial^2 u}{\partial x^2} \qquad (7.3.8)$$

with the initial condition $u(0, \cdot) = f$. **Exercise:** Prove that the operator $\mathbf{i}A$ is self-adjoint.

(ii) Another variant of the Schrödinger equation is associated to the operator $A : \operatorname{dom}(A) \to L^2(\mathbb{R}, \mathbb{C})$, defined by

(7.3.9)

$$
\begin{aligned}
\operatorname{dom}(A) &:= \left\{ f : \mathbb{R} \to \mathbb{C} \;\middle|\; \begin{array}{l} f \text{ is absolutely continuous and} \\ \frac{df}{dx} \text{ is absolutely continuous and} \\ \int_{-\infty}^{\infty} (|f|^2 + |-\hbar^2 \frac{d^2f}{dx^2} + x^2 f|^2)\, dx < \infty \end{array} \right\}, \\
(Af)(x) &:= \mathbf{i}\hbar \frac{d^2 f}{dx^2}(x) + \frac{x^2}{\mathbf{i}\hbar} f(x) \quad \text{for } f \in W^{2,2}(\mathbb{R}, \mathbb{C}) \text{ and } x \in \mathbb{R}.
\end{aligned}
$$

The operator $\mathbf{i}A$ is again self-adjoint and hence the operator A generates a unitary group $U : \mathbb{R} \to \mathcal{L}^c(L^2(\mathbb{R}, \mathbb{C}))$. If $f : \mathbb{R} \to \mathbb{C}$ is a smooth function with compact support and $u : \mathbb{R}^2 \to \mathbb{C}$ is defined by $u(t,x) := (U(t)f)(x)$, then u satisfies the **Schrödinger equation with quadratic potential**

$$\mathbf{i}\hbar \frac{\partial u}{\partial t}(t,x) = -\hbar^2 \frac{\partial^2 u}{\partial x^2}(t,x) + x^2 u(t,x) \qquad (7.3.10)$$

with the initial condition $u(0, \cdot) = f$. **Exercise:** Prove that the operator $\mathbf{i}A$ is self-adjoint.

Corollary 7.3.9 (Groups of Isometries). *Let H be a real Hilbert space and suppose that $A : \mathrm{dom}(A) \to H$ is a linear operator with a dense domain $\mathrm{dom}(A) \subset H$. Then the following are equivalent.*

(i) *A is the infinitesimal generator of a group of isometries.*

(ii) *If $\lambda \in \mathbb{R} \setminus \{0\}$, then $\lambda\mathbb{1} - A$ is bijective and $\left\|(\lambda\mathbb{1} - A)^{-1}\right\| \le |\lambda|^{-1}$.*

(iii) $\mathrm{dom}(A^*) = \mathrm{dom}(A)$ *and* $A^*x + Ax = 0$ *for all* $x \in \mathrm{dom}(A)$.

Proof. By Theorem 7.2.4, a map $S : \mathbb{R} \to \mathcal{L}(H)$ is a strongly continuous group of isometries if and only if both $[0, \infty) \to \mathcal{L}(H) : t \mapsto S(t)$ and $[0, \infty) \to \mathcal{L}(H) : t \mapsto S(-t)$ are contraction semigroups. Hence the equivalence of (i) and (ii) follows from the Lumer–Phillips Theorem 7.2.11. The equivalence of (i) and (iii) follows from Theorem 7.3.6 for the complexified operator $A^c : \mathrm{dom}(A^c) := \mathrm{dom}(A)^c \to H^c$. □

Example 7.3.10 (Shift Group). (i) The formula $(L(t)f)(s) := f(s + t)$ for $s, t \in \mathbb{R}$ and $f \in H := L^2(\mathbb{R})$ defines a shift group $L : \mathbb{R} \to \mathcal{L}(H)$ of isometries. Its infinitesimal generator $A : \mathrm{dom}(A) = W^{1,2}(\mathbb{R}) \to L^2(\mathbb{R})$ is given by $Af = f'$ for $f \in W^{1,2}(\mathbb{R})$ and satisfies $A^* = -A$. (See Example 7.1.4.)

(ii) The formulas $(R(t)f)(s) := f(s - t)$ for $s \ge t \ge 0$ and $(R(t)f)(s) := 0$ for $t > s \ge 0$ and $f \in H := L^2([0, \infty))$ define a semigroup $R : [0, \infty) \to \mathcal{L}(H)$ of isometric embeddings. The infinitesimal generator $B : \mathrm{dom}(B) \to H$ has the domain $\mathrm{dom}(B) = W_0^{1,2}([0, \infty)) := \{f \in W^{1,2}([0, \infty)) \,|\, f(0) = 0\}$ and is given by $Bf = -f'$. Its adjoint has the domain $\mathrm{dom}(B^*) = W^{1,2}([0, \infty))$ and satisfies $Bf + B^*f = 0$ for $f \in \mathrm{dom}(B) \subsetneq \mathrm{dom}(B^*)$.

Example 7.3.11 (Wave Equation). (i) The group $\mathscr{S} : \mathbb{R} \to \mathcal{L}(\mathscr{H})$ on the Hilbert space $\mathscr{H} = L^2(\mathbb{R}, \mathbb{R}^2)$, given by (7.1.15) in Example 7.1.7, consists of isometries and has the infinitesimal generator $\mathscr{A} = -\mathscr{A}^*$ on $\mathscr{H}$, given by $\mathrm{dom}(\mathscr{A}) = W^{1,2}(\mathbb{R}, \mathbb{R}^2)$ and $\mathscr{A}(f, g) = (g', f')$.

(ii) Fix real numbers $a < b$ and consider the wave equation

$$\frac{\partial^2 u}{\partial t^2} = \frac{\partial^2 u}{\partial x^2}, \qquad u(t, a) = u(t, b) = 0, \tag{7.3.11}$$

on the compact interval $I := [a, b]$. Equation (7.3.11) gives rise to a strongly continuous group of isometries on the Hilbert space $H := W_0^{1,2}(I) \times L^2(I)$, where $W_0^{1,2}(I) := \left\{f \in W^{1,2}(I) \,|\, f(a) = f(b) = 0\right\}$ and

$$\|(f, g)\|_H := \sqrt{\int_a^b \left(|f'(x)|^2 + |g(x)|^2\right) dx}$$

for $f \in W_0^{1,2}(I)$ and $g \in L^2(I)$. Its infinitesimal generator is the operator

$$\mathrm{dom}(A) = (W^{2,2}(I) \cap W_0^{1,2}(I)) \times W_0^{1,2}(I), \qquad A(f, g) = (g, f'').$$

7.4. Analytic Semigroups

7.4.1. Properties of Analytic Semigroups. For a strongly continuous semigroup

$$S : [0, \infty) \to \mathcal{L}^c(X)$$

on a complex Banach space X an important question is whether the function $t \mapsto S(t)x$ extends to a holomorphic function on a neighborhood of the positive real axis for all $x \in X$. A necessary condition for the existence of such an extension is *instant regularity*, i.e. the image of the operator $S(t)$ must be contained in the domain of the infinitesimal generator for all $t > 0$. The formal definition involves the sectors

$$\begin{aligned} U_\delta &:= \left\{ z \in \mathbb{C} \setminus \{0\} \,\middle|\, |\arg(z)| < \delta \right\} \\ &= \left\{ re^{\mathbf{i}\theta} \,\middle|\, r > 0 \text{ and } -\delta < \theta < \delta \right\} \end{aligned} \tag{7.4.1}$$

for $0 < \delta < \pi/2$.

Definition 7.4.1 (Analytic Semigroups). Let X be a complex Banach space. A strongly continuous semigroup $S : [0, \infty) \to \mathcal{L}^c(X)$ is called **analytic** if there exist a number $0 < \delta < \pi/2$ and an extension of S to $\overline{U}_\delta$, still denoted by

$$S : \overline{U}_\delta \to \mathcal{L}^c(X),$$

such that, for every $x \in X$, the function

$$\overline{U}_\delta \to X : z \mapsto S(z)x$$

is continuous and its restriction to the interior $U_\delta \subset \mathbb{C}$ is holomorphic.

The next theorem summarizes the basic properties of analytic semigroups. In particular, it shows that the map $S_\theta : [0, \infty) \to \mathcal{L}^c(X)$, defined by

$$S_\theta(t) := S(te^{\mathbf{i}\theta}) \qquad \text{for } t \geq 0, \tag{7.4.2}$$

is a strongly continuous semigroup for $-\delta \leq \theta \leq \delta$, and that its infinitesimal generator is the operator $A_\theta : \mathrm{dom}(A) \to X$ defined by

$$A_\theta x := e^{\mathbf{i}\theta} A x \qquad \text{for } x \in \mathrm{dom}(A). \tag{7.4.3}$$

It also shows that the semigroups S_θ satisfy an exponential estimate of the form $\|S_\theta(t)\| \leq M e^{\omega \cos(\theta) t}$, where the constants $\omega \in \mathbb{R}$ and $M \geq 1$ can be chosen independent of θ. Let ω_0 be the infimum of all $\omega \in \mathbb{R}$ for which such an estimate exists. Then the spectrum of A is contained in the sector

$$C_\delta := \left\{ \omega_0 + re^{\mathbf{i}\theta} \,\middle|\, r \geq 0,\ \pi/2 - \delta \leq |\theta| \leq \pi \right\} \tag{7.4.4}$$

(see Figure 7.4.1).

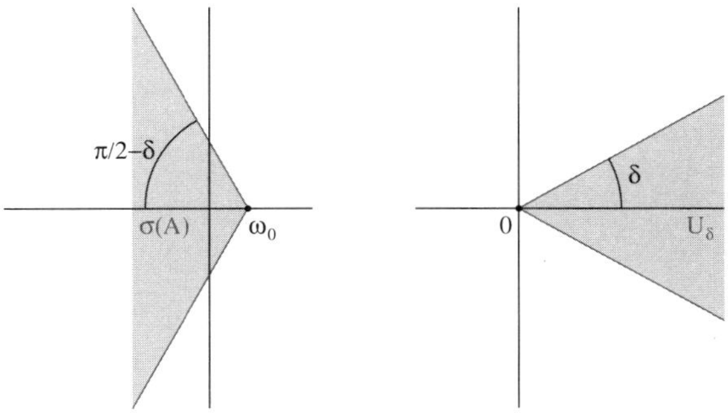

Figure 7.4.1. The spectrum of the generator of an analytic semigroup.

Theorem 7.4.2 (Analytic Semigroups). *Let X be a complex Banach space, let $0 < \delta < \pi/2$, let $S : \overline{U}_\delta \to \mathcal{L}^c(X)$ be an analytic semigroup, and let A be its infinitesimal generator. Then the following hold.*

(i) *$S(t+z) = S(t)S(z)$ for all $t, z \in \overline{U}_\delta$.*

(ii) *If $z \in U_\delta$, then $\operatorname{im}(S(z)) \subset \operatorname{dom}(A)$, $AS(z) \in \mathcal{L}^c(X)$, and*

$$\lim_{\substack{h\to 0\\ h\in\mathbb{C}\setminus\{0\}}} \left\| \frac{S(z+h)-S(z)}{h} - AS(z) \right\| = 0. \tag{7.4.5}$$

Moreover, the function $U_\delta \to \mathcal{L}^c(X) : z \mapsto AS(z)$ is holomorphic.

(iii) *If $x \in \operatorname{dom}(A)$ and $z \in \overline{U}_\delta$, then*

$$S(z)x \in \operatorname{dom}(A), \qquad AS(z)x = S(z)Ax.$$

(iv) *If $z \in U_\delta$, then $\operatorname{im}(S(z)) \subset \operatorname{dom}(A^\infty)$.*

(v) *For each*

$$\omega > \omega_0 := \inf_{r>0} r^{-1} \sup\{\log\|S(z)\| \mid z \in \overline{U}_\delta,\ \operatorname{Re}(z) = r\} \tag{7.4.6}$$

there exists a constant $M \geq 1$ such that $\|S(z)\| \leq Me^{\omega \operatorname{Re}(z)}$ for all $z \in \overline{U}_\delta$.

(vi) *Let $x \in X$ and $z_0 \in U_\delta$. Choose $r > 0$ such that $B_r(z_0) \subset U_\delta$. Then*

$$S(z)x = \sum_{n=0}^{\infty} \frac{(z-z_0)^n}{n!} A^n S(z_0)x \qquad \text{for all } z \in B_r(z_0). \tag{7.4.7}$$

The power series in (7.4.7) converges absolutely and uniformly on every compact subset of $B_r(z_0)$.

(vii) *For $-\delta \leq \theta \leq \delta$ the map S_θ in (7.4.2) is a strongly continuous semigroup whose infinitesimal generator is the operator A_θ in (7.4.3).*

(viii) *If ω_0 is as in (v), then $\sigma(A) \subset C_\delta$ (see equation (7.4.4)).*

Proof. We prove part (i). Fix a number $t > 0$ and two elements $x \in X$ and $x^* \in X^*$. Define functions $u, v, w : U_\delta \to \mathbb{C}$ by

$$u(z) := \langle x^*, S(t+z)x \rangle,$$
$$v(z) := \langle x^*, S(z)S(t)x \rangle,$$
$$w(z) := \langle x^*, S(t)S(z)x \rangle = \langle S(t)^* x^*, S(z)x \rangle$$

for $z \in U_\delta$. By assumption these functions are holomorphic and agree on the positive real axis. Hence they agree on all of U_δ by unique continuation. This shows that

$$S(t+z) = S(z)S(t) = S(t)S(z)$$

for all $t > 0$ and all $z \in \overline{U}_\delta$. Repeat the argument with $t \in U_\delta$ to obtain

$$S(t+z) = S(t)S(z)$$

for all $t, z \in \overline{U}_\delta$. This proves part (i).

We prove part (ii). Let $x \in X$ and define $f : U_\delta \to X$ by

$$f(z) := S(z)x$$

for $z \in U_\delta$. This function is holomorphic by assumption and

$$\frac{f(z+h) - f(z)}{h} = \frac{S(h)S(z)x - S(z)x}{h} \qquad \text{for all } h > 0$$

by part (i). The difference quotient on the left converges to $f'(z)$ as h tends to zero because f is holomorphic. Hence it follows from the definition of the infinitesimal generator that

$$S(z)x \in \mathrm{dom}(A), \qquad AS(z)x = f'(z)$$

for all $z \in U_\delta$. Since f' is holomorphic by Exercise 5.1.13, and every weakly holomorphic operator valued function is holomorphic by Lemma 5.1.12, this proves part (ii).

We prove part (iii). Let $x \in \mathrm{dom}(A)$ and define $f, g : U_\delta \to X$ by

$$f(z) := S(z)Ax, \qquad g(z) := AS(z)x \qquad \text{for } z \in U_\delta.$$

Then f is holomorphic by assumption and g is holomorphic by part (ii). Moreover, the functions agree on the positive real axis by Lemma 7.1.13. Hence they agree on all of U_δ by unique continuation. This proves part (iii) for $z \in U_\delta$. Now let $z \in \overline{U}_\delta$ and choose a sequence $z_n \in U_\delta$ that converges to z. Then it follows from the strong continuity of the map $S : \overline{U}_\delta \to \mathcal{L}^c(X)$ and from what we have just proved that

$$\lim_{n\to\infty} S(z_n)x = S(z)x, \qquad \lim_{n\to\infty} AS(z_n)x = \lim_{n\to\infty} S(z_n)Ax = S(z)Ax.$$

Since A is closed, it follows that $S(z)x \in \mathrm{dom}(A)$ and $AS(z)x = S(z)Ax$. This proves part (iii).

We prove part (iv). We prove by induction on n that $S(z)x \in \mathrm{dom}(A^n)$ for all $z \in U_\delta$ and all $x \in X$. For $n = 1$ this was established in part (ii). Assume by induction that $S(z)x \in \mathrm{dom}(A^n)$ for all $z \in U_\delta$ and all $x \in X$. Fix two elements $x \in X$ and $z \in U_\delta$. Then it follows from parts (i), (ii), (iii) and the induction hypothesis that

$$AS(z)x = AS(z/2)S(z/2)x = S(z/2)AS(z/2)x \in \mathrm{dom}(A^n)$$

and hence $S(z)x \in \mathrm{dom}(A^{n+1})$. This completes the induction argument and the proof of part (iv).

We prove part (v). The function $\overline{U}_\delta \to [0, \infty) : z \mapsto \|S(z)x\|$ is bounded on every compact subset of U_δ and for every $x \in X$ by strong continuity. Hence it follows from the Uniform Boundedness Theorem 2.1.1 and the analyticity of the semigroup that, for every real number $r > 0$, there exists a constant $c \geq 1$ such that $c^{-1} \leq \|S(z)\| \leq c$ for all $z \in \overline{U}_\delta$ with $\mathrm{Re}(z) \leq r$. Define

$$\omega_0 := \inf_{r>0} \frac{\omega(r)}{r}, \qquad \omega(r) := \sup\left\{\log\|S(z)\| \,\middle|\, z \in \overline{U}_\delta,\ \mathrm{Re}(z) = r\right\}, \tag{7.4.8}$$

and define the functions $g : \overline{U}_\delta \to \mathbb{R}$ and $M : [0, \infty) \to [0, \infty)$ by

$$g(z) := \log\|S(z)\|, \qquad M(r) := \sup_{z\in\overline{U}_\delta,\, \mathrm{Re}(z)\leq r} |g(z)| \tag{7.4.9}$$

for $z \in \overline{U}_\delta$ and $r \geq 0$. Then it follows from part (i) that

$$g(t + z) \leq g(t) + g(z)$$

for all $t, z \in \overline{U}_\delta$. Fix a real number $r > 0$ and let $z \in \overline{U}_\delta \setminus \{0\}$. Then there exist an integer $k \geq 0$ and a number $0 \leq s < r$ such that $\mathrm{Re}(z) = kr + s$. Define $\zeta := \mathrm{Re}(z)^{-1}z$. Then $g(z) = g(kr\zeta + s\zeta)$ and hence

$$\frac{g(z)}{\mathrm{Re}(z)} \leq \frac{kg(r\zeta) + g(s\zeta)}{\mathrm{Re}(z)} = \frac{g(r\zeta)}{r} - \frac{sg(r\zeta)}{r\mathrm{Re}(z)} + \frac{g(s\zeta)}{\mathrm{Re}(z)} \leq \frac{\omega(r)}{r} + \frac{2M(r)}{\mathrm{Re}(z)}.$$

Now fix a constant $\omega > \omega_0$, choose $r > 0$ such that $r^{-1}\omega(r) < \omega$, and then choose $R > 0$ such that $r^{-1}\omega(r) + 2R^{-1}M(r) \leq \omega$. Then each $z \in \overline{U}_\delta$ with $|z| \geq R$ satisfies $|z|^{-1}g(z) \leq \omega$ and hence $\|S(z)\| = e^{g(z)} \leq e^{\omega\mathrm{Re}(z)}$. This proves part (v) with $M := \sup_{z\in\overline{U}_\delta,\, \mathrm{Re}(z)\leq R} e^{-\omega\mathrm{Re}(z)}\|S(z)\|$.

We prove part (vi). Let $x \in X$ and $x^* \in X^*$ and define $f : U_\delta \to \mathbb{C}$ by $f(z) := \langle x^*, S(z)x\rangle$. By parts (ii), (iii), and (iv) the derivatives of f are given by $f^{(n)}(z) = \langle x^*, A^nS(z)x\rangle$ for $n \in \mathbb{N}$ and $z \in U_\delta$. Hence part (vi) follows by carrying over the familiar result in complex analysis about the convergence of power series (e.g. [**1**, p. 179] or [**74**, Thm. 3.43]) to operator valued holomorphic functions. (See also Exercises 5.1.13 and 5.1.14.) This proves part (vi).

We prove part (vii). Fix a real number $-\delta \le \theta \le \delta$. That S_θ is strongly continuous follows directly from the definition and that it is a semigroup follows from part (i). We must prove that its infinitesimal generator is the operator $A_\theta = e^{\mathbf{i}\theta}A : \mathrm{dom}(A) \to X$ in (7.4.3). To see this, fix an element $x_0 \in \mathrm{dom}(A)$ and define the function

$$x : [0,\infty) \to X$$

by

$$x(t) := S_\theta(t)x_0 = S(te^{\mathbf{i}\theta})x_0 \qquad \text{for } t \ge 0.$$

This function is continuous by assumption and takes values in the subspace $\mathrm{dom}(A_\theta) = \mathrm{dom}(A)$ by part (ii). Moreover, it follows from part (ii) that x is differentiable and

$$\begin{aligned} \frac{d}{dt}S_\theta(t)x &= \lim_{h\to 0} \frac{S(te^{\mathbf{i}\theta} + he^{\mathbf{i}\theta})x - S(te^{\mathbf{i}\theta})x}{h} \\ &= e^{\mathbf{i}\theta}AS(te^{\mathbf{i}\theta})x \\ &= S_\theta(t)A_\theta x \end{aligned}$$

for all $t \ge 0$. Here the last equality follows from part (iii). Thus x is continuously differentiable and satisfies the differential equation $\dot{x} = A_\theta x$. Hence S_θ and A_θ satisfy condition (iii) in Lemma 7.1.17 and so A_θ is the infinitesimal generator of S_θ. This proves part (vii)

We prove part (viii). Recall the definition of the spectrum of a closed unbounded operator in (6.1.9). Let $\lambda \in \sigma(A)$ and fix a real number $-\delta \le \theta \le \delta$. Then $e^{\mathbf{i}\theta}\lambda \in \sigma(A_\theta)$. Let $\omega > \omega_0$. Then part (v) asserts that there is a constant $M \ge 1$ such that $\|S_\theta(t)\| \le Me^{\omega\cos(\theta)t}$ for all $t \ge 0$. By Theorem 7.2.5 this implies that $\mathrm{Re}(e^{\mathbf{i}\theta}\lambda) \le \omega\cos(\theta)$. Since $\omega > \omega_0$ was chosen arbitrarily, this implies $\mathrm{Re}(e^{\mathbf{i}\theta}\lambda) \le \omega_0\cos(\theta)$, i.e.

$$\mathrm{Re}(e^{\mathbf{i}\theta}(\lambda - \omega_0)) \le 0 \qquad \text{for } -\delta \le \theta \le \delta.$$

Thus $\lambda \in C_\delta$. This proves part (viii) and Theorem 7.4.2. □

Example 7.4.3. This elementary example shows that the number ω_0 in (7.4.6) may depend on the domain U_δ on which the semigroup is (chosen to be) defined. Let $\lambda \in \mathbb{C}$ and consider the analytic semigroup $S : \overline{U}_\delta \to \mathcal{L}^c(X)$ on one-dimensional complex Banach space $X = \mathbb{C}$, given by

$$S(z)x = e^{\lambda z}x$$

for $z \in \overline{U}_\delta$ and $x \in X = \mathbb{C}$. This semigroup extends to a holomorphic function on the entire complex plane, so the number $0 < \delta < \pi/2$ can be chosen arbitrarily. We have $\log\|S(z)\| = \log|e^{\lambda z}| = \mathrm{Re}(\lambda z)$ for all $z \in \overline{U}_\delta$ and hence

$$\sup\left\{\log\|S(z)\| \,\middle|\, z \in \overline{U}_\delta,\ \mathrm{Re}(z) = r\right\} = r\big(\mathrm{Re}(\lambda) + \tan(\delta)\,|\mathrm{Im}(\lambda)|\big).$$

Thus $\omega_0 = \mathrm{Re}(\lambda) + \tan(\delta)\,|\mathrm{Im}(\lambda)|$ and so $\sigma(A) = \{\lambda\} \subset C_\delta$.

7.4.2. Generators of Analytic Semigroups. The next theorem is the main result of this section. It characterizes the infinitesimal generators of analytic semigroups.

Theorem 7.4.4 (Generators of Analytic Semigroups). *Let X be a complex Banach space and let $A : \mathrm{dom}(A) \to X$ be a complex linear operator with a dense domain and a closed graph. Fix a real number ω_0. Then the following are equivalent.*

(i) *There exists a number $0 < \delta < \pi/2$ such that A generates an analytic semigroup $S : \overline{U}_\delta \to \mathcal{L}^c(X)$ that satisfies*

$$\lim_{t\to\infty} \frac{\log\|S(t)\|}{t} = \inf_{t>0} \frac{\log\|S(t)\|}{t} \le \omega_0. \tag{7.4.10}$$

(ii) *For each $\omega > \omega_0$ there exists a constant $M \ge 1$ such that*

$$\|(\lambda 1\!\!1 - A)^{-1}\| \le \frac{M}{|\lambda - \omega|} \qquad \text{for all } \lambda \in \mathbb{C} \text{ with } \mathrm{Re}\lambda > \omega. \tag{7.4.11}$$

If these equivalent conditions are satisfied, then $\mathrm{im}(S(t)) \subset \mathrm{dom}(A)$ for all $t > 0$ and, for each $\omega > \omega_0$, there exists a constant $M \ge 1$ such that

$$\|AS(t)x\| \le \frac{M}{t} e^{\omega t}\|x\| \qquad \text{for all } t > 0 \text{ and all } x \in X. \tag{7.4.12}$$

Proof. We prove that (i) implies the last assertion. Thus assume part (i). Then $\mathrm{im}(S(t)) \subset \mathrm{dom}(A)$ for all $t > 0$ by Theorem 7.4.2. Let $\omega > \omega_0$ and assume $\omega_1 := \inf_{r>0} \sup\{\frac{\log\|S(z)\|}{r} \,|\, z \in \overline{U}_\delta,\, \mathrm{Re}(z) = r\} < \omega$. (Shrink δ if necessary.) Choose $r > 0$ so small that

$$\overline{B_r(1)} \subset U_\delta, \qquad \omega_1 < \frac{\omega}{1+r}, \qquad \omega_1 < \frac{\omega}{1-r}.$$

(Note that ω might be negative.) Let $t > 0$ and define $\gamma_t : [0,1] \to U_\delta$ by $\gamma_t(s) := t + rte^{2\pi \mathbf{i} s}$ for $0 \le s \le 1$. Fix an element $x \in X$. Then $AS(t)x$ is the derivative at $z = t$ of the holomorphic function $U_\delta \to X : z \mapsto S(z)x$ by Theorem 7.4.2. Hence the Cauchy integral formula asserts that

$$AS(t)x = \frac{1}{2\pi\mathbf{i}} \int_{\gamma_t} \frac{S(z)x}{(z-t)^2}\, dz = \frac{1}{rt}\int_0^1 e^{-2\pi\mathbf{i}s} S(t + rte^{2\pi\mathbf{i}s})x\, ds.$$

Choose $M \ge 1$ such that $\|S(z)\| \le Me^{\frac{\omega\mathrm{Re}(z)}{1+r}}$ and $\|S(z)\| \le Me^{\frac{\omega\mathrm{Re}(z)}{1-r}}$ for $z \in \overline{U}_\delta$. Since $(1-r)t \le \mathrm{Re}(t + rte^{2\pi\mathbf{i}s}) \le (1+r)t$ this implies

$$\|AS(t)x\| \le \frac{1}{rt} \sup_{s\in\mathbb{R}} \left\| S(t + rte^{2\pi\mathbf{i}s})x \right\| \le \frac{M}{rt} e^{\omega t}\|x\|.$$

This shows that (i) implies (7.4.12).

We prove that (i) implies (ii). Thus assume part (i). Let $\omega > \omega_0$ and assume

$$\omega_1 := \inf_{r>0} \sup \left\{ \frac{\log\|S(z)\|}{r} \,\middle|\, z \in \overline{U}_\delta,\ \mathrm{Re}(z) = r \right\} < \omega.$$

(Shrink δ if necessary.) Then, by part (v) of Theorem 7.4.2, there exists a constant $M \geq 1$ such that

$$\|S(z)\| \leq M e^{\omega \mathrm{Re}(z)} \qquad \text{for all } z \in \overline{U}_\delta.$$

Thus the semigroup $S_{-\delta}$ in (7.4.2) satisfies the inequality

$$\|S_{-\delta}(t)\| = \|S(te^{-\mathbf{i}\delta})\| \leq M e^{\omega \cos(\delta) t}$$

for all $t \geq 0$. Since the operator $A_{-\delta} = e^{-\mathbf{i}\delta} A$ in (7.4.3) is the infinitesimal generator of $S_{-\delta}$, it follows from Corollary 7.2.8 that every complex number λ' with $\mathrm{Re}(\lambda') > \omega \cos(\delta)$ belongs to the resolvent set of $A_{-\delta}$ and satisfies

$$\|(\lambda' \mathbb{1} - e^{-\mathbf{i}\delta} A)^{-1}\| \leq \frac{M}{\mathrm{Re}(\lambda') - \omega\cos(\delta)}. \tag{7.4.13}$$

Define

$$c := \sqrt{\frac{1}{\sin^2(\delta)} + \frac{1}{\cos^2(\delta)}}. \tag{7.4.14}$$

Let $\lambda \in \mathbb{C}$ such that $\mathrm{Re}(\lambda) > \omega$ and $\mathrm{Im}\lambda \geq 0$. Define $\lambda' := e^{-\mathbf{i}\delta}\lambda$. Then

$$\mathrm{Re}(\lambda') - \omega\cos(\delta) = \cos(\delta)(\mathrm{Re}(\lambda) - \omega) + \sin(\delta)\mathrm{Im}(\lambda) > 0,$$

hence

$$\mathrm{Re}(\lambda) - \omega \leq \frac{\mathrm{Re}(\lambda') - \omega\cos(\delta)}{\cos(\delta)}, \qquad \mathrm{Im}(\lambda) \leq \frac{\mathrm{Re}(\lambda') - \omega\cos(\delta)}{\sin(\delta)},$$

and so

$$|\lambda - \omega| \leq c\left(\mathrm{Re}(\lambda') - \omega\cos(\delta)\right). \tag{7.4.15}$$

Since $\mathrm{Re}\lambda' > \omega\cos(\delta)$, the operator $\lambda\mathbb{1} - A = e^{\mathbf{i}\delta}(\lambda'\mathbb{1} - e^{-\mathbf{i}\delta}A)$ is invertible and, by (7.4.13), (7.4.14), and (7.4.15), it satisfies the estimate

$$\|(\lambda\mathbb{1} - A)^{-1}\| = \|(\lambda'\mathbb{1} - e^{-\mathbf{i}\delta}A)^{-1}\| \leq \frac{M}{\mathrm{Re}\lambda' - \omega\cos(\delta)} \leq \frac{cM}{|\lambda - \omega|}.$$

This shows that A satisfies (7.4.11) whenever $\mathrm{Im}(\lambda) \geq 0$. When $\mathrm{Im}(\lambda) \leq 0$ repeat this argument with $A_{-\delta}$ replaced by A_δ and $\lambda' := e^{\mathbf{i}\delta}\lambda$ to obtain that A satisfies (7.4.11). This shows that (i) implies (ii).

We prove that (ii) implies (i). Thus assume part (ii). We prove in eight steps that A generates an analytic semigroup satisfying (7.4.10).

Step 1. *Let $\omega > \omega_0$ and choose $M \ge 1$ such that (7.4.11) holds. Choose the real number $0 < \varepsilon_0 \le \pi/2$ such that $\sin(\varepsilon_0) = 1/M$ and define*

$$M_\varepsilon := \frac{M}{1 - M\sin(\varepsilon)} \qquad \text{for } 0 < \varepsilon < \varepsilon_0. \tag{7.4.16}$$

Then

$$\sigma(A) \subset \{\omega + re^{\mathbf{i}\theta} \,|\, r \ge 0,\, \pi/2 + \varepsilon_0 \le |\theta| \le \pi\}$$

and, if $0 < \varepsilon < \varepsilon_0$, then

$$\|(\lambda\mathbb{1} - A)^{-1}\| \le \frac{M_\varepsilon}{|\lambda - \omega|} \tag{7.4.17}$$

for all $\lambda = \omega + re^{\mathbf{i}\theta}$ with $r > 0$ and $|\theta| \le \pi/2 + \varepsilon$.

We prove first that, for all $\lambda \in \mathbb{C}$,

$$\mathrm{Re}\lambda \ge \omega,\ \lambda \ne \omega \quad \Longrightarrow \quad \|(\lambda\mathbb{1} - A)^{-1}\| \le \frac{M}{|\lambda - \omega|}. \tag{7.4.18}$$

If $\mathrm{Re}\lambda > \omega$, this holds by assumption. Thus assume $\lambda = \omega + \mathbf{i}t$ for $t \in \mathbb{R}\setminus\{0\}$ and define $\lambda_s := \omega + s + \mathbf{i}t$ for $s > 0$. Then $\|(\lambda_s\mathbb{1} - A)^{-1}\| \le M/|t|$ for all $s > 0$. With $0 < s < |t|/M$ this implies

$$|\lambda - \lambda_s|\|(\lambda_s\mathbb{1} - A)^{-1}\| \le \frac{sM}{|t|} < 1$$

and so it follows from Lemma 6.1.10 that $\lambda \in \rho(A)$ and $\|(\lambda\mathbb{1} - A)^{-1}\| \le \frac{M}{|t| - sM}$. Take the limit $s \to 0$ to obtain the estimate (7.4.18).

Now let $0 < \varepsilon < \varepsilon_0$ and let $\lambda = \omega \pm \mathbf{i}re^{\pm\mathbf{i}\theta}$ with $r > 0$ and $0 < \theta \le \varepsilon$. Consider the number $\mu := \omega \pm \mathbf{i}r/\cos(\theta)$. It satisfies $|\lambda - \mu| = r\tan(\theta)$ and

$$\|(\mu\mathbb{1} - A)^{-1}\| \le \frac{M}{|\mu - \omega|} = \frac{M\cos(\theta)}{r} \le \frac{M}{r}$$

by (7.4.18). Hence

$$|\lambda - \mu|\|(\mu\mathbb{1} - A)^{-1}\| \le \frac{M\cos(\theta)}{r}|\lambda - \mu| = M\sin(\theta) \le M\sin(\varepsilon) < 1.$$

Thus $\lambda \in \rho(A)$ and

$$(\lambda\mathbb{1} - A)^{-1} = \sum_{k=0}^{\infty}(\mu - \lambda)^k(\mu\mathbb{1} - A)^{-k-1}$$

by Lemma 6.1.10. Hence

$$\|(\lambda\mathbb{1} - A)^{-1}\| \le \frac{\|(\mu\mathbb{1} - A)^{-1}\|}{1 - |\lambda - \mu|\|(\mu\mathbb{1} - A)^{-1}\|} \le \frac{M/r}{1 - M\sin(\varepsilon)} = \frac{M_\varepsilon}{|\lambda - \omega|}.$$

Here the last step uses the equation $|\lambda - \omega| = r$. This proves Step 1.

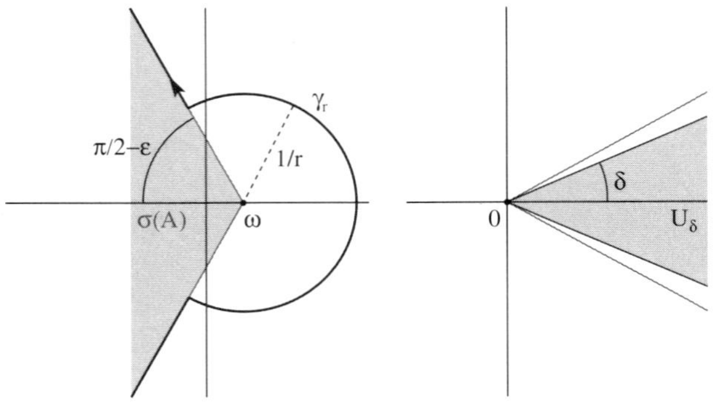

Figure 7.4.2. Integration along γ_r.

Step 2. *Let $\omega > \omega_0$ and $0 < \varepsilon < \varepsilon_0 \le \pi/2$ be as in Step 1. For $r > 0$ define the curve $\gamma_r = \gamma_{r,\varepsilon} : \mathbb{R} \to \mathbb{C}$ by*

$$\gamma_r(t) := \begin{cases} \omega + \frac{1}{r} e^{\mathbf{i}rt(\frac{\pi}{2}+\varepsilon)}, & \text{for } -1/r \le t \le 1/r, \\ \omega + \mathbf{i}te^{-\mathbf{i}\varepsilon}, & \text{for } t \le -1/r, \\ \omega + \mathbf{i}te^{\mathbf{i}\varepsilon}, & \text{for } t \ge 1/r \end{cases} \tag{7.4.19}$$

(see Figure 7.4.2). Then the formula

$$S(z) := \frac{1}{2\pi\mathbf{i}} \int_{\gamma_r} e^{z\zeta}(\zeta\mathbb{1} - A)^{-1}\, d\zeta \qquad \text{for } z \in U_\varepsilon \tag{7.4.20}$$

defines a holomorphic map $S : U_\varepsilon \to \mathcal{L}^c(X)$, which is independent of r.

Step 1 asserts that $\omega + \mathbf{i}te^{\mathbf{i}\varepsilon} \in \rho(A)$ and $\omega - \mathbf{i}te^{-\mathbf{i}\varepsilon} \in \rho(A)$ for $t > 0$ and

$$\left\| ((\omega \pm \mathbf{i}te^{\pm\mathbf{i}\varepsilon})\mathbb{1} - A)^{-1} \right\| \le \frac{M_\varepsilon}{t} \qquad \text{for all } t > 0.$$

Let $z = |z|e^{\mathbf{i}\theta} \in U_\varepsilon$ with $|\theta| < \varepsilon$. Then

$$\mathrm{Re}(z\mathbf{i}e^{\mathbf{i}\varepsilon}) = -|z|\sin(\varepsilon + \theta) < 0, \qquad \mathrm{Re}(-z\mathbf{i}e^{-\mathbf{i}\varepsilon}) = -|z|\sin(\varepsilon - \theta) < 0.$$

Hence

$$\left\| \frac{e^{\pm\mathbf{i}\varepsilon}}{2\pi} e^{z(\omega \pm \mathbf{i}te^{\pm\mathbf{i}\varepsilon})}((\omega \pm \mathbf{i}te^{\pm\mathbf{i}\varepsilon})\mathbb{1} - A)^{-1} \right\| \le \frac{M_\varepsilon e^{|z|\omega\cos(\theta)}}{2\pi} \frac{e^{-t|z|\sin(\varepsilon\pm\theta)}}{t}$$

for all $t \ge 1/r$. This shows that the integrals

$$S^{\pm}(z) := \frac{e^{\pm\mathbf{i}\varepsilon}}{2\pi} \int_{1/r}^{\infty} e^{z(\omega \pm \mathbf{i}te^{\pm\mathbf{i}\varepsilon})}((\omega \pm \mathbf{i}te^{\pm\mathbf{i}\varepsilon})\mathbb{1} - A)^{-1}\, dt$$

converge in $\mathcal{L}^c(X)$. That the map $S : U_\varepsilon \to \mathcal{L}^c(X)$ is holomorphic follows from the definition and the convergence of the integrals. That it is independent of the choice of r follows from Step 1 and the Cauchy integral formula. This proves Step 2.

Step 3. *Let ε and S be as in Step 2 and let $0 < \delta < \varepsilon$. Then there exists a constant $M_{\delta,\varepsilon} \geq 1$ such that*

$$\|S(z)\| \leq M_{\delta,\varepsilon} e^{\omega \mathrm{Re}(z)}$$

for all $z \in \overline{U}_\delta \setminus \{0\}$.

Let $z \in \overline{U}_\delta \setminus \{0\}$ and choose $r := |z|$ in (7.4.19). Then $z = re^{\mathbf{i}\theta}$ with $|\theta| \leq \delta$. Hence, by Step 2,

$$\begin{aligned}
S(z) &= \frac{1}{2\pi\mathbf{i}} \int_{\gamma_r} e^{z\zeta}(\zeta\mathbb{1} - A)^{-1}\, d\zeta \\
&= \frac{1}{2\pi\mathbf{i}} \int_{-\infty}^{\infty} e^{z\gamma_r(t)} \dot{\gamma}_r(t)(\gamma_r(t)\mathbb{1} - A)^{-1}\, dt \\
&= \frac{\pi + 2\varepsilon}{4\pi} \int_{-1/r}^{1/r} e^{z(\omega + \frac{1}{r}e^{\mathbf{i}rt(\frac{\pi}{2}+\varepsilon)})} e^{\mathbf{i}rt(\frac{\pi}{2}+\varepsilon)} \left(\left(\omega + \frac{e^{\mathbf{i}rt(\frac{\pi}{2}+\varepsilon)}}{r} \right)\mathbb{1} - A \right)^{-1} dt \\
&\quad + \frac{e^{-\mathbf{i}\varepsilon}}{2\pi} \int_{-\infty}^{-1/r} e^{z(\omega + \mathbf{i}te^{-\mathbf{i}\varepsilon})}((\omega + \mathbf{i}te^{-\mathbf{i}\varepsilon})\mathbb{1} - A)^{-1}\, dt \\
&\quad + \frac{e^{\mathbf{i}\varepsilon}}{2\pi} \int_{1/r}^{\infty} e^{z(\omega + \mathbf{i}te^{\mathbf{i}\varepsilon})}((\omega + \mathbf{i}te^{\mathbf{i}\varepsilon})\mathbb{1} - A)^{-1}\, dt \\
&=: S^0(z) + S^-(z) + S^+(z).
\end{aligned}$$

By Step 1, $\|((\omega + r^{-1}e^{\mathbf{i}rt(\frac{\pi}{2}+\varepsilon)})\mathbb{1} - A)^{-1}\| \leq M_\varepsilon r$ and hence

$$\|S^0(z)\| \leq \frac{\pi + 2\varepsilon}{2\pi r} e^{\omega r\cos(\theta)+1} M_\varepsilon r \leq M_\varepsilon e^{\omega r\cos(\theta)+1}.$$

Now use the fact that $\mathrm{Re}(\pm z\mathbf{i}e^{\pm\mathbf{i}\varepsilon}) = -r\sin(\varepsilon \pm \theta) < 0$ to obtain

$$\begin{aligned}
\|S^\pm(z)\| &\leq \frac{M_\varepsilon e^{\omega r\cos(\theta)}}{2\pi} \int_{1/r}^{\infty} \frac{e^{-tr\sin(\varepsilon\pm\theta)}}{t}\, dt \\
&\leq \frac{M_\varepsilon e^{\omega r\cos(\theta)}}{2\pi} \int_{1/r}^{\infty} \frac{e^{-tr\sin(\varepsilon-\delta)}}{t}\, dt \\
&= \frac{M_\varepsilon e^{\omega r\cos(\theta)}}{2\pi} \int_{1}^{\infty} \frac{e^{-s\sin(\varepsilon-\delta)}}{s}\, ds \\
&\leq \frac{M_\varepsilon e^{\omega r\cos(\theta)}}{2\pi\sin(\varepsilon - \delta)}.
\end{aligned}$$

Since $r\cos(\theta) = \mathrm{Re}(z)$, the last two estimates imply

$$\|S(z)\| \leq M_\varepsilon \left(e + \frac{1}{\pi\sin(\varepsilon - \delta)} \right) e^{\omega\mathrm{Re}(z)} \qquad \text{for all } z \in \overline{U}_\delta \setminus \{0\}.$$

This proves Step 3.

Step 4. *Let $0 < \delta < \varepsilon < \pi/2$ and let $z \in \overline{U}_\delta$. Choose a real number $r > 0$ and let*

$$\gamma_r = \gamma_{r,\varepsilon} : \mathbb{R} \to \mathbb{C}$$

be given by (7.4.19) *as in Step 2. Then*

$$\frac{1}{2\pi\mathbf{i}} \int_{\gamma_r} \frac{e^{z\zeta}}{\zeta - \omega}\, d\zeta = e^{\omega z}.$$

The loop obtained from $\gamma_r|_{[-T,T]}$ by joining the endpoints with a straight line encircles the number ω with winding number one for $T \geq 1/r$. Moreover, the straight line

$$\beta_T : [-1, 1] \to \mathbb{C}$$

joining the endpoints (from top to bottom) is given by

$$\beta_T(s) := \omega - T\sin(\varepsilon) - \mathbf{i}sT\cos(\varepsilon)$$

and so

$$\begin{aligned}
\left| \frac{1}{2\pi\mathbf{i}} \int_{\beta_T} \frac{e^{z\zeta}}{\zeta - \omega}\, d\zeta \right| &= \left| \frac{-T\cos(\varepsilon)}{2\pi} \int_{-1}^{1} \frac{e^{z(\omega - T\sin(\varepsilon) - \mathbf{i}sT\cos(\varepsilon))}}{-T\sin(\varepsilon) - \mathbf{i}sT\cos(\varepsilon)}\, ds \right| \\
&\leq \frac{\cos(\varepsilon) e^{\omega \mathrm{Re}(z)}}{\sin(\varepsilon)\pi} e^{-T\sin(\varepsilon)\mathrm{Re}(z) + T\cos(\varepsilon)|\mathrm{Im}(z)|}.
\end{aligned}$$

Since $z \in \overline{U}_\delta$, the last factor is bounded above by $e^{-|z|T\sin(\varepsilon - \delta)}$ and so converges exponentially to zero as T tends to infinity. This proves Step 4.

Step 5. *For $0 < \delta < \varepsilon < \varepsilon_0$ the map*

$$S : \overline{U}_\delta \setminus \{0\} \to \mathcal{L}^c(X)$$

in Step 2 satisfies

$$\lim_{r \to 0} \sup \left\{ \|S(z)x - x\| \,\middle|\, z \in \overline{U}_\delta,\, |z| = r \right\} = 0$$

for all $x \in X$.

Assume first that $x \in \mathrm{dom}(A)$. Let $z \in \overline{U}_\delta \setminus \{0\}$, define $r := |z|$, and let the curve $\gamma_r : \mathbb{R} \to \mathbb{C}$ be given by equation (7.4.19). Then, by Step 2 and Step 4,

$$\begin{aligned}
S(z)x - e^{\omega z}x &= \frac{1}{2\pi\mathbf{i}} \int_{\gamma_r} e^{z\zeta} \left((\zeta\mathbb{1} - A)^{-1}x - (\zeta - \omega)^{-1}x \right) d\zeta \\
&= \frac{1}{2\pi\mathbf{i}} \int_{\gamma_r} \frac{e^{z\zeta}}{\zeta - \omega} (\zeta\mathbb{1} - A)^{-1}(Ax - \omega x)\, d\zeta \\
&= \frac{1}{2\pi\mathbf{i}} \int_{-\infty}^{\infty} \frac{\dot\gamma_r(t) e^{z\gamma_r(t)}}{\gamma_r(t) - \omega} (\gamma_r(t)\mathbb{1} - A)^{-1}(Ax - \omega x)\, dt.
\end{aligned}$$

Since

$$\|(\gamma_r(t)\mathbb{1} - A)^{-1}\| \le \frac{M_\varepsilon}{|\gamma_r(t) - \omega|}$$

by Step 1 and

$$|\gamma_r(t) - \omega| \ge \frac{1}{r}$$

by (7.4.19), it follows that

$$\begin{aligned}\|S(z)x - e^{\omega z}x\| &\le \frac{M_\varepsilon}{2\pi}\int_{-\infty}^{\infty} \frac{|\dot\gamma_r(t)|e^{\mathrm{Re}(z\gamma_r(t))}}{|\gamma_r(t) - \omega|^2}\,dt\,\|Ax - \omega x\| \\ &\le \frac{M_\varepsilon}{2\pi}\int_{-\infty}^{\infty} |\dot\gamma_r(t)|e^{\mathrm{Re}(z\gamma_r(t))}\,dt\,\|Ax - \omega x\|\,r^2.\end{aligned}$$

Now

$$\gamma_r(t) - \omega = \frac{1}{r}e^{\mathbf{i}rt(\frac{\pi}{2}+\varepsilon)} \qquad \text{for } |t| \le \frac{1}{r}$$

and

$$\gamma_r(t) - \omega = te^{\mathbf{i}(\frac{\pi}{2}+\varepsilon)} \qquad \text{for } t \ge \frac{1}{r}$$

and

$$\gamma_r(t) - \omega = -te^{-\mathbf{i}(\frac{\pi}{2}+\varepsilon)} \qquad \text{for } t \le -\frac{1}{r}.$$

Thus $|\dot\gamma_r(t)| \le \pi$ for $|t| < 1/r$ and $|\dot\gamma_r(t)| = 1$ for $|t| > 1/r$. Write $z = re^{\mathbf{i}\theta}$ with $|\theta| \le \delta < \varepsilon$ and use the inequality

$$\mathrm{Re}\left(tze^{\mathbf{i}(\frac{\pi}{2}+\varepsilon)}\right) = tr\cos\left(\frac{\pi}{2} + \varepsilon + \theta\right) = -tr\sin(\varepsilon+\theta) \le -tr\sin(\varepsilon - \delta)$$

to obtain

$$\begin{aligned}\int_{-\infty}^{\infty} |\dot\gamma_r(t)|\,e^{\mathrm{Re}(z\gamma_r(t))}\,dt &= e^{\omega\mathrm{Re}(z)}\int_{-1/r}^{1/r} |\dot\gamma_r(t)|\,e^{\mathrm{Re}(\frac{z}{r}e^{\mathbf{i}rt(\frac{\pi}{2}+\varepsilon)})}\,dt \\ &\quad + 2e^{\omega\mathrm{Re}(z)}\int_{1/r}^{\infty} e^{\mathrm{Re}(tze^{\mathbf{i}(\frac{\pi}{2}+\varepsilon)})}\,dt \\ &\le e^{\omega\mathrm{Re}(z)}\left(\frac{2\pi e}{r} + 2\int_{1/r}^{\infty} e^{-tr\sin(\varepsilon-\delta)}\,dt\right) \\ &\le e^{\omega\mathrm{Re}(z)}\left(\frac{2\pi e}{r} + \frac{2}{r\sin(\varepsilon-\delta)}\right).\end{aligned}$$

Combine these inequalities to obtain

$$\begin{aligned}\|S(z)x - e^{\omega z}x\| &\le \frac{M_\varepsilon}{2\pi}\int_{-\infty}^{\infty} |\dot\gamma_r(t)|e^{\mathrm{Re}(z\gamma_r(t))}\,dt\,\|Ax - \omega x\|\,r^2 \\ &\le M_\varepsilon e^{\omega\mathrm{Re}(z)}\left(e + \frac{1}{\pi\sin(\varepsilon-\delta)}\right)\|Ax - \omega x\|\,r\end{aligned}$$

for all $z \in \overline{U}_\delta \setminus \{0\}$ with $|z| = r$. This proves Step 5 in the case $x \in \mathrm{dom}(A)$. The general case follows from the special case by Step 3 and Theorem 2.1.5.

Step 6. *Let $0 < \varepsilon < \varepsilon_0$, let S be as in Step 2, and let $0 < \delta < \varepsilon$. Extend the map $S : \overline{U}_\delta \setminus \{0\} \to \mathcal{L}^c(X)$ to all of $\overline{U}_\delta$ by setting $S(0) := \mathbb{1}$. Then*

$$S : \overline{U}_\delta \to \mathcal{L}^c(X)$$

is strongly continuous and satisfies

$$\frac{S(z+h)x - S(z)x}{h} = \int_0^1 S(z+th)Ax\,dt \tag{7.4.21}$$

for all $x \in \mathrm{dom}(A)$ and all $z, h \in \overline{U}_\delta$.

Strong continuity follows from Step 5. To prove (7.4.21), let $x \in \mathrm{dom}(A)$ and $z, h \in \overline{U}_\delta$. Assume first that $z \neq 0$. Define the curve $\gamma_r = \gamma_{r,\varepsilon} : \mathbb{R} \to \mathbb{C}$ by (7.4.19) as in Step 2. Then

$$\begin{aligned}
\int_0^1 S(z+th)Ax\,dt &= \frac{1}{2\pi\mathbf{i}} \int_0^1 \int_{\gamma_r} e^{(z+th)\zeta}(\zeta\mathbb{1} - A)^{-1}Ax\,d\zeta\,dt \\
&= \frac{1}{2\pi\mathbf{i}} \int_{\gamma_r} \int_0^1 e^{(z+th)\zeta}\,dt(\zeta\mathbb{1} - A)^{-1}Ax\,d\zeta \\
&= \frac{1}{2\pi\mathbf{i}} \int_{\gamma_r} \frac{e^{(z+h)\zeta} - e^{z\zeta}}{h\zeta}(\zeta\mathbb{1} - A)^{-1}Ax\,d\zeta \\
&= \frac{1}{2\pi\mathbf{i}} \int_{\gamma_r} \frac{e^{(z+h)\zeta} - e^{z\zeta}}{h}\left((\zeta\mathbb{1} - A)^{-1}x - \frac{x}{\zeta}\right)d\zeta \\
&= \frac{S(z+h)x - S(z)x}{h}.
\end{aligned}$$

Here the last assertion follows from the fact that, by the same argument as in Step 4, we have

$$\frac{1}{2\pi\mathbf{i}} \int_{\gamma_{r,\varepsilon}} \frac{e^{z\zeta}(e^{h\zeta} - 1)}{h\zeta} d\zeta = 1$$

whenever $r > 0$, $0 < \delta < \varepsilon < \pi/2$, and $z, h \in \overline{U}_\delta$. This proves (7.4.21) in the case $z \neq 0$. In the case $z = 0$ the equation then follows from strong continuity. This proves Step 6.

Step 7. *The map $S : \overline{U}_\delta \to \mathcal{L}^c(X)$ in Step 2 and Step 6 satisfies*

$$S(w+z) = S(w)S(z) \tag{7.4.22}$$

for all $z, w \in \overline{U}_\delta$.

By strong continuity it suffices to prove equation (7.4.22) for $z, w \in U_\delta$. Fix two elements $w, z \in U_\delta$. Choose two numbers $0 < \rho < r$, define the curve $\gamma = \gamma_{r,\varepsilon} : \mathbb{R} \to \mathbb{C}$ by equation (7.4.19) as in Step 2, and define the curve $\beta := \beta_{\rho,\delta} : \mathbb{R} \to \mathbb{C}$ by the same formula with ε replaced by δ and r replaced by ρ (see Figure 7.4.3).

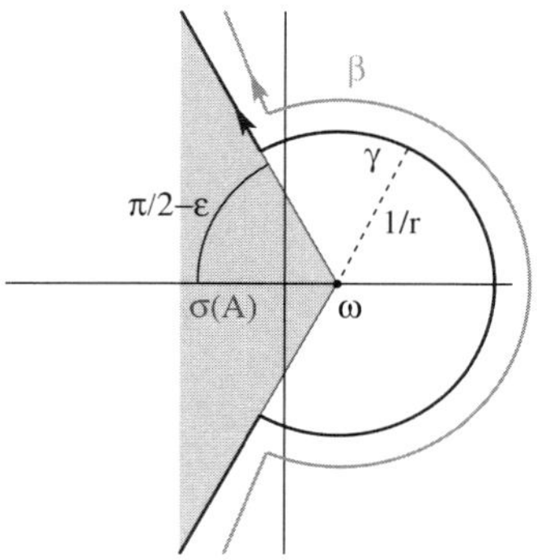

Figure 7.4.3. Integration along β and γ.

With this notation in place, the argument in the proof of Step 4 yields

$$\frac{1}{2\pi\mathbf{i}}\int_\gamma \frac{e^{z\eta}}{\eta-\beta(s)}d\eta = 0, \qquad \frac{1}{2\pi\mathbf{i}}\int_\beta \frac{e^{w\xi}}{\xi-\gamma(t)}d\xi = e^{w\gamma(t)}$$

for all $s,t\in\mathbb{R}$ and all $z,w\in\mathbb{C}$. The key observation is that the integrals along the relevant vertical straight lines converge to zero as in Step 4, and that in the first case the resulting γ-loops have winding number zero about $\beta(s)$, while in the second case the resulting β-loops have winding number one about $\gamma(t)$ for sufficiently large T. Hence

$$\begin{aligned}
S(w)S(z) &= \frac{1}{2\pi\mathbf{i}}\int_\beta e^{w\xi}\left(\xi\mathbb{1}-A\right)^{-1}S(z)\,d\xi \\
&= \frac{1}{2\pi\mathbf{i}}\int_\beta e^{w\xi}\left(\xi\mathbb{1}-A\right)^{-1}\left(\frac{1}{2\pi\mathbf{i}}\int_\gamma e^{z\eta}\left(\eta\mathbb{1}-A\right)^{-1}d\eta\right)d\xi \\
&= \frac{1}{2\pi\mathbf{i}}\int_\beta \left(\frac{1}{2\pi\mathbf{i}}\int_\gamma e^{w\xi+z\eta}\left(\xi\mathbb{1}-A\right)^{-1}\left(\eta\mathbb{1}-A\right)^{-1}d\eta\right)d\xi \\
&= \frac{1}{2\pi\mathbf{i}}\int_\beta \left(\frac{1}{2\pi\mathbf{i}}\int_\gamma \frac{e^{w\xi+z\eta}}{\eta-\xi}\left(\left(\xi\mathbb{1}-A\right)^{-1}-\left(\eta\mathbb{1}-A\right)^{-1}\right)d\eta\right)d\xi \\
&= \frac{1}{2\pi\mathbf{i}}\int_\beta \left(\frac{1}{2\pi\mathbf{i}}\int_\gamma \frac{e^{w\xi+z\eta}}{\eta-\xi}\,d\eta\right)\left(\xi\mathbb{1}-A\right)^{-1}d\xi \\
&\quad+\frac{1}{2\pi\mathbf{i}}\int_\gamma \left(\frac{1}{2\pi\mathbf{i}}\int_\beta \frac{e^{w\xi+z\eta}}{\xi-\eta}\,d\xi\right)\left(\eta\mathbb{1}-A\right)^{-1}d\eta \\
&= \frac{1}{2\pi\mathbf{i}}\int_\gamma e^{(w+z)\eta}\left(\eta\mathbb{1}-A\right)^{-1}d\eta \\
&= S(w+z).
\end{aligned}$$

This proves Step 7.

Step 8. *The map* $S : \overline{U}_\delta \to \mathcal{L}^c(X)$ *is an analytic semigroup. It satisfies* (7.4.10) *and its infinitesimal generator is the operator* A.

That S is an analytic semigroup follows from Step 6 and Step 7, and the estimate (7.4.10) follows from Step 3 by taking the limit $\omega \to \omega_0$. Now let $x \in \mathrm{dom}(A)$ and $t > 0$. Then the integral

$$S(t)x = \frac{1}{2\pi\mathbf{i}} \int_{\gamma_r} e^{t\zeta}(\zeta\mathbb{1} - A)^{-1}x\, d\zeta$$

in (7.4.20) converges in the Banach space $\mathrm{dom}(A)$ with the graph norm. Hence we have $S(t)x \in \mathrm{dom}(A)$ and

$$AS(t)x = \frac{1}{2\pi\mathbf{i}} \int_{\gamma_r} e^{t\zeta}(\zeta\mathbb{1} - A)^{-1}Ax\, d\zeta = S(t)Ax.$$

Moreover,

$$S(t)x - x = \int_0^t S(s)Ax\, ds$$

by Step 6. Hence A and S satisfy condition (ii) in Lemma 7.1.17 and so A is the infinitesimal generator of S. This proves Step 8 and Theorem 7.4.4. □

7.4.3. Examples of Analytic Semigroups. By Theorem 7.4.2 an analytic semigroup $S : [0, \infty) \to \mathcal{L}^c(X)$ on a complex Banach space X with infinitesimal generator $A : \mathrm{dom}(A) \to X$ satisfies $\mathrm{im}(S(t)) \subset \mathrm{dom}(A)$ for all $t > 0$. Hence a group of operators $S : \mathbb{R} \to \mathcal{L}^c(X)$ cannot be analytic unless its infinitesimal generator is a bounded operator (see Lemma 7.1.18 and Theorem 7.2.4).

Example 7.4.5 (Self-Adjoint Semigroups). Let H be a complex Hilbert space and let $A : \mathrm{dom}(A) \to H$ be a self-adjoint operator such that

$$\omega_0 := \sup_{x \in \mathrm{dom}(A)\setminus\{0\}} \frac{\langle x, Ax\rangle}{\|x\|^2} < \infty.$$

By Theorem 7.3.4 the operator A is the infinitesimal generator of a strongly continuous self-adjoint semigroup $S : [0, \infty) \to \mathcal{L}^c(H)$. Moreover, if $\lambda \in \mathbb{C}$ satisfies $\mathrm{Re}\lambda > \omega_0$, then $\lambda \in \rho(A)$ and

$$|\lambda - \omega_0|\, \|x\|^2 = \left|\lambda\|x\|^2 - \omega_0\|x\|^2\right| \le \left|\lambda\|x\|^2 - \langle x, Ax\rangle\right| \le \|x\|\|\lambda x - Ax\|$$

for all $x \in X$. This implies

$$\|(\lambda\mathbb{1} - A)^{-1}\| \le \frac{1}{|\lambda - \omega_0|} \qquad \text{for all } \lambda \in \mathbb{C} \text{ with } \mathrm{Re}\lambda > \omega_0.$$

Hence it follows from Theorem 7.4.4 that S is an analytic semigroup. In fact, the proof of Theorem 7.4.4 with $M = 1$ and $\varepsilon_0 = \pi/2$ shows that S extends to a holomorphic function $S : \{z \in \mathbb{C} \,|\, \mathrm{Re}z > \omega_0\} \to \mathcal{L}^c(H)$ on an open half-space and that the spectrum of A is contained in the half-axis $(-\infty, \omega_0]$.

Example 7.4.6 (Heat Equation). The solutions of the heat equation

$$\partial_t u = \Delta u, \qquad \Delta := \sum_{i=1}^{n} \frac{\partial^2}{\partial x_i^2}, \tag{7.4.23}$$

determine a contraction semigroup on $L^p(\mathbb{R}^n)$ for $1 < p < \infty$, given by

$$S(t)f := K_t * f, \qquad K_t(x) := \frac{1}{(4\pi t)^{n/2}} e^{-|x|^2/4t}, \tag{7.4.24}$$

for $t > 0$ and $f \in L^2(\mathbb{R}^n)$ (see Example 7.1.6). Its infinitesimal generator is the Laplace operator $\Delta : W^{2,p}(\mathbb{R}^n) \to L^p(\mathbb{R}^n)$ in Example 6.1.6. In the case $p = 2$ the semigroup S is self-adjoint, and so is analytic by Example 7.4.5. In general, one can verify directly that the formula (7.4.24) is well-defined for every complex number t with positive real part and defines a holomorphic function on the right half-plane.

Example 7.4.7. This example shows that every closed subset of a sector of the form C_δ in (7.4.4) is the spectrum of the infinitesimal generator of an analytic semigroup on a Hilbert space. Let H be a separable complex Hilbert space, let $(e_i)_{i \in \mathbb{N}}$ be a complex orthonormal basis of H, and let $(\lambda_i)_{i \in \mathbb{N}}$ be a sequence of complex numbers. Define the operator $A_\lambda : \mathrm{dom}(A_\lambda) \to H$ by

$$\begin{aligned} \mathrm{dom}(A_\lambda) &:= \left\{ x \in H \,\middle|\, \sum_{i=1}^{\infty} |\lambda_i|^2 |\langle e_i, x\rangle|^2 < \infty \right\}, \\ A_\lambda x &:= \sum_{i=1}^{\infty} \lambda_i \langle e_i, x\rangle e_i \qquad \text{for } x \in \mathrm{dom}(A_\lambda). \end{aligned} \tag{7.4.25}$$

By Example 7.1.12 this operator generates a strongly continuous semigroup if and only if $\sup_{i \in \mathbb{N}} \mathrm{Re}\lambda_i < \infty$. In this case the semigroup is given by

$$S_\lambda(t)x := \sum_{i=1}^{\infty} e^{\lambda_i t} \langle e_i, x\rangle e_i \qquad \text{for } t \geq 0 \text{ and } x \in H. \tag{7.4.26}$$

(See Example 7.1.3.) The semigroup (7.4.26) is analytic if and only if

$$\sup_{i \in \mathbb{N}} \frac{|\mathrm{Im}\lambda_i|}{\omega - \mathrm{Re}\lambda_i} < \infty \qquad \text{for } \omega > \omega_0 := \sup_{i \in \mathbb{N}} \mathrm{Re}\lambda_i. \tag{7.4.27}$$

Exercise: Show that this condition holds for some $\omega > \omega_0$ if and only if it holds for all $\omega > \omega_0$. Assuming (7.4.27), let $\omega > \omega_0$, choose $0 < \delta < \pi/2$ such that $\sin(\delta)|\mathrm{Im}\lambda_i| \leq \cos(\delta)(\omega - \mathrm{Re}\lambda_i)$ for all i, and define $M := 1/\sin(\delta)$. Show that, for all $\mu \in \mathbb{C}$,

$$\mathrm{Re}(\mu) \geq \omega \quad \Longrightarrow \quad \left\|(\mu\mathbb{1} - A_\lambda)^{-1}\right\| = \sup_{i \in \mathbb{N}} \frac{1}{|\mu - \lambda_i|} \leq \frac{M}{|\mu - \omega|}.$$

Show that $\sigma(A_\lambda) = \overline{\{\lambda_i \,|\, i \in \mathbb{N}\}} \subset \{\omega + re^{\mathrm{i}\theta} \,|\, r \geq 0,\ \pi/2 + \delta \leq |\theta| \leq \pi\}$.

7.5. Banach Space Valued Measurable Functions

This is a preparatory section. It studies measurable functions on an interval with values in a Banach space, a subject with many applications and of interest in its own right. The first subsection introduces the concept of a strongly measurable function and proves Pettis' Theorem. The next four subsections deal with the Banach space $L^p(I, X)$, the Radon–Nikodým property of a Banach space, the dual space of $L^p(I, X)$, and the Sobolev space $W^{1,p}(I, X)$. All these results will be used in Section 7.6 on the inhomogeneous equation $\dot{x} = Ax + f$ associated to a semigroup.

7.5.1. Measurable Functions. The following definition summarizes the different notions of measurability for functions with values in a Banach space. Although these definitions and many of the results carry over to functions on general measure spaces, in this book we will only use Banach space valued functions on an interval and restrict the discussion to that case.

Definition 7.5.1. Let X be a real Banach space and let $I \subset \mathbb{R}$ be an interval. A function $f : I \to X$ is called

• **weakly continuous** if the function

$$\langle x^*, f \rangle : I \to \mathbb{R}$$

is continuous for all $x^* \in X^*$,

• **weakly measurable** if the function

$$\langle x^*, f \rangle : I \to \mathbb{R}$$

is Borel measurable for all $x^* \in X^*$,

• **measurable** if $f^{-1}(B) \subset I$ is a Borel set for every Borel set $B \subset X$,

• a **measurable step function** if it is measurable and $f(I)$ is a finite set,

• **strongly measurable** if there exists a sequence of measurable step functions $f_n : I \to X$ such that $\lim_{n\to\infty} f_n(t) = f(t)$ for almost all $t \in I$.

The basic example, which illustrates the subtlety of this story is the function $[0, 1] \to L^\infty([0, 1]) : t \mapsto f_t$, defined by $f_t := \chi_{[0,t]}$, i.e. $f_t(s) = 1$ for $0 \le s \le t$ and $f_t(s) = 0$ for $t < s \le 1$. This function is weakly measurable, but not strongly measurable and is everywhere discontinuous. The same function, understood with values in the Banach space $L^1([0, 1])$, is an example of a Lipschitz continuous function which is nowhere differentiable.

It follows directly from the definition that the image of a strongly measurable function $f : I \to X$ is contained in a separable subspace of X. Example 7.5.3 below shows that weakly measurable functions need not satisfy this condition.

Theorem 7.5.2 (Pettis). *Let X be a real Banach space. Fix two numbers $a < b$ and a function $f : [a,b] \to X$. Then the following hold.*

(i) *Assume X is separable and let $E \subset X^*$ be a linear subspace such that*

$$\|x\| = \sup_{x^* \in E \setminus \{0\}} \frac{|\langle x^*, x \rangle|}{\|x^*\|} \qquad \text{for all } x \in X. \tag{7.5.1}$$

If $\langle x^*, f \rangle$ is measurable for all $x^* \in E$, then f is strongly measurable.

(ii) *If X is separable and f is weakly measurable, then f is strongly measurable.*

(iii) *If f is weakly continuous, then f is strongly measurable.*

(iv) *If f is strongly measurable, then the function $[a,b] \to \mathbb{R} : t \mapsto \|f(t)\|$ is Borel measurable.*

Proof. We prove part (i). Thus assume X is separable and $E \subset X^*$ is a linear subspace that satisfies (7.5.1). Abbreviate $I := [a,b]$ and let $f : I \to X$ be a function such that $\langle x^*, f \rangle : I \to \mathbb{R}$ is measurable for all $x^* \in E$. We prove in three steps that f is strongly measurable.

Step 1. *Let $\xi \in X$ and $r > 0$. Then $f^{-1}(\overline{B_r(\xi)})$ is a Borel subset of I.*

Choose a dense sequence $x_n \in X \setminus \overline{B_r(\xi)}$ and define

$$\varepsilon_n := \frac{1}{2}\big(\|x_n - \xi\| - r\big) > 0 \qquad \text{for } n \in \mathbb{N}.$$

Then $X \setminus \overline{B_r(\xi)} = \bigcup_{n=1}^\infty B_{\varepsilon_n}(x_n)$. For $n \in \mathbb{N}$ choose $x_n^* \in E$ such that

$$\|x_n^*\| = 1, \qquad \langle x_n^*, x_n - \xi \rangle > \|x_n - \xi\| - \varepsilon_n.$$

Then, for all $n \in \mathbb{N}$, all $\eta \in \overline{B_r(\xi)}$, and all $x \in B_{\varepsilon_n}(x_n)$, we have

$$\langle x_n^*, \eta \rangle \le \langle x_n^*, \xi \rangle + r = \langle x_n^*, \xi \rangle + \|x_n - \xi\| - 2\varepsilon_n < \langle x_n^*, x_n \rangle - \varepsilon_n < \langle x_n^*, x \rangle.$$

This implies $\overline{B_r(\xi)} = \bigcap_{n=1}^\infty \{y \in X \,|\, \langle x_n^*, y \rangle \le \langle x_n^*, \xi \rangle + r\}$. Hence

$$f^{-1}(\overline{B_r(\xi)}) = \bigcap_{n=1}^\infty \big\{t \in I \,\big|\, \langle x_n^*, f(t) \rangle \le \langle x_n^*, \xi \rangle + r\big\}$$

is a Borel set. This proves Step 1.

Step 2. *f is measurable.*

Let $U \subset X$ be open. Since X is separable, there exist a sequence $x_n \in X$ and a sequence of real numbers $\varepsilon_n > 0$ such that $U = \bigcup_{n=1}^\infty \overline{B_{\varepsilon_n}(x_n)}$. Hence it follows from Step 1 that $f^{-1}(U) = \bigcup_{n=1}^\infty f^{-1}(\overline{B_{\varepsilon_n}(x_n)})$ is a Borel subset of I. This shows that f is Borel measurable by [**75**, Thm. 1.20].

Step 3. *f is strongly measurable.*

Since X is separable there exists a dense sequence $x_k \in X$. For $k, n \in \mathbb{N}$ define the set

$$\Sigma_{k,n} := \left\{ t \in I \,\middle|\, \begin{array}{l} \|f(t) - x_k\| < 1/n \text{ and} \\ \|f(t) - x_i\| \geq 1/n \text{ for } i = 1, \dots, k-1 \end{array} \right\}. \tag{7.5.2}$$

This is a Borel subset of I by Step 2. Moreover $\Sigma_{k,n} \cap \Sigma_{\ell,n} = \emptyset$ for $k \neq \ell$ and $\bigcup_{k=1}^{\infty} \Sigma_{k,n} = I$. Hence, for each $n \in \mathbb{N}$, there is an $N_n \in \mathbb{N}$ such that

$$\mu\left(\bigcup_{k=N_n+1}^{\infty} \Sigma_{k,n} \right) < 2^{-n}. \tag{7.5.3}$$

Here μ denotes the restriction of the Lebesgue measure to the Borel σ-algebra of I. Define the functions $f_n : I \to X$ by

$$f_n(t) := \begin{cases} x_k, & \text{for } t \in \Sigma_{k,n} \text{ and } k = 1, \dots, N_n, \\ 0, & \text{for } t \in \bigcup_{k=N_n+1}^{\infty} \Sigma_{k,n}. \end{cases} \tag{7.5.4}$$

These are measurable step functions. Define

$$\Omega := \bigcap_{m=1}^{\infty} \bigcup_{n=m}^{\infty} \bigcup_{k=N_n+1}^{\infty} \Sigma_{k,n}, \qquad I \setminus \Omega = \bigcup_{m=1}^{\infty} \bigcap_{n=m}^{\infty} \bigcup_{k=1}^{N_n} \Sigma_{k,n}.$$

Then $\mu(\Omega) = 0$ by (7.5.3) and $\|f_n(t) - f(t)\| < 1/n$ for all $t \in \bigcup_{k=1}^{N_n} \Sigma_{k,n}$ by (7.5.2) and (7.5.4). If $t \in I \setminus \Omega$, then there exists an integer $m \in \mathbb{N}$ such that $t \in \bigcap_{n=m}^{\infty} \bigcup_{k=1}^{N_n} \Sigma_{k,n}$ and hence $\|f_n(t) - f(t)\| < 1/n$ for every integer $n \geq m$. Thus

$$\lim_{n \to \infty} f_n(t) = f(t) \qquad \text{for all } t \in I \setminus \Omega.$$

This proves Step 3 and part (i).

Part (ii) follows from (i) with $E = X^*$ and the Hahn–Banach Theorem (Corollary 2.3.4).

We prove part (iii). Assume $f : I \to X$ is weakly continuous and define

$$X_0 := \overline{\operatorname{span}\{f(t) \,|\, t \in I \cap \mathbb{Q}\}}.$$

If $t \in I$ and $x^* \in X_0^{\perp}$, then $\langle x^*, f(t) \rangle = 0$ by weak continuity. Hence it follows from Corollary 2.3.24 that $f(I) \subset X_0$. Since X_0 is separable by definition, it follows from (ii) that f is strongly measurable.

We prove part (iv). Assume $f : I \to X$ is strongly measurable and choose a sequence of measurable step functions $f_n : I \to X$ that converges almost everywhere to f. Then the sequence $\|f_n\| : I \to \mathbb{R}$ of measurable step functions converges almost everywhere to $\|f\| : I \to \mathbb{R}$ and hence the function $\|f\| : I \to \mathbb{R}$ is measurable. This proves Theorem 7.5.2. $\square$

The next example shows that the hypothesis that X is separable cannot be removed in part (ii) of Theorem 7.5.2.

Example 7.5.3. **(i)** Let H be a nonseparable real Hilbert space, equipped with an uncountable orthonormal basis

$$\{e_t\}_{0\le t\le 1}.$$

Thus the vectors $e_t \in H$ are parametrized by the elements of the unit interval $[0,1]\subset\mathbb{R}$ and satisfy $\langle e_s, e_t\rangle = 0$ for $s\ne t$ and $\|e_t\| = 1$ for all t. The function $f:[0,1]\to H$ defined by $f(t) := e_t$ is not strongly measurable because every Borel set $\Omega\subset[0,1]$ of measure zero has an uncountable complement, so $f([0,1]\setminus\Omega)$ is not contained in a separable subspace of H. However, the function f is weakly measurable because each $x\in H$ has the form $x = \sum_{i=1}^\infty \lambda_i e_{s_i}$ for a sequence $\lambda_i\in\mathbb{R}$ such that $\sum_{i=1}^\infty \lambda_i^2 < \infty$ and a sequence of pairwise distinct elements $s_i\in[0,1]$; thus $\langle x, f(t)\rangle = \lambda_i$ for $t = s_i$ and $\langle x, f(t)\rangle = 0$ for $t\notin\{s_i\,|\,i\in\mathbb{N}\}$.

(ii) Let $X := L^\infty([0,1])$ and define the function $f:[0,1]\to L^\infty([0,1])$ by

$$(f(t))(x) := f(t,x) := \left\{\begin{array}{ll} 1, & \text{if } 0\le x\le t,\\ 0, & \text{if } t < x \le 1.\end{array}\right.$$

This function satisfies $\|f(s)-f(t)\|_{L^\infty} = 1$ for all $s\ne t$ and the same argument as in part (i) shows that f is not strongly measurable. However, when the same function is considered with values in the Banach space $L^p([0,1])$ for $1\le p<\infty$, it is continuous and hence strongly measurable.

Theorem 7.5.4. *Let X be a Banach space. Fix real numbers $1\le p<\infty$ and $a<b$ and a function $f: I := [a,b]\to X$. The following are equivalent.*

(i) *f is strongly measurable and*

$$\int_a^b \|f(t)\|^p\,dt < \infty.$$

(ii) *For every $\varepsilon>0$ there exists a measurable step function $g: I\to X$ such that the function $I\to\mathbb{R}: t\mapsto \|f(t)-g(t)\|$ is Borel measurable and*

$$\int_a^b \|f(t)-g(t)\|^p\,dt < \varepsilon.$$

(iii) *For every $\varepsilon>0$ there exists a continuous function $g: I\to X$ such that the function $I\to\mathbb{R}: t\mapsto\|f(t)-g(t)\|$ is Borel measurable and*

$$\int_a^b \|f(t)-g(t)\|^p\,dt < \varepsilon.$$

Proof. We prove that (i) implies (ii). Choose a sequence of measurable step functions $g_n : I \to X$ that converges almost everywhere to f. For $n \in \mathbb{N}$ define the function $f_n : I \to X$ by

$$f_n(t) := \begin{cases} g_n(t), & \text{if } \|g_n(t)\| < \|f(t)\| + 1, \\ 0, & \text{if } \|g_n(t)\| \geq \|f(t)\| + 1, \end{cases} \qquad \text{for } t \in I.$$

Then f_n is a measurable step function for every $n \in \mathbb{N}$ by part (iv) of Theorem 7.5.2. Moreover, $\lim_{n\to\infty} \|f_n(t) - f(t)\|^p = 0$ for almost all $t \in I$ and

$$\|f(t) - f_n(t)\|^p \leq (2\|f(t)\| + 1)^p \leq 4^p \|f(t)\|^p + 2^p$$

for all $t \in I$ and all $n \in \mathbb{N}$. The function on the right is integrable by (i). Hence $\lim_{n\to\infty} \int_a^b \|f(t) - f_n(t)\|^p \, dt = 0$ by the Lebesgue Dominated Convergence Theorem. This shows that (i) implies (ii).

We prove that (ii) implies (i). Choose a sequence of measurable step functions $f_n : I \to X$ such that the function $\|f - f_n\| : I \to \mathbb{R}$ is Borel measurable and $\lim_{n\to\infty} \int_a^b \|f(t) - f_n(t)\|^p \, dt = 0$. Then there exists a subsequence f_{n_i} such that $\lim_{i\to\infty} \|f(t) - f_{n_i}(t)\| = 0$ for almost every $t \in I$ by [**75**, Cor. 4.10]. Hence f is strongly measurable. Now choose an integer n such that

$$\int_a^b \|f(t) - f_n(t)\|^p \, dt < 1.$$

Then, by Minkowski's inequality,

$$\left(\int_a^b \|f(t)\|^p \, dt\right)^{1/p} \leq \left(\int_a^b \|f_n(t)\|^p \, dt\right)^{1/p} + 1 < \infty.$$

Hence (ii) implies (i) and the same argument shows that (iii) implies (i).

We prove that (i) implies (iii). For this it suffices to assume that f is a measurable step function with precisely one nonzero value. Let $B \subset I$ be a Borel set and let $x \in X \setminus \{0\}$ and assume $f = \chi_B x$. Fix a constant $\varepsilon > 0$. Since the Lebesgue measure is regular by [**75**, Thm. 2.13], there exists a compact set $K \subset I$ and an open set $U \subset I$ such that

$$K \subset B \subset U, \qquad \mu(U \setminus K) < \frac{\varepsilon}{\|x\|^p}.$$

By Urysohn's Lemma there exists a continuous function $\psi : I \to [0,1]$ such that $\psi(t) = 1$ for all $t \in K$ and $\psi(t) = 0$ for all $t \in I \setminus U$. Define the function $g : I \to X$ by $g := \psi x$. Then $|\psi - \chi_B| \leq \chi_{U\setminus K}$ and hence

$$\int_a^b \|f(t) - g(t)\|^p \, dt \leq \int_{U\setminus K} \|x\|^p \, dt = \mu(U \setminus K)\|x\|^p < \varepsilon.$$

This proves Theorem 7.5.4. □

The next lemma is a direct consequence of Theorem 7.5.4. It will play a central role in Exercise 7.7.11.

Lemma 7.5.5. *Let X be a Banach space and fix real numbers $1 \leq p < \infty$ and $a < b$. Let $f : [a,b] \to X$ be a strongly measurable function such that*

$$\int_a^b \|f(t)\|^p \, dt < \infty.$$

Then, for every $\varepsilon > 0$, there exists a $\delta > 0$ such that, for all $h \in \mathbb{R}$,

$$0 < h < \delta \qquad \Longrightarrow \qquad \int_a^{b-h} \|f(t+h) - f(t)\|^p \, dt < \varepsilon.$$

Proof. Exercise. **Hint:** Prove this first when f is continuous and then use Theorem 7.5.4. □

7.5.2. The Banach Space $L^p(I, X)$. The remainder of this section begins with a discussion of Banach space valued L^p functions on an interval, and then moves on to the Radon–Nikodým property, the dual space of L^p, and the Sobolev space $W^{1,p}$. These are important topics with many applications. In particular, this material will be used in Section 7.6 on the inhomogeneous equation associated to a semigroup.

Let X be a real Banach space, fix real numbers $1 \leq p < \infty$ and $a < b$, and abbreviate $I := [a,b]$. Define $L^p(I,X) := \mathcal{L}^p(I,X)/\sim$, where

$$\mathcal{L}^p(I,X) := \left\{ f : I \to X \;\middle|\; \begin{array}{l} f \text{ is strongly measurable} \\ \text{and } \int_a^b \|f(t)\|^p \, dt < \infty \end{array} \right\} \tag{7.5.5}$$

and the equivalence relation is equality almost everywhere. It is often convenient to abuse notation and use f to denote an equivalence class in $L^p(I,X)$ as well as a representative of this class in $\mathcal{L}^p(I,X)$. For $f \in \mathcal{L}^p(I,X)$ define

$$\|f\|_{L^p} := \left(\int_a^b \|f(t)\|^p \, dt \right)^{1/p}. \tag{7.5.6}$$

By the Minkowski inequality $L^p(I,X)$ is a normed vector space. For $p = \infty$ we define $L^\infty(I,X) := \mathcal{L}^\infty(I,X)/\sim$, where

$$\mathcal{L}^\infty(I,X) := \left\{ f : I \to X \;\middle|\; \begin{array}{l} f \text{ is strongly measurable} \\ \text{and bounded} \end{array} \right\} \tag{7.5.7}$$

and the equivalence relation is again given by equality almost everywhere. The norm on $L^\infty(I,X)$ is the essential supremum

$$\|f\|_{L^\infty} := \inf \left\{ \sup_{t \in I \setminus E} \|f(t)\| \;\middle|\; \begin{array}{l} E \subset I \text{ is a Borel set} \\ \text{of Lebesgue measure zero} \end{array} \right\} \tag{7.5.8}$$

for $f \in \mathcal{L}^\infty(I,X)$. We emphasize that these definitions have been chosen such that the functions in $\mathcal{L}^p(I,X)$ are all strongly measurable.

Theorem 7.5.6. *Let X be a Banach space, let $I \subset \mathbb{R}$ be a compact interval, and let $1 \le p \le \infty$. Then the following hold.*

(i) *Let $(f_n)_{n\in\mathbb{N}}$ be a Cauchy sequence in $L^p(I,X)$. If $p = \infty$, then the sequence $(f_n(t))_{n\in\mathbb{N}}$ converges in X for almost every $t \in I$. If $1 \le p < \infty$, then there exists a subsequence $(f_{n_i})_{i\in\mathbb{N}}$ such that the sequence $(f_{n_i}(t))_{i\in\mathbb{N}}$ converges in X for almost every $t \in I$.*

(ii) *$L^p(I,X)$ is a Banach space.*

(iii) *For $1 \le p < \infty$, the subspace $C_0^\infty(I,X)$ of smooth functions $f : I \to X$ that vanish near the boundary is a dense subset of $L^p(I,X)$.*

(iv) *There exists a unique linear operator*

$$L^p(I,X) \to X : f \mapsto \int_a^b f(t)\,dt,$$

called the **integral**, *such that*

$$\left\langle x^*, \int_a^b f(t)\,dt \right\rangle = \int_a^b \langle x^*, f(t)\rangle\,dt \tag{7.5.9}$$

for all $f \in \mathcal{L}^p(I,X)$ and all $x^* \in X^*$.

Proof. We prove the assertions only for $p < \infty$. The case $p = \infty$ is left to the reader. Let $f_n \in L^p(I,X)$ be a Cauchy sequence. Choose a subsequence f_{n_i} such that $\|f_{n_i} - f_{n_{i+1}}\|_{L^p} < 2^{-i}$ for all $i \in \mathbb{N}$. Then the same argument as in [**75**, p. 139] shows that f_{n_i} converges almost everywhere to a function $f : I \to X$. Namely, the sequence of Borel measurable functions $\phi_k := \sum_{i=1}^k \|f_{n_{i+1}} - f_{n_i}\| : I \to [0,\infty)$ is monotonically increasing and satisfies $\|\phi_k\|_{L^p} < 1$ for all k. Hence, by the Lebesgue Monotone Convergence Theorem, the sequence $\phi_k^p : I \to [0,\infty)$ converges to a Borel measurable function $\psi : I \to [0,\infty]$ and

$$\int_a^b \psi(t)\,dt = \lim_{k\to\infty} \int_a^b \phi_k(t)^p\,dt \le 1.$$

Thus there is a Borel set $E \subset I$ of Lebesgue measure zero such that $\psi(t) < \infty$ for all $t \in I \setminus E$ (see [**75**, Lemma 1.47]). Hence the sequence $(f_{n_i}(t))_{i\in\mathbb{N}}$ converges in X for all $t \in I \setminus E$ by Lemma 1.5.1. Define $f : I \to X$ by

$$f(t) := \begin{cases} \lim_{i\to\infty} f_{n_i}(t), & \text{for } t \in I \setminus E, \\ 0, & \text{for } t \in E. \end{cases}$$

By Theorem 7.5.4 and the axiom of countable choice, there exists a sequence of measurable step functions $g_i : I \to X$ such that $\|g_i - f_{n_i}\|_{L^p} < 2^{-i}$ for all $i \in \mathbb{N}$. Use the same argument as above, pass to a further subsequence, and enlarge the Borel set E of Lebesgue measure zero, if necessary, to obtain that the sequence $(g_i(t) - f_{n_i}(t))_{i\in\mathbb{N}}$ converges to zero for every $t \in I \setminus E$. Then g_i converges to f almost everywhere, and so f is strongly measurable.

We must prove that $f \in \mathcal{L}^p(I, X)$ and $\lim_{n\to\infty} \|f - f_n\|_{L^p} = 0$. To see this, fix a constant $\varepsilon > 0$ and choose $n_0 \in \mathbb{N}$ such that $\|f_n - f_m\|_{L^p} < \varepsilon$ for all integers $n, m \geq n_0$. Then, by the Lemma of Fatou [**75**, Thm. 1.41],

$$\begin{aligned}\int_a^b \|f_n(t) - f(t)\|^p \, dt &= \int_a^b \liminf_{k\to\infty} \left\|f_n(t) - f_{n_k}(t)\chi_{I\setminus E}(t)\right\|^p dt \\ &\leq \liminf_{k\to\infty} \int_a^b \left\|f_n(t) - f_{n_k}(t)\chi_{I\setminus E}(t)\right\|^p dt \\ &= \liminf_{k\to\infty} \int_a^b \|f_n(t) - f_{n_k}(t)\|^p \, dt \\ &\leq \varepsilon^p\end{aligned}$$

for all $n \geq n_0$. Hence $\|f\|_{L^p} \leq \|f_{n_0}\|_{L_p} + \varepsilon < \infty$, and so $f \in L^p(I, X)$ and the sequence $(f_n)_{n\in\mathbb{N}}$ converges to f in $L^p(I, X)$. Hence $L^p(I, X)$ is a Banach space and this proves (i) and (ii).

We prove (iii). That $C(I, X)$ is dense in $L^p(I, X)$ follows directly from Theorem 7.5.4. Hence multiplication with smooth cutoff functions that vanish near the boundary shows that the space $C_c(I, X)$ of continuous functions with support in the interior of I is also dense in $L^p(I, X)$. Now fix a function $f \in C_c(I, X)$ and choose a smooth function $\rho : \mathbb{R} \to [0, \infty)$ with support in the interval $[-1, 1]$ and mean value 1, and define $\rho_\delta(t) := \delta^{-1}\rho(\delta^{-1}t)$ for $\delta > 0$ and $t \in \mathbb{R}$. Then the function $f_\delta : I \to \mathbb{R}$, defined by

$$f_\delta(t) := (\rho_\delta * f)(t) := \int_{\mathbb{R}} \rho_\delta(t - s) f(s) \, ds$$

for $t \in \mathbb{R}$, is smooth for every $\delta > 0$ and vanishes near the boundary of I for $\delta > 0$ sufficiently small. Moreover, f_δ converges to f uniformly, because

$$\begin{aligned}\sup_{t\in I} \|f_\delta(t) - f(t)\| &= \sup_{t\in I} \left\| \int_{\mathbb{R}} \rho_\delta(t - s)(f(s) - f(t)) \, ds \right\| \\ &\leq \sup \{\|f(s) - f(t)\| \mid s, t \in I, \, |s - t| \leq \delta\}\end{aligned}$$

and f is uniformly continuous. Since $\|f_\delta - f\|_{L^p} \leq |I|^{1/p} \|f_\delta - f\|_{L^\infty}$, this implies $\lim_{\delta\to 0} \|f_\delta - f\|_{L^p} = 0$ and this proves part (iii).

Next observe that the operator $C(I, X) \to X : f \mapsto \int_a^b f(t) \, dt$ in Lemma 5.1.8 is bounded with respect to the L^p norm on $C(I, X)$ by part (vi) of Lemma 5.1.10 and the Hölder inequality. Since the subspace $C(I, X)$ is dense in $L^p(I, X)$ by part (i), the integral extends uniquely to a bounded linear functional on $L^p(I, X)$. Since every linear operator satisfying (7.5.9) is necessarily bounded, this proves part (iv) and Theorem 7.5.6. □

7.5.3. The Radon–Nikodým Property. The next goal is to examine the dual space of $L^p(I, X)$. This is a surprisingly delicate topic and many mathematicians have worked on this problem, starting with Bochner [**15, 16**]. It has led to the question of whether an absolutely continuous function on an interval with values in a Banach space is almost everywhere differentiable. We begin this discussion by examining the derivative of a continuous function on the domain where it exists.

Lemma 7.5.7. *Let X be a Banach space, let $I = [0, 1]$ be the unit interval, and let $F : I \to X$ be a continuous function. Then the set*

$$Z := \{t \in I \,|\, f \text{ is not differentiable at } t\} \tag{7.5.10}$$

is a Borel set, and the function $f : I \to X$ defined by

$$f(t) := \begin{cases} 0, & \text{for } t \in Z, \\ F'(t), & \text{for } t \in I \setminus Z, \end{cases} \tag{7.5.11}$$

is strongly measurable.

Proof. Let $\varepsilon > 0$. Then the set

$$E(\varepsilon, h, h') := \left\{ t \in I \,\middle|\, \begin{array}{l} \text{if } t + h \in I \text{ and } t + h' \in I, \text{ then} \\ \left| \frac{F(t+h)-F(t)}{h} - \frac{F(t+h')-F(t)}{h'} \right| \le \varepsilon \end{array} \right\}$$

is a Borel set for all $h, h' \in \mathbb{R} \setminus \{0\}$ and hence so is the set

$$E_{\varepsilon,\delta} := \bigcap_{\substack{h,h' \in \mathbb{Q} \\ 0<|h|,|h'|<\delta}} E(\varepsilon, h, h') = \bigcap_{\substack{h,h' \in \mathbb{R} \\ 0<|h|,|h'|<\delta}} E(\varepsilon, h, h')$$

for all $\delta > 0$. Here the second equality holds because F is continuous. Thus

$$E := \bigcap_{\substack{\varepsilon \in \mathbb{Q} \\ \varepsilon > 0}} \bigcup_{\substack{\delta \in \mathbb{Q} \\ \delta > 0}} E_{\varepsilon,\delta} = \bigcap_{\varepsilon > 0} \bigcup_{\delta > 0} E_{\varepsilon,\delta}$$

is a Borel set. Now the function F is differentiable at an element $t \in I$ if and only if $t \in E$. Thus $Z = I \setminus E$ is a Borel set.

For each $n \in \mathbb{N}$ define the function $f_n : I \to X$ by

$$f_n(t) := \begin{cases} 0, & \text{if } t \in Z, \\ 2^n\big(F(t + 2^{-n}) - F(t)\big), & \text{if } t \in E \text{ and } 0 \le t \le 1/2, \\ 2^n\big(F(t) - F(t - 2^{-n})\big), & \text{if } t \in E \text{ and } 1/2 < t \le 1. \end{cases} \tag{7.5.12}$$

Let $X_0 \subset X$ be the smallest closed subspace that contains the image of F. Then X_0 is a separable subspace of X. For each n the function f_n takes values in X_0 and is weakly measurable, and hence is strongly measurable by part (ii) of Theorem 7.5.2. Moreover, $f(t) = \lim_{t \to \infty} f_n(t)$ for every $t \in I$. Hence f takes values in X_0 and is weakly measurable, and so is strongly measurable by part (ii) of Theorem 7.5.2. This proves Lemma 7.5.7. □

Let $I \subset \mathbb{R}$ be a compact interval and let $F : I \to X$ be a continuous function with values in a Banach space. Recall that F is called **Lipschitz continuous** if there exists a $c \geq 0$ such that $\|F(s) - F(t)\| \leq c\,|s-t|$ for all $s, t \in I$. Recall that F is called **absolutely continuous** if for every $\varepsilon > 0$ there exists a $\delta > 0$ such that every sequence $s_1 \leq t_1 \leq \cdots \leq s_N \leq t_N$ in I with $\sum_i |s_i - t_i| < \delta$ satisfies $\sum_i \|F(s_i) - F(t_i)\| < \varepsilon$.

Lemma 7.5.8. *Let X be a Banach space, let $I = [0,1]$, and let $F : I \to X$ be a Lipschitz continuous function that is almost everywhere differentiable. Then the function $f : I \to X$ defined by (7.5.10) and (7.5.11) is bounded and strongly measurable and satisfies*

$$F(t) - F(0) = \int_0^t f(s)\,ds \qquad \text{for all } t \in I.$$

Proof. Choose $c > 0$ such that $\|F(s) - F(t)\| \leq c\,|s-t|$ for all $s, t \in I$. Then the functions $f_n : I \to X$ in (7.5.12) satisfy $\|f_n(t)\| \leq c$ for all $t \in I$ and all $n \in \mathbb{N}$. Hence $\|f(t)\| \leq c$ for all $t \in I$. Second, f is strongly measurable by Lemma 7.5.7. Third, the set $Z \subset I$ in Lemma 7.5.7 has Lebesgue measure zero by assumption. Hence for each $x^* \in X^*$ the function $\langle x^*, F\rangle : I \to \mathbb{R}$ is absolutely continuous and its derivative agrees almost everywhere with the function $\langle x^*, f\rangle : I \to \mathbb{R}$. By [**75**, Thm. 6.19], this implies

$$\langle x^*, F(t) - F(0)\rangle = \int_0^t \langle x^*, f(s)\rangle\,ds$$

for all $t \in I$ and all $x^* \in X^*$. This proves Lemma 7.5.8. □

Lemma 7.5.9. *Let X be a Banach space, let $I = [0,1]$, let $f : I \to X$ be a strongly measurable function with $\int_0^1 \|f(t)\|\,dt < \infty$, and define $F : I \to X$ by $F(t) := \int_0^t f(s)\,ds$ for $t \in I$. Then F is absolutely continuous and almost everywhere differentiable with $F'(t) = f(t)$ for almost every $t \in I$.*

Proof. The absolute continuity of F follows as in [**75**, Thm. 6.29]. That F is almost everywhere differentiable with $F' = f$ follows from the Lebesgue Differentiation Theorem [**75**, Thm. 6.14] whose proof carries over verbatim to Banach space valued functions. This proves Lemma 7.5.9. □

With these preparations in place we are now ready to formulate the main problem of this subsection, namely whether or not every Lipschitz continuous function with values in a given Banach space X is almost everywhere differentiable. If it is, then Lemma 7.5.7 shows that its derivative is strongly measurable and Lemma 7.5.8 shows that it is the integral of its derivative. Lemma 7.5.9 shows that the integrals of bounded measurable functions are necessarily almost everywhere differentiable. The next lemma relates this problem to the differentiability of absolutely continuous functions.

Lemma 7.5.10. *Let X be a Banach space and let $I := [0,1]$ be the unit interval. Then the following are equivalent.*

(i) *Every Lipschitz continuous function $F : I \to X$ is almost everywhere differentiable.*

(ii) *Every absolutely continuous function $F : I \to X$ is almost everywhere differentiable.*

If these equivalent conditions are satisfied, and $F : I \to X$ is an absolutely continuous function, then its derivative $f := F' : I \to X$ is strongly measurable, $\int_0^1 \|f(s)\|\, ds < \infty$, and $F(t) - F(0) = \int_0^t f(s)\, ds$ for all $t \in I$.

Proof. That (ii) implies (i) is obvious, because every Lipschitz continuous function is absolutely continuous. Hence assume (i) and let $F : I \to X$ be an absolutely continuous function. Define $\Phi : [0,1] \to [0,\infty)$ by

$$\Phi(t) := \mathrm{Var}(F|_{[0,t]}) = \sup_{0=t_0<t_1<\cdots<t_N=t} \sum_{i=1}^{N} \|F(t_i) - F(t_{i-1})\| .$$

Then Φ is absolutely continuous and monotone. Denote

$$c := \Phi(1) = \mathrm{Var}(F).$$

Since $\|F(t) - F(s)\| \le \Phi(t) - \Phi(s)$ for all $0 \le s \le t \le 1$, there is a unique function $G : [0,c] \to X$ such that $G(\Phi(t)) = F(t)$ for all $t \in [0,1]$, and G is Lipschitz continuous with Lipschitz constant 1. Hence, by part (i), G is almost everywhere differentiable and so, by Lemma 7.5.8, there exists a strongly measurable $g : I \to X$ such that

$$\sup_{0\le\tau\le c} \|g(\tau)\| \le 1, \qquad G(\theta) = G(0) + \int_0^\theta g(\tau)\, d\tau$$

for all $\theta \in [0,c]$. Moreover, by [**75**, Thm. 6.19], there exists a Borel measurable function $\phi : I \to [0,\infty)$ with $\int_0^1 |\phi(s)|\, ds < \infty$ such that

$$\Phi(t) = \int_0^t \phi(s)\, ds$$

for all $t \in I$. Hence the function $f := \phi(g \circ \Phi) : I \to X$ is strongly measurable and satisfies $\int_0^1 \|f(s)\|\, ds \le \int_0^1 \phi(s)\, ds < \infty$ and

$$\int_0^t f(s)\, ds = \int_0^t \phi(s) g(\Phi(s))\, ds = \int_0^{\Phi(t)} g(\tau)\, d\tau = F(t) - F(0)$$

for all $t \in I$. Here the second step uses the fact that $C(I)$ is dense in $L^1(I)$ and so there exists a sequence of continuous functions $\phi_i : [0,1] \to [0,\infty)$ with $\int_0^1 \phi_i(t)\, dt = c$ and $\lim_{i\to\infty} \int_0^1 |\phi_i(t) - \phi(t)| = 0$. Now it follows from Lemma 7.5.9 that F is differentiable almost everywhere and $F' = f$. This proves Lemma 7.5.10. □

Definition 7.5.11. A Banach space X is said to have the **Radon–Nikodým property** if every Lipschitz continuous function $f : [0,1] \to X$ is almost everywhere differentiable or, equivalently, every absolutely continuous function $f : [0,1] \to X$ is almost everywhere differentiable.

Remark 7.5.12. The reason for this terminology lies in the fact that a Banach space X has the Radon–Nikodým property if and only if it satisfies the following for every measurable space $(M, \mathcal{A})$. *Let* $\nu : \mathcal{A} \to X$ *be a* **countably additive map**, *i.e. if* $A_i \in \mathcal{A}$ *is a sequence of pairwise disjoint measurable sets, then* $\nu\left(\bigcup_{i=1}^{\infty} A_i\right) = \lim_{n\to\infty} \sum_{i=1}^{n} \nu(A_i) = \sum_{i=1}^{\infty} \nu(A_i)$. *Assume* ν *has* **bounded variation**, *i.e.*

$$\mu(A) := \sup \left\{ \sum_{i=1}^{N} \|\nu(A_i)\| \;\middle|\; \begin{array}{l} A_1, \dots, A_N \in \mathcal{A}, \\ A_i \cap A_j = \emptyset \textit{ for } i \neq j, \\ A_1 \cup \cdots \cup A_N = A \end{array} \right\} < \infty$$

for all $A \in \mathcal{A}$. *Then there exists a strongly* $\mathcal{A}$*-measurable map* $f : M \to X$ *with* $\int_M \|f\|\, d\mu < \infty$ *and* $\nu(A) = \int_A f\, d\mu$ *for all* $A \in \mathcal{A}$.

That this condition is indeed equivalent to the Radon–Nikodým property in Definition 7.5.11 was proved by Bochner–Taylor [**17**] in the late 1930s. For other expositions see [**10, 21, 22**].

Theorem 7.5.13 (Dunford–Pettis). (i) *If* X *is a Banach space and its dual space* X^* *is separable, then* X^* *has the Radon–Nikodým property.*

(ii) *Every reflexive Banach space has the Radon–Nikodým property.*

Proof. See page 416. □

Remark 7.5.14. (i) Part (i) of Theorem 7.5.13 was proved by Gelfand [**29**] using the notion in Definition 7.5.11, and then by Dunford–Pettis [**25**] using the notion in Remark 7.5.12. That Hilbert spaces have the Radon–Nikodým property was first proved by Birkhoff [**14**], and this was extended to all reflexive spaces by Dunford–Pettis [**25**].

(ii) By part (i) of Theorem 7.5.13 the Banach space $X = \ell^1$ has the Radon–Nikodým property. This was first noted by Clarkson [**20**] and was extended by Dunford–Morse [**24**] to all Banach spaces with boundedly complete Schauder bases. Clarkson [**20**] also proved that all uniformly convex Banach spaces have the Radon–Nikodým property.

(iii) Banach spaces that do not have the Radon–Nikodým property include the examples $X = L^\infty([0,1])$ (Bochner [**16**]) and $X = c_0$ and $X = L^1([0,1])$ (Clarkson [**20**]). Hence c_0 and $L^1([0,1])$ cannot be isomorphic to the dual space of any Banach space.

Proof of Theorem 7.5.13. We prove part (i), following the exposition by Kreuter [**50**]. Let X be a real Banach space with a separable dual space X^* and let $G : I = [0,1] \to X^*$ be a Lipschitz continuous function with $G(0) = 0$ and Lipschitz constant 1. Since X^* is separable, so is X by Theorem 2.4.6. Hence there exists a linearly independent sequence $(x_k)_{k\in\mathbb{N}}$ in X such that

$$X = \overline{Y}, \qquad Y := \left\{ \sum_{k=1}^N \lambda_k x_k \,\middle|\, N \in \mathbb{N},\, \lambda_1, \ldots, \lambda_N \in \mathbb{R} \right\}.$$

For each $x \in X$ the function $\langle G, x\rangle : I \to \mathbb{R}$ is Lipschitz continuous with Lipschitz constant $\|x\|$. Hence $\langle G, x\rangle$ is almost everywhere differentiable. For each $k \in \mathbb{N}$ let $Z_k \subset I$ be the set of all $t \in I$ such that $\langle G, x_k\rangle$ is not differentiable at t. Then Z_k is a Borel set by Lemma 7.5.7 and it has Lebesgue measure zero. Hence the Borel set $Z := \bigcup_{k=1}^\infty Z_k$ has Lebesgue measure zero and $\langle G, y\rangle$ is differentiable on $I \setminus Z$ for each $y \in Y$. For $y \in Y$ define the function $g_y : I \to \mathbb{R}$ by

$$g_y(t) := \begin{cases} \lim_{h\to 0} h^{-1}\langle G(t+h) - G(t), y\rangle, & \text{if } t \in I \setminus Z, \\ 0, & \text{if } t \in Z. \end{cases} \tag{7.5.13}$$

This function is measurable and satisfies

$$\langle G(t), y\rangle = \int_0^t g_y(s)\, ds, \qquad \|g_y(t)\| \le \|y\| \qquad \text{for all } t \in I. \tag{7.5.14}$$

Moreover, for each $t \in I$ the functional $Y \to \mathbb{R} : y \mapsto g_y(t)$ is linear and bounded by (7.5.14), and so extends uniquely to a bounded linear functional on all of X. Thus there exists a unique function $g : I \to X^*$ such that

$$\langle g(t), y\rangle = g_y(t) \quad \text{for } t \in I \text{ and } y \in Y, \qquad \sup_{0\le t\le 1} \|g(t)\| \le 1. \tag{7.5.15}$$

By (7.5.14) we have

$$\langle G(t), x\rangle = \int_0^t \langle g(s), x\rangle\, ds \tag{7.5.16}$$

for all $x \in Y$ and all $t \in I$. By continuity in x the function $\langle g, x\rangle$ is Borel measurable for all $x \in X$ and equation (7.5.16) continues to hold for all $x \in X$ and all $t \in I$. Since X^* is separable, it follows from part (i) of Theorem 7.5.2 (with X replaced by X^* and $E := \iota(X) \subset X^{**}$) that the function $g : I \to X^*$ is strongly measurable. Hence $G(t) = \int_0^t g(s)\, ds$ for all $t \in I$ by (7.5.16), and so it follows from the Lebesgue Differentiation Theorem that G is almost everywhere differentiable (see Lemma 7.5.9). This proves part (i).

We prove part (ii). Let X be a reflexive Banach space and let $G : I \to X$ be a Lipschitz continuous function. Denote by $Y \subset X$ the smallest closed subspace of X that contains the image of G. Then Y is separable and is reflexive by Theorem 2.4.4. Hence it follows from part (i) that G is almost everywhere differentiable, and this proves Theorem 7.5.13. □

7.5.4. The Dual Space of $L^p(I,X)$. It is a natural question to ask how the dual space of $L^p(I,X)$ can be characterized. The obvious candidate for the dual space is $L^q(I,X^*)$ with $1/p+1/q=1$.

Lemma 7.5.15. *Let X be a real Banach space, let $I=[a,b]$ be a compact interval, let $1\le p,q\le\infty$ with $1/p+1/q=1$, and let $g\in\mathcal{L}^q(I,X^*)$. Then the map $\Lambda_g:L^p(I,X)\to\mathbb{R}$, defined by*

$$\Lambda_g(f):=\int_a^b\langle g(t),f(t)\rangle\,dt\qquad\text{for } f\in\mathcal{L}^p(I,X),\tag{7.5.17}$$

is a bounded linear functional with $\|\Lambda_g\|=\|g\|_{L^q}$.

Proof. The function $I\to\mathbb{R}:t\mapsto\langle g(t),f(t)\rangle$ is measurable because f and g are strongly measurable. Moreover, by the Hölder inequality, this function is integrable and satisfies $|\int_a^b\langle g(t),f(t)\rangle\,dt|\le\|g\|_{L^q}\|f\|_{L^p}$. Hence the map $\Lambda_g:L^p(I,X)\to\mathbb{R}$ is a bounded linear functional with $\|\Lambda_g\|\le\|g\|_{L^q}$. Thus the map $L^q(I,X^*)\to L^p(I,X)^*$ is a bounded linear operator of norm less than or equal to one. To prove that it is an isometry, it suffices to prove the equation $\|\Lambda_g\|=\|g\|_{L^q}$ for all elements g of a dense subset of $L^q(I,X^*)$. Such a dense subset is the set of measurable step functions by Theorem 7.5.4, provided that $q<\infty$. Here we focus on the case $1<p,q<\infty$ and leave the remaining cases to the reader. Consider a function of the form

$$g=\sum_{i=1}^m\chi_{B_i}x_i^*$$

for $x_1^*,\dots,x_m^*\in X^*\setminus\{0\}$ and pairwise disjoint Borel sets $B_1,\dots,B_m\subset I$. Fix a number $\varepsilon>0$ and choose elements $x_1,\dots,x_m\in X$ such that $\|x_i\|=1$ and $\langle x_i^*,x_i\rangle>(1-\varepsilon)\|x_i^*\|$ for all i. Define the function $f:I\to X$ by

$$f:=\sum_i\chi_{B_i}\|x_i^*\|^{q-1}x_i.$$

Then

$$\int_I\langle g,f\rangle=\sum_i\mu(B_i)\|x_i^*\|^{q-1}\langle x_i^*,x_i\rangle>(1-\varepsilon)\|g\|_{L^q}^q$$

and

$$\|f\|_{L^p}=\left(\sum_i\mu(B_i)\|x_i^*\|^{p(q-1)}\right)^{1/p}=\left(\sum_i\mu(B_i)\|x_i^*\|^q\right)^{1-1/q}=\|g\|_{L^q}^{q-1}.$$

This implies $\|\Lambda_g\|\ge\|f\|_{L_p}^{-1}\int_I\langle g,f\rangle>(1-\varepsilon)\|g\|_{L_q}$. Since $\varepsilon>0$ was chosen arbitrarily, we find that $\|\Lambda_g\|=\|g\|_{L_q}$ for every measurable step function $g:I\to X^*$ and this proves Lemma 7.5.15. □

The central question is now under which conditions the isometric embedding $L^q(I, X^*) \to L^p(I, X)^*$ in Lemma 7.5.15 is surjective. The answer depends on the Banach space X and is surprisingly subtle. It was first noted by Bochner [**15, 16**] that a positive answer requires that every absolutely continuous function with values in the dual space X^* is almost everywhere differentiable.

Theorem 7.5.16 (Bochner). *Let X be a Banach space, let $I := [0, 1]$, and let $p, q > 1$ with $1/p + 1/q = 1$. Then the following are equivalent.*

(i) *The isometric embedding $L^q(I, X^*) \to L^p(I, X)^*$ is surjective.*

(ii) *The isometric embedding $L^\infty(I, X^*) \to L^1(I, X)^*$ is surjective.*

(iii) *The dual space X^* has the Radon–Nikodým property.*

Proof. We prove that (i) implies (ii). Let $\Lambda : L^1(I, X) \to \mathbb{R}$ be a bounded linear functional and denote

$$c := \|\Lambda\|.$$

Then Λ restricts to a bounded linear functional on $L^p(I, X)$. Hence by part (i) there is a function $g \in \mathcal{L}^q(I, X^*)$ such that

$$\int_I \langle g, f\rangle = \Lambda(f) \le c\,\|f\|_{L^1}$$

for all $f \in \mathcal{L}^p(I, X)$. We claim that $\|g\|_{L^\infty} \le c$. Otherwise, there exists a constant $\delta > 0$ such that the set

$$A := \{t \in I \mid \|g(t)\| > c + \delta\}$$

has positive measure. By Theorem 7.5.6 there is a sequence of measurable step functions $g_i : I \to X^* \setminus \{0\}$ that converges in L^q and almost everywhere to g. For each i let $f_i : I \to X$ be a measurable step function that satisfies $\langle g_i(t), f_i(t)\rangle \ge (1 - \frac{1}{i})\,\|g_i(t)\|$ and $\|f_i(t)\| = 1$ for all i and t. Then

$$\langle g(t), f_i(t)\rangle \ge \left(1 - \frac{1}{i}\right)\|g_i(t)\| - \|g_i(t) - g(t)\|$$

for all i and t, and hence

$$\liminf_{i\to\infty} \int_I \langle g, \chi_A f_i\rangle \ge \lim_{i\to\infty} \int_A \|g_i\| = \int_A \|g\| \ge (c + \delta)\mu(A).$$

Thus

$$\int_I \langle g, \chi_A f_i\rangle > c\mu(A) = c\,\|\chi_A f_i\|_{L^1}$$

for i sufficiently large. This contradiction shows that $\|g\|_{L^\infty} \le c$ as claimed. This proves that (i) implies (ii).

We prove that (ii) implies (iii). Let $G : I \to X^*$ be a Lipschitz continuous function with Lipschitz constant c so that

$$\|G(s) - G(t)\| \le c\,|s-t|$$

for all $s, t \in I$. For a step function $f : I \to X$ of the form

$$f = \sum_{i=0}^{N} \chi_{[t_{i-1},t_i)} x_i$$

with $0 = t_0 < t_1 < \cdots < t_N = 1$ and $x_i \in X$ define

$$\Lambda(f) := \sum_{i=1}^{N} \langle G(t_i) - G(t_{i-1}), x_i \rangle.$$

Then

$$|\Lambda(f)| \le c \sum_{i=1}^{N} (t_i - t_{i-1}) \|x_i\| = c \|f\|_{L^1}.$$

Thus Λ is a bounded linear functional on a dense subset of $L^1(I, X)$, by Theorem 7.5.6, and hence extends uniquely to a bounded linear functional on $L^1(I, X)$ which will still be denoted by

$$\Lambda : L^1(I, X) \to \mathbb{R}.$$

By part (ii), there exists a bounded strongly measurable function $g : I \to X^*$ such that

$$\int_0^1 \langle g(t), f(t) \rangle \, dt = \Lambda(f)$$

for all $f \in \mathcal{L}^1(I, X)$. Take $f := \chi_{[0,t)} x$ to obtain

$$\begin{aligned} \left\langle \int_0^t g(s)\, ds, x \right\rangle &= \int_0^t \langle g(s), x \rangle \, ds \\ &= \Lambda(\chi_{[0,t)} x) \\ &= \langle G(t) - G(0), x \rangle \end{aligned}$$

for all $t \in I$ and all $x \in X$. This implies

$$\int_0^t g(s)\, ds = G(t) - G(0)$$

for all $t \in I$. Hence it follows from the Lebesgue Differentiation Theorem (see for example [**75**, Thm. 6.14]) that the function G is almost everywhere differentiable and

$$G' = g.$$

This shows that X^* has the Radon–Nikodým property.

We prove that (iii) implies (i). Let $\Lambda : L^p(I,X) \to \mathbb{R}$ be a bounded linear functional and let $\mathcal{B} \subset 2^I$ be the Borel σ-algebra. Define the map $\nu : \mathcal{B} \to X^*$ by

$$\langle \nu(B), x\rangle := \Lambda(\chi_B x) \qquad \text{for } B \in \mathcal{B} \text{ and } x \in X. \tag{7.5.18}$$

More precisely, the linear functional $X \to \mathbb{R} : x \mapsto \Lambda(\chi_B x)$ is bounded because $|\Lambda(\chi_B x)| \le \|\Lambda\| \, \|\chi_B x\|_{L^1} \le \|\Lambda\| \, \mu(B)^{1/p} \, \|x\|$. We prove that every finite sequence of pairwise disjoint Borel sets $B_1, \dots, B_N \in \mathcal{B}$ satisfies

$$\sum_{i=1}^N \|\nu(B_i)\| \le \|\Lambda\| \, \mu\left(\bigcup_{i=1}^N B_i\right)^{1/p}. \tag{7.5.19}$$

To see this, fix a constant $\varepsilon > 0$ and, for each i, choose a vector $x_i \in X$ such that $\|x_i\| = 1$ and $\langle \nu(B_i), x_i\rangle \ge (1-\varepsilon)\|\nu(B_i)\|$. Define $f := \sum_i \chi_{B_i} x_i$. Then

$$\sum_i \|\nu(B_i)\| \le \sum_i \frac{\langle \nu(B_i), x_i\rangle}{1-\varepsilon} = \frac{\Lambda(f)}{1-\varepsilon} \le \frac{\|\Lambda\| \, \|f\|_{L^p}}{1-\varepsilon} = \frac{\|\Lambda\| \, \mu(\bigcup_i B_i)^{1/p}}{1-\varepsilon}.$$

This proves (7.5.19).

Now define the function $G : I \to X^*$ by

$$G(t) := \nu([0,t]) \qquad \text{for } t \in I. \tag{7.5.20}$$

This function satisfies $G(0) = 0$ and is absolutely continuous by (7.5.19). Hence, by (iii) there exists a function $g \in \mathcal{L}^1(I, X^*)$ such that

$$G(t) = \int_0^t g(s)\, ds \qquad \text{for all } t \in I. \tag{7.5.21}$$

For each $x \in X$ consider the bounded linear functional $\Lambda_x : L^p(I) \to \mathbb{R}$ defined by $\Lambda_x(\phi) := \Lambda(\phi x)$ for $\phi \in \mathcal{L}^p(I)$. By [**75**, Thm. 4.35] there exists a function $g_x \in \mathcal{L}^q(I)$ such that

$$\int_I g_x \phi = \Lambda_x(\phi) = \Lambda(\phi x) \qquad \text{for all } \phi \in \mathcal{L}^p(I). \tag{7.5.22}$$

Then, for each $t \in I$ and each $x \in X$, we have

$$\int_0^t g_x(s)\, ds = \Lambda(\chi_{[0,t]} x) = \langle \nu([0,t]), x\rangle = \langle G(t), x\rangle = \int_0^t \langle g(s), x\rangle\, ds.$$

Here the first equality follows from the definition of g_x in (7.5.22), the second from the definition of ν in (7.5.18), the third from the definition of G in (7.5.20), and the last from (7.5.21). This shows that

$$g_x(t) = \langle g(t), x\rangle \tag{7.5.23}$$

for every $x \in X$ and almost every $t \in I$.

We prove that every $f \in \mathcal{L}^p(I,X)$ satisfies

$$\langle g, f\rangle \in \mathcal{L}^1(I), \qquad \int_I \langle g, f\rangle = \Lambda(f), \qquad \int_I |\langle g, f\rangle| \le \|\Lambda\| \, \|f\|_{L^p}. \tag{7.5.24}$$

First let $f : I \to X$ be a measurable step function of the form $f = \sum_i \chi_{B_i} x_i$, where the $B_i \subset I$ are pairwise disjoint Borel sets and $x_i \in X \setminus \{0\}$. Then

$$\langle g, f\rangle = \sum_i \chi_{B_i}\langle g, x_i\rangle \overset{a.e.}{=} \sum_i \chi_{B_i} g_{x_i},$$

where the last equation follows from (7.5.23). Thus $\langle g, f\rangle$ is integrable because χ_{B_i} is bounded and $g_{x_i} \in \mathcal{L}^q(I)$ for each i. Moreover, it follows from the definition of the functions g_{x_i} in (7.5.22) that

$$\int_I \langle g, f\rangle = \sum_i \int_I g_{x_i}\chi_{B_i} = \sum_i \Lambda(\chi_{B_i} x_i) = \Lambda(f).$$

Now define the function $\phi_i : I \to \mathbb{R}$ by $\phi_i(t) := 0$ for $t \in I \setminus B_i$, by $\phi_i(t) := 1$ for $t \in B_i$ with $g_{x_i}(t) \ge 0$, and by $\phi_i(t) := -1$ for $t \in B_i$ with $g_{x_i}(t) < 0$. Let $\widetilde{f} := \sum_i \chi_{B_i}\phi_i x_i$. Then

$$|\langle g, f\rangle| = \sum_i \chi_{B_i}|g_{x_i}| = \sum_i \chi_{B_i}\phi_i g_{x_i} \overset{a.e.}{=} \left\langle g, \sum_i \chi_{B_i}\phi_i x_i\right\rangle = \left\langle g, \widetilde{f}\right\rangle$$

and hence

$$\int_I |\langle g, f\rangle| = \int_I \left\langle g, \widetilde{f}\right\rangle = \Lambda(\widetilde{f}) \le \|\Lambda\| \, \|\widetilde{f}\|_{L^p} = \|\Lambda\| \, \|f\|_{L^p}.$$

This proves (7.5.24) for measurable step functions $f : I \to X$.

Now let $f \in \mathcal{L}^p(I,X)$ and choose a sequence of measurable step functions $f_i : I \to X$ that converges in L^p and almost everywhere to f. Then

$$\int_I |\langle g, f_i\rangle - \langle g, f_j\rangle| = \int_I |\langle g, f_i - f_j\rangle| \le \|\Lambda\| \, \|f_i - f_j\|_{L^p}$$

and so $(\langle g, f_i\rangle)_{i\in\mathbb{N}}$ is a Cauchy sequence in $L^1(I)$. Thus it converges to a function $h \in L^1(I)$. Passing to a suitable subsequence, we may assume the sequence converges almost everywhere to h. Hence

$$h(t) \overset{a.e.}{=} \lim_{i\to\infty} \langle g(t), f_i(t)\rangle \overset{a.e.}{=} \langle g(t), f(t)\rangle.$$

Hence $\langle g, f\rangle$ is integrable and

$$\int_I \langle g, f\rangle = \int_I h = \lim_{i\to\infty} \int_I \langle g, f_i\rangle = \lim_{i\to\infty} \Lambda(f_i) = \Lambda(f).$$

Moreover,

$$\|\langle g, f\rangle\|_{L^1} = \|h\|_{L^1} = \lim_{i\to\infty} \|\langle g, f_i\rangle\|_{L^1} \le \lim_{i\to\infty} \|\Lambda\| \, \|f_i\|_{L^p} = \|\Lambda\| \, \|f\|_{L^p}$$

and this proves (7.5.24).

With the help of (7.5.24) we are now able to prove that $g \in \mathcal{L}^q(I, X^*)$. For $n \in \mathbb{N}$ define the function $g_n : I \to X^*$ by

$$g_n(t) := \begin{cases} g(t), & \text{if } \|g(t)\| \le n, \\ 0, & \text{if } \|g(t)\| > n, \end{cases} \qquad \text{for } t \in I.$$

These functions are strongly measurable and satisfy $\lim_{n\to\infty} g_n(t) = g(t)$ for all $t \in I$. Moreover, it follows from (7.5.24) that

$$\int_I |\langle g_n, f\rangle| \le \int_I |\langle g, f\rangle| \le \|\Lambda\| \, \|f\|_{L^p}$$

for every $n \in \mathbb{N}$ and every $f \in \mathcal{L}^p(I, X)$. Since each function g_n is bounded, and hence an element of the space $\mathcal{L}^q(I, X^*)$, this implies

$$\|g_n\|_{L^q} = \sup_{f \in L^p(I,X)\setminus\{0\}} \frac{\left|\int_I \langle g_n, f\rangle\right|}{\|f\|_{L^p}} \le \sup_{f \in L^p(I,X)\setminus\{0\}} \frac{\int_I |\langle g_n, f\rangle|}{\|f\|_{L^p}} \le \|\Lambda\| \, .$$

Here the equality follows from Lemma 7.5.15. By the Lebesgue Monotone Convergence Theorem, this implies

$$\int_I \|g(t)\|^q \, dt = \lim_{n\to\infty} \int_I \|g_n(t)\|^q \, dt \le \|\Lambda\|^q \, .$$

Thus $g \in \mathcal{L}^q(I, X^*)$, $\|g\|_{L^q} \le \|\Lambda\|$, and $\int_I \langle g, f\rangle = \Lambda(f)$ for all $f \in \mathcal{L}^p(I, X)$ by (7.5.24). This completes the proof of Theorem 7.5.16. □

The following result was proved by R. S. Phillips [**66**] in 1943.

Corollary 7.5.17 (Phillips). *Fix a constant $1 < p < \infty$, let X be a reflexive Banach space, and let $I \subset \mathbb{R}$ be a compact interval. Then the Banach space $L^p(I, X)$ is reflexive.*

Proof. Choose the real number $1 < q < \infty$ such that $1/p + 1/q = 1$. The dual space X^* is reflexive by Theorem 2.4.4, and so has the Radon–Nikodým property by part (ii) of Theorem 7.5.13. Hence Theorem 7.5.16 asserts that the isometric embeddings

$$L^q(I, X^*) \to L^p(I, X)^* \tag{7.5.25}$$

and

$$L^p(I, X^{**}) \to L^q(I, X^*)^* \tag{7.5.26}$$

are isomorphisms. Now the canonical inclusion $\iota : L^p(I, X) \to L^p(I, X)^{**}$ is the composition

$$L^p(I, X) \to L^p(I, X^{**}) \to L^q(I, X^*)^* \to L^p(I, X)^{**}, \tag{7.5.27}$$

where the first map is induced by the canonical isomorphism $\iota : X \to X^{**}$, the second map is the isomorphism (7.5.26), and the third map is the inverse of the dual operator of (7.5.25) (see Corollary 4.1.18). This proves Corollary 7.5.17. □

7.5.5. The Sobolev Space $W^{1,p}(I,X)$. Let X be a Banach space, fix a real number $1 \le p < \infty$, and let $I = [a,b] \subset \mathbb{R}$ be a compact interval. The **Sobolev space** $W^{1,p}(I,X)$ can be defined as the completion of the space of continuously differentiable functions $f : I \to X$ with respect to the norm

$$\|f\|_{W^{1,p}} := \left(\int_a^b \left(\|f(t)\|^p + \|f'(t)\|^p\right) dt\right)^{1/p}. \tag{7.5.28}$$

Alternatively, $W^{1,p}(I,X)$ is the space of all functions $f : I \to \mathbb{R}$ that can be expressed as the integrals of L^p functions, i.e.

$$W^{1,p}(I,X) := \left\{ f : I \to X \;\middle|\; \begin{array}{l} \text{there exists a strongly} \\ \text{measurable function } g : I \to X \\ \text{such that } \int_a^b \|g(t)\|^p \, dt < \infty \\ \text{and } f(t) - f(a) = \int_a^t g(s)\, ds \\ \text{for all } t \in I \end{array} \right\}. \tag{7.5.29}$$

The Lebesgue Differentiation Theorem asserts that the function $g : I \to X$ in (7.5.29) is uniquely determined by f up to equality almost everywhere and agrees with the derivative of f (Lemma 7.5.9). The norm is again given by equation (7.5.28). With this definition the functions in $W^{1,p}(I,X)$ are absolutely continuous, are almost everywhere differentiable, have derivatives in $\mathcal{L}^p(I,X)$, and can be expressed as the integrals of their derivatives. If X has the Radon–Nikodým property, then every absolutely continuous function $f : I \to X$ is almost everywhere differentiable and we have

$$W^{1,p}(I,X) := \left\{ f : I \to X \;\middle|\; \begin{array}{l} f \text{ is absolutely continuous} \\ \text{and } f' \in \mathcal{L}^p(I,X) \end{array} \right\}.$$

If X does not have the Radon–Nikodým property, this last definition does not even make sense, because absolutely continuous functions need not be differentiable. Thus we will work with the definition (7.5.29). However, in all the relevant examples in this book the Banach space in question is reflexive and therefore does have the Radon–Nikodým property by Theorem 7.5.13. The next theorem asserts that the Sobolev space $W^{1,p}(I,X)$ is a Banach space and that the space $C^\infty(I,X)$ of smooth functions $f : I \to X$ is dense in $W^{1,p}(I,X)$.

Theorem 7.5.18. *Let X be a Banach space, let $I = [a,b] \subset \mathbb{R}$ be a compact interval, and fix a constant $1 \le p < \infty$. Then the following hold.*

(i) *There exists a $c > 0$ such that $\|f\|_{L^\infty} \le c\,\|f\|_{W^{1,p}}$ for all $f \in W^{1,p}(I,X)$.*

(ii) *The Sobolev space $W^{1,p}(I,X)$ is complete with the norm* (7.5.28).

(iii) *The subspace $C^\infty(I,X)$ is dense in $W^{1,p}(I,X)$.*

(iv) *If X is reflexive and $1 < p < \infty$, then $W^{1,p}(I,X)$ is reflexive.*

Proof. We prove part (i). Let $f \in W^{1,p}$ and choose $g \in \mathcal{L}^p(I,X)$ such that $\int_a^t g(s)\,ds = f(t) - f(a)$ for all $t \in I$. Then, by Hölder's inequality,

$$\|f(t) - f(s)\| = \left\|\int_s^t g(r)\,dr\right\| \le \left|\int_s^t \|g(r)\|\,dr\right| \le (b-a)^{1/q}\,\|g\|_{L^p}$$

for all $s,t \in [a,b]$. Here $1 < q \le \infty$ is chosen such that $1/p + 1/q = 1$. In the case $q = \infty$ we use the standard convention $(b-a)^{1/q} = (b-a)^0 := 1$. Now raise this inequality to the power p and integrate to obtain

$$\int_a^b \|f(t) - f(s)\|^p\,ds \le (b-a)^{1+p/q}\,\|g\|_{L^p}^p \le (b-a)^p\,\|g\|_{L^p}^p .$$

Take the pth root of this estimate to obtain

$$\left(\int_a^b \|f(t) - f(s)\|^p\,ds\right)^{1/p} \le (b-a)\,\|g\|_{L^p} .$$

Hence $(b-a)^{1/p}\,\|f(t)\| \le \|f\|_{L^p} + (b-a)\,\|g\|_{L^p}$ for all $t \in I$ by Minkowski's inequality. This proves part (i).

We prove part (ii). Let $f_n : I \to X$ be a Cauchy sequence in $W^{1,p}(I,X)$ and choose a sequence $g_n \in \mathcal{L}^p(I,X)$ such that $\int_a^t g_n(s)\,ds = f_n(t) - f_n(a)$ for all $t \in I$ and all $n \in \mathbb{N}$. Then $(f_n)_{n\in\mathbb{N}}$ is a Cauchy sequence in the Banach space $C(I,X)$ of continuous functions with the supremum norm, and $(g_n)_{n\in\mathbb{N}}$ is a Cauchy sequence in the Banach space $L^p(I,X)$. Hence the sequence f_n converges uniformly to a continuous function $f : I \to X$ and g_n converges to a function $g \in L^p(I,X)$ by Theorem 7.5.6. The limit functions satisfy

$$f(t) - f(a) = \lim_{n\to\infty}(f_n(t) - f_n(a)) = \lim_{n\to\infty}\int_a^t g_n(s)\,ds = \int_a^t g(s)\,ds$$

for all $t \in I$. Thus $f \in W^{1,p}(I,X)$ and

$$\lim_{n\to\infty}\|f - f_n\|_{W^{1,p}} = \lim_{n\to\infty}\left(\|f - f_n\|_{L^p}^p + \|g - g_n\|_{L^p}^p\right)^{1/p} = 0.$$

This proves part (ii).

We prove part (iii) by a standard mollifier argument. Let $f \in W^{1,p}(I,X)$ and extend f to all of $\mathbb{R}$ by $f(t) := f(b)$ for $t > b$ and $f(t) := f(a)$ for $t < a$. Choose a smooth function $\rho : \mathbb{R} \to [0,\infty)$ with compact support and mean value 1 and define $\rho_\delta(t) := \delta^{-1}\rho(\delta^{-1}t)$ for $\delta > 0$ and $t \in \mathbb{R}$. Then the function $f_\delta : I \to \mathbb{R}$, defined by $f_\delta(t) := (\rho_\delta * f)(t) := \int_{\mathbb{R}}\rho_\delta(t-s)f(s)\,ds$ for $t \in \mathbb{R}$, is smooth for every $\delta > 0$, and f_δ converges to f uniformly, and hence also in the L^p-norm. Moreover, $f_\delta' = \rho_\delta * f'$ converges to f' in the L^p-norm and thus $\lim_{\delta\to 0}\|f - f_\delta\|_{W^{1,p}} = 0$. This proves part (iii).

We prove part (iv). The map $W^{1,p}(I,X) \to L^p(I, X \times X) : f \mapsto (f, f')$ is an isometric embedding. The target space is reflexive by Corollary 7.5.17, so $W^{1,p}(I,X)$ is reflexive by Theorem 2.4.4. This proves Theorem 7.5.18. □

7.6. Inhomogeneous Equations

Let X be a real Banach space and let $A : \mathrm{dom}(A) \to X$ be the infinitesimal generator of a strongly continuous semigroup $S : [0, \infty) \to \mathcal{L}(X)$. This section is devoted to the study of the solutions of the inhomogeneous equation

$$\dot{x} = Ax + f, \qquad x(0) = x_0. \tag{7.6.1}$$

Here we assume that the function $f : [0, \infty) \to \mathbb{R}$ is strongly measurable and locally integrable. In this situation we consider the function $x : [0, \infty) \to X$, defined by the variation of constants formula

$$x(t) := S(t)x_0 + \int_0^t S(t-s)f(s)\,ds \tag{7.6.2}$$

for $t \geq 0$. If $x_0 \in \mathrm{dom}(A)$ and $f : [0, \infty) \to X$ is continuously differentiable, then, by Lemma 7.1.14, the function $x : [0, \infty) \to X$ in (7.6.2) is continuously differentiable, takes values in the domain of A, and satisfies equation (7.6.1). While this is a rather crude general observation, it is the starting point for any more refined study of the solutions of (7.6.1).

7.6.1. Weak Solutions. As a first step we use the concepts developed in Section 7.5 to introduce the notion of a weak solution. This notion uses test functions $g : I \to X^*$ on a compact interval $I \subset \mathbb{R}$ that take values in $\mathrm{dom}((A^*)^\infty)$ and have the property that the function $(A^*)^k g : I \to X^*$ is smooth for every $k \in \mathbb{N}$. The space of such functions will be denoted by $C^\infty(I, \mathrm{dom}(A^*)^\infty)$.

Definition 7.6.1 (Weak Solution). Let X be a real Banach space and let $A : \mathrm{dom}(A) \to X$ be the infinitesimal generator of a strongly continuous semigroup $S : [0, \infty) \to \mathcal{L}(X)$. Fix a compact interval $I = [0, T]$, a strongly measurable function $f : I \to X$ with $\int_0^T \|f(t)\|\,dt < \infty$, and an element $x_0 \in X$. A **weak solution of equation** (7.6.1) is a strongly measurable function $x : I \to X$ with $\int_0^T \|x(t)\|\,dt < \infty$ that satisfies the condition

$$\langle g(0), x_0 \rangle + \int_0^T \langle g(s), f(s) \rangle\,ds + \int_0^T \langle \dot{g}(s) + A^*g(s), x(s) \rangle\,ds = 0 \tag{7.6.3}$$

for every test function $g \in C^\infty(I, \mathrm{dom}(A^*)^\infty)$ with $g(T) = 0$.

The next theorem shows that equation (7.6.1) admits a unique (almost everywhere) weak solution and that it is given by (7.6.2).

Theorem 7.6.2 (Existence and Uniqueness). *Let X, I, S, A, f, x_0 be as in Definition* 7.6.1 *and let $x \in \mathcal{L}^1(I, X)$. The following are equivalent.*

(i) *x is a weak solution of* (7.6.1).

(ii) *x is given by* (7.6.2) *for almost every $t \in I$.*

Proof. We prove that (ii) implies (i). Let $g \in C^\infty(I, \mathrm{dom}(A^*)^\infty)$ be a test function with $g(T) = 0$. Recall from Theorem 7.3.1 that the restriction of the dual semigroup $S^*(t)$ to the strong closure $E \subset X^*$ of the domain of A^* is a strongly continuous semigroup, whose infinitesimal generator is the restriction of the operator A^* to the subspace $\{x^* \in \mathrm{dom}(A^*) \,|\, A^*x^* \in E\}$. This implies that the function $I \to X^* : t \mapsto S^*(t)g(t)$ is continuously differentiable with the derivative

$$\frac{d}{dt} S^*(t)g(t) = S^*(t)(\dot{g}(t) + A^*g(t))$$

for $t \in I$. Hence the function $x_0(t) := S(t)x_0$ satisfies

$$\begin{aligned}
\int_0^T \langle \dot{g}(t) + A^*g(t), x_0(t) \rangle \, dt &= \int_0^T \langle \dot{g}(t) + A^*g(t), S(t)x_0 \rangle \, dt \\
&= \int_0^T \langle S^*(t)(\dot{g}(t) + A^*g(t)), x_0 \rangle \, dt \\
&= \int_0^T \frac{d}{dt} \langle S^*(t)g(t), x_0 \rangle \, dt \\
&= \langle S^*(T)g(T), x_0 \rangle - \langle g(0), x_0 \rangle \\
&= -\langle g(0), x_0 \rangle
\end{aligned}$$

and for $x_1(t) := \int_0^t S(t-s)f(s)\, ds$ we obtain

$$\begin{aligned}
&\int_0^T \langle \dot{g}(t) + A^*g(t), x_1(t) \rangle \, dt \\
&= \int_0^T \int_0^t \langle \dot{g}(t) + A^*g(t), S(t-s)f(s) \rangle \, dsdt \\
&= \int_0^T \int_0^t \langle S^*(t-s)(\dot{g}(t) + A^*g(t)), f(s) \rangle \, dsdt \\
&= \int_0^T \int_s^T \langle S^*(t-s)(\dot{g}(t) + A^*g(t)), f(s) \rangle \, dtds \\
&= \int_0^T \int_0^{T-s} \langle S^*(t)(\dot{g}(s+t) + A^*g(s+t)), f(s) \rangle \, dtds \\
&= \int_0^T \int_0^{T-s} \frac{d}{dt} \langle S^*(t)g(s+t), f(s) \rangle \, dtds \\
&= \int_0^T \Big(\langle S^*(T-s)g(T), f(s) \rangle - \langle g(s), f(s) \rangle \Big) \, ds \\
&= -\int_0^T \langle g(s), f(s) \rangle \, ds.
\end{aligned}$$

Take the sum of these equations to obtain that $x := x_0 + x_1 : I \to X$ is a weak solution of (7.6.1).

We prove that (i) implies (ii). Thus assume that $x : I \to X$ is a weak solution of (7.6.1) and define the function $y : I \to X$ by

$$y(t) := x(t) - S(t)x_0 - \int_0^t S(t-s)f(s)\,ds \qquad \text{for } 0 \le t \le T.$$

Then, by what we have just proved, y is a weak solution of equation (7.6.1) with $x_0 = 0$ and $f = 0$, i.e. $y \in \mathcal{L}^1(I, X)$ and

$$\int_0^T \langle \dot{g}(s) + A^*g(s), y(s)\rangle\,ds = 0 \tag{7.6.4}$$

for all $g \in C^\infty(I, \mathrm{dom}(A^*)^\infty)$ with $g(T) = 0$. We must prove that $y(t) = 0$ for almost every $t \in I$. To see this, fix an element $x^* \in \mathrm{dom}((A^*)^\infty)$ and a smooth function $\phi : I \to \mathbb{R}$, and define the function $g : I \to X^*$ by

$$g(s) := \int_0^{T-s} \phi(r)S^*(T-s-r)x^*\,dr \qquad \text{for } 0 \le s \le T.$$

Then $g(T) = 0$. Moreover, it follows from Theorem 7.3.1 by induction that, for $k \in \mathbb{N}_0$, the restriction of $S^*(t)$ to $E_k := \left\{\xi^* \in \mathrm{dom}(A^*)^k \,|\, (A^*)^k \in E\right\}$ is a strongly continuous semigroup whose infinitesimal generator is the restriction $B_k := A^*|_{E_{k+1}} : E_{k+1} \to E_k$. Apply Lemma 7.1.14 to this semigroup to deduce that, for every integer $k \ge 0$, the function $g : I \to X^*$ takes values in E_{k+1}, is continuously differentiable as a function with values in E_k, and satisfies $\frac{d}{ds}g(T-s) = A^*g(T-s) + \phi(T-s)x^*$ or equivalently

$$\dot{g}(s) + A^*g(s) = \phi(s)x^* \qquad \text{for } 0 \le s \le T. \tag{7.6.5}$$

This implies $g \in C^\ell(I, E_k)$ for all $k, \ell \in \mathbb{N}_0$ and so $g \in C^\infty(I, \mathrm{dom}((A^*)^\infty))$. Thus it follows from (7.6.4) and (7.6.5) that

$$\int_0^T \phi(s)\langle x^*, y(s)\rangle\,ds = 0 \tag{7.6.6}$$

for all $x^* \in \mathrm{dom}((A^*)^\infty)$ and all $\phi \in C^\infty(I)$. Choose a sequence of smooth functions $\phi_i : [0,1] \to [0,1]$ converging pointwise to the characteristic function of the subinterval $[0,t]$ and use Lebesgue dominated convergence and equation (7.6.6) to obtain $\int_0^t \langle x^*, y(s)\rangle\,ds = 0$ and hence

$$\left\langle x^*, \int_0^t y(s)\,ds\right\rangle = 0 \tag{7.6.7}$$

for all $x^* \in \mathrm{dom}((A^*)^\infty)$ and all $t \in I$. Since $\mathrm{dom}((A^*)^\infty)$ is dense in E by Lemma 7.1.16, equation (7.6.7) continues to hold for all $x^* \in E$ and all $t \in I$. Since E contains the domain of A^*, it is weak* dense in X^* by part (iii) of Theorem 6.2.2. This implies $\int_0^t y(s)\,ds = 0$ for $0 \le t \le T$. Now it follows from Lebesgue differentiation that $y(t) = 0$ for almost every $t \in I$ (Lemma 7.5.9). This proves Theorem 7.6.2. □

7.6.2. Regular Solutions. The next theorem examines the properties of weak solutions of (7.6.1) that belong to the Sobolev space $W^{1,1}(I,X)$.

Theorem 7.6.3 (Regular Solutions). *Let X, I, S, A, f, x_0 be as in Definition 7.6.1 and let $x : I \to X$ be a strongly measurable function. Then the following are equivalent.*

(i) *$x \in W^{1,1}(I,X)$ and x is a weak solution of equation (7.6.1).*

(ii) *There exists a Borel set $Z \subset I$ of Lebesgue measure zero such that*

- *$x(t) \in \mathrm{dom}(A)$ for every $t \in I \setminus Z$,*
- *the function $x : I \to X$ is differentiable on $I \setminus Z$ and*
$$\dot{x}(t) = Ax(t) + f(t) \qquad \text{for all } t \in I \setminus Z,$$
- *the function $y : I \to X$, defined by*
$$y(t) := \begin{cases} \dot{x}(t), & \text{for } t \in I \setminus Z, \\ 0, & \text{for } t \in Z, \end{cases}$$
is strongly measurable and satisfies $\int_0^T \|y(s)\|\, ds < \infty$ and
$$x(t) = x_0 + \int_0^t y(s)\, ds \qquad \text{for all } t \in I.$$

Proof. We prove that (i) implies (ii). Thus assume that $x \in W^{1,1}(I,X)$ is a weak solution of equation (7.6.1). Then x is continuous and

$$x(t) = S(t)x_0 + \int_0^t S(t-s)f(s)\, ds \qquad \text{for } 0 \le t \le T$$

by Theorem 7.6.2. For $0 \le t < t+h \le T$ this implies

$$\begin{aligned} \frac{S(h)x(t) - x(t)}{h} &= \frac{x(t+h) - x(t)}{h} - \frac{1}{h}\int_0^h S(s)f(t)\, ds \\ &\quad - \frac{1}{h}\int_0^h S(h-s)\Big(f(t+s) - f(t)\Big)\, ds. \end{aligned} \tag{7.6.8}$$

Moreover, by definition of $W^{1,1}(I,X)$ there exists a function $\xi \in \mathcal{L}^1(I,X)$ such that $x(t) = x_0 + \int_0^t \xi(s)\, ds$ for all $t \in I$. Hence, by Lebesgue differentiation, there exists a Borel set $Z \subset I$ of Lebesgue measure zero such that

- x is differentiable on $I \setminus Z$ and $\dot{x}(t) = \xi(t)$ for all $t \in I \setminus Z$,
- $\lim_{h \searrow 0} \frac{1}{h}\int_0^h \|f(t+s) - f(t)\|\, ds = 0$ for all $t \in I \setminus Z$.

For $t \in I \setminus Z$ this implies that the right-hand side of (7.6.8) converges to $\xi(t) - f(t)$ as h tends to zero. Thus $x(t) \in \mathrm{dom}(A)$ and $Ax(t) = \xi(t) - f(t)$ for all $t \in I \setminus Z$. This shows that x satisfies (ii) with this Borel set Z and the function $y : I \to X$ defined by $y(t) := \xi(t) = \dot{x}(t)$ for $t \in I \setminus Z$ and by $y(t) = 0$ for $t \in Z$.

We prove that (ii) implies (i). Thus assume that $Z \subset I$ is a Borel set of Lebesgue measure zero that satisfies the requirements of part (ii). Let $E \subset X^*$ be the strong closure of the domain of the dual operator A^*. Fix an element $x^* \in \mathrm{dom}(A^*)$ with $A^*x^* \in E$ and a real number $0 < t \leq T$. Then, by Theorem 7.3.1, the function $[0,t] \to X^* : s \mapsto S^*(t-s)x^*$ is continuously differentiable and has the derivative

$$\frac{d}{ds}S^*(t-s)x^* = -S^*(t-s)A^*x^* = -A^*S^*(t-s)x^*.$$

Moreover, by assumption, the function $x : I \to X$ is absolutely continuous and differentiable in $I \setminus Z$. This implies that the function

$$[0,t] \to \mathbb{R} : s \mapsto \langle S^*(t-s)x^*, x(s)\rangle \tag{7.6.9}$$

is absolutely continuous and differentiable in $[0,t] \setminus Z$. Since $x(t) \in \mathrm{dom}(A)$ and $\dot{x}(t) = Ax(t) + f(t)$ for $t \in I \setminus Z$, the function (7.6.9) has the derivative

$$\begin{aligned}\frac{d}{ds}\langle S^*(t-s)x^*, x(s)\rangle &= \langle S^*(t-s)x^*, \dot{x}(s)\rangle - \langle A^*S^*(t-s)x^*, x(s)\rangle \\ &= \langle S^*(t-s)x^*, \dot{x}(s) - Ax(s)\rangle \\ &= \langle x^*, S(t-s)f(s)\rangle\end{aligned}$$

for $s \in [0,t] \setminus Z$. Since Z has measure zero and (7.6.9) is absolutely continuous, this implies $\langle x^*, x(t)\rangle - \langle S^*(t)x^*, x(0)\rangle = \int_0^t \langle x^*, S(t-s)f(s)\rangle\, ds$ and hence

$$\left\langle x^*, x(t) - S(t)x_0 - \int_0^t S(t-s)f(s)\, ds \right\rangle = 0 \tag{7.6.10}$$

for all $x^* \in \mathrm{dom}(A^*)$ with $A^*x^* \in E$. By Theorem 7.3.1 the set of all such x^* is the domain of the infinitesimal generator of the strongly continuous semigroup $[0,\infty) \to \mathcal{L}(E) : t \mapsto S^*(t)|_E$ and so is dense in E by Lemma 7.1.16. Hence equation (7.6.10) continues to hold for all $x^* \in E$ and hence, in particular, for all $x^* \in \mathrm{dom}(A^*)$. Since the domain of A^* is weak* dense in X^*, by Theorem 6.2.2, it follows that $x(t)$ is given by equation (7.6.2) for every $t \in [0,T]$. This proves Theorem 7.6.3. □

Theorem 7.6.3 leads to the question under which conditions on x_0 and f the weak solution (7.6.2) of (7.6.1) belongs to the Sobolev space $W^{1,1}(I,X)$. This is the fundamental **regularity problem for semigroups**. It has two parts, one for the inhomogeneous term f when $x_0 = 0$ (see Subsection 7.6.3) and one for the initial condition x_0 when $f = 0$ (see Subsection 7.6.4). By Lemma 7.1.14, the weak solution (7.6.2) belongs to $W^{1,1}(I,X)$ whenever $x_0 \in \mathrm{dom}(A)$ and $f : I \to X$ is continuously differentiable. By Exercise 7.7.12 this continues to hold for $f \in W^{1,1}(I,X)$.

7.6.3. Maximal Regularity. In applications one is interested in a refined regularity problem associated to a number $1 \le q < \infty$, which asks for weak solutions in the Sobolev space $W^{1,q}(I,X)$ when $f \in L^q(I,X)$. The sharp answer would be that, for every $f \in L^q(I,X)$, the formula (7.6.2) with $x_0 = 0$ defines a weak solution $x : I \to X$ of (7.6.1) in the Sobolev space $W^{1,q}(I,X)$, i.e. both $\dot{x}$ and Ax, and not just their difference, belong to the space $L^q(I,X)$. This property is called *maximal q-regularity.*

Definition 7.6.4 (Maximal Regularity). Let X be a Banach space, let $A : \mathrm{dom}(A) \to X$ be the infinitesimal generator of a strongly continuous semigroup $S : [0,\infty) \to \mathcal{L}(X)$, and fix a real number $q \ge 1$. The semigroup S is called **maximal q-regular** if, for every $T > 0$, there exists a $c_T > 0$ such that every continuously differentiable function $f : [0,T] \to X$ satisfies

$$\left(\int_0^T \left\| A\int_0^t S(t-s)f(s)\,ds \right\|^q dt\right)^{1/q} \le c_T \left(\int_0^T \|f(t)\|^q\,dt\right)^{1/q}. \tag{7.6.11}$$

This condition is independent of T. The semigroup S is called **uniformly maximal q-regular** if it is maximal q-regular and the constant in (7.6.11) can be chosen independent of T.

Lemma 7.6.5 (Maximal Regularity). *Let X be a Banach space and let $A : \mathrm{dom}(A) \to X$ be the infinitesimal generator of a strongly continuous semigroup $S : [0,\infty) \to \mathcal{L}(X)$. Fix two real numbers $q \ge 1$ and $T > 0$ and abbreviate $I := [0,T]$. Then the following are equivalent.*

(i) *For every strongly measurable function $f : I \to X$ with $\int_I \|f\|^q < \infty$ equation (7.6.1) has a weak solution $x \in W^{1,q}(I,X)$ with $x(0) = x_0 = 0$.*

(ii) *For every strongly measurable function $f : I \to X$ with $\int_I \|f\|^q < \infty$ the continuous function $x : I \to X$ defined by (7.6.2) with $x_0 = 0$ belongs to the Sobolev space $W^{1,q}(I,X)$.*

(iii) *The semigroup S is maximal q-regular.*

Proof. That (i) implies (ii) follows directly from Theorem 7.6.2. To prove that (ii) implies (iii), denote by $\iota : W^{1,q}(I,X) \to C(I,X)$ the obvious inclusion and define the linear operator $\mathscr{S} : L^q(I,X) \to C(I,X)$ by

$$(\mathscr{S}f)(t) := \int_0^t S(t-s)f(s)\,ds$$

for $f \in L^q(I,X)$ and $t \in I$. Then ι is a bounded linear operator by part (i) of Theorem 7.5.18. To prove that $\mathscr{S}$ is a bounded linear operator, choose a constant $M \ge 1$ such that $\|S(t)\| \le M$ for $0 \le t \le T$ (Lemma 7.1.8). Then

$$\|(\mathscr{S}f)(t)\| \le \int_0^t \|S(t-s)f(s)\|\,ds \le M\int_0^T \|f(s)\|\,ds \le MT^{1-1/q}\|f\|_{L^q}$$

for all $t \in I$ and all $f \in \mathcal{L}^q(I,X)$. Moreover, $\mathrm{im}(\mathscr{S}) \subset \mathrm{im}(\iota)$ by (ii). Since ι is injective, Corollary 2.2.17 (Douglas factorization) asserts that the linear operator $\iota^{-1} \circ \mathscr{S} : L^q(I,X) \to W^{1,q}(I,X)$ is bounded. Thus there exists a constant $C > 0$ such that $\|\mathscr{S}f\|_{W^{1,q}} \le C\|f\|_{L^q}$ for all $f \in L^q(I,X)$. For $f \in C^1(I,X)$ this is equivalent to the estimate (7.6.11). Thus S is maximal q-regular.

We prove that (iii) implies (i). Assume S is maximal q-regular and let $f : I \to X$ be a strongly measurable function with $\int_0^T \|f(t)\|^q\,dt < \infty$. By part (iii) of Theorem 7.5.4, there exists a sequence of smooth functions $f_i : I \to X$ such that $\lim_{i\to\infty}\|f_i(t) - f(t)\|_{L^q} = 0$. Define the functions $x : I \to X$ and $x_i : I \to X$, $i \in \mathbb{N}$, by

$$x(t) := \int_0^t S(t-s)f(s)\,ds, \qquad x_i(t) := \int_0^t S(t-s)f_i(s)\,ds$$

for $t \in I$. Then $\lim_{i\to\infty}\sup_{t\in I}\|x_i(t) - x(t)\| = 0$. By Lemma 7.1.14, we have $x_i(t) \in \mathrm{dom}(A)$ and $\dot{x}_i(t) = Ax_i(t) + f_i(t) =: y_i(t)$ for all t and i. Moreover, $\|Ax_i - Ax_j\|_{L^q} \le c_T\|f_i - f_j\|_{L^q}$ for all i,j by maximal regularity. Thus $(y_i)_{i\in\mathbb{N}}$ is a Cauchy sequence in $L^q(I,X)$ and so, by Theorem 7.5.6, there exists a function $y \in \mathcal{L}^q(I,X)$ with $\lim_{i\to\infty}\|y_i - y\|_{L^q} = 0$. Hence

$$x(t) = \lim_{i\to\infty} x_i(t) = \lim_{i\to\infty}\int_0^t y_i(s)\,ds = \int_0^t y(s)\,ds$$

for all $t \in I$ and so $x \in W^{1,q}(I,X)$. Since x is a weak solution of (7.6.1) by Theorem 7.6.2, this proves Lemma 7.6.5. □

The next lemma shows that there are many semigroups that cannot be maximal q-regular for any $q \ge 1$. Such examples include all strongly continuous groups generated by unbounded operators.

Lemma 7.6.6. *Let $S : [0,\infty) \to \mathcal{L}(X)$ be a strongly continuous semigroup on a Banach space X that is maximal q-regular for some $q \ge 1$. Then*

$$\mathrm{im}(S(t)) \subset \mathrm{dom}(A) \qquad \textit{for all } t > 0.$$

Proof. Assume that there exists a $T > 0$ such that $\mathrm{im}(S(T)) \not\subset \mathrm{dom}(A)$, abbreviate $I := [0,T]$, and choose $\xi \in X$ such that $S(T)\xi \in X \setminus \mathrm{dom}(A)$. Define the function $f : I \to X$ by $f(t) := S(t)\xi$ for $0 \le t \le T$. Then we have $f \in C(I,X) \subset L^q(I,X)$ and

$$x(t) := \int_0^t S(t-s)f(s)\,ds = tS(t)\xi \in X \setminus \mathrm{dom}(A)$$

for $0 < t \le T$. Hence the function $x : I \to X$ cannot belong to the Sobolev space $W^{1,1}(I,X)$ by Theorem 7.6.3. This shows that the semigroup S violates condition (i) in Lemma 7.6.5 for any $q \ge 1$ and hence cannot be maximal q-regular. This proves Lemma 7.6.6. □

Remark 7.6.7. Let X be a reflexive Banach space. Let $S : [0, \infty) \to \mathcal{L}(X)$ be a strongly continuous semigroup with infinitesimal generator A such that

$$\operatorname{im}(S(t)) \subset \operatorname{dom}(A^2), \qquad \left\|A^2 S(t)x\right\| \leq ct^{-2}\left\|x\right\| \tag{7.6.12}$$

for all $t > 0$, all $x \in X$, and some $c > 0$. Under these assumptions it was proved by Benedek–Calderón–Panzone [**9**] that S is (uniformly) maximal q-regular for some $q > 1$ if and only if it is (uniformly) maximal q-regular for all $q > 1$. Another exposition of their theorem can be found in [**76**]. Note that analytic contraction semigroups satisfy (7.6.12) by Theorem 7.4.4.

Remark 7.6.8. Let $(M, \mathcal{A}, \mu)$ be a measure space and let

$$S : [0, \infty) \to \mathcal{L}(L^2(\mu))$$

be an analytic semigroup that satisfies the estimate

$$\|S(t)f\|_{L^p} \leq \|f\|_{L^p} \tag{7.6.13}$$

for all $p \geq 1$, all $t \geq 0$, and all $f \in L^p(\mu) \cap L^2(\mu)$. Under this assumption a theorem of Lamberton [**54**] asserts that the induced contraction semigroup on $L^p(\mu)$ is uniformly maximal q-regular for all $p, q > 1$. For the heat flow in Example 7.1.6 an exposition can be found in [**76**]. The proof goes far beyond the scope of the present book. However, for $p = q = 2$ the result follows from an elementary abstract observation that is explained below.

For the study of maximal regularity it is convenient to introduce a Banach space that contains all the regular solutions of equation (7.6.1). For each $q \geq 1$ and each interval $I = [0, T]$ this is the space

$$\begin{aligned}
\mathscr{W}_A^{1,q}(I, X) &:= W^{1,q}(I, X) \cap L^q(I, \operatorname{dom}(A)) \\
&:= \left\{ x \in W^{1,q}(I, X) \;\middle|\; \begin{array}{l} \text{there is a Borel set } Z \subset I \\ \text{of measure zero such that} \\ x(t) \in \operatorname{dom}(A) \text{ for } t \in I \setminus Z, \\ \text{the function } Ax : I \to X \\ \text{is strongly measurable,} \\ \text{and } \int_0^T \|Ax(t)\|^q \, dt < \infty \end{array} \right\}
\end{aligned} \tag{7.6.14}$$

equipped with the norm

$$\|x\|_{\mathscr{W}_A^{1,q}} := \left(\int_0^T \left(\|x(t)\|^q + \|\dot{x}(t)\|^q + \|Ax(t)\|^q \right) dt \right)^{1/q}. \tag{7.6.15}$$

In this definition the function $Ax : I \to X$ is understood to be zero for $t \in Z$. The next lemma summarizes some basic properties of this space. In particular, it is a Banach space and is reflexive when X is reflexive and $1 < q < \infty$.

Lemma 7.6.9. *Let $A : \mathrm{dom}(A) \to X$ be the infinitesimal generator of a strongly continuous semigroup on a Banach space X, let $1 \le q < \infty$, and let $T > 0$ and $I := [0, T]$. Then the following hold.*

(i) *Let $x : I \to \mathrm{dom}(A)$ be any function. Then x is strongly measurable in the Banach space $\mathrm{dom}(A)$ with the graph norm if and only if both functions $x : I \to X$ and $Ax : I \to X$ are strongly measurable in X.*

(ii) *$\mathscr{W}_A^{1,q}(I, X)$ is a Banach space with the norm (7.6.15).*

(iii) *The space*

$$C^\infty(I, \mathrm{dom}(A)) := \left\{ x : I \to \mathrm{dom}(A) \,\middle|\, \begin{array}{l} \text{the functions } x : I \to X \\ \text{and } Ax : I \to X \text{ are smooth} \end{array} \right\}$$

is dense in $\mathscr{W}_A^{1,q}(I, X)$.

(iv) *If X is reflexive and $1 < q < \infty$, then $\mathscr{W}_A^{1,q}(I, X)$ is reflexive.*

Proof. We prove part (i). Let us temporarily denote by $\iota : \mathrm{dom}(A) \to X$ the obvious inclusion and think of $x : I \to \mathrm{dom}(A)$ solely as a function with values in the Banach space $\mathrm{dom}(A)$, equipped with the graph norm. Then the operator $\lambda\iota - A : \mathrm{dom}(A) \to X$ is invertible for $\lambda > 0$ sufficiently large and for such a λ we have $x = (\lambda\iota - A)^{-1} \circ (\lambda\iota \circ x - A \circ x)$. Thus, if $x : I \to \mathrm{dom}(A)$ is strongly measurable, so are $\iota \circ x, A \circ x : I \to X$, and conversely if those two are strongly measurable, so is $\lambda\iota \circ x - A \circ x$ and hence also x.

We prove part (ii). Let $(x_i)_{i\in\mathbb{N}}$ be a Cauchy sequence in $\mathscr{W}_A^{1,q}(I, X)$. Then $(x_i)_{i\in\mathbb{N}}$ is a Cauchy sequence in $W^{1,q}(I, X)$ and hence converges to a function $x \in W^{1,q}(I, X)$, both with respect to the $W^{1,q}$-norm and with respect to the supremum norm by Theorem 7.5.18. Moreover, the functions $y_i := Ax_i : I \to X$ form a Cauchy sequence in $L^q(I, X)$. Hence Theorem 7.5.6 asserts that there exists a strongly measurable function $y : I \to X$ such that $\int_I \|y\|^q < \infty$ and $\lim_{i\to\infty} \|y_i - y\|_{L^q} = 0$, and that a subsequence of y_i converges almost everywhere to y. Since A is closed, this implies that there exists a Borel set $Z \subset I$ of measure zero such that $x(t) \in \mathrm{dom}(A)$ and $Ax(t) = y(t)$ for all $t \in I \setminus Z$. Hence $x \in \mathscr{W}_A^{1,q}(I, X)$ and

$$\lim_{i\to\infty} \|x - x_i\|_{\mathscr{W}_A^{1,q}} = \lim_{i\to\infty} \left(\|x - x_i\|_{W^{1,q}}^q + \|y - y_i\|_{L^q}^q \right)^{1/q} = 0.$$

This proves part (ii). Part (iii) follows from the same mollifier argument as in the proof of Theorem 7.5.18, and part (iv) follows from the fact that the map $\mathscr{W}_A^{1,q}(I, X) \to W^{1,q}(I, X \times X \times X) : x \mapsto (x, \dot{x}, Ax)$ is an isometric embedding, by definition, and the target space is reflexive whenever X is reflexive and $1 < q < \infty$, by Corollary 7.5.17. This proves Lemma 7.6.9. □

It follows from Theorem 7.6.3 that the weak $W^{1,q}$ solutions of equation (7.6.1) with $f \in L^q(I,X)$ are elements of the space $\mathscr{W}_A^{1,q}(I,X)$ and that the inhomogeneous term in the equation can be recovered from the element $x \in \mathscr{W}_A^{1,q}(I,X)$ via the formula $f = \dot{x} - Ax$. Thus the semigroup generated by A is maximal q-regular if and only if the map

$$\left\{x \in \mathscr{W}_A^{1,q}(I,X)\,\middle|\, x(0)=0\right\} \to L^q(I,X) : x \mapsto \dot{x} - Ax \tag{7.6.16}$$

is a Banach space isomorphism. If that holds, then the bounded linear operator $\iota^{-1} \circ \mathscr{S} : L^q(I,X) \to W^{1,q}(I,X)$ in the proof of Lemma 7.6.5 is the inverse of the operator (7.6.16).

7.6.4. Regular Initial Conditions. With these preparations we are ready to formulate the second regularity problem for equation (7.6.1). The question is, which initial conditions $x_0 \in X$ give rise to solutions of the homogeneous equation in the space $\mathscr{W}_A^{1,q}(I,X)$. Define the normed vector space

$$X_{A,q} := \left\{ x \in X \,\middle|\, \begin{array}{l} S(t)x \in \mathrm{dom}(A) \text{ for all } t > 0 \\ \text{and } \int_0^T \|AS(t)x\|_X^q \, dt < \infty \end{array} \right\}, \tag{7.6.17}$$

$$\|x\|_{A,q} := \|x\|_X + \left(\int_0^T \|AS(t)x\|_X^q \, dt\right)^{1/q} \qquad \text{for } x \in X_{A,q}. \tag{7.6.18}$$

Lemma 7.6.10. *Let $A : \mathrm{dom}(A) \to X$ be the infinitesimal generator of a strongly continuous semigroup $S : [0,\infty) \to \mathcal{L}(X)$ on a Banach space X, let $I = [0,T]$, and let $1 \le q < \infty$. Then the following hold.*

(i) *$X_{A,q}$ is a Banach space with the norm* (7.6.18) *and* $\mathrm{dom}(A) \subset X_{A,q} \subset X$ *with continuous dense inclusions.*

(ii) *The subspace $X_{A,q} \subset X$ is invariant under the operator $S(t)$ for all $t \ge 0$ and $S(t)$ restricts to a strongly continuous semigroup on the space $X_{A,q}$.*

(iii) *Let $x_0 \in X$ and define the function $x : I \to X$ by $x(t) := S(t)x_0$ for $0 \le t \le T$. Then $x_0 \in X_{A,q}$ if and only if $x \in \mathscr{W}_A^{1,q}(I,X)$.*

(iv) *Assume S is maximal q-regular. Then there exists a $c > 0$ such that every continuously differentiable function $x : I \to \mathrm{dom}(A)$ satisfies*

$$\sup_{0 \le t \le T} \|x(t)\|_{A,q} \le c \|x\|_{\mathscr{W}_A^{1,q}}. \tag{7.6.19}$$

Thus there is a continuous inclusion $\mathscr{W}_A^{1,q}(I,X) \hookrightarrow C(I,X_{A,q})$.

(v) *Assume S is maximal q-regular. Then the map*

$$\mathscr{W}_A^{1,q}(I,X) \to X_{A,q} \times L^q(I,X) : x \mapsto (x(0), \dot{x} - Ax) \tag{7.6.20}$$

is a Banach space isomorphism. If this holds, then the inverse of (7.6.20) *is the operator $X_{A,q} \times L^q(I,X) \to \mathscr{W}_A^{1,q}(I,X) : (x_0, f) \mapsto x$ defined by* (7.6.2).

Proof. Let $(x_i)_{i\in\mathbb{N}}$ be a Cauchy sequence in $X_{A,q}$ and define the functions $y_i : I \to X$ by $y_i(0) := 0$ and

$$y_i(t) := AS(t)x_i$$

for $0 < t \le T$ and $i \in \mathbb{N}$. Then $(x_i)_{i\in\mathbb{N}}$ is a Cauchy sequence in X and $(y_i)_{i\in\mathbb{N}}$ is a Cauchy sequence in $L^q(I,X)$. Thus there exist an element $x \in X$ and, by Theorem 7.5.6, a strongly measurable function $y : I \to X$ with

$$\int_0^T \|y(t)\|_X^q \, dt < \infty$$

such that

$$\lim_{i\to\infty} \|x - x_i\|_X = 0, \qquad \lim_{i\to\infty} \int_0^T \|y(t) - y_i(t)\|_X^q \, dt = 0.$$

Passing to a subsequence, if necessary, we may also assume that the sequence $(y_i)_{i\in\mathbb{N}}$ converges almost everywhere to y by part (i) of Theorem 7.5.6. Thus there is a Borel set $Z \subset I$ of measure zero such that

$$y(t) = \lim_{i\to\infty} AS(t)x_i \qquad \text{for all } t \in I \setminus Z.$$

Since A is closed and

$$\lim_{i\to\infty} S(t)x_i = S(t)x,$$

this implies

$$S(t)x \in \operatorname{dom}(A), \qquad AS(t)x = y(t) \qquad \text{for all } t \in I \setminus Z.$$

Since Z has measure zero, we obtain $S(t)x \in \operatorname{dom}(A)$ for all $t > 0$ and

$$\int_0^T \|AS(t)x\|_X^q \, dt = \int_0^T \|y(t)\|_X^q \, dt < \infty.$$

Hence $x \in X_{A,q}$ and

$$\begin{aligned}\lim_{i\to\infty} \|x - x_i\|_{A,q} &= \lim_{i\to\infty} \|x - x_i\|_X + \lim_{i\to\infty} \left(\int_0^T \|y(t) - AS(t)x_i\|_X^q \, dt\right)^{1/q} \\ &= 0.\end{aligned}$$

This shows that $X_{A,q}$ is a Banach space. That the obvious inclusions

$$\operatorname{dom}(A) \hookrightarrow X_{A,q}, \qquad X_{A,q} \hookrightarrow X$$

are continuous follows directly from the definition of the norms. That $X_{A,q}$ is dense in X follows from Lemma 7.1.16 and the fact that $\operatorname{dom}(A) \subset X_{A,q}$. To prove that $\operatorname{dom}(A)$ is dense in $X_{A,q}$, one can use part (ii), which is an easy exercise left to the reader, and observe that the domain of the infinitesimal generator of the restricted semigroup $S(t)|_{X_{A,q}}$ contains $\operatorname{dom}(A)$. This proves parts (i) and (ii). Part (iii) follows directly from the definitions.

To prove part (iv), assume that S is maximal q-regular, and define the bounded linear operator

$$\mathscr{S} : L^q(I,X) \to \mathscr{W}_A^{1,q}(I,X)$$

by

$$(\mathscr{S}f)(t) := \int_0^t S(t-s)f(s)\,ds$$

for $f \in L^q(I,X)$ and $0 \le t \le T$. Composing $\mathscr{S}$ with the bounded linear operator

$$\mathscr{W}_A^{1,q}(I,X) \to L^q(I,X) : x \mapsto \dot{x} - Ax$$

we obtain a bounded linear operator

$$\mathscr{T} : \mathscr{W}_A^{1,q}(I,X) \to \mathscr{W}_A^{1,q}(I,X)$$

given by

$$(\mathscr{T}x)(t) := \int_0^t S(t-s)\big(\dot{x}(s) - Ax(s)\big)\,ds$$

for $x \in \mathscr{W}_A^{1,q}(I,X)$ and $0 \le t \le T$. For $x \in C^1(I,\mathrm{dom}(A))$ and $0 \le t \le T$ we obtain from Lemma 7.1.14 the equation

$$(\mathscr{T}x)(t) - x(t) = S(t)x(0).$$

This implies the inequality

$$\|x(0)\|_{A,q} \le \|S(\cdot)x(0)\|_{\mathscr{W}_A^{1,q}} \le \Big(1 + \|\mathscr{T}\|\Big)\,\|x\|_{\mathscr{W}_A^{1,q}}$$

for all $x \in C^1(I,\mathrm{dom}(A))$. Since $C^1(I,\mathrm{dom}(A))$ is dense in $\mathscr{W}_A^{1,q}(I,X)$, this inequality continues to hold for all $x \in \mathscr{W}_A^{1,q}(I,X)$. Similar estimates, with a constant independent of t, for all the evaluation maps

$$\mathscr{W}_A^{1,q}(I,X) \to X_{A,q} : x \mapsto x(t)$$

can be obtained by shortening the interval for $0 \le t \le T/2$ and in addition reversing time for $T/2 \le t \le T$. Here one must use the fact that in the definition of the norm (7.6.18) on the space $X_{A,q}$, the number T can be chosen arbitrarily. Different choices of T give rise to equivalent norms. This proves part (iv). Part (v) follows directly from part (iv) and this completes the proof of Lemma 7.6.10. □

The preceding discussion sets up a general abstract framework for suitable Banach spaces of initial conditions and solutions for linear Cauchy problems. Under the assumption of maximal q-regularity these spaces can be used to obtain well-posed Cauchy problems for PDEs with nonlinearities in the highest order terms (see Remark 7.6.14 below).

7.6.5. Regularity in Hilbert Spaces. For self-adjoint semigroups on Hilbert spaces maximal q-regularity is easy to verify for $q = 2$.

Theorem 7.6.11. *Every self-adjoint semigroup on a Hilbert space is maximal* 2*-regular.*

Proof. Let H be a Hilbert space and let $S : [0, \infty) \to \mathcal{L}(H)$ be a strongly continuous semigroup of self-adjoint operators with infinitesimal generator $A : \mathrm{dom}(A) \to H$. Then, by Theorem 7.3.4, we have

$$\omega := \sup_{x \in \mathrm{dom}(A) \setminus \{0\}} \frac{\langle x, Ax \rangle_H}{\|x\|_H^2} < \infty. \tag{7.6.21}$$

Let $V \subset H$ be the completion of $\mathrm{dom}(A)$ with respect to the norm

$$\|x\|_V := \sqrt{\langle x, cx - Ax \rangle}, \qquad c := \omega + 1. \tag{7.6.22}$$

Now let $x_0 \in \mathrm{dom}(A)$, let $f : [0, T] \to H$ be a continuously differentiable function, and define the function $x : [0, T] \to H$ by

$$x(t) := S(t)x_0 + \int_0^t S(t - s)f(s)\, ds \qquad \text{for } 0 \le t \le T. \tag{7.6.23}$$

Then $x(t) \in \mathrm{dom}(A)$ for all t and the function $x : [0, T] \to H$ is continuously differentiable and satisfies $\dot{x}(t) = Ax(t) + f(t)$ for all t (Lemma 7.1.14). Thus the function $t \mapsto \frac{1}{2}\|x(t)\|_V^2$ is continuously differentiable and

$$\begin{aligned}
\frac{d}{dt}\frac{1}{2}\|x(t)\|_V^2 &= \langle \dot{x}(t), cx(t) - Ax(t) \rangle_H \\
&= \langle f(t) + Ax(t), cx(t) - Ax(t) \rangle_H \\
&\le \|f(t)\|_H \|cx(t)\|_H + \|Ax(t)\|_H \|cx(t)\|_H \\
&\quad + \|f(t)\|_H \|Ax(t)\|_H - \|Ax(t)\|_H^2 \\
&\le \frac{3}{2}\|f(t)\|_H^2 + \frac{3c^2}{2}\|x(t)\|_H^2 - \frac{1}{2}\|Ax(t)\|_H^2.
\end{aligned}$$

Integrate this inequality over the interval $[0, T]$ to obtain

$$\|x(T)\|_V^2 + \int_0^T \|Ax(t)\|_H^2\, dt \le \|x_0\|_V^2 + 3\int_0^T \left(\|f(t)\|_H^2 + c^2 \|x(t)\|_H^2 \right) dt.$$

Now take $x_0 = 0$ and define $c_T := (2\omega)^{-1}(e^{2\omega T} - 1)$ when $\omega \ne 0$ and $c_T := T$ when $\omega = 0$. Then $\int_0^T \|x(t)\|_H^2\, dt \le Tc_T \int_0^T \|f(t)\|_H^2\, dt$ and so

$$\int_0^T \|Ax(t)\|_H^2\, dt \le 3\left(1 + c^2 Tc_T\right) \int_0^T \|f(t)\|_H^2\, dt.$$

This proves Theorem 7.6.11. □

Remark 7.6.12. Let A and ω be as in the proof of Theorem 7.6.11. Let $B : \mathrm{dom}(B) \to H$ be the unique self-adjoint operator with $\langle x, Bx \rangle \geq 0$ for all $x \in \mathrm{dom}(B)$ that satisfies $B^2 = \omega\mathbb{1} - A$ (see Exercise 6.5.8). Then the space V in (7.6.22) is the domain of B, equipped with the graph norm of B. Moreover, V agrees with the space $X_{A,2}$ in (7.6.17) and hence there is a canonical inclusion $\mathscr{W}_A^{1,2}(I, H) \hookrightarrow C(I, V)$ (see part (iv) of Lemma 7.6.10).

Remark 7.6.13. For parabolic (second order) equations in an L^p-space, the question of finding the space of initial conditions that give rise to solutions in $W^{1,q}(I, L^p) \cap L^q(I, W^{2,p})$ has been studied by many mathematicians (see [**11, 12, 32, 46, 63, 64, 84, 85**]). For the heat equation a theorem of Grigor'yan–Liu [**32**], which is based on work of Triebel [**84, 85**], asserts that the initial conditions in the Besov space

$$B_q^{s,p}(\mathbb{R}^n), \qquad s = 2 - \frac{2}{q},$$

give rise to solutions in the space

$$\mathscr{W}^{1,q,p} := W^{1,q}([0,T], L^p(\mathbb{R}^n)) \cap L^q([0,T], W^{2,p}(\mathbb{R}^n)).$$

For $p = q = 2$ the relevant Besov space is the Hilbert space $W^{1,2}(\mathbb{R}^n)$ and the proof reduces to the simple abstract argument in Theorem 7.6.11. For $p \neq 2$ the Grigor'yan–Liu Theorem is a deep result which goes far beyond the scope of the present book. Another exposition is given in [**76**].

Remark 7.6.14. One reason for the importance of such results is that one can reformulate the existence and uniqueness problem for nonlinear parabolic equations of the form

$$\partial_t u = \Delta u + f(u), \qquad u(0, \cdot) = u_0, \tag{7.6.24}$$

as a fixed point problem for the map $\mathscr{W}^{1,q,p} \to \mathscr{W}^{1,q,p} : u \mapsto \mathcal{F}(u)$ given by

$$(\mathcal{F}(u))(t) := S(t)u_0 + \int_0^t S(t-s) f(u(s))\, ds. \tag{7.6.25}$$

Here $S : [0, \infty) \to \mathcal{L}(L^p(\mathbb{R}^n))$ is the heat semigroup and f can be a map from $W^{2,p}(\mathbb{R}^n)$ to $L^p(\mathbb{R}^n)$. Thus one can deal with nonlinearities in the highest order terms. Moreover, by standard regularity arguments, one can replace the Laplace operator by a general second order elliptic operator. In this situation it is sometimes important to choose $p > n/2$ to obtain the relevant nonlinear estimates, and so the easy case $p = 2$ may not suffice. Many important geometric PDEs, such as the Ricci flow, the mean curvature flow, the Yang-Mills flow, the harmonic map flow, or the Donaldson geometric flow for symplectic four-manifolds [**51, 52**] can be formulated in this manner.

7.7. Problems

Exercise 7.7.1 (Semigroups on Complex Banach Spaces). Let X be a complex Banach space and let $A : \mathrm{dom}(A) \to X$ be the infinitesimal generator of a strongly continuous semigroup $S : [0, \infty) \to \mathcal{L}(X)$. Suppose that $\mathrm{dom}(A)$ is a complex subspace of X and that A is complex linear. Prove that $S(t) \in \mathcal{L}^c(X)$ for all $t \geq 0$. **Hint:** Define the operator $T(t) \in \mathcal{L}(X)$ by

$$T(t)x := -\mathbf{i}S(t)\mathbf{i}x$$

for $x \in X$ and $t \geq 0$. Show that T is a strongly continuous semigroup with infinitesimal generator A and use Corollary 7.2.3.

Exercise 7.7.2 (Contraction Semigroups). Let X be a complex Banach space and let $A : \mathrm{dom}(A) \to X$ be a complex linear operator with a dense domain $\mathrm{dom}(A) \subset X$. Consider the following conditions.

(i) A generates a contraction semigroup.

(ii) A has a closed graph and both A and A^* are dissipative.

Prove that (ii) implies (i). If X is reflexive prove that (i) is equivalent to (ii). Find an example of an operator on a nonreflexive Banach space that satisfies (i) but not (ii). **Hint:** Definition 7.2.10.

Exercise 7.7.3 (Dual Semigroup). Prove that the domain of the infinitesimal generator A of the group on $L^1(\mathbb{R})$ in Example 7.3.3 is the space of absolutely continuous real valued functions on $\mathbb{R}$ with integrable derivative. Prove that the domain of the dual operator A^* on $L^\infty(\mathbb{R})$ is the space of bounded Lipschitz continuous functions from $\mathbb{R}$ to itself. Prove that $\sigma(A) = \sigma(A^*) = \mathbf{i}\mathbb{R}$. Prove that the operator A^* does not satisfy the requirements of the Hille–Yosida–Phillips Theorem 7.2.5 because its domain is not dense.

Exercise 7.7.4 (Infinitesimal Generators of Unitary Groups). Let H be a complex Hilbert space and let $A : \mathrm{dom}(A) \to H$ be an unbounded complex linear operator with a dense domain $\mathrm{dom}(A) \subset H$. Prove that the following are equivalent.

(i) If $\lambda \in \mathbb{R} \setminus \{0\}$, then $\lambda\mathbb{1} - A$ is bijective and $\left\|(\lambda\mathbb{1} - A)^{-1}\right\| \leq |\lambda|^{-1}$.

(ii) If $\lambda \in \mathbb{C} \setminus \mathbf{i}\mathbb{R}$, then $\lambda\mathbb{1} - A$ is bijective and $\left\|(\lambda\mathbb{1} - A)^{-1}\right\| \leq |\mathrm{Re}\lambda|^{-1}$.

(iii) $\mathrm{dom}(A^*) = \mathrm{dom}(A)$ and $A^*x + Ax = 0$ for all $x \in \mathrm{dom}(A)$.

Hint: Each of these conditions is equivalent to the assertion that A generates a unitary group, by Theorem 7.2.11 and Theorem 7.3.6. The exercise is to establish their equivalence without using semigroup theory. Show that (i) $\Longrightarrow$ (iii) $\Longrightarrow$ (ii) $\Longrightarrow$ (i).

Exercise 7.7.5 (The Sobolev Space $W^{1,2}(\mathbb{R})$). Prove that the space of smooth functions $f : \mathbb{R} \to \mathbb{R}$ with compact support and mean value zero is dense in $L^2(\mathbb{R})$. Deduce that the completion of $C_0^\infty(\mathbb{R})$ with respect to the norm $f \mapsto \|f'\|_{L^2}$ in Example 7.1.7 can be identified with the space of equivalence classes of absolutely continuous function $f : \mathbb{R} \to \mathbb{R}$ with $\frac{df}{dx} \in L^2(\mathbb{R})$ under the equivalence relation $f_1 \sim f_2$ iff $f_1 - f_2$ is constant.

Exercise 7.7.6 (Maximal Regularity). Let $S : [0, \infty) \to \mathcal{L}(X)$ be a strongly continuous semigroup on a Banach space X with infinitesimal generator A and let $q > 1$. If the estimate (7.6.11) holds for some $T > 0$, prove that it holds for all $T > 0$ with a constant depending on T.

Exercise 7.7.7 (The Banach Space $L^p(I, X)$ and its Dual).

(a) Verify the assertions of Theorem 7.5.6 for $p = \infty$.

(b) Verify the assertions of Lemma 7.5.15 for $p = 1$ and $p = \infty$.

(c) Prove that the composition (7.5.27) in the proof of Corollary 7.5.17 is the canonical inclusion $\iota : L^p(I, X) \to L^p(I, X)^{**}$.

Exercise 7.7.8 (The Radon–Nikodým Property). Let $I := [0, 1]$ be the unit interval and $1 \le p \le \infty$. Define the function $f : [0, 1] \to L^p(I)$ by

$$(f(t))(s) := \begin{cases} 1, & \text{if } 0 \le s \le t, \\ 0, & \text{if } t < s \le 1, \end{cases} \qquad \text{for } s, t \in I.$$

When $p = \infty$, prove that f is everywhere discontinuous. When $1 < p < \infty$, prove that f is Hölder continuous. When $p = 1$, prove that f is Lipschitz continuous and nowhere differentiable. Deduce that $L^1(I)$ is not isomorphic to the dual space of any Banach space. **Hint:** Theorem 7.5.13.

Exercise 7.7.9 (Lebesgue Differentiation). Let X be a Banach space and let $f : I := [0, 1] \to X$ be a strongly measurable function such that $\int_0^1 \|f(t)\| \, dt < \infty$. Define the function $F : [0, 1] \to X$ by

$$F(t) := \int_0^t f(s) \, ds \qquad \text{for } 0 \le t \le 1.$$

Prove that F is absolutely continuous and that there is a Borel set $Z \subset I$ of Lebesgue measure zero such that F is differentiable on $I \setminus Z$ and

$$F'(t) = \lim_{h \to 0} \frac{F(t + h) - F(t)}{h} = f(t) \qquad \text{for every } t \in I \setminus Z.$$

Hint: The proof of the Lebesgue Differentiation Theorem in [**75**, Thm. 6.14] carries over verbatim to Banach space valued functions.

Exercise 7.7.10 (Bounded Lipschitz Continuous Functions). Prove that the closure of the space of bounded Lipschitz continuous functions in $L^\infty(\mathbb{R})$ is the space of bounded uniformly continuous functions on $\mathbb{R}$.

Exercise 7.7.11 (Weak and Strong Continuity). Let X be a real Banach space and let $S : [0,\infty) \to \mathcal{L}(X)$ be a one-parameter semigroup. Prove that the following are equivalent.

(i) The function $[0,\infty) \to X : t \mapsto S(t)x$ is continuous for all $x \in X$.

(ii) The function $[0,\infty) \to \mathbb{R} : t \mapsto \langle x^*, S(t)x\rangle$ is continuous for all $x \in X$ and all $x^* \in X^*$.

Hint: To prove that (ii) implies (i), show first that

$$\sup_{0\le t\le T} \|S(t)\| < \infty \qquad \text{for all } T > 0, \tag{7.7.1}$$

using the Uniform Boundedness Theorem 2.1.1. Second, use part (iii) of Theorem 7.5.2 and Lemma 7.5.5 to prove that

$$\lim_{\substack{h\to 0\\ h>0}} \int_0^{T-h} \|S(t+h)x - S(t)x\|\, dt = 0. \tag{7.7.2}$$

Third, fix a constant $\varepsilon > 0$, define

$$M := \sup_{0\le s\le \varepsilon} \|S(s)\|,$$

prove the estimate

$$\|S(t+h)x - S(t)x\| \le \frac{M}{\varepsilon} \int_{t-\varepsilon}^{t} \|S(s+h)x - S(s)x\|\, ds \tag{7.7.3}$$

for $x \in X$ and $0 < |h| < \varepsilon < t/2$, and use this estimate to show that the function $[0,\infty) \to X : t \mapsto S(t)x$ is continuous for $t > 0$. Fourth, prove that

$$\lim_{t\to 0} \|S(t)x - x\| = 0 \tag{7.7.4}$$

for all $x \in X$, by observing that x belongs to the closure of the linear subspace

$$Z := \operatorname{span}\left\{ S(t)x \,\middle|\, 0 < t < 1 \right\}$$

and using $\lim_{t\to 0} \|S(t)z - z\| = 0$ for all $z \in Z$.

Exercise 7.7.12 (Regularity of Weak Solutions). Let X be a Banach space and let $S : [0,\infty) \to \mathcal{L}(X)$ be a strongly continuous semigroup with infinitesimal generator $A : \operatorname{dom}(A) \to X$. Let $I := [0,T]$ and $f \in \mathcal{L}^1(I,X)$ and define $x(t) := \int_0^t S(t-s)f(s)\, ds$ for $0 \le t \le T$.

(a) If $f \in W^{1,1}(I,X)$, prove that $x \in W^{1,1}(I,X)$.

(b) If $f(t) \in \operatorname{dom}(A)$ for all t and $Af \in \mathcal{L}^1(I,X)$, prove that $x \in W^{1,1}(I,X)$.

Hint: For part (a) use Lemma 7.1.14 and approximation. For part (b) assume first that $Af : I \to X$ is continuous and then use approximation.

Exercise 7.7.13 (Semigroups and Compact Operators). Let $I = [0, 1]$ be the unit interval, let U, X, Y be real Banach spaces, and let $[0, \infty) \to \mathcal{L}(X) : t \mapsto S(t)$ be a strongly continuous semigroup.

(a) Let $I \to \mathcal{L}(X, Y) : t \mapsto K(t)$ be a strongly continuous family of operators. Prove that the operator

$$X \to C(I, Y) : x \mapsto K(\cdot)x \tag{7.7.5}$$

is compact if and only if the operator $K(t) \in \mathcal{L}(X, Y)$ is compact for every $t \in I$ and the map $K : I \to \mathcal{L}(X, Y)$ is continuous with respect to the operator norm on $\mathcal{L}(X, Y)$. **Hint:** Consider the set $\mathscr{F} \subset C(I, Y)$ whose elements are the functions $f_x := K(\cdot)x$ for all $x \in X$ with $\|x\| \leq 1$. Prove that $\mathscr{F}$ is equi-continuous if and only if the map $K : I \to \mathcal{L}(X, Y)$ is continuous with respect to the operator norm. Use Theorem 1.1.11.

(b) For $t \in I$ let $K(t) \in \mathcal{L}(X, Y)$ be a compact operator and suppose that the function $K : I \to \mathcal{L}(X, Y)$ is continuous with respect to the norm topology. Prove that the operator

$$L^1(I, X) \to Y : f \mapsto \int_0^1 K(t)f(t)\,dt \tag{7.7.6}$$

is compact. **Hint:** Show first that the function $I \to Y : t \mapsto K(t)f(t)$ is strongly measurable whenever $f : I \to X$ is strongly measurable. Second, use part (a) to prove that the operator $Y^* \to C(I, X^*) : y^* \mapsto K^*(\cdot)y^*$ is compact. Third, show that the composition of this operator with the canonical isometric inclusion $C(I, X^*) \to L^1(I, X)^*$ (Lemma 7.5.15) is the dual operator of (7.7.6). Then use Theorem 4.2.10.

(c) Let $B \in \mathcal{L}(U, X)$ be a compact operator. Prove that the operator

$$L^1(I, U) \to X : f \mapsto \int_0^1 S(t)Bf(t)\,dt \tag{7.7.7}$$

is compact. **Hint:** Show that the map $I \to \mathcal{L}(U, X) : t \mapsto S(t)B$ is continuous in the norm topology and use part (b).

(d) Let $C \in \mathcal{L}(X, Y)$ be a compact operator. If X is reflexive, prove that the operator

$$X \to C(I, Y) : x \mapsto CS(\cdot)x \tag{7.7.8}$$

is compact. Find an example of a semigroup on a nonreflexive Banach space X and a compact operator $C : X \to Y$ such that the operator (7.7.8) is not compact. **Hint:** Consider the shift semigroup on $X = L^1([0, 1])$ and let $C : X \to \mathbb{R}$ be the bounded linear functional $x \mapsto \int_0^1 x(t)\,dt$. Relate this to the fact that the inclusion of $W^{1,1}(I)$ into $C(I)$ is not a compact operator. (See Exercise 4.5.16.)

Exercise 7.7.14 (Semigroups and Functional Calculus). Let H be a complex Hilbert space, let $A : \mathrm{dom}(A) \to H$ be an unbounded self-adjoint operator on H with spectrum

$$\Sigma := \sigma(A) \subset (-\infty, 0],$$

and let $\Psi_A : C_b(\Sigma) \to \mathcal{L}^c(H)$ be the functional calculus in Theorem 6.4.1. Let $\mathcal{B} \subset 2^\Sigma$ be the Borel σ-algebra and, for $x, y \in H$, define the signed Borel measure $\mu_{x,y} : \mathcal{B} \to \mathbb{R}$ by $\mu_{x,y}(\Omega) := \mathrm{Re}\langle x, \Psi_A(\chi_\Omega)y\rangle$ for all $\Omega \in \mathcal{B}$ as in Definition 6.4.3 and Theorem 6.4.4. For $z \in \mathbb{C}$ with $\mathrm{Re}(z) \geq 0$ define the linear operator $S(z) \in \mathcal{L}^c(H)$ by

$$\mathrm{Re}\langle x, S(z)y\rangle := \int_\Sigma e^{z\lambda} d\mu_{x,y}(\lambda) \tag{7.7.9}$$

for $x, y \in H$ (see Theorem 5.6.2).

(a) Verify that $S(z) = \Psi_A(f_z)$ for all $z \in \mathbb{C}$ with $\mathrm{Re}(z) \geq 0$, where the function $f_z : \Sigma \to \mathbb{C}$ is defined by $f_z(\lambda) := e^{\lambda z}$ for $\lambda \in \Sigma$.

(b) Verify the formulas $S(0) = \mathrm{id}$ and $S(z + w) = S(w)S(z)$ for all $w, z \in \mathbb{C}$ with $\mathrm{Re}(z) \geq 0$ and $\mathrm{Re}(w) \geq 0$.

(c) Show that the map $z \mapsto S(z)$ is continuous in the norm topology on the open right half-plane and is strongly continuous on the closed right half-plane.

(d) Show that the map $z \mapsto S(z)$ is the analytic semigroup generated by A (see Example 7.4.5 and Theorem 7.4.4).

(e) Show that the map $\mathbb{R} \to \mathcal{L}^c(H) : t \mapsto S(\mathbf{i}t)$ is the unitary group generated by $\mathbf{i}A$ (see Theorem 7.3.6).

(f) By considering the Laplace operator on $\mathbb{R}^n$, deduce from (e) that the heat equation in Example 7.1.6 and the Schrödinger equation in Example 7.3.8 (adapted to dimension n) fit into a single strongly continuous semigroup on the right half-plane.

Appendix A

Zorn and Tychonoff

This appendix establishes the equivalence of the axiom of choice and the Lemma of Zorn and gives a self-contained proof of Tychonoff's Theorem.

A.1. The Lemma of Zorn

Our proof of the equivalence of the axiom of choice and the Lemma of Zorn follows the exposition by Imre Leader [**56**] and is based on the *Bourbaki–Witt Fixed Point Theorem*. Here are some basic definitions.

Definition A.1.1. A relation $\preccurlyeq$ on a set P is called a **partial order** if it is reflexive, anti-symmetric, and transitive, i.e. if it satisfies the conditions

- $p \preccurlyeq p$,
- if $p \preccurlyeq q$ and $q \preccurlyeq p$, then $p = q$,
- if $p \preccurlyeq q$ and $q \preccurlyeq r$, then $p \preccurlyeq r$

for all $p, q, r \in P$. A **partially ordered set** is a pair $(P, \preccurlyeq)$ consisting of a set P and a partial order $\preccurlyeq$ on P.

Definition A.1.2. Let $(P, \preccurlyeq)$ be a partially ordered set.

(i) An element $m \in P$ is called **maximal** if $m \not\preccurlyeq p$ for all $p \in P \setminus \{m\}$.

(ii) A **chain** in P is a **totally ordered** subset $C \subset P$, i.e. any two distinct elements $p, q \in C$ satisfy either $p \preccurlyeq q$ or $q \preccurlyeq p$.

(iii) Let $C \subset P$ be a nonempty chain. An element $a \in P$ is called an **upper bound of** C if every element $p \in C$ satisfies $p \preccurlyeq a$. It is called a **supremum of** C if it is an upper bound of C and every upper bound $b \in P$ of C satisfies $a \preccurlyeq b$. The supremum, if it exists, is unique and denoted by $\sup C$.

The Lemma of Zorn. *Let $(P, \preccurlyeq)$ be a partially ordered set such that every nonempty chain $C \subset P$ admits an upper bound. Let $p \in P$. Then there exists a maximal element $m \in P$ such that $p \preccurlyeq m$.*

The Axiom of Choice. *Let I and X be two nonempty sets and, for each element $i \in I$, let $X_i \subset X$ be a nonempty subset. Then there exists a map $g : I \to X$ such that every $i \in I$ satisfies $g(i) \in X_i$.*

Theorem A.1.3. *The axiom of choice is equivalent to the Lemma of Zorn.*

Proof. See page 448. □

Theorem A.1.4 (Bourbaki–Witt). *Let $(P, \preccurlyeq)$ be a nonempty partially ordered set such that every nonempty chain $C \subset P$ admits a supremum and let $f : P \to P$ be a map such that*

$$p \preccurlyeq f(p) \qquad \text{for all } p \in P.$$

Then there exists an element $p \in P$ such that $f(p) = p$.

Proof. Fix any element $p_0 \in P$ and denote by $\mathcal{A} \subset 2^P$ the set of all subsets $A \subset P$ that satisfy the following three conditions.

(I) $p_0 \in A$.

(II) If $p \in A$, then $f(p) \in A$.

(III) If $C \subset A$ is a nonempty chain, then $\sup C \in A$.

Then $\mathcal{A}$ is nonempty because $P \in \mathcal{A}$. Now let

$$E := \bigcap_{A \in \mathcal{A}} A \subset P$$

be the intersection of all subsets $A \in \mathcal{A}$. Then the set E also satisfies the conditions (I), (II), and (III) and hence is itself an element of $\mathcal{A}$. In particular, E is nonempty. We prove in five steps that E is a chain.

Step 1. *Every element $p \in E$ satisfies $p_0 \preccurlyeq p$.*

The set $P_0 := \{p \in P \,|\, p_0 \preccurlyeq p\}$ satisfies the conditions (I), (II), and (III), and hence is an element of $\mathcal{A}$. Thus $E \subset P_0$ and this proves Step 1.

Step 2. *Let $F \subset E$ be the subset*

$$F := \left\{ q \in E \,\middle|\, \begin{array}{l} \text{every element } p \in E \setminus \{q\} \\ \text{with } p \preccurlyeq q \text{ also satisfies } f(p) \preccurlyeq q \end{array} \right\}.$$

Then $p_0 \in F$.

By Step 1 there is no element $p \in E \setminus \{p_0\}$ with $p \preccurlyeq p_0$. Hence $p_0 \in F$.

Step 3. *Let $p \in E$ and $q \in F$. Then $p \preccurlyeq q$ or $f(q) \preccurlyeq p$.*

Fix an element $q \in F$ and consider the set

$$E_q := \{p \in E \,|\, p \preccurlyeq q\} \cup \{p \in E \,|\, f(q) \preccurlyeq p\}.$$

We will prove that $E_q \in \mathcal{A}$. Since $q \in F \subset E$ we have $p_0 \preccurlyeq q$ by Step 1. Since $p_0 \in E$, this implies $p_0 \in E_q$ and so E_q satisfies condition (I).

We prove that E_q satisfies (II). Fix an element $p \in E_q$. Then $f(p) \in E$ because E satisfies (II). If $p \preccurlyeq q$ and $p \neq q$, then $f(p) \preccurlyeq q$, because q is an element of F, and this implies $f(p) \in E_q$. If $p = q$, then $f(q) \preccurlyeq f(p)$ and this implies $f(p) \in E_q$. If $p \not\preccurlyeq q$, then we must have $f(q) \preccurlyeq p$, because $p \in E_q$, and this implies again $f(q) \preccurlyeq f(p)$ and therefore $f(p) \in E_q$. This shows that E_q satisfies condition (II).

We prove that E_q satisfies (III). Thus let $C \subset E_q$ be a nonempty chain and $s := \sup C$. Then $s \in E$ because E satisfies (III). If $p \preccurlyeq q$ for all $p \in C$, then $s \preccurlyeq q$ and therefore $s \in E_q$. Otherwise there exists an element $p \in C$ with $p \not\preccurlyeq q$. Since $p \in E_q$, we must have $f(q) \preccurlyeq p \preccurlyeq s$ and therefore $s \in E_q$. This shows that E_q satisfies condition (III).

Thus we have proved that $E_q \in \mathcal{A}$ and thus $E \subset E_q$, by definition of the set E. This proves Step 3.

Step 4. $F = E$.

We will prove that $F \in \mathcal{A}$. By Step 2 we have $p_0 \in F$ and so F satisfies (I).

We prove that F satisfies (II). Fix an element $q \in F$. We must prove that $f(q) \in F$. To see this, note first that $f(q) \in E$ because E satisfies (II). Now let $p \in E \setminus \{f(q)\}$ with $p \preccurlyeq f(q)$. Under these assumptions we must show that $f(p) \preccurlyeq f(q)$. Since $f(q) \not\preccurlyeq p$, we have $p \preccurlyeq q$ by Step 3. If $p \neq q$, then it follows from the definition of F that $f(p) \preccurlyeq q \preccurlyeq f(q)$. If $p = q$, then we also have $f(p) \preccurlyeq f(q)$. Thus we have shown that $f(p) \preccurlyeq f(q)$ for every element $p \in E \setminus \{f(q)\}$ with $p \preccurlyeq f(q)$. Hence $f(q) \in F$ and this shows that F satisfies condition (II).

We prove that F satisfies (III). Let $C \subset F$ be a nonempty chain and define $s := \sup C$. We must prove that $s \in F$. To see this, note first that $s \in E$ because E satisfies (III). Now let $p \in E \setminus \{s\}$ with $p \preccurlyeq s$. Under these assumptions we must show that $f(p) \preccurlyeq s$. Since $s \neq p$, we have $s \not\preccurlyeq p$. Thus there exists an element $q \in C$ with $q \not\preccurlyeq p$, and hence also $f(q) \not\preccurlyeq p$. Since $q \in C \subset F$, this implies $p \preccurlyeq q$ by Step 3. Since $p \neq q$ and $q \in F$, this implies $f(p) \preccurlyeq q$. Since $q \in C$ and $s = \sup C$, this implies $f(p) \preccurlyeq s$. Thus we have proved that $s \in F$ and so F satisfies condition (III).

Thus we have proved that $F \in \mathcal{A}$, hence $E \subset F$, and therefore $E = F$. This proves Step 4.

Step 5. *E is a chain.*

Let $p, q \in E$. Then $q \in F$ by Step 4, and so $p \preccurlyeq q$ or $f(q) \preccurlyeq p$ by Step 3. Thus $p \preccurlyeq q$ or $q \preccurlyeq p$ and this proves Step 5.

By Step 5, the set E has a supremum $s := \sup E \in P$. Since E satisfies condition (III) we have $s \in E$. Since E also satisfies (II), this implies $f(s) \in E$ and hence $f(s) \preccurlyeq s$. Since $s \preccurlyeq f(s)$ by assumption, we have $f(s) = s$ and this proves Theorem A.1.4. □

We remark that the Lemma of Zorn implies the existence of a maximal element $m \in P$ under the assumptions of Theorem A.1.4, and that any such maximal element must be a fixed point of f. However, the above proof of the Bourbaki–Witt Theorem does not use the Lemma of Zorn (nor does it use the axiom of choice) and so the result can be used to show that the axiom of choice implies the Lemma of Zorn.

Proof of Theorem A.1.3. First assume the Lemma of Zorn. Let I and X be nonempty sets and, for each $i \in I$, let $X_i \subset X$ be a nonempty subset, as in the assumptions of the axiom of choice. Define

$$\mathscr{P} := \left\{ (J, g) \;\middle|\; \begin{array}{l} \emptyset \neq J \subset I,\, g : J \to X, \\ g(i) \in X_i \text{ for all } i \in J \end{array} \right\}.$$

This set is partially ordered by the relation

$$(J, g) \preccurlyeq (K, h) \qquad \stackrel{\text{def}}{\iff} \qquad J \subset K \quad \text{and} \quad h|_J = g$$

for $(J, g), (K, h) \in \mathscr{P}$. It is nonempty, because each pair (i_0, x_0) with $i_0 \in I$ and $x_0 \in X_{i_0}$ determines a pair $(J_0, g_0) \in \mathscr{P}$ with $J_0 := \{i_0\}$, $g_0(i_0) := x_0$. Moreover, each nonempty chain $\mathscr{C} \subset \mathscr{P}$ has a supremum

$$(K, h) = \sup \mathscr{C} \in \mathscr{P}$$

given by

$$K := \bigcup_{(J,g) \in \mathscr{C}} J, \qquad h(i) := g(i) \quad \text{for } (J, g) \in \mathscr{C} \text{ and } i \in J.$$

Hence it follows from the Lemma of Zorn that there exists a maximal element $(J, g) \in \mathscr{P}$. This element must satisfy $J = I$. Otherwise $J \subsetneq I$, hence there exist an element $i_0 \in I \setminus J$ and an element $x_0 \in X_{i_0}$, and then the pair $(K, h) \in \mathscr{P}$ with

$$K := J \cup \{i_0\}, \qquad h(i) := \begin{cases} g(i), & \text{if } i \in J, \\ x_0, & \text{if } i = i_0, \end{cases} \qquad \text{for } i \in K,$$

satisfies $(J, g) \preccurlyeq (K, h)$ and $(J, g) \neq (K, h)$, in contradiction to maximality. This shows that there exists a map $g : I \to X$ that satisfies $g(i) \in X_i$ for all $i \in I$, and so the axiom of choice holds.

Conversely, assume the Axiom of Choice. Under this assumption we prove the Lemma of Zorn in two steps.

Step 1. *Let $(P, \preccurlyeq)$ be a nonempty partially ordered set such that every nonempty chain $C \subset P$ has a supremum. Then P has a maximal element.*

Assume, by contradiction, that P does not have a maximal element. Then the set

$$S(p) := \{q \in P \,|\, p \preccurlyeq q,\ p \neq q\} \subset P$$

is nonempty for every element $p \in P$. Hence the axiom of choice asserts that there exists a map $f : P \to P$ such that $f(p) \in S(p)$ for every $p \in P$. This map f satisfies the condition

$$p \preccurlyeq f(p) \qquad \text{for all } p \in P$$

but does not have a fixed point, in contradiction to Theorem A.1.4. This contradiction shows that our assumption, that P does not have a maximal element, must have been wrong. This proves Step 1.

Step 2. *Let $(P, \preccurlyeq)$ be a partially ordered set such that every nonempty chain $C \subset P$ admits an upper bound. Let $p \in P$. Then there exists a maximal element $m \in P$ with $p \preccurlyeq m$.*

Let $\mathscr{P} \subset 2^P$ be the set of all chains $C \subset P$ that contain the point p. Then $\mathscr{P}$ is a nonempty set, partially ordered by inclusion. Now let $\mathscr{C} \subset \mathscr{P}$ be a nonempty chain in $\mathscr{P}$ and define the set

$$S := \bigcup_{C \in \mathscr{C}} C.$$

This set contains the point p and we claim that it is a chain in P. To see this, let $p_0, p_1 \in S$ and choose chains $C_0, C_1 \in \mathscr{C}$ such that $p_0 \in C_0$ and $p_1 \in C_1$. Since $\mathscr{C}$ is a chain we have $C_0 \subset C_1$ or $C_1 \subset C_0$. Hence $C := C_0 \cup C_1 \in \mathscr{C}$ is a chain in P that contains both p_0 and p_1, and thus $p_0 \preccurlyeq p_1$ or $p_1 \preccurlyeq p_0$. This shows that S is an element of $\mathscr{P}$ and therefore is the supremum of the chain of chains $\mathscr{C} \subset \mathscr{P}$. Thus we have proved that every nonempty chain in $\mathscr{P}$ has a supremum. Hence Step 1 asserts that there exists a maximal chain $M \subset P$ that contains the point p. Let $m \in P$ be an upper bound of M. Then $p \preccurlyeq m$. Moreover, m is a maximal element of P, because otherwise there would exist an element $q \in P$ with $m \preccurlyeq q$ and $m \neq q$, so $q \notin M$, and then $M' := M \cup \{q\}$ would be a larger chain containing p, in contradiction to the maximality of M. This proves Step 2 and Theorem A.1.3. □

A.2. Tychonoff's Theorem

The purpose of this appendix is to state and prove Tychonoff's Theorem. It plays a central role in the proof of the Banach–Alaoglu Theorem for non-separable Banach spaces (Theorem 3.2.4).

Theorem A.2.1 (Tychonoff). *Let I be any set and, for each $i \in I$, let K_i be a compact topological space. Then the product*

$$K := \prod_{i \in I} K_i = \left\{ x = (x_i)_{i \in I} \,\middle|\, x_i \in K_i \text{ for all } i \in I \right\}$$

is compact with respect to the product topology (i.e. the weakest topology on K such that the obvious projection $\pi_i : K \to K_i$ is continuous for every $i \in I$).

Proof. See page 451. □

The proof of Theorem A.2.1 uses the characterization of compactness in terms of the finite intersection property in part (i) of Remark A.2.3 below.

Definition A.2.2. Let K be a set. A collection $\mathcal{A} \subset 2^K$ of subsets of K is said to have the **finite intersection property** if

$$\mathcal{A} \neq \emptyset$$

and

$$n \in \mathbb{N}, \quad A_1, \ldots, A_n \in \mathcal{A} \qquad \Longrightarrow \qquad A_1 \cap \cdots \cap A_n \neq \emptyset.$$

A collection $\mathcal{A} \subset 2^K$ with the finite intersection property is called **maximal** if every collection $\mathcal{B} \subset 2^K$ that has the finite intersection property and contains $\mathcal{A}$ is equal to $\mathcal{A}$.

The significance of this definition rests on the following observations.

Remark A.2.3. **(i)** A topological space K is compact if and only if every collection $\mathcal{A} \subset 2^K$ of closed subsets of K with the finite intersection property has a nonempty intersection, i.e. there is an element $x \in K$ such that $x \in A$ for all $A \in \mathcal{A}$.

(ii) Let K be any set and let $\mathcal{A} \subset 2^K$ be a collection of subsets of K that has the finite intersection property. Then, by the Lemma of Zorn, there exists a maximal collection $\mathcal{B} \subset 2^K$ with the finite intersection property that contains $\mathcal{A}$.

(iii) Let $\mathcal{B} \subset 2^K$ be a maximal collection with the finite intersection property. Then

$$n \in \mathbb{N}, \quad B_1, \ldots, B_n \in \mathcal{B} \qquad \Longrightarrow \qquad B_1 \cap \cdots \cap B_n \in \mathcal{B}$$

and, for every subset $C \subset K$,

$$C \cap B \neq \emptyset \text{ for all } B \in \mathcal{B} \qquad \Longrightarrow \qquad C \in \mathcal{B}.$$

Proof of Theorem A.2.1. Let

$$K = \prod_{i \in I} K_i$$

be a product of compact topological spaces and denote the canonical projections by $\pi_i : K \to K_i$ for $i \in I$. Let $\mathcal{A} \subset 2^K$ be a collection of closed subsets of K that has the finite intersection property. Then, by part (ii) of Remark A.2.3, there exists a maximal collection $\mathcal{B} \subset 2^K$ of subsets of K that has the finite intersection property and contains $\mathcal{A}$. We prove that there exists an $x \in X$ such that $x \in \overline{B}$ for all $B \in \mathcal{B}$. To see this, define

$$\mathcal{B}_i := \left\{ \overline{\pi_i(B)} \,\middle|\, B \in \mathcal{B} \right\} \subset 2^{K_i}$$

for $i \in I$. Then $\mathcal{B}_i$ is a collection of closed subsets of K_i that has the finite intersection property. Since K_i is compact, it follows from part (i) of Remark A.2.3 that

$$\bigcap_{B \in \mathcal{B}} \overline{\pi_i(B)} \neq \emptyset$$

for all $i \in I$. Hence it follows from the axiom of choice that there exists an element $x = (x_i)_{i \in I} \in K$ such that

$$x_i \in \overline{\pi_i(B)} \qquad \text{for all } i \in I \text{ and all } B \in \mathcal{B}.$$

We claim that $x \in \overline{B}$ for every $B \in \mathcal{B}$. To see this, let $U \subset K$ be an open set containing x. Then, by definition of the product topology, there exists a finite set $J \subset I$ and a collection of open sets $U_j \subset K_j$ for $j \in J$ such that

$$x \in \bigcap_{j \in J} \pi_j^{-1}(U_j) \subset U.$$

Hence

$$x_j = \pi_j(x) \in U_j \cap \overline{\pi_j(B)} \qquad \text{for all } j \in J \text{ and all } B \in \mathcal{B}.$$

Since U_j is open, this implies $U_j \cap \pi_j(B) \neq \emptyset$ and hence

$$\pi_j^{-1}(U_j) \cap B \neq \emptyset \qquad \text{for all } j \in J \text{ and all } B \in \mathcal{B}.$$

By part (iii) of Remark A.2.3 this implies $\pi_j^{-1}(U_j) \in \mathcal{B}$ for all $j \in J$. Use part (iii) of Remark A.2.3 again to deduce that $\bigcap_{j \in J} \pi_j^{-1}(U_j) \in \mathcal{B}$, and hence

$$\bigcap_{j \in J} \pi_j^{-1}(U_j) \cap B \neq \emptyset \qquad \text{for all } B \in \mathcal{B}.$$

This shows that $U \cap B \neq \emptyset$ for every $B \in \mathcal{B}$ and every open set $U \subset K$ containing x. Thus $x \in \overline{B}$ for all $B \in \mathcal{B}$ and therefore $x \in A$ for all $A \in \mathcal{A}$. Hence K is compact, by part (i) of Remark A.2.3, and this proves Theorem A.2.1. □

Bibliography

[1] L. V. Ahlfors, *Complex analysis: An introduction to the theory of analytic functions of one complex variable*, 3rd ed., International Series in Pure and Applied Mathematics, McGraw-Hill Book Co., New York, 1978. MR510197

[2] N. I. Akhiezer and I. M. Glazman, *Theory of linear operators in Hilbert space. Vols. I & II*, Translated from the third Russian edition by E. R. Dawson; Translation edited by W. N. Everitt, Monographs and Studies in Mathematics, Vols. 9 & 10, Pitman (Advanced Publishing Program), Boston, Mass.-London, 1981. MR615737

[3] F. Albiac and N. J. Kalton, *Topics in Banach space theory*, Graduate Texts in Mathematics, vol. 233, Springer, New York, 2006. MR2192298

[4] S. A. Argyros and R. G. Haydon, *A hereditarily indecomposable $\mathcal{L}_\infty$-space that solves the scalar-plus-compact problem*, Acta Math. **206** (2011), no. 1, 1–54. MR2784662

[5] M. F. Atiyah, *Algebraic topology and operators in Hilbert space*, Lectures in Modern Analysis and Applications. I, Springer, Berlin, 1969, pp. 101–121. MR0248803

[6] M. F. Atiyah, *Bott periodicity and the index of elliptic operators*, Quart. J. Math. Oxford Ser. (2) **19** (1968), 113–140. MR0228000

[7] M. F. Atiyah, *K-theory*, Lecture notes by D. W. Anderson, W. A. Benjamin, Inc., New York-Amsterdam, 1967. MR0224083

[8] S. Banach, *Théorie des opérations linéaires*, Monografje Matewatyczne, Warsaw, 1932.

[9] A. Benedek, A.-P. Calderón, and R. Panzone, *Convolution operators on Banach space valued functions*, Proc. Nat. Acad. Sci. U.S.A. **48** (1962), 356–365. MR0133653

[10] Y. Benyamini and J. Lindenstrauss, *Geometric nonlinear functional analysis. Vol. 1*, American Mathematical Society Colloquium Publications, vol. 48, American Mathematical Society, Providence, RI, 2000. MR1727673

[11] O. V. Besov, *On some families of functional spaces. Imbedding and extension theorems* (Russian), Dokl. Akad. Nauk SSSR **126** (1959), 1163–1165. MR0107165

[12] O. V. Besov, *Equivalent normings of spaces of functions of variable smoothness* (Russian, with Russian summary), Tr. Mat. Inst. Steklova **243** (2003), 87–95; English transl., Proc. Steklov Inst. Math. **4(243)** (2003), 80–88. MR2049464

[13] G. D. Birkhoff, *Proof of the ergodic theorem*, Proc. Nat. Acad. Sci. U.S.A. **17** (1931), 656–660.

[14] G. D. Birkhoff, *Integration of functions with values in a Banach space*, Trans. Amer. Math. Soc. **38** (1935), no. 2, 357–378. MR1501815

[15] Salomon Bochner, *Integration von Funkionen, deren Werte die Elemente eines * Vectorraumes sind*, Fund. Math. **20** (1933), 262–276. * *

[16] S. Bochner, *Absolut-additive abstrakte Mengenfunctionen*, Fund. Math. **21** (1933), 211–213.

[17] S. Bochner and A. E. Taylor, *Linear functionals on certain spaces of abstractly-valued functions*, Ann. of Math. (2) **39** (1938), no. 4, 913–944. MR1503445

[18] B. Bockelman, *Analysis notes: The Stone–Weierstraß Theorem*, October 2005. `http://www.math.unl.edu/~s-bbockel1/922-notes/Stone_Weierstrass_Theorem.html`

[19] H. Brezis, *Functional analysis, Sobolev spaces and partial differential equations*, Universitext, Springer, New York, 2011. MR2759829

[20] J. A. Clarkson, *Uniformly convex spaces*, Trans. Amer. Math. Soc. **40** (1936), no. 3, 396–414. MR1501880

[21] J. Diestel and J. J. Uhl Jr., *The Radon-Nikodým theorem for Banach space valued measures*, Rocky Mountain J. Math. **6** (1976), no. 1, 1–46. MR0399852

[22] J. Diestel and J. J. Uhl Jr., *Vector measures*, Mathematical Surveys, No. 15, American Mathematical Society, Providence, R.I., 1977. MR0453964

[23] R. G. Douglas, *On majorization, factorization, and range inclusion of operators on Hilbert space*, Proc. Amer. Math. Soc. **17** (1966), 413–415. MR0203464

[24] N. Dunford and A. P. Morse, *Remarks on the preceding paper of James A. Clarkson: "Uniformly convex spaces" [Trans. Amer. Math. Soc.* **40** *(1936), no. 3, 396–414; MR1501880*, Trans. Amer. Math. Soc. **40** (1936), no. 3, 415–420. MR1501881

[25] N. Dunford and B. J. Pettis, *Linear operations on summable functions*, Trans. Amer. Math. Soc. **47** (1940), 323–392. MR0002020

[26] N. Dunford and J. T. Schwartz, *Linear operators. Part I: General theory*, Wiley Classics Library, John Wiley & Sons, Inc., New York, 1988.

[27] P. Enflo, *A counterexample to the approximation problem in Banach spaces*, Acta Math. **130** (1973), 309–317. MR0402468

[28] A. Friedman, *Partial differential equations of parabolic type*, Prentice-Hall, Inc., Englewood Cliffs, N.J., 1964. MR0181836

[29] I. M. Gelfand, *Abstrakte Functionen und lineare Operatoren*, Math. Sbornik N.S. **46** (1938), 235–286.

[30] D. Gilbarg and N. S. Trudinger, *Elliptic partial differential equations of second order*, 2nd ed., Grundlehren der Mathematischen Wissenschaften [Fundamental Principles of Mathematical Sciences], vol. 224, Springer-Verlag, Berlin, 1983. MR737190

[31] W. T. Gowers, *A solution to Banach's hyperplane problem*, Bull. London Math. Soc. **26** (1994), no. 6, 523–530. MR1315601

[32] A. Grigor'yan and L. Liu, *Heat kernel and Lipschitz-Besov spaces*, Forum Math. **27** (2015), no. 6, 3567–3613. MR3420351

[33] A. Grothendieck, *Produits tensoriels topologiques et espaces nucléaires* (French), Mem. Amer. Math. Soc. **No. 16** (1955), 140. MR0075539

[34] H. Herrlich, *Axiom of choice*, Lecture Notes in Mathematics, vol. 1876, Springer-Verlag, Berlin, 2006. MR2243715

[35] E. Hille, *Functional analysis and semi-groups*, American Mathematical Society Colloquium Publications, vol. 31, American Mathematical Society, New York, 1948. MR0025077

[36] R. B. Holmes, *Geometric functional analysis and its applications*, Graduate Texts in Mathematics, No. 24, Springer-Verlag, New York-Heidelberg, 1975. MR0410335

[37] R. C. James, *Bases and reflexivity of Banach spaces*, Ann. of Math. (2) **52** (1950), 518–527. MR0039915

[38] R. C. James, *A non-reflexive Banach space isometric with its second conjugate space*, Proc. Nat. Acad. Sci. U.S.A. **37** (1951), 174–177. MR0044024

[39] R. C. James, *Weakly compact sets*, Trans. Amer. Math. Soc. **113** (1964), 129–140. MR0165344

[40] R. C. James, *Characterizations of reflexivity*, Studia Math. **23** (1963/1964), 205–216. MR0170192

[41] R. C. James, *A separable somewhat reflexive Banach space with nonseparable dual*, Bull. Amer. Math. Soc. **80** (1974), 738–743. MR0417763

[42] K. Jänich, *Vektorraumbündel und der Raum der Fredholm-Operatoren*, Math. Ann. **161** (1965), 129–142. MR0190946

[43] K. Jarosz, *Any Banach space has an equivalent norm with trivial isometries*, Israel J. Math. **64** (1988), no. 1, 49–56. MR981748

[44] T. Kato, *Perturbation theory for linear operators*, Die Grundlehren der mathematischen Wissenschaften, Band 132, Corrected printing of the second edition, Springer-Verlag New York, Inc., New York, 1980. MR0203473

[45] J. L. Kelley, *General topology*, Graduate Texts in Mathematics, No. 27, Springer-Verlag, New York-Berlin, 1975. Reprint of the 1955 edition [Van Nostrand, Toronto, Ont.] MR0370454

[46] G. Kerkyacharian and P. Petrushev, *Heat kernel based decomposition of spaces of distributions in the framework of Dirichlet spaces*, Trans. Amer. Math. Soc. **367** (2015), no. 1, 121–189. MR3271256

[47] M. Krein and D. Milman, *On extreme points of regular convex sets* (English., with Ukrainian summary), Studia Math. **9** (1940), 133–138. MR0004990

[48] A. W. Knapp, *Basic real analysis*, Cornerstones, Birkhäuser Boston, Inc., Boston, MA, 2005. Digital second edition 2016. MR2155259

[49] A. W. Knapp, *Advanced real analysis*, Cornerstones, Birkhäuser Boston, Inc., Boston, MA, 2005. Digital second edition 2017. MR2155260

[50] M. Kreuter, *Sobolev spaces of vector valued functions*, Master Thesis, Universität Ulm, 2015.

[51] R. Krom, *The Donaldson geometric flow is a local smooth semiflow*, Preprint, December 2015. `http://arxiv.org/abs/1512.09199`

[52] R. Krom, D. A. Salamon, *The Donaldson geometric flow for symplectic four-manifolds*, Preprint, December 2015. `http://arxiv.org/abs/1512.09198`

[53] N. H. Kuiper, *The homotopy type of the unitary group of Hilbert space*, Topology **3** (1965), 19–30. MR0179792

[54] D. Lamberton, *Équations d'évolution linéaires associées à des semi-groupes de contractions dans les espaces L^p* (French), J. Funct. Anal. **72** (1987), no. 2, 252–262. MR886813

[55] P. D. Lax, *Functional analysis*, Pure and Applied Mathematics (New York), Wiley-Interscience [John Wiley & Sons], New York, 2002. MR1892228

[56] I. Leader, *Logic and set theory*, University of Cambridge, Michaelmas 2003.

[57] L. Liu, D. Yang, and W. Yuan, *Besov-type and Triebel-Lizorkin-type spaces associated with heat kernels*, Collect. Math. **67** (2016), no. 2, 247–310. MR3484021

[58] G. Lumer and R. S. Phillips, *Dissipative operators in a Banach space*, Pacific J. Math. **11** (1961), 679–698. MR0132403

[59] R. E. Megginson, *An introduction to Banach space theory*, Graduate Texts in Mathematics, vol. 183, Springer-Verlag, New York, 1998. MR1650235

[60] D. Mil′man, *Characteristics of extremal points of regularly convex sets* (Russian), Doklady Akad. Nauk SSSR (N.S.) **57** (1947), 119–122. MR0022313

[61] J. R. Munkres, *Topology: a first course, second edition*, Prentice-Hall, Inc., Englewood Cliffs, N.J., 2000. MR0464128

[62] J. von Neumann, *Proof of the quasi-ergodic hypothesis*, Proc. Natl. Acad. Sci. U.S.A. **18** (1932), 70–82.

[63] J. Peetre, *On spaces of Triebel-Lizorkin type*, Ark. Mat. **13** (1975), 123–130. MR0380394

[64] J. Peetre, *New thoughts on Besov spaces*, Mathematics Department, Duke University, Durham, N.C., 1976. Duke University Mathematics Series, No. 1. MR0461123

[65] R. S. Phillips, *On linear transformations*, Trans. Amer. Math. Soc. **48** (1940), 516–541. MR0004094

[66] R. S. Phillips, *On weakly compact subsets of a Banach space*, Amer. J. Math. **65** (1943), 108–136. MR0007938

[67] R. S. Phillips, *On the generation of semigroups of linear operators*, Pacific J. Math. **2** (1952), 343–369. MR0050797

[68] R. S. Phillips, *A note on the abstract Cauchy problem*, Proc. Nat. Acad. Sci. U.S.A. **40** (1954), 244–248. MR0061759

[69] R. S. Phillips, *The adjoint semi-group*, Pacific J. Math. **5** (1955), 269–283. MR0070976

[70] J. D. Pryce, *Weak compactness in locally convex spaces*, Proc. Amer. Math. Soc. **17** (1966), 148–155. MR0190695

[71] D. Quillen, *Determinants of Cauchy-Riemann operators on Riemann surfaces* (Russian), Funktsional. Anal. i Prilozhen. **19** (1985), no. 1, 37–41, 96; English transl., Funct. Anal. Appl. **19** (1985), 31–34. MR783704

[72] M. Reed, B. Simon, *Methods of modern mathematical physics I: Functional analysis*, Revised and Enlarged Edition, Academic Press, 1980.

[73] W. Rudin, *Functional analysis*, 2nd ed., International Series in Pure and Applied Mathematics, McGraw-Hill, Inc., New York, 1991. MR1157815

[74] D. A. Salamon, *Funktionentheorie*, Birkhäuser, 2012. `http://www.math.ethz.ch/~salamon/PREPRINTS/cxana.pdf`

[75] D. A. Salamon, *Measure and integration*, EMS Textbooks in Mathematics, European Mathematical Society (EMS), Zürich, 2016. MR3469972

[76] D. A. Salamon, *Parabolic $L^p - L^q$ estimates*, ETH Preprint, 2016. `http://www.math.ethz.ch/~salamon/PREPRINTS/parabolic.pdf`

[77] D. A. Salamon, *Notes on the universal determinant bundle*, ETH, Preprint, 2013. `http://www.math.ethz.ch/~salamon/PREPRINTS/det.pdf`

[78] L. Schwartz, *Functional analysis*, Notes by Paul Rabinowitz. Courant Institute, NYU, 1964.

[79] B. Simon, *A Comprehensive Course in Analysis*, five vols, American Mathematical Society, Providence, RI, 2015. MR3364090

[80] A. D. Sokal, *A really simple elementary proof of the uniform boundedness theorem*, Amer. Math. Monthly **118** (2011), no. 5, 450–452. MR2805031

[81] M. H. Stone, *On one-parameter unitary groups in Hilbert space*, Ann. of Math. (2) **33** (1932), no. 3, 643–648. MR1503079

[82] A. Szankowski, *Subspaces without the approximation property*, Israel J. Math. **30** (1978), no. 1-2, 123–129. MR508257

[83] A. Szankowski, *$B(\mathcal{H})$ does not have the approximation property*, Acta Math. **147** (1981), no. 1-2, 89–108. MR631090

[84] H. Triebel, *Interpolation theory, function spaces, differential operators*, North-Holland Mathematical Library, vol. 18, North-Holland Publishing Co., Amsterdam-New York, 1978. MR503903

[85] H. Triebel, *Theory of function spaces. II*, Monographs in Mathematics, vol. 84, Birkhäuser Verlag, Basel, 1992. MR1163193

[86] D. Werner, *Funktionalanalysis*, Springer-Verlag, Berlin, 2004. MR1787146

[87] K. Yosida, *On the differentiability and the representation of one-parameter semi-group of linear operators*, J. Math. Soc. Japan **1** (1948), 15–21. MR0028537

[88] K. Yosida, *Functional analysis*, 6th ed., Grundlehren der Mathematischen Wissenschaften [Fundamental Principles of Mathematical Sciences], vol. 123, Springer-Verlag, Berlin-New York, 1980. MR617913

Notation

Index